Human Milk and Lactation

Human Milk and Lactation

Special Issue Editor
Maria Lorella Gianni

MDPI • Basel • Beijing • Wuhan • Barcelona • Belgrade • Manchester • Tokyo • Cluj • Tianjin

Special Issue Editor
Maria Lorella Gianni
University of Milan
Italy

Editorial Office
MDPI
St. Alban-Anlage 66
4052 Basel, Switzerland

This is a reprint of articles from the Special Issue published online in the open access journal *Nutrients* (ISSN 2072-6643) (available at: https://www.mdpi.com/journal/nutrients/special_issues/Human_Milk_Lactation).

For citation purposes, cite each article independently as indicated on the article page online and as indicated below:

LastName, A.A.; LastName, B.B.; LastName, C.C. Article Title. *Journal Name* **Year**, *Article Number*, Page Range.

ISBN 978-3-03928-923-3 (Hbk)
ISBN 978-3-03928-924-0 (PDF)

© 2020 by the authors. Articles in this book are Open Access and distributed under the Creative Commons Attribution (CC BY) license, which allows users to download, copy and build upon published articles, as long as the author and publisher are properly credited, which ensures maximum dissemination and a wider impact of our publications.

The book as a whole is distributed by MDPI under the terms and conditions of the Creative Commons license CC BY-NC-ND.

Contents

About the Special Issue Editor ix

Maria Lorella Gianni, Daniela Morniroli, Maria Enrica Bettinelli and Fabio Mosca
Human Milk and Lactation
Reprinted from: *Nutrients* 2020, 12, 899, doi:10.3390/nu12040899 1

Sara Moukarzel, Alejandra M. Wiedeman, Lynda S. Soberanes, Roger A. Dyer, Sheila M. Innis and Yvonne Lamers
Variability of Water-Soluble Forms of Choline Concentrations in Human Milk during Storage, after Pasteurization, and among Women
Reprinted from: *Nutrients* 2019, 11, 3024, doi:10.3390/nu11123024 5

Maria Lorella Gianni, Maria Enrica Bettinelli, Priscilla Manfra, Gabriele Sorrentino, Elena Bezze, Laura Plevani, Giacomo Cavallaro, Genny Raffaeli, Beatrice Letizia Crippa, Lorenzo Colombo, Daniela Morniroli, Nadia Liotto, Paola Roggero, Eduardo Villamor, Paola Marchisio and Fabio Mosca
Breastfeeding Difficulties and Risk for Early Breastfeeding Cessation
Reprinted from: *Nutrients* 2019, 11, 2266, doi:10.3390/nu11102266 17

Maria Grunewald, Christian Hellmuth, Franca F. Kirchberg, Maria Luisa Mearin, Renata Auricchio, Gemma Castillejo, Ilma R. Korponay-Szabo, Isabel Polanco, Maria Roca, Sabine L. Vriezinga, Katharina Werkstetter, Berthold Koletzko and Hans Demmelmair
Variation and Interdependencies of Human Milk Macronutrients, Fatty Acids, Adiponectin, Insulin, and IGF-II in the European PreventCD Cohort
Reprinted from: *Nutrients*, 11, 2034, doi:10.3390/nu11092034 27

Michele R. Machado, Fernanda Kamp, Juliana C. Nunes, Tatiana El-Bacha and Alexandre G. Torres
Breast Milk Content of Vitamin A and E from Early- to Mid-Lactation Is Affected by Inadequate Dietary Intake in Brazilian Adult Women
Reprinted from: *Nutrients* 2019, 11, 2025, doi:10.3390/nu11092025 47

Aleksandra Wesolowska, Joanna Brys, Olga Barbarska, Kamila Strom, Jolanta Szymanska-Majchrzak, Katarzyna Karzel, Emilia Pawlikowska, Monika A. Zielinska, Jadwiga Hamulka and Gabriela Oledzka
Lipid Profile, Lipase Bioactivity, and Lipophilic Antioxidant Content in High Pressure Processed Donor Human Milk
Reprinted from: *Nutrients* 2019, 11, 1972, doi:10.3390/nu11091972 61

Jiahui Yu, Tinglan Yuan, Xinghe Zhang, Qingzhe Jin, Wei Wei and Xingguo Wang
Quantification of Nervonic Acid in Human Milk in the First 30 Days of Lactation: Influence of Lactation Stages and Comparison with Infant Formulae
Reprinted from: *Nutrients* 2019, 11, 1892, doi:10.3390/nu11081892 77

Magalie Sabatier, Clara L. Garcia-Rodenas, Carlos A. De Castro, Peter Kastenmayer, Mario Vigo, Stéphane Dubascoux, Daniel Andrey, Marine Nicolas, Janique Richoz Payot, Valentine Bordier, Sagar K. Thakkar, Lydie Beauport, Jean-François Tolsa, Céline J. Fischer Fumeaux and Michael Affolter
Longitudinal Changes of Mineral Concentrations in Preterm and Term Human Milk from Lactating Swiss Women
Reprinted from: *Nutrients* 2019, 11, 1855, doi:10.3390/nu11081855 91

Alessandra Mazzocchi, Maria Lorella Giannì, Daniela Morniroli, Ludovica Leone,
Paola Roggero, Carlo Agostoni, Valentina De Cosmi and Fabio Mosca
Hormones in Breast Milk and Effect on Infants' Growth: A Systematic Review
Reprinted from: *Nutrients* 2019, 11, 1845, doi:10.3390/nu11081845 105

Daniela Hampel, Setareh Shahab-Ferdows, Muttaquina Hossain, M. Munirul Islam,
Tahmeed Ahmed and Lindsay H. Allen
Validation and Application of Biocrates Absolute*IDQ*® p180 Targeted Metabolomics Kit Using
Human Milk
Reprinted from: *Nutrients* 2019, 11, 1733, doi:10.3390/nu11081733 117

Adekunle Dawodu, Khalil M. Salameh, Najah S. Al-Janahi, Abdulbari Bener and
Naser Elkum
The Effect of High-Dose Postpartum Maternal Vitamin D Supplementation Alone Compared
with Maternal Plus Infant Vitamin D Supplementation in Breastfeeding Infants in a High-Risk
Population. A Randomized Controlled Trial
Reprinted from: *Nutrients* 2019, 11, 1632, doi:10.3390/nu11071632 127

Agnieszka Bzikowska-Jura, Aneta Czerwonogrodzka-Senczyna, Edyta Jasińska-Melon,
Hanna Mojska, Gabriela Olędzka, Aleksandra Wesołowska and Dorota Szostak-Węgierek
The Concentration of Omega-3 Fatty Acids in Human Milk Is Related to Their Habitual but Not
Current Intake
Reprinted from: *Nutrients* 2019, 11, 1585, doi:10.3390/nu11071585 145

Veronique Demers-Mathieu, Robert K. Huston, Andi M. Markell, Elizabeth A. McCulley,
Rachel L. Martin and David C. Dallas
Antenatal Influenza A-Specific IgA, IgM, and IgG Antibodies in Mother's Own Breast Milk and
Donor Breast Milk, and Gastric Contents and Stools from Preterm Infants
Reprinted from: *Nutrients* 2019, 11, 1567, doi:10.3390/nu11071567 161

Céline J. Fischer Fumeaux, Clara L. Garcia-Rodenas, Carlos A. De Castro,
Marie-Claude Courtet-Compondu, Sagar K. Thakkar, Lydie Beauport,
Jean-François Tolsa and Michael Affolter
Longitudinal Analysis of Macronutrient Composition in Preterm and Term Human Milk:
A Prospective Cohort Study
Reprinted from: *Nutrients* 2019, 11, 1525, doi:10.3390/nu11071525 173

Sean Austin, Carlos A. De Castro, Norbert Sprenger, Aristea Binia, Michael Affolter,
Clara L. Garcia-Rodenas, Lydie Beauport, Jean-François Tolsa and Céline J. Fischer Fumeaux
Human Milk Oligosaccharides in the Milk of Mothers Delivering Term versus Preterm Infants
Reprinted from: *Nutrients* 2019, 11, 1282, doi:10.3390/nu11061282 185

Malgorzata Witkowska-Zimny, Ewa Kamińska-El-Hassan and Edyta Wróbel
Milk Therapy: Unexpected Uses for Human Breast Milk
Reprinted from: *Nutrients* 2019, 11, 944, doi:10.3390/nu11050944 201

Amanda de Sousa Rebouças, Ana Gabriella Costa Lemos da Silva, Amanda Freitas de
Oliveira, Lorena Thalia Pereira da Silva, Vanessa de Freitas Felgueiras, Marina Sampaio
Cruz, Vivian Nogueira Silbiger, Karla Danielly da Silva Ribeiro and Roberto Dimenstein
Factors Associated with Increased Alpha-Tocopherol Content in Milk in Response to Maternal
Supplementation with 800 IU of Vitamin E
Reprinted from: *Nutrients* 2019, 11, 900, doi:10.3390/nu11040900 213

Therese A. O'Sullivan, Joy Cooke, Chris McCafferty and Roslyn Giglia
Online Video Instruction on Hand Expression of Colostrum in Pregnancy is an Effective Educational Tool
Reprinted from: *Nutrients* **2019**, *11*, 883, doi:10.3390/nu11040883 227

Mohèb Elwakiel, Sjef Boeren, Jos A. Hageman, Ignatius M. Szeto, Henk A. Schols and Kasper A. Hettinga
Variability of Serum Proteins in Chinese and Dutch Human Milk during Lactation
Reprinted from: *Nutrients* **2019**, *11*, 499, doi:10.3390/nu11030499 239

Beatrice Letizia Crippa, Lorenzo Colombo, Daniela Morniroli, Dario Consonni, Maria Enrica Bettinelli, Irene Spreafico, Giulia Vercesi, Patrizio Sannino, Paola Agnese Mauri, Lidia Zanotta, Annalisa Canziani, Paola Roggero, Laura Plevani, Donatella Bertoli, Stefania Zorzan, Maria Lorella Giannì and Fabio Mosca
Do a Few Weeks Matter? Late Preterm Infants and Breastfeeding Issues
Reprinted from: *Nutrients* **2019**, *11*, 312, doi:10.3390/nu11020312 253

Ida Emilie Ingvordsen Lindahl, Virginia M. Artegoitia, Eimear Downey, James A. O'Mahony, Carol-Anne O'Shea, C. Anthony Ryan, Alan L. Kelly, Hanne C. Bertram and Ulrik K. Sundekilde
Quantification of Human Milk Phospholipids: The Effect of Gestational and Lactational Age on Phospholipid Composition
Reprinted from: *Nutrients* **2019**, *11*, 222, doi:10.3390/nu11020222 263

Monika A. Zielinska, Jadwiga Hamulka and Aleksandra Wesolowska
Carotenoid Content in Breastmilk in the 3rd and 6th Month of Lactation and Its Associations with Maternal Dietary Intake and Anthropometric Characteristics
Reprinted from: *Nutrients* **2019**, *11*, 193, doi:10.3390/nu11010193 277

Sagar K. Thakkar, Carlos Antonio De Castro, Lydie Beauport, Jean-François Tolsa, Céline J. Fischer Fumeaux, Michael Affolter and Francesca Giuffrida
Temporal Progression of Fatty Acids in Preterm and Term Human Milk of Mothers from Switzerland
Reprinted from: *Nutrients* **2019**, *11*, 112, doi:10.3390/nu11010112 293

Jing Zhu and Kelly A. Dingess
The Functional Power of the Human Milk Proteome
Reprinted from: *Nutrients* **2019**, *11*, 1834, doi:10.3390/nu11081834 305

Andrea Gila-Diaz, Silvia M. Arribas, Alba Algara, María A. Martín-Cabrejas, Ángel Luis López de Pablo, Miguel Sáenz de Pipaón and David Ramiro-Cortijo
A Review of Bioactive Factors in Human Breastmilk: A Focus on Prematurity
Reprinted from: *Nutrients* **2019**, *11*, 1307, doi:10.3390/nu11061307 333

About the Special Issue Editor

Maria Lorella Gianni earned her medical degree in 1993 and completed her residency training in pediatrics in 1998 at the University of Milan, San Paolo Hospital, where she began working on inborn errors of metabolism and infant nutrition and later health. Since 2002, she has been working at Fondazione IRCCS "Cà Granda" Ospedale Maggiore Policlinico, University of Milan, Italy. She is involved in monitoring the health, nutritional status, and neurodevelopmental outcome of high-risk infants.

Editorial

Human Milk and Lactation

Maria Lorella Gianni [1,2,*], Daniela Morniroli [1], Maria Enrica Bettinelli [2] and Fabio Mosca [1,2]

[1] Fondazione IRCCS Ca' Granda Ospedale Maggiore Policlinico, NICU, Via Commenda 12, 20122 Milan, Italy; daniela.morniroli@gmail.com (D.M.); fabio.mosca@unimi.it (F.M.)
[2] Department of Clinical Science and Community Health, University of Milan, Via Commenda 19, 20122 Milan, Italy; mebettinelli@gmail.com
* Correspondence: maria.gianni@unimi.it; Tel.: +39-0255-032-483

Received: 9 March 2020; Accepted: 17 March 2020; Published: 26 March 2020

Human milk is uniquely tailored to meet infants' specific nutritional requirements [1]. However, it is more than just "milk" since it has emerged as an evolutionary strategy to promote human well-being [2]. This dynamic and bioactive fluid allows mother–infant signalling over lactation, guiding the infant in the developmental and physiological processes. Human milk exerts protection and life-long biological effects, playing a crucial role in promoting healthy growth and optimal cognitive development [3,4]. For this evidence, the promotion of breastfeeding initiation and duration becomes paramount in all healthcare settings [5]. The latest scientific advances have provided insight into different components of human milk and their dynamic and flexible changes over time in response to several biological and environmental triggers. However, the complexity of human milk composition and the synergistic mechanisms responsible for its beneficial health effects have not yet been unravelled [4]. This special issue has brought together a variety of articles, including original works and literature reviews, further exploring the complexity of the human milk biofluid and the mechanisms underlying the beneficial effects associated with breastfeeding. In this issue, the mounting amount of data regarding human milk proteome and metabolome, gathered using advanced technological achievements such as "omics" techniques, has been reviewed, describing the multitude of bioactive components and their relationship with infants' cognitive development, growth and immune functions [6,7]. Changes in human milk protein content over the first months of lactation in mothers from different geographical and ethnic origins have been investigated [8]. The high abundance of immune active proteins reflects the well-known immunological properties of mothers' milk [8]. Authors enhance the importance of passive immunisation through mothers' antibodies transfer from breast milk, which has a key role for infant immune protection in the first months [9]. Differences in oligosaccharides content between term and preterm milk have also been examined in view of the potential implications for preterm infants' clinical outcomes with special regard to their increased vulnerability to infections [10]. Given the widely known anti-inflammatory and antimicrobial properties of human milk, authors have also explored its implementation as a powerful therapeutic agent for skin issues, suggesting its potential use in settings with limited access to medicine [11].

In this special edition, attention has been focused on the variability of human milk compounds depending on individual differences among mothers and, far more significant, on mothers' nutritional status and anthropometric characteristics. Authors outline the importance of a healthy lifestyle and a correct micro and macronutrient intake, before and during pregnancy and lactation, in order to promote adequate levels of vitamins and other components in human milk [12–16]. Moreover, author recommendations indicate the need for identifying women at risk for a deficiency, who could, therefore, benefit from an appropriate supplementation aimed at increasing breastmilk micronutrient content [12–16].

The more the exceptional qualities of human milk are brought up, the more the support of breastfeeding initiation and duration becomes fundamental [5]. However, breastfeeding rates are still lower than recommended, especially in developed countries. Authors highlight the association among breastfeeding difficulties in the first months of lactation and early breastfeeding cessation and advocate the provision of continued tailored breastfeeding support also after hospital discharge [17]. Within this context, the effectiveness of online sources including an expert instructional video in improving maternal knowledge and confidence regarding antenatal colostrum expressing, which in turn may promote long term breastfeeding, has been explored [18].

In this issue, authors have investigated the potential relationship between the presence of unique components of human milk and the positive long-life beneficial effects associated with breastfeeding. In view of the crucial role of neuronic acid in white matter development, its content in human milk through the first month of lactation has been quantified and compared with that of formula milk from three fat sources [19]. Human milk's hormonal content, which seems to be involved in infants' metabolic pathways, including appetite and energy balance, has been also examined in light of the reduced risk of developing overweight and metabolic syndrome in human milk-fed infants [20,21].

Benefits of human milk feeding are indeed even more critical among specific populations at high risk of developing adverse outcomes, as preterm infants [22]. This value is highlighted not only by the positive effects that human milk has in modulating preterms' outcomes at every level but also by the results of studies in this issue demonstrating the higher levels of bioactive, micro and macronutrient contents in preterm milk, compared to full-term [10,23–26]. Within this context, however, authors have underlined the potential lack of mineral content of preterm milk that should be taken into consideration in the approach to the fortification of milk for the preterm population [27].

Since human milk feeding is associated with several life-long important beneficial health effects, in a dose-dependent relation, its promotion and support should be considered as a public health issue [2]. Unfortunately, the authors underline that breastfeeding initiation and duration are even more challenging in preterm infants [28]. Therefore, donor human milk has been studied for its role as a fresh mother's milk substitute. Even though donor milk has to be processed through pasteurisation for microbiological safety reasons and supplemented with fortifiers, it has been demonstrated to be a better feeding alternative for preterm infants, compared to formula milk, when the own mother's milk is not available [29]. The refrigeration, freezing, and pasteurisation of donor milk have a variable impact on vitamin, enzymes and nutrients concentration, resulting in a diminished bioactive function of donor milk [30]. In this issue, changes in concentrations after pasteurization of water-soluble forms of choline, which is crucial for infants' development, have been investigated together with the potential for reducing the loss of donor human milk compounds by using innovative techniques including high-pressure processing [31,32].

As the diverse articles in this special issue highlight, commitment towards filling the knowledge gap on the complex and highly dynamic human milk composition and the strictly interrelated mechanisms underpinning its positive long-life biological effects is crucial for a deeper understanding of the biology of the developing infant and the optimisation of infant feeding, particularly that of the most vulnerable infants.

Author Contributions: M.L.G., D.M. wrote the editorial, M.E.B., F.M. reviewed and revised the editorial. All authors have read and agreed to the published version of the manuscript.

Funding: This research received no external funding.

Conflicts of Interest: The authors declare no conflict of interest.

References

1. WHO/UNICEF. *Global Strategy for Infant and Young Child Feeding*; World Health Organization: Geneva, Switzerland, 2003.
2. Goldman, A.S. Evolution of immune functions of the mammary gland and protection of the infant. *Breastfeed. Med.* **2012**, *7*, 132–142. [CrossRef] [PubMed]

3. Mosca, F.; Giannì, M.L. Human Milk: Composition and Health Benefits. *Pediatr. Med. Chir.* **2017**, *39*, 155. [CrossRef] [PubMed]
4. Bardanzellu, F.; Peroni, D.G.; Fanos, V. Human Breast Milk: Bioactive Components, from Stem Cells to Health Outcomes. *Curr. Nutr. Rep.* **2020**, *9*, 1–13. [CrossRef] [PubMed]
5. Rollins, N.C.; Bhandari, N.; Hajeebhoy, N.; Horton, S.; Lutter, C.K.; Martines, J.C.; Piwoz, E.G.; Richter, L.M.; Victoria, C.G. The Lancet Breastfeeding Series Group. Why invest, and what it will take to improve breastfeeding practices? *Lancet* **2016**, *387*, 491–504. [CrossRef]
6. Zhu, J.; Dingess, K.A. The Functional Power of the Human Milk Proteome. *Nutrients* **2019**, *11*, 1834. [CrossRef] [PubMed]
7. Hampel, D.; Shahab-Ferdows, S.; Hossain, M.; Islam, M.M.; Ahmed, T.; Allen, L.H. Validation and Application of Biocrates AbsoluteIDQ® p180 Targeted Metabolomics Kit Using Human Milk. *Nutrients* **2019**, *11*, 1733. [CrossRef] [PubMed]
8. Elwakiel, M.; Boeren, S.; Hageman, J.A.; Szeto, I.M.; Schols, H.A.; Hettinga, K.A. Variability of Serum Proteins in Chinese and Dutch Human Milk during Lactation. *Nutrients* **2019**, *11*, 499. [CrossRef]
9. Demers-Mathieu, V.; Huston, R.K.; Markell, A.M.; McCulley, E.A.; Martin, R.L.; Dallas, D.C. Antenatal Influenza A-Specific IgA, IgM, and IgG Antibodies in Mother's Own Breast Milk and Donor Breast Milk, and Gastric Contents and Stools from Preterm Infants. *Nutrients* **2019**, *11*, 1567. [CrossRef]
10. Austin, S.; De Castro, C.A.; Sprenger, N.; Binia, A.; Affolter, M.; Garcia-Rodenas, C.L.; Beauport, L.; Tolsa, J.F.; Fischer Fumeaux, C.J. Human Milk Oligosaccharides in the Milk of Mothers Delivering Term versus Preterm Infants. *Nutrients* **2019**, *11*, 1282. [CrossRef]
11. Witkowska-Zimny, M.; Kamińska-El-Hassan, E.; Wróbel, E. Milk Therapy: Unexpected Uses for Human Breast Milk. *Nutrients* **2019**, *11*, 944. [CrossRef]
12. Machado, M.R.; Kamp, F.; Nunes, J.C.; El-Bacha, T.; Torres, A.G. Breast Milk Content of Vitamin A and E from Early- to Mid-Lactation Is Affected by Inadequate Dietary Intake in Brazilian Adult Women. *Nutrients* **2019**, *11*, 2025. [CrossRef] [PubMed]
13. Zielinska, M.A.; Hamulka, J.; Wesolowska, A. Carotenoid Content in Breastmilk in the 3rd and 6th Month of Lactation and Its Associations with Maternal Dietary Intake and Anthropometric Characteristics. *Nutrients* **2019**, *11*, 193. [CrossRef] [PubMed]
14. Bzikowska-Jura, A.; Czerwonogrodzka-Senczyna, A.; Jasińska-Melon, E.; Mojska, H.; Olędzka, G.; Wesołowska, A.; Szostak-Węgierek, D. The Concentration of Omega-3 Fatty Acids in Human Milk Is Related to Their Habitual but Not Current Intake. *Nutrients* **2019**, *11*, 1585. [CrossRef] [PubMed]
15. Dawodu, A.; Salameh, K.M.; Al-Janahi, N.S.; Bener, A.; Elkum, N. The Effect of High-Dose Postpartum Maternal Vitamin D Supplementation Alone Compared with Maternal Plus Infant Vitamin D Supplementation in Breastfeeding Infants in a High-Risk Population. A Randomized Controlled Trial. *Nutrients* **2019**, *11*, 1632. [CrossRef]
16. de Sousa Rebouças, A.; Costa Lemos da Silva, A.G.; Freitas de Oliveira, A.; Thalia Pereira da Silva, L.; de Freitas Felgueiras, V.; Cruz, M.S.; Silbiger, V.N.; da Silva Ribeiro, K.D.; Dimenstein, R. Factors Associated with Increased Alpha-Tocopherol Content in Milk in Response to Maternal Supplementation with 800 IU of Vitamin E. *Nutrients* **2019**, *11*, 900. [CrossRef]
17. Gianni, M.L.; Bettinelli, M.E.; Manfra, P.; Sorrentino, G.; Bezze, E.; Plevani, L.; Cavallaro, G.; Raffaeli, G.; Crippa, B.L.; Colombo, L.; et al. Breastfeeding Difficulties and Risk for Early Breastfeeding Cessation. *Nutrients* **2019**, *11*, 2266. [CrossRef]
18. O'Sullivan, T.A.; Cooke, J.; McCafferty, C.; Giglia, R. Online Video Instruction on Hand Expression of Colostrum in Pregnancy is an Effective Educational Tool. *Nutrients* **2019**, *11*, 883. [CrossRef]
19. Yu, J.; Yuan, T.; Zhang, X.; Jin, Q.; Wei, W.; Wang, X. Quantification of Nervonic Acid in Human Milk in the First 30 Days of Lactation: Influence of Lactation Stages and Comparison with Infant Formulae. *Nutrients* **2019**, *11*, 1892. [CrossRef]
20. Mazzocchi, A.; Giannì, M.L.; Morniroli, D.; Leone, L.; Roggero, P.; Agostoni, C.; De Cosmi, V.; Mosca, F. Hormones in Breast Milk and Effect on Infants' Growth: A Systematic Review. *Nutrients* **2019**, *11*, 1845. [CrossRef]
21. Grunewald, M.; Hellmuth, C.; Kirchberg, F.F.; Mearin, M.L.; Auricchio, R.; Castillejo, G.; Korponay-Szabo, I.R.; Polanco, I.; Roca, M.; Vriezinga, S.L.; et al. Variation and Interdependencies of Human Milk Macronutrients,

Fatty Acids, Adiponectin, Insulin, and IGF-II in the European PreventCD Cohort. *Nutrients* **2019**, *11*, 2034. [CrossRef]
22. Verduci, E.; Giannì, M.L.; Di Benedetto, A. Human Milk Feeding in Preterm Infants: What Has Been Done and What Is to Be Done. *Nutrients* **2019**, *12*, 44. [CrossRef] [PubMed]
23. Fischer Fumeaux, C.J.; Garcia-Rodenas, C.L.; De Castro, C.A.; Courtet-Compondu, M.C.; Thakkar, S.K.; Beauport, L.; Tolsa, J.F.; Affolter, M. Longitudinal Analysis of Macronutrient Composition in Preterm and Term Human Milk: A Prospective Cohort Study. *Nutrients* **2019**, *11*, 1525. [CrossRef] [PubMed]
24. Thakkar, S.K.; De Castro, C.A.; Beauport, L.; Tolsa, J.F.; Fischer Fumeaux, C.J.; Affolter, M.; Giuffrida, F. Temporal Progression of Fatty Acids in Preterm and Term Human Milk of Mothers from Switzerland. *Nutrients* **2019**, *11*, 112. [CrossRef] [PubMed]
25. Ingvordsen Lindahl, I.; Artegoitia, V.M.; Downey, E.; O'Mahony, J.A.; O'Shea, C.A.; Ryan, C.A.; Kelly, A.L.; Bertram, H.C.; Sundekilde, U.K. Quantification of Human Milk Phospholipids: The Effect of Gestational and Lactational Age on Phospholipid Composition. *Nutrients* **2019**, *11*, 222. [CrossRef]
26. Gila-Diaz, A.; Arribas, S.M.; Algara, A.; Martín-Cabrejas, M.A.; López de Pablo, Á.L.; Sáenz de Pipaón, M.; Ramiro-Cortijo, D. A Review of Bioactive Factors in Human Breastmilk: A Focus on Prematurity. *Nutrients* **2019**, *11*, 1307. [CrossRef]
27. Sabatier, M.; Garcia-Rodenas, C.L.; Castro, C.A.; Kastenmayer, P.; Vigo, M.; Dubascoux, S.; Andrey, D.; Nicolas, M.; Payot, J.R.; Bordier, V.; et al. Longitudinal Changes of Mineral Concentrations in Preterm and Term Human Milk from Lactating Swiss Women. *Nutrients* **2019**, *11*, 1855. [CrossRef]
28. Crippa, B.L.; Colombo, L.; Morniroli, D.; Consonni, D.; Bettinelli, M.E.; Spreafico, I.; Vercesi, G.; Sannino, P.; Mauri, P.A.; Zanotta, L.; et al. Do a Few Weeks Matter? Late Preterm Infants and Breastfeeding Issues. *Nutrients* **2019**, *11*, 312. [CrossRef]
29. American Academy of Pediatrics. Section on Breastfeeding. Breastfeeding and the use of human milk. *Pediatrics* **2012**, *129*, e827–e841. [CrossRef]
30. Peila, C.; Emmerik, N.E.; Giribaldi, M.; Stahl, B.; Ruitenberg, J.E.; van Elburg, R.M.; Moro, G.E.; Bertino, E.; Coscia, A.; Cavallarin, L. Human Milk Processing: A Systematic Review of Innovative Techniques to Ensure the Safety and Quality of Donor Milk. *J. Pediatr. Gastroenterol. Nutr.* **2017**, *64*, 353–361. [CrossRef]
31. Moukarzel, S.; Wiedeman, A.M.; Soberanes, L.S.; Dyer, R.A.; Innis, S.M.; Lamers, Y. Variability of Water-Soluble Forms of Choline Concentrations in Human Milk during Storage, after Pasteurization, and among Women. *Nutrients* **2019**, *11*, 3024. [CrossRef]
32. Wesolowska, A.; Brys, J.; Barbarska, O.; Strom, K.; Szymanska-Majchrzak, J.; Karzel, K.; Pawlikowska, E.; Zielinska, M.A.; Hamulka, J.; Oledzka, G. Lipid Profile, Lipase Bioactivity, and Lipophilic Antioxidant Content in High Pressure Processed Donor Human Milk. *Nutrients* **2019**, *11*, 1972. [CrossRef] [PubMed]

© 2020 by the authors. Licensee MDPI, Basel, Switzerland. This article is an open access article distributed under the terms and conditions of the Creative Commons Attribution (CC BY) license (http://creativecommons.org/licenses/by/4.0/).

Article

Variability of Water-Soluble Forms of Choline Concentrations in Human Milk during Storage, after Pasteurization, and among Women

Sara Moukarzel [1,‡], Alejandra M. Wiedeman [2,3,‡], Lynda S. Soberanes [2,4], Roger A. Dyer [2,3], Sheila M. Innis [2,3,†] and Yvonne Lamers [2,4,*]

1. Larsson-Rosenquist Foundation Mother-Milk-Infant Center of Research Excellence, University of California, San Diego, CA 92093, USA; smoukarzel@ucsd.edu
2. British Columbia Children's Hospital Research Institute, Vancouver, BC V5Z 4H4, Canada; awiedeman@bcchr.ca (A.M.W.); lynda.soberanes@gmail.com (L.S.S.); radyer@mail.ubc.ca (R.A.D.); sheila.innis@ubc.ca (S.M.I.)
3. Department of Pediatrics, Faculty of Medicine, The University of British Columbia, Vancouver, BC V5Z 4H4, Canada
4. Food, Nutrition, and Health Program, Faculty of Land and Food Systems, The University of British Columbia, Vancouver, BC V6T 1Z4, Canada
* Correspondence: yvonne.lamers@ubc.ca; Tel.: +1-604-827-1776
† Deceased.
‡ Shared first authorship.

Received: 30 October 2019; Accepted: 6 December 2019; Published: 11 December 2019

Abstract: Choline is critical for infant development and mother's milk is the sole source of choline for fully breastfed infants until six months of age. Human milk choline consists to 85% of water-soluble forms of choline including free choline (FC), phosphocholine (PhosC), and glycerophosphocholine (GPC). Donor milk requires safe handling procedures such as cold storage and pasteurization. However, the stability of water-soluble forms of choline during these processes is not known. The objectives of this research were to determine the effect of storage and pasteurization on milk choline concentration, and the diurnal intra- and inter-individual variability of water-soluble choline forms. Milk samples were collected from healthy women who were fully breastfeeding a full-term, singleton infant <6 months. Milk total water-soluble forms of choline, PhosC, and GPC concentrations did not change during storage at room temperature for up to 4 h. Individual and total water-soluble forms of choline concentrations did not change after storage for 24 h in the refrigerator or for up to one week in the household freezer. Holder pasteurization decreased PhosC and GPC, and thereby total water-soluble choline form concentrations by <5%. We did not observe diurnal variations in PhosC and total water-soluble forms of choline concentrations, but significant differences in FC and GPC concentrations across five sampling time points throughout one day. In conclusion, these outcomes contribute new knowledge for the derivation of evidence-informed guidelines for the handling and storage of expressed human milk as well as the development of optimized milk collection and storage protocols for research studies.

Keywords: human milk; donor milk; choline; phosphocholine; storage; pasteurization; milk banking; pumping; breastfeeding; lactation

1. Introduction

Choline is an essential nutrient with crucial roles in brain function, neurodevelopment, and growth [1,2]. Biological roles of choline include neurogenesis and synapse formation in the form of acetylcholine, membrane biogenesis, cell division, lipid transports, and myelination in the form of choline phospholipids, and as a methyl donor in the form of its oxidation product betaine [3–6]. Betaine contributes to the generation of S-adenosylmethionine, which is the main methyl donor involved in creatine and phosphatidylcholine synthesis, and DNA methylation, among other biochemical reactions [7]. Choline adequacy in early infancy and rapid stages of growth is critical to support membrane formation, cell proliferation, and parenchymal growth [6].

Exclusive breastfeeding ad libitum is the recommended feeding practice for the first six months of life, with continued breastfeeding up to two years of age [8]. Human milk contains various forms of the essential nutrient choline. The three water-soluble forms of choline, i.e., free choline (FC), phosphocholine (PhosC), and glycerophosphocholine (GPC), contribute to approximately 85% of total choline in human milk; the lipid-soluble phosphatidylcholine and sphingomyelin account for the remaining 15% [9–13].

The practice of human milk pumping and storage for later use is on the rise both at home and in clinical settings to accommodate various situations when feeding human milk at the breast is not possible, e.g., mothers returning to work; insufficient volume of mother's own milk, and need for donor milk [14,15]. Current guidelines for the safe handling of expressed human milk focus largely on microbiological safety, whereby Holder pasteurization (at 62.5 °C for 30 min) of donor milk is mandatory in all North American hospitals [16]. Additionally, it is recommended that milk be stored in the refrigerator at ≤4 °C for no more than eight days, and in the freezer at −17 °C for up to 12 months [17]. Whilst evidence suggests that refrigeration, freezing and pasteurization impact human milk concentration of folate, vitamin B6, vitamin C, and other nutrients to various degrees [18–21], the effect of pasteurization and short- and long-term storage at different temperatures on water-soluble forms of choline in expressed human milk is not known.

Water-soluble forms of choline concentrations in human milk may vary within and between women. Higher maternal dietary choline intake increases milk concentrations of water-soluble, yet not fat-soluble, forms of choline, as shown in a 12-week dose-response feeding study and a supplementation trial from 18 gestational weeks to 90 days postpartum [11,13]. No consistent changes in total concentration of water-soluble forms of choline in human milk were observed in six women over a time period of 72 h [22]. The diurnal changes and variability of individual water-soluble choline forms among women have not been studied to date. Understanding these variabilities has important implications for designing study protocols (e.g., sampling techniques) that focus on identifying the determinants of choline in human milk and infant outcomes due to milk choline consumption.

The objectives of this research were to determine the effect of cold storage and pasteurization on the concentrations of water-soluble forms of choline in human milk, and to determine the intra- and inter-individual variability of water-soluble forms of choline concentrations in human milk within a day. Water-soluble forms of choline seem stable during short-term cold storage and most forms remained unchanged after four hours of storage at room temperature and six months of storage at ultra-low freezing temperatures, as employed in research settings. Holder pasteurization significantly impacted milk concentrations of water-soluble forms of choline, but to a small extent that is outweighed by the microbiological safety benefits of pasteurization. We observed intra-individual variability of individual but not total water-soluble forms of choline throughout one day, and recommend standardized sampling protocols for research studies.

2. Materials and Methods

2.1. Participants and Study Design

The research consisted of two cross-sectional studies, i.e., the stability and variability study, as well as the secondary analysis of bio-banked milk samples for the pasteurization study. For all studies, healthy women who were exclusively breastfeeding a healthy, full-term and singleton infant <6 months of age were eligible to participate. Exclusion criteria included: suffering from diabetes mellitus Type 1 or 2, or any chronic disease involving fat metabolism, taking routine medications known to influence fat metabolism, or consumption of more than 1 alcoholic drink per day.

2.1.1. Stability Study

A convenience sample of 6 mothers of 2- to 6- month old infants was included in this study. Women were recruited through active and passive recruitment methods in the Greater Vancouver area in 2014. Signed informed consent was obtained prior to enrolment. Information on maternal and infant age, sociodemographic status, supplement use, and general infant health, was collected using a self-administered questionnaire. Ethical approval was granted by the University of British Columbia and the British Columbia Children's and Women's (C&W) Hospital Research Ethics Board (H12-03191).

Women provided a fresh complete milk expression collected in the morning (between 9:00–10:00 a.m.) using a commercial pump (Medela); samples were immediately transferred on ice to the lab to be aliquoted and analyzed. One milk aliquot was analyzed immediately, with the rest of the samples being stored at different temperatures for different durations (see Supplemental Table S1), prior to analysis.

2.1.2. Pasteurization Study

Mid-feed human milk samples (milk collected after breastfeeding or pumping milk for approximately 3 min) from mothers of 1-month old infants were used in the pasteurization study; these bio-banked samples were derived from a completed, randomized controlled trial studying the effect of DHA supplementation during pregnancy on the cognitive and visual outcomes on infants [23]. The study was approved by C&W REB (H03-70242). A subset of milk samples was randomly selected from the bio-banked samples that had been stored for 5–9 years at −80 °C. In the original study, milk samples were immediately frozen at home for a maximum time of 3 days, and transferred on ice to the lab where they were frozen at −80 °C until analysis. Original milk samples obtained from the participants were thawed on ice, homogenized by gentle mixing, and divided in two aliquots. The first aliquot was stored at −20 °C for 1–2 h, and the second aliquot underwent Holder pasteurization in a water bath. Water-soluble forms of choline were quantified before and after pasteurization on the same day.

2.1.3. Variability Study

Twenty women were enrolled in this cross-sectional study. Recruitment and enrolment procedures were similar to those described for the stability study. Sociodemographic, health and supplement use data were collected using the same self-administered questionnaire as in the stability study. Ethical approval was granted by the University of British Columbia and the British Columbia Children's and Women's Hospital Research Ethics Board (H12-03191).

Each participant provided 5 mid-feed milk samples at different times throughout one day. For feasibility purposes, the five different time points were flexible as follows: before breakfast, before lunch, 45–60 min after lunch, 45–60 min after dinner, and before bedtime. Participants were instructed to place the vials in their home freezers immediately after pumping. Frozen samples were transferred to the lab on ice the day after collection and stored at −80 °C for 2 weeks, when choline analysis was completed.

2.2. Human Milk Choline Quantitation

Concentrations of water-soluble forms of choline in milk were determined using isotope dilution liquid chromatography tandem mass spectrometry as previously described [24]. In brief, aliquots of 20 μL of human milk were transferred to Eppendorf tubes containing 10 μL of deuterium-labeled internal standards (choline-d9, PhosC-d9, GPC-d9) and vortexed. Protein was precipitated with 30 μL of methanol with 0.1% formic acid. The supernatant was recovered after centrifugation at $18{,}000 \times g$ at 4 °C for 10 min, transferred to an autosampler vial and mixed with acetonitrile with 0.1% formic acid in dilutions of 1:5. The inter-assay and intra-assay coefficient of variation (CV) based on 5 replicates were as follows: For FC, 5.5% and 4.1%; for PhosC, 6.4% and 5.2%; and for GPC 9.5% and 2.3%; respectively.

2.3. Statistical Analysis

Participant characteristics are presented using descriptive statistics. Normality of data distribution was assessed using Shapiro–Wilk test. Differences in milk concentrations of water-soluble forms of choline under different storage conditions were determined using the related-samples Friedman test, followed by Wilcoxon signed-ranks test as post-hoc analysis and adjusted using the Bonferroni correction. Differences in water-soluble forms of choline concentrations after pasteurization were determined using the Wilcoxon signed-ranks test. Intra-individual variability in water-soluble forms of choline concentrations was determined using the related-samples Friedman test, followed by Wilcoxon signed-ranks test as post-hoc analysis and adjusted using the Bonferroni correction. Analyses were performed using the IBM SPSS statistics software (IBM SPSS Statistics for Windows, Version 25.0. SPSS Inc., Chicago, IL, USA). Level of significance was set at p values < 0.05.

3. Results

3.1. Participant Characteristics

The characteristics of women included in each study are summarized in Table 1.

Table 1. Participant characteristics of each study.

Characteristics Total Sample Size	Stability Study ($n = 6$)	Pasteurization Study ($n = 33$)	Variability Study ($n = 20$)
Age, y	35 ± 2 [1]	34 ± 4	32 ± 4
Postpartum, mo	4.0 (3.0) [2]	1	4.5 (3.0)
First-time breastfeeding, n (%)	6 (100)	20 (61)	10 (50)
Ethnic background, n (%)			
European	2 (33)	26 (79)	11 (55)
Latin American	2 (33)	2 (6)	4 (20)
Middle Eastern	0	3 (9)	3 (15)
First Nation	1 (17)	0	1 (5)
Chinese Asian	1 (17)	2 (6)	1 (5)

[1] Mean ± SD; [2] median (IQR).

3.2. Stability of Water-Soluble Choline Forms at Different Storage Conditions

The concentration of total water-soluble forms of choline, PhosC, and GPC did not significantly change during storage at room temperature for up to 4 h (Figure 1, Supplemental Table S2). Compared to fresh milk samples, i.e., baseline values, only FC concentration significantly increased after 3 h and 4 h of storage at room temperature.

We observed no changes in total water-soluble forms of choline and PhosC concentrations independent of condition and duration of cold storage, including in the refrigerator at 4 °C for 24 h, in the freezer at −20 °C for up to 1 week, and in the ultra-low freezer at −80 °C for up to 6 months (Figure 1, Supplemental Table S2). The concentration of FC and GPC did not change after storage at 4 °C for 24 h or at −20 °C for up to 1 week, but significantly increased between baseline and 6 months of storage at −80 °C.

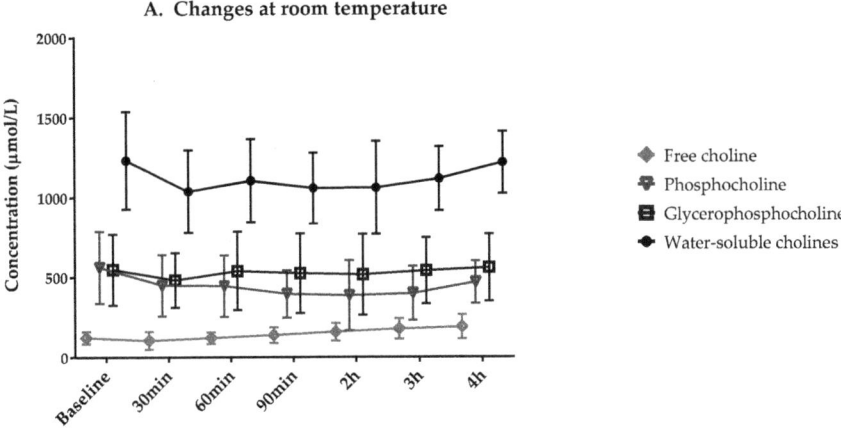

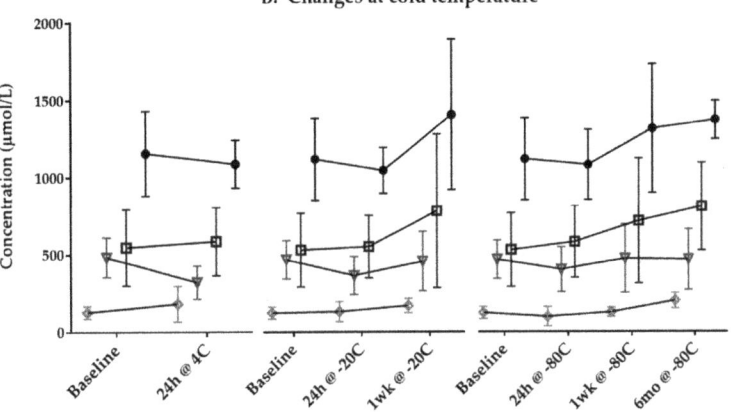

Figure 1. Changes in water-soluble forms of choline concentrations in human milk at (**A**) room temperature ($n = 6$) and (**B**) under different cold storage conditions ($n = 5$). Data presented as mean ± SD.

3.3. Stability of Water-Soluble Choline Forms during Pasteurization

Total water-soluble forms of choline concentrations significantly decreased by approximately 5% after Holder pasteurization in 33 milk samples (Table 2). This decrease seems to be largely driven by the decrease in the main water-soluble forms of choline in human milk, i.e., GPC and PhosC. The concentrations of GPC and PhosC showed a mean decrease of 4%–5%. Median FC concentration remained constant and did not seem to be impacted by exposure to pasteurization.

Table 2. Water-soluble forms of choline concentrations in human milk before and after Holder pasteurization [1].

Choline Form	Before, µmol/L	After, µmol/L	Difference, µmol/L	Difference, %	p-Value [2]
Free choline					
Mean ± SD	124 ± 60	124 ± 58	0.6 ± 14	1.3 ± 10	0.893
Median (min; max)	117 (47.9; 293)	120 (44.6; 275)	−1 (−29; 46)	−0.5 (−18; 33.0)	
Phosphocholine					
Mean ± SD	675 ± 220	632 ± 189	−43 ± 73	−4.7 ± 10	0.003
Median (min; max)	676 (230; 1131)	621 (261; 960)	−41 (−238; 111)	−6.5 (−21; 25)	
Glycerophosphocholine					
Mean ± SD	442 ± 181	425 ± 181	−17 ± 43	−3.6 ± 10	0.015
Median (min; max)	381 (243; 904)	387 (211; 994)	−20 (−122; 97)	-5.6 (−21; 26)	
Total water-soluble choline					
Mean ± SD	1241 ± 249	1186 ± 201	−56 ± 124	−3.4 ± 10	0.017
Median (min; max)	1229 (789; 1794)	1136 (792; 1566)	−63 (358; 231)	−4.8 (−20; 27)	

[1] $n = 33$ samples; [2] Wilcoxon signed-rank test to test for differences between before and after Holder pasteurization.

3.4. Intra- and Inter-Individual Variability of Water-Soluble Forms of Choline Concentrations

Total water-soluble forms of choline and PhosC concentrations did not vary significantly within a woman during the day based on analysis of five milk samples per mother (Table 3). However, significant changes in FC and GPC were found. Posthoc analysis showed a 22.7% increase in FC from T1 to T4 ($p = 0.027$) and a 12.2% decrease in GPC from T1 to T3 ($p = 0.027$).

Additionally, as shown in Figure 2, intra-individual variability in FC and GPC concentrations varied between women, whereby milk FC and GPC concentrations showed a large variability in some, i.e., 2 or 3 mothers, with milk samples of most other mothers reflecting minimal variability.

Table 3. Concentrations of water-soluble forms of choline in mid-feed milk samples collected at five different time points within one day [1].

Choline Form	T1	T2	T3	T4	T5	p [2]
Free choline						
Median (IQR)	119 (73.5)	125 (63.7)	131 (75.4)	146 (119)	132 (81.1)	0.029
Range	51.9–156	60.4–253	69.2–474	59.9–453	505–332	
Phosphocholine						
Median (IQR)	490 (328)	488 (270)	441 (359)	457 (330)	482 (335)	0.545
Range	138–906	82.1–1113	115–1120	78.9–1188	88.9–958	
Glycerophosphocholine						
Median (IQR)	625 (194)	559 (280)	549 (320)	533 (275)	682 (408)	0.008
Range	187–1363	107–1115	104–1188	99.1–1268	155–1445	
Total water-soluble choline						
Median (IQR)	1727 (366)	1219 (410)	1200 (308)	1289 (404)	1230 (344)	0.224
Range	955–2117	768–1817	809–1894	752–1896	652–2729	

[1] $n = 20$ mothers exclusively breastfeeding 2–6 months old infants. T1, milk collected before breakfast; T2, before lunch; T3, 45–60 min after lunch; T4, 45–60 min after dinner; T5, before bedtime. [2] p values for repeated-measures Friedman test for difference across time points within a woman.

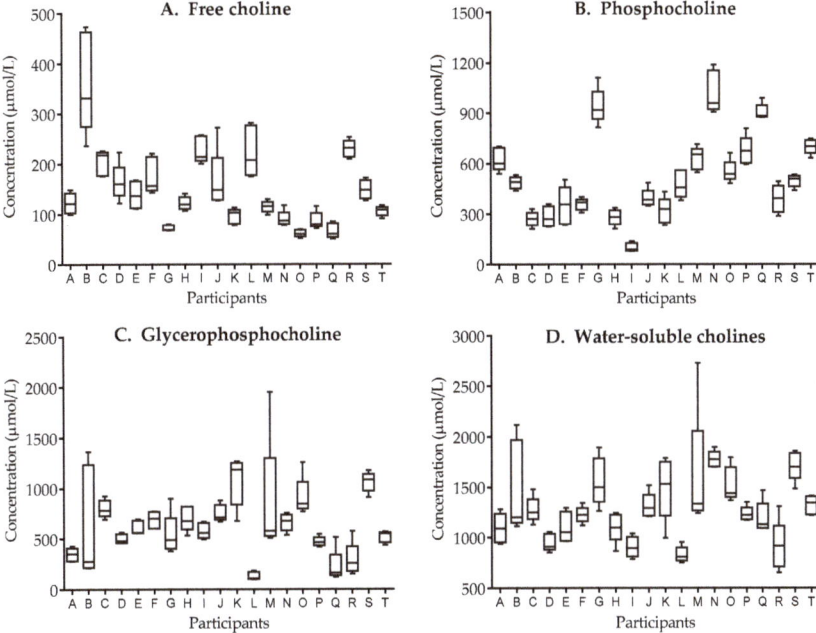

Figure 2. Intra-individual variability of human milk concentration of (**A**) free choline, (**B**) phosphocholine, (**C**) glycerophosphocholine, and (**D**) total water-soluble forms of choline. Boxes represent median and 25th–75th percentile; whiskers represent minimum and maximum values. Data for each participant (n = 20, represented on the x-axis) includes five mid-feed milk samples collected at separate feeds on a single day.

4. Discussion

In this study, we report three findings on the stability of water-soluble forms of choline in expressed human milk: First, storing human milk for 3 or 4 h at room temperature significantly increases milk FC concentration. Second, thawing human milk after 6 months of freezing at −80 °C significantly increases FC and GPC concentrations. Third, Holder pasteurization decreases PhosC and GPC concentrations, and thereby lowers total concentration of water-soluble forms of choline. Additionally, we report that total concentration of water-soluble forms of choline does not underlie diurnal variations within a woman; however, significant diurnal variations in FC and GPC concentrations may occur among some, but not all, women. These findings contribute to the literature being harnessed to develop evidence-informed guidelines for the handling and storage of expressed human milk as well as the development of optimized milk collection and storage protocols for research studies.

Milk expression for in-home use or human milk banking have become increasingly recognized as a first best alternative to direct breastfeeding. While ensuring microbiological safety of expressed milk remains a top priority, maintaining milk's bioactive and nutritional quality is an important weighing factor in developing milk handling and storage guidelines. The impact of storage and pasteurization on the immunological properties, digestive enzymes, antioxidant capacity, and macro- and micronutrient composition in human milk has been recently reviewed [19,25–29], with limited data on water-soluble forms of choline. In an older study, Zeisel et al. [22] compared fresh milk to either samples incubated for 15 min at 37 °C or to samples frozen for 72 h at −10 °C followed by incubation at 37 °C for 15 min. Using radioisotope labeled choline compounds, the authors showed that neither of the two milk handling and storage conditions resulted in significant changes in FC concentration. These findings combined with ours suggest that following current recommendation of keeping freshly expressed milk up to 4 h at room temperature should not significantly alter the composition of water-soluble forms of choline, except for minor possible increases in FC. The possible mechanism(s) underlying the increase in FC concentration cannot be determined in our study as designed.

The increase in FC concentration may be explained by the enzymatic breakdown of the lipid-soluble choline compounds phosphatidylcholine and sphingomyelin. The presence of these phospholipids within the milk fat globule membrane, a complex tri-layer of proteins and lipids, poses an analytical challenge for their accurate quantification [30]. In addition to identifying factors influencing the variability of choline forms in human milk in future studies, it is critical to also investigate whether changes in the composition of choline forms in milk, not total choline concentration per se, have functional implications for infant health and development. Choline has a wide array of functions that support infant growth and development. As an essential component of phosphatidylcholine and sphingomyelin, choline is involved in the maintenance of cell membrane structural integrity and signaling pathways, as well as in parenchymal growth, cell proliferation, and membrane formation [3–6]. Choline, via its oxidized form betaine, also functions as a methyl group donor in the generation of S-adenosylmethionine [7]. Additionally, choline is crucial for brain function as precursor of the neurotransmitter acetylcholine [4]. To date, the contribution of different choline forms in human milk to the variety in choline functions in the developing infant is largely unknown and merits further investigation.

The cold storage conditions we tested are relevant to home and clinical use (refrigeration temperature typically around 4 °C and freezing at −20 °C) and to storage conditions in research settings (ultra-low freeze storage at −80 °C). Our findings suggest that short-term storage of milk aliquots in the refrigerator (at 4 °C) for 1 day and in the home freezer (at −20 °C) for up to one week and long-term storage at −80 °C for up to 6 months does not alter the total concentration of water-soluble forms of choline in expressed human milk samples. However, milk FC and GPC concentrations increased in samples that were thawed after six months of ultra-low freeze storage (i.e., at −80 °C). The increase in FC and GPC concentrations after prolonged freezer storage may be explained by the breakdown of lipid-soluble forms of choline, similar to the increase in FC concentration in milk stored at room temperature for 4 h. Similar changes were observed in serum samples that were inappropriately processed and stored, with increasing concentration of total water-soluble choline concentration [31]. We were limited to six participants for the stability study as it was logistically challenging to recruit participants willing to commute to our research site, provide a fresh milk sample, and provide the sample as a complete milk expression. We however compensated in our statistical analysis for this limitation (i.e., possible high variability across samples due to small sample size) by using related-samples analyses rather than independent comparisons across conditions. Future research is warranted to enhance the stability data by testing more milk samples as well as more frequent and longer duration intervals to identify whether and when water-soluble forms of choline concentration or composition may change in different cold storage conditions.

To our knowledge, this study is the first to test the effect of pasteurization on water-soluble forms of choline concentrations in human milk. We showed that Holder pasteurization lowered milk PhosC and GPC concentrations by about 5%, and thereby total water-soluble forms of choline concentration by 3%. Limitation of our study was the use of previously frozen milk samples, for 5–9 years at −80 °C, which is not consistent with clinical practice of pooling human milk and it undergoing pasteurization. We used biobanked milk samples because of low milk volumes collected in the stability and variability study, as well as to reduce participant burden. Additionally, for this preliminary study, we were mainly focusing on the comparison of choline concentrations between before and after pasteurization. We acknowledge that the long-term storage at −80 °C may have affected the milk concentrations of the water-soluble forms of choline; however, the total water-soluble choline concentration in the pre-pasteurized samples was similar to that of the fresh milk samples in the stability study (1241 µmol/L versus 1231 µmol/L, respectively), as well as compared to those reported in the literature [11,24,32]. This seems to reflect that the total water-soluble choline concentration was not affected by the long-term storage at −80 °C and that our findings may contribute to the literature of how human milk is altered by pasteurization.

Considering the crucial benefits of pasteurization for microbiological safety, we evaluate the <5% decrease in milk concentration of water-soluble forms of choline as minor yet recommend the confirmation of our findings in a second and larger study. The impact of pasteurization on lipid-soluble choline compounds should also be investigated in future studies. If confirmed, our finding of a 5% decrease should reassure clinicians that the benefits of providing pasteurized human milk continue to outweigh the risk of nutrient losses. Indeed, despite previous reports of a decrease in several nutrients, including folate, vitamin C and B6, due to pasteurization [18,25,26], the use of pasteurization techniques continues to ensure the biologically-safe provision of human milk at hospital settings around the globe.

In regards to the variability of water-soluble forms of choline concentration in expressed human milk, we observed diurnal changes in milk FC and GPC concentrations, but not in PhosC and total water-soluble choline forms. The total concentration of water-soluble forms of choline in expressed human milk we found was similar to the concentrations previously reported by our team for Canadian lactating women [24,32] and by Fischer et al. for US women [11]. Because no substantial diurnal changes were observed and the total concentration of water-soluble forms of choline in human milk seems stable, we conclude that the time point of milk sample collection in studies on water-soluble forms of choline may not influence the study outcomes. However, because of the diverse intra-individual variability of FC and GPC concentration within a day, we recommend to standardize the time of milk sample collection across study participants. The influencing factors of the intra-individual variability for some of the water-soluble choline forms, when present, are not fully elucidated but may be related to dietary intake of choline and/or genetic variants related to choline absorption, distribution, and metabolism [11]. The effect of acute versus long-term dietary choline intake on milk composition of water-soluble forms of choline merits further investigation.

In conclusion, we provide new information on the stability of water-soluble forms of choline concentration that will help in the development of evidence-based guidelines for the safe handling and storage of expressed milk samples. Because breast milk is the recommended sole source of nutrients, including that of choline, for infants under the age of 6 months, the handling of expressed milk needs to address microbiological safety as well as nutrient-protective needs. Further research is warranted on the effect of acute versus long-term dietary choline intake on the composition of choline forms in human milk as well as the metabolic and functional significance of individual choline forms on infant growth and development.

Supplementary Materials: The following are available online at http://www.mdpi.com/2072-6643/11/12/3024/s1: Supplementary Table S1: Storage duration and temperature condition for stability testing of human milk choline composition; Supplementary Table S2: Changes in water-soluble forms of choline concentrations in human milk under different storage conditions.

Author Contributions: The authors' contributions were as follows: S.M.I., L.S.S., and A.M.W. designed the research; L.S.S., S.M., and A.M.W. coordinated and conducted the studies from which the milk samples were obtained; R.A.D. contributed to laboratory analysis; A.M.W. and Y.L. analyzed the data; L.S.S. prepared the initial draft of the manuscript; S.M., A.M.W., and Y.L. provided critical input into the writing and editing of the manuscript. All authors have read and approved the final version of the manuscript.

Funding: This research received no external funding. A.M.W. acknowledges funding support through Doctoral Becas-Chile Scholarship provided through Comisión Nacional de Investigación Científica y Tecnológica (CONICYT) from the Government of Chile and Nutritional Research Fellowship from The University of British Columbia, Canada. L.S.S. acknowledges funding in form of a graduate student scholarship from the Consejo Nacional de Ciencia y Tecnología (CONACYT), Mexico. Y.L. acknowledges funding from the Canada Research Chair Program/Canadian Institutes of Health Research.

Acknowledgments: We thank all study participants for their time and efforts.

Conflicts of Interest: The authors declare no conflict of interest.

References

1. Zeisel, S.; Niculescu, M. Perinatal choline influences brain structure and function. *Nutr. Rev.* **2006**, *64*, 197–203. [CrossRef] [PubMed]
2. Zeisel, S.H. Nutritional Importance of Choline for Brain Development. *J. Am. Coll. Nutr.* **2004**, *23*, 621S–626S. [CrossRef] [PubMed]
3. Zeisel, S.H. Choline: Critical Role During Fetal Development and Dietary Requirements in Adults. *Annu. Rev. Nutr.* **2006**, *26*, 229–250. [CrossRef]
4. Zeisel, S.H.; Blusztajn, J.K. Choline and Human Nutrition. *Rev. Nutr* **1994**, *14*, 269–296. [CrossRef]
5. Zeisel, S.H. Biochemistry, physiology, and pharmacology. *Ann. Rev. Nutr.* **1981**, *1*, 121. [CrossRef]
6. Jiang, X.; West, A.A.; Caudill, M.A. Maternal choline supplementation: A nutritional approach for improving offspring health? *Trends Endocrinol. Metab.* **2014**, *25*, 263–273. [CrossRef]
7. Craig, S.A. Betaine in human nutrition. *Am. J. Clin. Nutr.* **2004**, *80*, 539–549. [CrossRef]
8. WHO Exclusive Breastfeeding for Six Months Best for Babies Everywhere. Available online: http://www.who.int/mediacentre/news/statements/2011/breastfeeding_20110115/en (accessed on 17 July 2019).
9. Holmes, H.C.; Snodgrass, G.J.A.I.; Iles, R.A. Changes in the choline content of human breast milk in the first 3 weeks after birth. *Eur. J. Pediatr.* **2000**, *159*, 198–204. [CrossRef]
10. Ilcol, Y.O.; Ozbek, R.; Hamurtekin, E.; Ulus, I.H. Choline status in newborns, infants, children, breast-feeding women, breast-fed infants and human breast milk. *J. Nutr. Biochem.* **2005**, *16*, 489–499. [CrossRef]
11. Fischer, L.M.; Da Costa, K.A.; Galanko, J.; Sha, W.; Stephenson, B.; Vick, J.; Zeisel, S.H. Choline intake and genetic polymorphisms influence choline metabolite concentrations in human breast milk and plasma. *Am. J. Clin. Nutr.* **2010**, *92*, 336–346. [CrossRef]
12. Holmes-McNary, M.Q.; Cheng, W.L.; Mar, M.H.; Fussell, S.; Zeisel, S.H. Choline and choline esters in human and rat milk and in infant formulas. *Am. J. Clin. Nutr.* **1996**, *64*, 572–576. [CrossRef] [PubMed]
13. Davenport, C.; Yan, J.; Taesuwan, S.; Shields, K.; West, A.A.; Jiang, X.; Perry, C.A.; Malysheva, O.V.; Stabler, S.P.; Allen, R.H.; et al. Choline intakes exceeding recommendations during human lactation improve breast milk choline content by increasing PEMT pathway metabolites. *J. Nutr. Biochem.* **2015**, *26*, 903–911. [CrossRef] [PubMed]
14. Jaeger, M.C.; Lawson, M.; Filteau, S. The impact of prematurity and neonatal illness on the decision to breast-feed. *J. Adv. Nurs.* **1997**, *25*, 729–737. [CrossRef] [PubMed]
15. Johns, H.M.; Forster, D.A.; Amir, L.H.; McLachlan, H.L. Prevalence and outcomes of breast milk expressing in women with healthy term infants: A systematic review. *BMC Pregnancy Childbirth* **2013**, *13*, 212. [CrossRef]
16. Jones, F. *Best Practice for Expressing, Storing and Handling Human Milk in Hospitals, Homes, and Child Care Settings*, 3rd ed.; Human Milk Banking Association of North America, Inc.: Fort Worth, TX, USA, 2011.
17. Eglash, A.; Simon, L.; Brodribb, W.; Reece-Stremtan, S.; Noble, L.; Brent, N.; Bunik, M.; Harrel, C.; Lawrence, R.A.; LeFort, Y.; et al. ABM Clinical Protocol #8: Human Milk Storage Information for Home Use for Full-Term Infants, Revised 2017. *Breastfeed. Med.* **2017**, *12*, 390–395.
18. Donnelly-Vanderloo, M.; O'Connor, D.L.; Shoukri, M. Impact of pasteurization and procedures commonly used to rethermalize stored human milk on folate content. *Nutr. Res.* **1994**, *14*, 1305–1316. [CrossRef]
19. Peila, C.; Moro, G.; Bertino, E.; Cavallarin, L.; Giribaldi, M.; Giuliani, F.; Cresi, F.; Coscia, A. The Effect of Holder Pasteurization on Nutrients and Biologically-Active Components in Donor Human Milk: A Review. *Nutrients* **2016**, *8*, 477. [CrossRef]
20. García-Lara, N.R.; Escuder-Vieco, D.; García-Algar, O.; De la Cruz, J.; Lora, D.; Pallás-Alonso, C. Effect of Freezing Time on Macronutrients and Energy Content of Breastmilk. *Breastfeed. Med.* **2012**, *7*, 295–301. [CrossRef]
21. Ahrabi, A.F.; Handa, D.; Codipilly, C.N.; Shah, S.; Williams, J.E.; McGuire, M.A.; Potak, D.; Aharon, G.G.; Schanler, R.J. Effects of Extended Freezer Storage on the Integrity of Human Milk. *J. Pediatr.* **2016**, *177*, 140–143. [CrossRef]
22. Zeisel, S.H.; Char, D.; Sheard, N.F. Choline, phosphatidylcholine and sphingomyelin in human and bovine milk and infant formulas. *J. Nutr.* **1986**, *116*, 50–58. [CrossRef]
23. Innis, S.M.; Friesen, R.W. Essential n-3 fatty acids in pregnant women and early visual acuity maturation in term infants. *Am. J. Clin. Nutr.* **2008**, *87*, 548–557. [CrossRef]

24. Moukarzel, S.; Soberanes, L.; Dyer, R.A.; Albersheim, S.; Elango, R.; Innis, S.M.; Moukarzel, S.; Soberanes, L.; Dyer, R.A.; Albersheim, S.; et al. Relationships among Different Water-Soluble Choline Compounds Differ between Human Preterm and Donor Milk. *Nutrients* **2017**, *9*, 369. [CrossRef]
25. Peters, M.D.J.; McArthur, A.; Munn, Z. Safe management of expressed breast milk: A systematic review. *Women Birth* **2016**, *29*, 473–481. [CrossRef]
26. Nessel, I.; Khashu, M.; Dyall, S.C. The effects of storage conditions on long-chain polyunsaturated fatty acids, lipid mediators, and antioxidants in donor human milk—A review. *Prostaglandins Leukot. Essent. Fat. Acids* **2019**, *149*, 8–17. [CrossRef]
27. Loikas, S.; Lopponen, M.; Suominen, P.; Moller, J.; Irjala, K.; Isoaho, R.; Kivela, S.L.; Koskinen, P.; Pelliniemi, T.T. RIA for serum holo-transcobalamin: Method evaluation in the clinical laboratory and reference interval. *Clin. Chem.* **2003**, *49*, 455–462. [CrossRef]
28. O'Connor, D.L.; Ewaschuk, J.B.; Unger, S. Human milk pasteurization: Benefits and risks. *Curr. Opin. Clin. Nutr. Metab. Care* **2015**, *18*, 269–275. [CrossRef]
29. Gao, C.; Miller, J.; Middleton, P.F.; Huang, Y.C.; McPhee, A.J.; Gibson, R.A. Changes to breast milk fatty acid composition during storage, handling and processing: A systematic review. *Prostaglandins Leukot. Essent. Fat. Acids* **2019**, *146*, 1–10. [CrossRef]
30. Moukarzel, S.; Dyer, R.A.; Keller, B.O.; Elango, R.; Innis, S.M. Human milk plasmalogens are highly enriched in long-chain PUFAs. *J. Nutr.* **2016**, *146*, 2412–2417. [CrossRef]
31. Holm, P.I.; Ueland, P.M.; Kvalheim, G.; Lien, E.A. Determination of choline, betaine, and dimethylglycine in plasma by a high-throughput method based on normal-phase chromatography-tandem mass spectrometry. *Clin. Chem.* **2003**, *49*, 286–294. [CrossRef]
32. Wiedeman, A.M.; Whitfield, K.C.; March, K.M.; Chen, N.N.; Kroeun, H.; Sokhoing, L.; Sophonneary, P.; Dyer, R.A.; Xu, Z.; Kitts, D.D.; et al. Concentrations of Water-Soluble Forms of Choline in Human Milk from Lactating Women in Canada and Cambodia. *Nutrients* **2018**, *10*, 381. [CrossRef]

© 2019 by the authors. Licensee MDPI, Basel, Switzerland. This article is an open access article distributed under the terms and conditions of the Creative Commons Attribution (CC BY) license (http://creativecommons.org/licenses/by/4.0/).

Article

Breastfeeding Difficulties and Risk for Early Breastfeeding Cessation

Maria Lorella Gianni [1,2,*], Maria Enrica Bettinelli [2], Priscilla Manfra [1], Gabriele Sorrentino [1], Elena Bezze [1], Laura Plevani [1], Giacomo Cavallaro [1], Genny Raffaeli [1], Beatrice Letizia Crippa [1,2], Lorenzo Colombo [1], Daniela Morniroli [1,2], Nadia Liotto [1,2], Paola Roggero [1,2], Eduardo Villamor [3], Paola Marchisio [4,5] and Fabio Mosca [1,2]

1. Fondazione IRCCS Ca' Granda Ospedale Maggiore Policlinico, NICU, via Commenda 12, 20122 Milan, Italy; priscilla.manfra@gmail.com (P.M.); gabriele.sorrentino@mangiagalli.it (G.S.); elena.bezze@policlinico.mi.it (E.B.); laura.plevani@mangiagalli.it (L.P.); giacomo.cavallaro@mangiagalli.it (G.C.); genny.raffaeli@gmail.com (G.R.); beatriceletizia.crippa@gmail.com (B.L.C.); lorenzo.colombo@policlinico.mi.it (L.C.); daniela.morniroli@gmail.com (D.M.); nadia.liotto@mangiagalli.it (N.L.); paola.roggero@unimi.it (P.R.); fabio.mosca@mangiagalli.it (F.M.)
2. Department of Clinical Sciences and Community Health, University of Milan, Via San Barnaba 8, 20122 Milan, Italy; maria.bettinelli@unimi.it
3. Department of Pediatrics, Maastricht University Medical Center (MUMC+), School for Oncology and Developmental Biology (GROW), 6202 AZ Maastricht, The Netherlands; e.villamor@mumc.nl
4. Fondazione IRCCS Ca' Granda Ospedale Maggiore Policlinico, 20122 Milan, Italy; paola.marchisio@unimi.it
5. Department of Pathophysiology and Transplantation, University of Milan, 20122 Milan, Italy
* Correspondence: maria.gianni@unimi.it; Tel.: +39-0255032483

Received: 25 August 2019; Accepted: 17 September 2019; Published: 20 September 2019

Abstract: Although breast milk is the normative feeding for infants, breastfeeding rates are lower than recommended. We investigated breastfeeding difficulties experienced by mothers in the first months after delivery and their association with early breastfeeding discontinuation. We conducted a prospective observational study. Mothers breastfeeding singleton healthy term newborns at hospital discharge were enrolled and, at three months post-delivery, were administered a questionnaire on their breastfeeding experience. Association among neonatal/maternal characteristics, breastfeeding difficulties and support after hospital discharge, and type of feeding at three months was assessed using multivariate binary logistic regression analysis. We enrolled 792 mothers, 552 completed the study. Around 70.3% of mothers experienced breastfeeding difficulties, reporting cracked nipples, perception of insufficient amount of milk, pain, and fatigue. Difficulties occurred mostly within the first month. Half of mothers with breastfeeding issues felt well-supported by health professionals. Maternal perception of not having a sufficient amount of milk, infant's failure to thrive, mastitis, and the return to work were associated with a higher risk of non-exclusive breastfeeding at three months whereas vaginal delivery and breastfeeding support after hospital discharge were associated with a decreased risk. These results underline the importance of continued, tailored professional breastfeeding support.

Keywords: breastfeeding difficulties; early breastfeeding cessation; term infants; breastfeeding support

1. Introduction

Breastfeeding is associated with improvement of infants' survival and significant health benefits both for infants and mothers in a dose-response manner [1–3]. Consequently, promotion and support of breastfeeding initiation, duration, and exclusivity is a public health issue. However, the worldwide rates of breastfeeding are lower than international recommendations, especially in high-income countries [4]. Therefore, there is a need for increasing the health care professionals' awareness of the intrinsic factors associated with early breastfeeding cessation and for gaining further insight into the related modifiable risk factors [5]. Several determinants of breastfeeding have been described within a complex framework, including structural settings and individual factors that are involved at multiple levels [6]. Among the individual factors, the experience of breastfeeding difficulties greatly contributes to early breastfeeding cessation and causes mothers to be less likely to breastfeed a future child [7]. However, "breastfeeding difficulties" includes a wide range of different biological, psychological, and social factors [8]. Unpacking this issue to gain further insight into the modifiable barriers mothers experience during breastfeeding may help health professionals in overcoming them and in refining community support [5].

The aim of the present study was to investigate the breastfeeding difficulties experienced by mothers of healthy, singleton term-born infants in the first months after delivery and their association with early breastfeeding discontinuation.

2. Materials and Methods

We conducted a prospective, observational study in the nursery of Fondazione IRCCS Ca' Granda Ospedale Maggiore Policlinico in Milan, Lombardy, Italy. The hospital is a Level III center for neonatal care that covers around 6000 deliveries per year, admitting pregnant women prevalently resident in Lombardy but also those resident in other Italian regions.

All subjects gave their informed consent for inclusion before they participated in the study. The study was conducted in accordance with the Declaration of Helsinki, and the protocol was approved by the Ethics Committee of Milano (Comitato Etico Milano Area 2, n. 0120, atti n. 1580/2018).

Mothers with a low risk for early breastfeeding cessation, that is having delivered singleton, healthy, term (gestational age ≥37 weeks) newborns with the birthweight ≥10th percentile for gestational age, according to the Bertino's neonatal growth chart [9], and breastfeeding were enrolled at hospital discharge, which occurred within the completion of the first 72 h after delivery. Exclusion criteria included exclusive formula feeding, multiple pregnancy, non-Italian speaking mothers due to fact that the language barrier could have interfered with the accuracy of the answers, and mothers whose newborns were admitted to Neonatal Intensive Care Unit and/or were affected by any condition that could interfere with breastfeeding, such as congenital diseases, chromosomal abnormalities, lung disease, brain disease, metabolic disease, cardiac disease, or gastrointestinal diseases. Breastfeeding was promoted and supported in all mother-infant pairs throughout the hospital stay, following the Ten Steps to Successful Breastfeeding [10]. Socio-demographic maternal variables (age, marital status, education, mode of delivery, parity), basic infants' characteristics (gestational age, birth weight, length, head circumference, Apgar score), and the infants' mode of feeding at hospital discharge were collected. At discharge, mothers were instructed to record in a diary their infant's mode of feeding at seven days, one month, and three months after delivery. The mode of feeding was categorized according to the World Health Organization definition [11] as exclusive breastfeeding (infants are fed only breast milk and no other food or drink; not even water; oral rehydration solutions, drops and syrups such as vitamins, minerals and medicines are permitted); predominant breastfeeding (breast milk is the infant's predominant source of nourishment but liquids such as water and water-based drinks are permitted); complementary feeding (infants are mainly breastfed but also consume formula milk and other liquid or non-dairy foods); and exclusive formula feeding.

At three months post-delivery, mothers were contacted by phone in order to collect the recorded infant feeding data, and were reminded to access and complete the online questionnaire investigating their breastfeeding experience following hospital discharge within the subsequent 48 h. Specifically, mothers were asked whether they had encountered any difficulty with regard to breastfeeding. If the mothers answered yes, they had to report which difficulties they had encountered during their breastfeeding experience, when the encountered difficulties had arisen (discharge–1st month after delivery, 1st month–2nd month after delivery, 2nd month–3rd month after delivery) and how they had been solved. Mothers were also required to rate the breastfeeding support they had received by health care professionals after hospital discharge (excellent, very good, satisfactory, poor, very poor, or unacceptable).

Statistical Analysis

Data are presented as mean (SD) or number of observations (%). For analysis, maternal age was divided into two categories based on the median value; maternal educational age was categorized as ≤13 years or >13 years, while breastfeeding support after discharge was considered positive if mothers rated it either as excellent, very good, or satisfactory and negative if the mothers rated it either as poor, very poor or unacceptable. Mode of feeding was categorized as exclusive breastfeeding vs. non-exclusive breastfeeding. The latter category included complementary feeding and exclusive formula feeding. Association between socio-demographic characteristics, the mode of delivery, parity, the occurrence of breastfeeding difficulties at any time point of the study, having been supported after hospital discharge and the mode of infant's feeding at three months (reference group: non-exclusive breastfeeding) were first assessed using univariate binary logistic regression analysis. A multivariate binary logistic regression analysis was then conducted in order to identify which breastfeeding difficulties arisen through the study period were independently associated with the type of feeding at three months. When adjusting the model, we included the items that showed a significant association with type of feeding at univariate analysis. Statistical significance was set at the $\alpha = 0.05$ level. The statistical analyses were performed using SPSS (version 12, SPSS, Inc., Chicago, IL, USA).

3. Results

Of the 1843 mothers who delivered during the study period, 868 were eligible for the study. A total of 76 mothers refused to participate and 792 mother-infant pairs were enrolled. Among these, 552 (70%) completed the study and the online questionnaire whereas the remaining 240 mothers did not complete either the study or the online questionnaire since it was not possible to reach them by telephone after hospital discharge.

Basic characteristics of mother-infant pairs which completed the study are summarized in Table 1. Mother-infant pairs that have not completed the study did not significantly differ from the ones completing the study.

Table 1. Basic characteristics of the mother-infant pairs that completed ($n = 552$) and that not completed (240) the study.

	Mothers that Completed the Study ($n = 552$)	Mothers Who did not Complete the Study ($n = 240$)
Maternal age, years (mean ± SD)	35.5 ± 4.6	34.9 ± 4.6
Marital status, n (%)		
Married or cohabitant	540 (98)	237 (99)
Single or divorced	12 (2)	3 (1)
Maternal education level, n (%)		
≤13 years	150 (27)	75 (31)
>13 years	402 (73)	165 (69)
Vaginal delivery, n (%)	369 (66.8)	157 (65.4)
Primiparous, n (%)	290 (52.5)	138 (57.5)

Table 1. *Cont.*

	Mothers that Completed the Study (*n* = 552)	Mothers Who did not Complete the Study (*n* = 240)
	Infants Born to Mothers that had Completed the Study (*n* = 552)	Infants Born to Mothers Who did not Complete the Study (*n* = 240)
Gestational age, weeks (mean ± SD)	39.2 ± 1.0	39.3 ± 0.9
Birth weight, g (mean ± SD)	3368 ± 350	3390 ± 332
Length, cm (mean ± SD)	50.1 ± 1.6	50.3 ± 1.5
Head circumference, cm (mean ± SD)	34.5 ± 1.4	34.3 ± 1.3

The mode of feeding at each time point of the study is reported in Table 2. At enrollment, 95% of the mothers practiced exclusive breastfeeding, whereas 5% of the mothers practiced complementary feeding. At one and three months, exclusive breastfeeding rates declined to 73% and 68%, respectively, whereas complementary feeding rates were 20% and 15%, respectively. Percentage of infants receiving exclusive formula feeding was 7% at one month, increasing up to 17% at three months.

Table 2. Mode of feeding at each time point of the study.

	Enrollment	Seven Days	One Month	Three Months
Exclusive breastfeeding	524 (95%)	447 (81%)	402 (73%)	375 (68%)
Predominant breastfeeding	0%	5 (1%)	5 (1%)	5 (1%)
Complementary feeding	28 (5%)	99 (18%)	105 (19%)	77 (14%)
Exclusive formula feeding	0%	0%	39 (7%)	94 (17%)

A total of 388 (70.3%) mothers experienced difficulties during breastfeeding. The difficulties most frequently reported by the mothers were cracked nipples, the perception of insufficient amount of milk, pain, and fatigue (Table 3).

Table 3. Breastfeeding difficulties arisen at any time point of the study according to mothers' experience.

Breastfeeding Difficulties	N (%)
Cracked nipples	159 (41.0)
Perception of an insufficient amount of milk	139 (35.8)
Pain not associated with cracked nipples	121 (31.2)
Fatigue	117 (30.2)
Breast engorgement	102 (26.3)
Infant's failure to thrive	79 (20.4)
Incorrect latching	74 (19.1)
Perception of own's milk limited nutritional value	68 (17.5)
Mastitis	27 (7.0)
Return to work	17 (4.4)
Prescription drugs	8 (2.1)

Most of the mothers (63%) reported the occurrence of difficulties within the first month after delivery whereas, in the second and third month after delivery, difficulties were experienced only by 9% and 10% of the enrolled mothers, respectively. A total of 189 (48.7%) mothers among those that have encountered difficulties in breastfeeding reported they were successfully supported by health professionals, whereas 78 (20.1%) mothers solved the difficulties by themselves and 45 (11.6%) mothers with the support of friends or relatives. Difficulties were not solved in 19.6% of cases; however, 7% of these latter mothers kept on breastfeeding.

After hospital discharge, the breastfeeding support received by health professionals was rated as either excellent, very good, or satisfactory in most cases (86.1%) whereas only in the 13.9% of cases the breastfeeding support was reported as either poor, very poor, or unacceptable. The mothers who rated the breastfeeding support after hospital discharge as negative were at higher risk of non-exclusive breastfeeding at three months than the mothers that rated the support after hospital discharge as positive (OR = 1.367, 95%CI 1.09–1.70, $p = 0.005$).

Univariate analysis showed that the absence of breastfeeding difficulties and having been supported in case of difficulties were significantly associated with a lower risk of non-exclusive breastfeeding at three months (OR = 0.051; 95% CI 0.022; 0.117, $p < 0.0001$; OR = 0.39; 95% CI 0.202–0.756, $p = 0.005$, respectively). When taking into account the type of breastfeeding difficulties, the perception of not having enough milk, pain perception, infant's failure to thrive, the perception of milk's limited nutritional value, the occurrence of mastitis, and the return to work were associated with a higher risk of non-exclusive breastfeeding at three months (Table 4). Primiparity and an incorrect latching tended to be associated with a higher risk of non-exclusive breastfeeding at three months whereas vaginal delivery resulted in being associated with a lower risk (Table 4). No significant association was found between maternal education level and age, breast engorgement, cracked nipples, fatigue, prescription drugs, and the infant's mode of feeding (Table 4).

Table 4. Association among maternal age and education, the mode of delivery, parity, and types of breastfeeding difficulties and the mode of infant's feeding at three months (univariate binary logistic regression analysis).

	Reference Group: Non-Exclusive Breastfeeding		
	OR	95%; CI	p
Maternal age (≤35 vs. >35 years)	1.02	0.711; 1.465	0.913
Maternal education (≤13 vs. >13 years)	0.67	0.237; 1.93	0.465
Mode of delivery (spontaneous vs. caesarean delivery)	0.60	0.415; 0.881	0.009
Parity (primiparous vs. multiparous)	1.42	0.988; 2.051	0.058
Cracked nipples (yes vs. no)	1.38	0.933; 2.042	0.107
Perception of not having enough milk (yes vs. no)	9.23	5.961; 14.301	<0.0001
Pain not associated with cracked nipples (yes vs. no)	1.62	1.066; 2.487	0.024
Fatigue (yes vs. no)	1.22	0.790; 1.903	0.363
Breast engorgement (yes vs. no)	0.87	0.545; 1,412	0.590
Infant's failure to thrive (yes vs. no)	5.136	3.094; 8.525	<0.0001
Incorrect latching (yes vs. no)	1.58	0.949; 2.635	0.078
Perception of milk's limited nutritional value (yes vs. no)	3.44	2.015; 5.898	<0.0001
Mastitis (yes vs. no)	2.49	1.144; 5.420	0.022
Return to work (yes vs. no)	7.65	2.457; 23.830	<0.0001
Prescription drugs (yes vs. no)	2.29	0.266; 19.761	0.452

Multivariate binary logistic regression showed that the maternal perception of not having a sufficient amount of milk, infant's failure to thrive, mastitis, and the return to work were associated with a higher risk of non-exclusive breastfeeding at three months whereas vaginal delivery and breastfeeding support after hospital discharge were associated with a decreased risk (Table 5).

Table 5. Association among the mode of delivery, having been supported after discharge, the types of breastfeeding difficulties and the mode of infant's feeding at three months (multivariate binary logistic regression analysis).

	Reference Group: Non-Exclusive Breastfeeding			
	B	OR	95%; CI	p
Mode of delivery (spontaneous vs. caesarean delivery)	−0.57	0.56	0.329; 0.961	0.035
Having been supported after hospital discharge (yes vs. no)	−1.28	0.27	0.130; 0.594	0.001
Perception of not having enough milk (yes vs. no)	1.96	7.15	4.096; 12.499	<0.0001
Pain not associated with cracked nipples (yes vs. no)	0.25	1.29	0.737; 2.265	0.37
Infant's failure to thrive (yes vs. no)	1.00	2.73	1.441; 5.180	0.002
Perception of milk's limited nutritional value (yes vs. no)	0.59	1.81	0.912; 3.607	0.089
Mastitis (yes vs. no)	1.07	2.92	1.166; 7.314	0.022
Return to work (yes vs. no)	1.63	5.136	1.046; 25.204	0.044

4. Discussion

Increasing awareness of the modifiable barriers experienced by mothers during breastfeeding may help health professionals in the detection of mothers at risk for early cessation of breastfeeding and the implementation of targeted breastfeeding support [12,13].

Our findings contribute to the understanding of the specific breastfeeding difficulties experienced by mothers with a low risk for early breastfeeding cessation, which appear to be related to several major areas, including lactational, nutritional, psychosocial, lifestyle, and medical factors, towards which breastfeeding promotion and support at the community level should be directed. Indeed, although in our study, the mother–infant dyads were enrolled in only one hospital, the present results reflect the primary care provided by the national "family pediatrics" network at the community level since, according to the Italian Public Health Care System, all patients aged 0–16 years must have an identified primary care provider among those available in the different regional health districts [14].

The perception of not having enough milk, the infant's failure to thrive, and mastitis are well-known factors acting negatively on breastfeeding [15–19], according to our results. Moreover, in this study, the return to work was associated with early exclusive breastfeeding failure. As previously described, balancing work and exclusive breastfeeding is challenging and requires a strong support in the short and long term [20,21]. In this scenario, employers could play a critical role in providing encouragement for working mothers to continue breastfeeding after returning to work and workplaces should establish dedicated breastfeeding rooms [22–26].

The perception of milk's limited nutritional value and pain during lactation was associated with a higher risk of exclusive breastfeeding discontinuation only in univariate analysis. It can be speculated that these factors might be closely related to the perception of reduced milk supply and often mentioned together. Incorrect latching showed a tendency even though it did not reach statistical significance, possibly reflecting the provision of adequate education and support both during the hospital stay and after hospital discharge with regard to the improvement of mothers' breastfeeding technique.

The findings of the present study are consistent with previous studies in the literature. Poor breastfeeding technique has been reported among the individual factors associated with unsuccessful breastfeeding [6,15,24,27], indicating that adequate breastfeeding support, including evaluation of latching, position, and feeding at the breast, could prevent nipple cracks and thus mastitis. Accordingly, the impact on breastfeeding cessation of acute pain, fever, and other typical mastitis symptoms presented by 8–10% of breastfeeding mothers has been broadly described in literature [28–30]. Mosca et al. [31] found that lactational and nutritional factors were the most cited by mothers as determinants for breastfeeding discontinuation, particularly during the first three months after delivery. Remarkably, the authors reported that the evaluation by a health care professional was rated as important only in 29% to 51% of cases whereas the maternal perception of inadequate milk or insufficient milk supply was cited as important by 40% up to 99% of mothers through the six months' study duration.

The present findings highlight the importance of educating mothers on the criteria that have to be taken into account when considering the adequateness of breast milk supply. Moreover, in this study, our results confirm the association between infant's failure to thrive and discontinuation of exclusive breastfeeding at three months. Accordingly, it has also been described how infant's failure to thrive, objectively evaluated by a healthcare professional, was one of the reasons of exclusive breastfeeding discontinuation, reported throughout the first 6 months of lactation [31]. Interestingly, a study by Flaherman et al. [32] has reported how early and limited administration of small quantities of formula milk during hospital stay could improve breastfeeding rates at three months. The authors speculated that limiting infants' weight loss during the first days of life may reduce maternal milk supply concern, which has been associated with breastfeeding discontinuation. It is then crucial to enhance maternal confidence in her own abilities, enabling mothers to get further insight into the lactation process and the peculiar characteristics of infant growth that often take place in spurts [16]. Within this context, it has to be underlined that a previous negative breastfeeding experience and difficulty negatively affect the likelihood of subsequent breastfeeding success, leading to a potential fear of breastfeeding secondary to prior breastfeeding trauma [7].

In agreement with previous data [15,27,31], in the present study, mothers reported psychosocial factors, in terms of pain and fatigue as breastfeeding difficulties in a relatively high number of cases. The occurrence of physical difficulty during breastfeeding has been associated with a greater risk for developing depressive symptoms in the postnatal period. Hence, it is crucial to provide mothers with early adequate breastfeeding support, including emotional [8].

Accordingly, antenatal and postnatal support including mothers' counseling and education positively affects breastfeeding success [6,12]. Consistently, in the present study, the availability of adequate support at the community level was associated with exclusive breastfeeding at three months post-delivery. Moreover, our results confirm that the mode of delivery modulates breastfeeding success [33], although it must be considered that caesarean section does not seem to negatively impact breastfeeding outcomes at six months, once adequate breastfeeding support is provided [34].

On the contrary, no mention about lifestyle factors, previously reported by other authors, regarding body image, such as wish to lose weight or dislike of breast appearance and breastfeeding convenience [8,15], have been reported, suggesting a positive breastfeeding attitude within the enrolled mothers.

Remarkably, most of the reported breastfeeding difficulties occurred within the first month after delivery, highlighting the importance of offering continuity of care after hospital discharge as underlined in the third guiding principle of the Ten Steps to Successful Breastfeeding [10]. Moreover, the largest decrease in exclusive breastfeeding in the present study was registered between enrollment and seven days after birth.

Literature shows how global breastfeeding rates are far below the international targets, particularly for high-income countries [4], although Italy has one of the highest rates of early initiation of breastfeeding. Moreover, according to the Italian National Statistics Institute [35], in Italy, 48.7% of infants are being exclusively breastfed in the first month, with a drop to 43.9% within the first three months. A survey conducted in 2012 in Lombardy [36] reported a progressive reduction of exclusive breastfeeding rates from 67.3% at hospital discharge to 47.3% and 27% within 120 and 180 days, respectively. Our rates are higher and reflect a particular local context of a high-income country where the breastfeeding benefits are well known and mothers are also supported at the community level. It must be acknowledged that this study focused on mothers with a low risk for breastfeeding cessation and did not include non-Italian speaking mothers due to the potential language barrier that could have interfered with the accuracy of the results, even though they could actually represent a subgroup particularly in need of breastfeeding support structures.

The strength of the present study is that it enrolled a relatively large sample of breastfeeding mother-infant pairs even though the duration of follow up was relatively limited and the dropout rate was 30%, thus partially limiting the generalizability of the present findings. However, it has to be taken into account that, with regard to cohort studies, although the maximum follow-up rate possible should be achieved, dropout rates ranging from 20% up to 50% have been suggested as acceptable [37].

5. Conclusions

Our findings provide further insight into breastfeeding difficulties experienced by mothers through the first three months after delivery in a high-income country with a positive breastfeeding culture and attitude. We underline the importance of providing continued tailored professional support in the community in the attempt to overcome maternal breastfeeding difficulties after discharge from the hospital.

Author Contributions: Conceptualization, M.L.G., M.E.B., G.C., P.R., E.V., P.M. and F.M.; methodology, M.L.G., G.C., P.R., E.V., F.M.; validation, M.L.G., M.E.B., E.B., L.P., P.R. and F.M.; formal analysis, B.L.C. and N.L.; investigation, P.M., G.S., G.R., B.L.C., D.M. and N.L.; supervision, M.L.G., E.B., L.P., G.C., P.R., E.V., P.M., F.M.; resources, M.L.G., G.S., and D.M.; data curation, P.M., G.S., G.R., B.L.C., D.M., N.L.; visualization, E.B., L.P.; writing—original draft preparation, M.L.G. and M.E.B.; writing—review and editing, G.C., G.R., L.C., E.V., P.M., F.M.

Funding: This research received no external funding.

Acknowledgments: We thank the mothers for participating in the study.

Conflicts of Interest: Maria Lorella Giannì is a Guest Editor of Nutrients. The other authors declare no conflict of interest.

Abbreviations

SD   Standard Deviation,
OR   Odd Ratio,
CI   Confidence Interval

References

1. Shamir, R. The Benefits of Breast Feeding. *Nestle Nutr. Inst. Workshop Ser.* **2016**, *86*, 67–76. [CrossRef] [PubMed]
2. Mosca, F.; Giannì, M.L. Human milk: Composition and health benefits. *Pediatr. Med. Chir.* **2017**, *39*, 155. [CrossRef] [PubMed]
3. Brown, A. Breastfeeding as a public health responsibility: A review of the evidence. *J. Hum. Nutr. Diet.* **2017**, *30*, 759–770. [CrossRef] [PubMed]
4. Victora, C.G.; Bahl, R.; Barros, A.J.; França, G.V.; Horton, S.; Krasevec, J.; Murch, S.; Sankar, M.J.; Walker, N.; Rollins, N.C.; et al. Breastfeeding in the 21st century: Epidemiology, mechanisms, and lifelong effect. *Lancet* **2016**, *387*, 475–490. [CrossRef]
5. Sayres, S.; Visentin, L. Breastfeeding: Uncovering barriers and offering solutions. *Curr. Opin. Pediatr.* **2018**, *30*, 591–596. [CrossRef]
6. Rollins, N.C.; Bhandari, N.; Hajeebhoy, N.; Horton, S.; Lutter, C.K.; Martines, J.C.; Piwoz, E.G.; Richter, L.M.; Victoria, C.G. The Lancet Breastfeeding Series Group. Why invest, and what it will take to improve breastfeeding practices? *Lancet* **2016**, *387*, 491–504. [CrossRef]
7. Palmér, L. Previous breastfeeding difficulties: An existential breastfeeding trauma with two intertwined pathways for future breastfeeding-fear and longing. *Int. J. Qual. Stud. Health Well Being* **2019**, *14*, 1588034. [CrossRef]
8. Brown, A.; Rance, J.; Bennett, P.J. Understanding the relationship between breastfeeding and postnatal depression: The role of pain and physical difficulties. *J. Adv. Nurs.* **2016**, *72*, 273–282. [CrossRef]

9. Bertino, E.; Di Nicola, P.; Varalda, A.; Occhi, L.; Giuliani, F.; Coscia, A. Neonatal growth charts. *J. Matern. Neonatal Med.* **2012**, *25*, 67–69. [CrossRef]
10. Ten Steps to Successful Breastfeeding. Available online: https://www.who.int/nutrition/bfhi/ten-steps/en/ (accessed on 2 April 2019).
11. The World Health Organization's Infant Feeding Recommendation. Available online: https://www.who.int/nutrition/en/ (accessed on 2 April 2019).
12. McFadden, A.; Gavine, A.; Renfrew, M.J.; Wade, A.; Buchanan, P.; Taylor, J.L.; MacGillivray, S.; Veitch, E.; Rennie, A.M.; Crowther, S.A.; et al. Support for healthy breastfeeding mothers with healthy term babies. *Cochrane Database Syst. Rev.* **2017**, *2*, CD001141. [CrossRef]
13. Heidari, Z.; Kohan, S.; Keshvari, M. Empowerment in breastfeeding as viewed by women: A qualitative study. *J. Educ. Health Promot.* **2017**, *6*, 33. [CrossRef]
14. Corsello, G.; Ferrara, P.; Chiamenti, G.; Nigri, L.; Campanozzi, A.; Pettoello-Mantovani, M. The Child Health Care System in Italy. *J. Pediatr.* **2016**, *177S*, S116–S126. [CrossRef]
15. Odom, E.C.; Li, R.; Scanlon, K.S.; Perrine, C.G.; Grummer-Strawn, L. Reasons for earlier than desired cessation of breastfeeding. *Pediatrics* **2013**, *131*, e726–e732. [CrossRef] [PubMed]
16. Li, R.; Fein, S.B.; Chen, J.; Grummer-Strawn, L.M. Why Mothers Stop Breastfeeding: Mothers' Self-reported Reasons for Stopping During the First Year. *Pediatrics* **2008**, *122*, S69–S76. [CrossRef] [PubMed]
17. Brown, C.R.L.; Dodds, L.; Legge, A.; Bryanton, J.; Semenic, S. Factors influencing the reasons why mothers stop breastfeeding. *Can. J. Public Health* **2014**, *105*, e179–e185. [CrossRef] [PubMed]
18. Kirkland, V.L.; Fein, S.B. Characterizing reasons for breastfeeding cessation throughout the first year postpartum using the construct of thriving. *J. Hum. Lact.* **2003**, *19*, 278–285. [CrossRef] [PubMed]
19. Ahluwalia, I.B.; Morrow, B.; Hsia, J. Why do women stop breastfeeding? Findings from the pregnancy risk assessment and monitoring system. *Pediatrics* **2005**, *116*, 1408–1412. [CrossRef] [PubMed]
20. Thomas-Jackson, S.C.; Bentley, G.E.; Keyton, K.; Reifman, A.; Boylan, M.; Hart, S.L. In-hospital breastfeeding and intention to return to work influence mothers' breastfeeding intentions. *J. Hum. Lact.* **2016**, *32*, NP76–NP83. [CrossRef]
21. Pounds, L.; Fisher, C.M.; Barnes-Josiah, D.; Coleman, J.D.; Lefebvre, R.C. The role of early maternal support in balancing full-time work and infant exclusive breastfeeding: A qualitative study. *Breastfeed. Med.* **2017**, *12*, 33–38. [CrossRef]
22. Tsai, S.Y. Employee perception of breastfeeding-friendly support and benefits of breastfeeding as a predictor of intention to use breast-pumping breaks after returning to work among employed mothers. *Breastfeed. Med.* **2014**, *9*, 16–23. [CrossRef] [PubMed]
23. Bettinelli, M.E. Breastfeeding policies and breastfeeding support programs in the mother's workplace. *J. Matern. Fetal Neonatal Med.* **2012**, *25*, 81–82. [CrossRef] [PubMed]
24. Maharlouei, N.; Pourhaghighi, A.; Raeisi Shahraki, H.; Zohoori, D.; Lankarani, K.B. Factors affecting exclusive breastfeeding, using adaptive LASSO regression. *Int. J. Community Based Nurs. Midwifery* **2018**, *6*, 260–271.
25. Mirkovic, K.R.; Perrine, C.G.; Scanlon, K.S.; Grummer-Strawn, L.M. Maternity leave duration and full-time/part-time work status are associated with US mothers' ability to meet breastfeeding intentions. *J. Hum. Lact.* **2014**, *30*, 416–419. [CrossRef]
26. Dinour, L.M.; Szaro, L.M. Employer-based programs to support breastfeeding among working mothers: A Systematic review. *Breastfeed. Med.* **2017**, *12*, 131–141. [CrossRef] [PubMed]
27. Colombo, L.; Crippa, B.; Consonni, D.; Bettinelli, M.; Agosti, V.; Mangino, G.; Plevani, L.; Bezze, E.N.; Mauri, P.A.; Zanotta, L.; et al. Breastfeeding determinants in healthy term newborns. *Nutrients* **2018**, *10*, 48. [CrossRef]
28. Khanal, V.; Scott, J.A.; Lee, A.H.; Binns, C.W. Incidence of mastitis in the neonatal period in a traditional breastfeeding society: Results of a cohort study. *Breastfeed. Med.* **2015**, *10*, 481–487. [CrossRef] [PubMed]
29. Spencer, J.P. Management of mastitis in breastfeeding women. *Am. Fam. Physician* **2008**, *78*, 727–731. [PubMed]
30. Schwartz, K.; D'Arcy, H.J.; Gillespie, B.; Bobo, J.; Longeway, M.; Foxman, B. Factors associated with weaning in the first 3 months postpartum. *J. Fam. Pract.* **2002**, *51*, 439–444.
31. Mosca, F.; Roggero, P.; Garbarino, F.; Morniroli, D.; Bracco, B.; Morlacchi, L.; Consonni, D.; Marlladi, D.; Gianni, M.L. Determinants of breastfeeding discontinuation in an Italian cohort of mother-infant dyads in the first six months of life: A randomized controlled trial. *Ital. J. Pediatr.* **2018**, *44*, 134. [CrossRef]

32. Flaherman, V.J.; Aby, J.; Burgos, A.E.; Lee, K.A.; Cabana, M.D.; Newman, T.B. Effect of early limited formula on duration and exclusivity of breastfeeding in at-risk infants: An RCT. *Pediatrics* **2013**, *131*, 1059–1065. [CrossRef]
33. Cato, K.; Sylvén, S.M.; Lindbäck, J.; Skalkidou, A.; Rubertsson, C. Risk factors for exclusive breastfeeding lasting less than two months-identifying women in need of targeted breastfeeding support. *PLoS ONE* **2017**, *12*, e0179402. [CrossRef] [PubMed]
34. Prior, E.; Santhakumaran, S.; Gale, C.; Philipps, L.H.; Modi, N.; Hyde, M.J. Breastfeeding after cesarean delivery: A systematic review and meta-analysis of world literature. *Am. J. Clin. Nutr.* **2012**, *95*, 1113–1135. [CrossRef] [PubMed]
35. Istituto Nazionale di Statistica. Available online: https://www.istat.it/it/archivio/141431 (accessed on 29 July 2019).
36. Regione Lombardia Sanità. Available online: http://www.epicentro.iss.it/argomenti/allattamento/pdf/Report%20allattamento%20RL%202012.pdf (accessed on 29 July 2019).
37. Fewtrell, M.S.; Kennedy, K.; Singhal, A.; Martin, R.M.; Ness, A.; Hadders-Algra, M.; Koletzko, B.; Lucas, A. How much loss to follow-up is acceptable in long-term randomised trials and prospective studies? *Arch. Dis. Child.* **2008**, *93*, 458–461. [CrossRef] [PubMed]

© 2019 by the authors. Licensee MDPI, Basel, Switzerland. This article is an open access article distributed under the terms and conditions of the Creative Commons Attribution (CC BY) license (http://creativecommons.org/licenses/by/4.0/).

Article

Variation and Interdependencies of Human Milk Macronutrients, Fatty Acids, Adiponectin, Insulin, and IGF-II in the European PreventCD Cohort

Maria Grunewald [1], Christian Hellmuth [1], Franca F. Kirchberg [1], Maria Luisa Mearin [2], Renata Auricchio [3], Gemma Castillejo [4], Ilma R. Korponay-Szabo [5], Isabel Polanco [6], Maria Roca [7], Sabine L. Vriezinga [2], Katharina Werkstetter [1], Berthold Koletzko [1,*] and Hans Demmelmair [1,*]

1. Ludwig-Maximilians-Universität, Division of Metabolic and Nutritional Medicine, Dr. von Hauner Children's Hospital, University of Munich Medical Center, 80337 Munich, Germany
2. Department of Paediatrics, Leiden University Medical Center, 2300 Leiden, The Netherlands
3. Department of Medical Translational Sciences and European Laboratory for the Investigation of Food-Induced Diseases, University Federico II, 80131 Naples, Italy
4. Department of Pediatric Gastroenterology Unit, Hospital Universitari Sant Joan de Reus, URV, IIPV, 43201 Reus, Spain
5. Celiac Disease Center, Heim Pál Children's Hospital, 1089 Budapest, Hungary
6. Department of Pediatric Gastroenterology and Nutrition, La Paz University Hospital, 28033 Madrid, Spain
7. U. Enfermedad Celiaca e Inmunopatología Digestiva, Instituto de Investigación Sanitaria La Fe, 46026 Valencia, Spain
* Correspondence: office.koletzko@med.uni-muenchen.de (B.K.); hans.demmelmair@med.uni-muenchen.de (H.D.); Tel.: +49-89-4400-52826 (B.K.); +49-89-4400-53692 (H.D.)

Received: 2 July 2019; Accepted: 23 August 2019; Published: 30 August 2019

Abstract: Human milk composition is variable. The identification of influencing factors and interdependencies of components may help to understand the physiology of lactation. In this study, we analyzed linear trends in human milk composition over time, the variation across different European countries and the influence of maternal celiac disease. Within a multicenter European study exploring potential prevention of celiac disease in a high-risk population (PreventCD), 569 human milk samples were donated by women from five European countries between 16 and 163 days postpartum. Some 202 mothers provided two samples at different time points. Protein, carbohydrates, fat and fatty acids, insulin, adiponectin, and insulin-like growth factor II (IGF-II) were analyzed. Milk protein and n-6 long chain polyunsaturated fatty acids decreased during the first three months of lactation. Fatty acid composition was significantly influenced by the country of residence. IGF-II and adiponectin concentrations correlated with protein content ($r = 0.24$ and $r = 0.35$), and IGF-II also correlated with fat content ($r = 0.36$), suggesting a possible regulatory role of IGF in milk macronutrient synthesis. Regarding the impact of celiac disease, only the level in palmitic acid was influenced by this disease, suggesting that breastfeeding by celiac disease mothers should not be discouraged.

Keywords: human milk; celiac disease; hormones; fatty acids; duration of lactation; country; carbohydrate; fat

1. Introduction

Breastfeeding supports physiological infant growth and development [1]. The importance of early life nutrition has been stimulated in studies investigating human milk composition and influencing factors [2–5]. A recent meta-analysis found that the average energy content in human milk of mothers with term born babies hardly changes from lactation week 2 to weeks 10–12 [6]. However, at both time points, the energy content shows large inter-individual variation. This primarily reflects a high variation of milk fat content, but also protein and to a lesser extent lactose are variable [7]. Colostrum and transitional milk are clearly different from mature milk. After the second week of lactation, changes associated with the duration of lactation, like the decrease in protein content, only partially explain the variation in milk composition and other influencing factors, for example, maternal diet, have to be considered [7].

The fatty acid (FA) composition of human milk fat is dependent on maternal diet. This has been demonstrated for essential FA and their long chain polyunsaturated derivatives (LC-PUFA) [8], as well as for medium chain FA (MCFA, C8.0 to C14.0) contents in milk, which are influenced by the ratio of dietary carbohydrates to fat [9]. Milk protein is composed of casein and whey, which is mainly comprised by α-lactalbumin and lactoferrin, but also includes a variety of lower concentrated proteins and peptides [10]. Insulin, insulin-like growth factors, and adipokines are metabolic regulators that might modulate infantile metabolism after milk feeding [11,12]. The hormones in milk may be derived from the maternal circulation, as suggested for insulin [13], or they could be synthesized in the mammary gland [11]; and their concentrations may be related to other human milk components. Co-variation of peptide hormone and macronutrient concentrations in human milk might complicate the identification of growth promoting or growth attenuating effects to individual compounds. This could also in part explain why studies observing the relationship between human milk composition and infant growth often yield ambiguous results [14–17].

Celiac disease (CD) is an intolerance of gluten, a protein present in various cereals. The disease is associated with atrophy of the intestinal villi, inflammation of the jejunal mucosa, and intestinal malabsorption [18]. A lifelong gluten free diet (GFD) is required to improve the histopathology and symptoms of CD, such as steatorrhea, diarrhea, and abdominal distension [18]. However, there is a risk that adherence to a GFD induces nutritional deficiencies, as GFDs have been found to be low in iron, calcium, B-vitamins, and some fatty acids [19]. There are ambiguous findings in relation to the effect of a GFD on fatty acid status biomarkers [20,21]. It is currently not known whether human milk fatty acid composition is affected by maternal CD. So far, it has only been shown that CD affects cytokines in milk [22]. Significant effects of CD or GFD on macronutrient contents or fatty acid composition could be of importance for the nutrition of breast fed infants of CD mothers and might require specific dietary recommendations.

In this study, we determined protein, fat, carbohydrate, individual FA, insulin, adiponectin, and insulin-like growth factor II (IGF-II) in milk samples collected in the large European PreventCD prospective cohort study. We aimed to compare milk composition between mothers with CD and healthy mothers, to investigate any effects by country of residence and duration of lactation on milk composition and to analyze the variation and interdependencies of the measured milk components.

2. Materials and Methods

Human milk samples were collected from 2007 to 2010 within the PreventCD study [23]. Details on the study population are reported in Vriezinga et al. [24]. Briefly, healthy newborns were enrolled if they had at least one first-degree family member with biopsy-confirmed celiac disease and were tested positive for the risk alleles *HLA-DQ2* and/or *HLA-DQ8*. Infants born preterm or with any congenital disorder were excluded. Infants were randomized to the introduction of either small amounts of gluten or to placebo at the age of 16 weeks.

The PreventCD study was approved by the medical ethics committee of each participating center and complied with Good Clinical Practice guidelines (ICH-GCP) regulations. The study was conducted according to the Declaration of Helsinki. The PreventCD Current Controlled Trials number is ICTRP CTRP NTR890.

Milk samples for this study were donated by mothers in five European countries between 16 days and 163 days postpartum. The included milk samples were collected in the Netherlands (Leiden, $n = 116$), Italy (Naples, $n = 68$), Spain (Madrid, Valencia, and Barcelona, $n = 138$), Hungary (Budapest, $n = 120$), and Germany (Munich, $n = 127$).

Mothers were asked to express milk manually or by pump once a month during the first six months after birth without further specification for fore- or hind-milk sampling and time of day. Milk samples were first frozen at −20 °C in home freezers, transferred to the hospital on ice, and then stored at −80 °C. Samples for the reported analyses were aliquoted (1–2 mL) and randomly selected, aiming for two samples from each mother, with one sample collected until 3 months postpartum (early samples), and one sample collected during months 4 or 5 (late samples).

2.1. Measurements

Analytical procedures were previously described in a publication observing the association between milk components and the infant metabolome [25]. Measurement of total fat and total carbohydrates was performed via mid-infrared spectroscopy with a Human Milk Analyzer (MIRIS AB, Uppsala, Sweden) [26]. Owing to limited available sample volumes, the samples were diluted 1:3 with water. Samples were sonicated and heated to 40 °C prior to analysis. Tests with a diluted reference milk sample revealed intra- and inter-assay coefficients of variation (CVs) (7 and 13 determinations) for fat (5.3% and 6.6%) and carbohydrates (4.8% and 4.5%), comparable to the inter-assay CVs of undiluted milk samples (fat: 5.6% and carbohydrates: 4.3%). The calibration curve of eight different diluted samples versus the same eight undiluted samples showed high correlations with R^2 of 0.99 for fat and 0.90 for carbohydrates, respectively.

As the protein measurement by infrared spectroscopy (MIRIS) led to unsatisfactory CVs, the protein content was measured with an adapted Bradford method [27]. The intra batch—and inter batch—assay CVs of 4 and 16 determinations were calculated with 4.3% and 9.7%, respectively, using samples with 1.3 g/dL protein. Spiking recovery was determined to be 99.1% ± 27.6% in eight low (+0.27 g/dL) and 105.8% ± 16.5% in eight high (+0.44 g/dL) spiked samples.

Analysis of the FA composition of milk lipids was performed as previously described using 20 μL of milk [28]. The lipid bound FAs were converted in situ with acidic catalysis into FA methyl esters, which were subsequently extracted into hexane and analyzed by gas chromatography. The method enabled quantification of FA with 8 to 24 carbon atoms, including the major LC-PUFA. The weight percentages of 35 FA were determined with a mean CV of 4.9%, as estimated from 31 analyses of control milk aliquots measured along with study samples.

For the analysis of hormones, milk aliquots were thawed overnight at 4 °C and skimmed by centrifugation at 4000× g and 4 °C for 30 min. Total adiponectin concentration was measured with a commercially available ELISA kit (Biovendor RD191023100 High Sensitivity Adiponectin, Brno, Czech Republic) in 50 μL skimmed milk with a 1:3 dilution following the protocol of the manufacturer. The intra-batch and inter-batch CVs of 4 and 8 determinations were 4.5% and 4.8%, respectively. Spiking recovery was found to be 105.1% ± 14.0% in eight low (+2 ng/mL) and 91.6% ± 4.0% in eight high (+10 ng/mL) spiked determinations.

Insulin was measured with the Mercodia Insulin ELISA kit 10-1113-01 (Mercodia, Uppsala, Sweden) from 25 μL of undiluted, skimmed human milk, according to the protocol of the manufacturer. The intra-batch and inter-batch CVs of 4 and 8 determinations were 3.4% and 11.0%, respectively. Spiking recovery was determined to be 92.3% ± 14.8% in seven low (+21 mU/L) and 85.9% ± 7.2% in seven high (+42 mU/L) spiked samples.

IGF-II was determined with a radioimmunoassay from 30 μL of full fat milk by Mediagnost (Reutlingen, Germany) using the R-30 IGF-II RIA kit, according to the protocol of the manufacturer. The kit had already successfully been applied for the analysis of IGF-II in human milk [29].

2.2. Data Analysis

In order to evaluate the effects of duration of lactation and country of residence, data were divided into subsets of early (day 16–100) and late (day 101–163) lactation. Statistical analyses were performed independently on both subsets, that is, separately on the early and late samples. We identified outliers by calculating the numeric distance to its nearest neighbor. If this distance (gap) was bigger than one standard deviation of the corresponding parameter, the observations more distant from the mean were excluded from further analyses.

Using univariate linear regression, we tested for effects of individual factors (mode of delivery, maternal age at delivery, duration of gestation, infants' gender, birth weight, maternal pre-pregnancy weight, maternal pre-pregnancy body mass index (BMI), maternal CD status, day of lactation, or country of residence) on measured milk analytes. As potentially significant predictors for the multiple regression analysis, we selected the variates that showed Bonferroni corrected p-values below 0.2 in both data sets [30].

Potentially significant factors were included in the multiple linear regression analysis to test for effects of these factors on the standardized analyte concentrations. Standardization, the transformation of the analytes to have a mean of 0 and standard deviation of 1, was done in order to obtain comparable model estimates. We used weighted effects coding for the categorical variable "country of residence" (each variable is coded such that the estimated effects for each category are to be interpreted as deviations from the weighted mean of the whole data set) to test whether milk components from individual countries differed significantly from the global mean. Subsequently, we utilized analysis of variance (ANOVA) to test for significant differences in the means of the measured analytes across countries.

For the determination of the relationships among selected analytes, correlation coefficients according to Pearson were calculated for the early and late dataset, respectively.

For the exploration of intra-individual stability of concentrations and percentages, we related data points in the early data set to the corresponding data points in the late data set for the 202 mothers who donated two samples. Intra-individual comparisons were done with paired t-tests and correlation coefficients were calculated according to Pearson.

All statistical analyses were performed with the software R (version 3.0.2., the R foundation for statistical computing). We adjusted the confidence intervals and p-values that we report here for multiple testing (41 milk compounds) using Bonferroni's method.

3. Results

A total of 569 samples from 367 mothers were available. After outlier removal, the early dataset (lactation days 16 to 100) contained results from 319 milk samples with a minimum of 307 values for each analyte. The late dataset (lactation days 101 to 163) with 250 milk samples provided a minimum of 233 values for individual analytes. Early samples were collected on lactation days 42 ± 21 (mean ± SD) and late samples were collected 120 ± 8 days postpartum. A total of 202 of the late samples had an earlier sampled counterpart in the first subset from the same mother. The characteristics of the mothers and their children are summarized in Table 1.

Table 1. Characteristics of participating mothers and their infants.

Variable	M	SD	N *
Age mother, years	33.4	±3.9	357
Gestational age, weeks	39.3	±1.4	366
Pre-pregnancy BMI mother, kg/m^2	22.4	±3.4	175
Birth weight, g	3373	±455	364
	n	%	N *
Mothers with celiac disease	184	50.1	367
Exclusive breastfeeding at 4 months	264	77.9	339
Infant gender female	182	49.6	367

* N corresponds to the number of participants with available information, BMI, body mass index.

Day of lactation, country of residence, and CD status were identified as potentially relevant variables for milk composition. Pre-pregnancy BMI showed a positive correlation with human milk insulin (Figure 1), but was not considered in other analyses as we have this information only from a small subset of mothers.

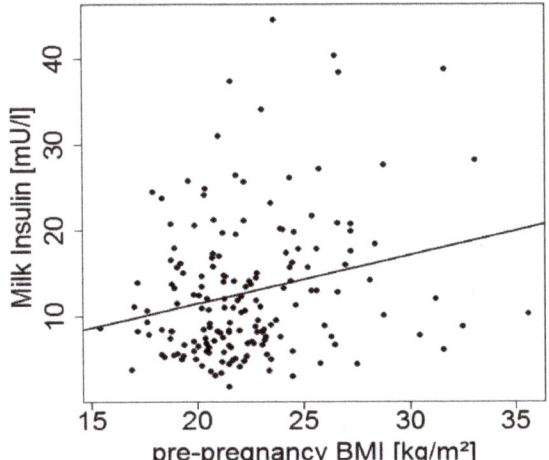

Figure 1. Milk insulin levels in early milk samples versus maternal pre-pregnancy body mass index (BMI) ($r = 0.24$, $p = 0.002$, $n = 175$).

3.1. Influence of Maternal CD Status, Day of Lactation, and Country of Residence

About half of the participating mothers were CD patients (Table 1). Five out of 184 mothers with CD did not follow a GFD. The early and late dataset showed that the milk of CD negative and positive women was not significantly different regarding the hormone and macronutrient concentrations. Among the FA percentages, only palmitic acid (C16:0) showed significantly decreased percentages in milk of mothers with CD compared with non-CD mothers. Taking all available data into account, palmitic acid contributed 22.3% ± 3.1% to total milk fatty acids in the healthy mothers and 22.0% ± 2.8% in mothers with CD.

Within the first three months of lactation, levels of protein, n-6 LC-PUFA percentages, n-3 eicosatrienoic acid (20:3n-3), capric acid (10:0), lauric acid (12:0), and the monounsaturated fatty acids (MUFA) C20:1n-9 and C24:1n-9 decreased significantly over time (Table 2). Day of lactation did not show significant effects on milk FA composition during months 4 and 5 postpartum (Table 3). During the first three months of lactation, most FA percentages differed significantly across the tested countries (Table 4, docosahexaenoic acid (DHA) in Figure 2A), and long-chain FA also differed by country in late samples (Table 5, DHA in Figure 2A). Comparisons of the individual FA between countries identified a huge number of differences, which were mostly similar in the early and the late data set (Tables 4 and 5). In the case of DHA, the mean value found in the early samples from Hungary was significantly lower than in the sample from all other countries, and in the late samples, values for Italy and Hungary were similarly low. This is also reflected in 57% and 73%, respectively, of Hungarian samples with DHA below 0.2%, while in the whole sample set, only 29% of the early and 51% of the late samples were below 0.2%. The highest DHA percentages were found in the samples from Spain and the Netherlands, where only 15% and 23%, respectively, of the early samples and 37% and 36%, respectively, of the late samples were below 0.2% DHA.

Table 2. Mean analyte concentrations (±SD) of early samples collected until day 100 of lactation.

	Mean ± SD	CD Mother β (CI: 0.06%; 99.94%)	Day of Lactation β (CI: 0.06%; 99.94%)	Country p
Hormones				
IGF-II, ng/mL	17.41 ± 6.09	−0.048 (−0.423; 0.326)	−0.005 (−0.014; 0.003)	<0.001
Insulin, mU/L	12.46 ± 8.19	−0.090 (−0.480; 0.300)	0.003 (−0.006; 0.012)	1
Adiponectin, ng/mL	19.28 ± 6.63	0.134 (−0.267; 0.534)	−0.005 (−0.014; 0.004)	1
macronutrients, g/dL				
Fat	2.2 ± 1.2	−0.202 (−0.600; 0.196)	−0.002 (−0.011; 0.007)	0.031
Carbohydrates	6.5 ± 0.4	−0.102 (−0.482; 0.278)	−0.002 (−0.010; 0.007)	0.701
Protein	1.16 ± 0.22	0.081 (−0.276; 0.439)	**−0.015 (−0.023; −0.007)**	0.049
Fatty Acids, wt %				
SFA				
C8:0	0.26 ± 0.10	0.129 (−0.258; 0.516)	−0.007 (−0.016; 0.001)	0.002
C10:0	1.36 ± 0.32	0.224 (−0.161; 0.609)	**−0.013 (−0.022; −0.004)**	0.122
C12:0	5.37 ± 1.69	0.244 (−0.150; 0.639)	**−0.01 (−0.019; −0.001)**	1
C13:0	0.04 ± 0.01	0.134 (−0.258; 0.526)	−0.002 (−0.011; 0.007)	0.005
C14:0	5.76 ± 1.62	0.131 (−0.268; 0.530)	−0.006 (−0.015; 0.003)	0.031
C15:0	0.32 ± 0.12	−0.068 (−0.404; 0.269)	0.001 (−0.007; 0.008)	<0.001
C16:0	22.16 ± 2.92	**−0.446 (−0.770; −0.122)**	0.002 (−0.005; 0.009)	<0.001
C17:0	0.31 ± 0.06	0.020 (−0.342; 0.382)	0.005 (−0.003; 0.013)	<0.001
C18:0	7.37 ± 1.33	−0.108 (−0.500; 0.284)	0.007 (−0.002; 0.015)	<0.001
C20:0	0.27 ± 0.10	−0.006 (−0.421; 0.409)	0.003 (−0.006; 0.013)	0.821
C22:0	0.10 ± 0.03	0.143 (−0.239; 0.525)	0.004 (−0.004; 0.013)	<0.001
C24:0	0.09 ± 0.04	0.181 (−0.184; 0.545)	−0.004 (−0.012; 0.004)	<0.001
MUFA				
C14:1	0.23 ± 0.11	−0.102 (−0.442; 0.237)	−0.004 (−0.011; 0.004)	<0.001
C15:1	0.07 ± 0.03	0.005 (−0.338; 0.348)	0.001 (−0.007; 0.008)	<0.001
C16:1 n-7	2.23 ± 0.69	−0.103 (−0.480; 0.274)	−0.006 (−0.015; 0.002)	<0.001
C18:1 n-9	35.33 ± 4.35	0.216 (−0.119; 0.550)	0.005 (−0.003; 0.012)	<0.001
C18:1 n-7	1.62 ± 0.25	0.185 (−0.223; 0.593)	−0.007 (−0.016; 0.002)	1
C20:1 n-9	0.46 ± 0.08	0.235 (−0.140; 0.610)	**−0.011 (−0.019; −0.003)**	0.272
C24:1 n-9	0.07 ± 0.02	0.306 (−0.042; 0.653)	**−0.015 (−0.022; −0.007)**	<0.001
PUFA				
n6				
C18:2 n-6	13.27 ± 4.16	−0.028 (−0.380; 0.323)	0.002 (−0.005; 0.010)	<0.001
C18:3 n-6	0.16 ± 0.05	−0.205 (−0.603; 0.193)	0.000 (−0.009; 0.009)	0.619
C20:2 n-6	0.30 ± 0.09	0.098 (−0.239; 0.435)	**−0.014 (−0.022; −0.007)**	<0.001
C20:3 n-6	0.44 ± 0.11	−0.010 (−0.374; 0.354)	**−0.019 (−0.027; −0.011)**	<0.001
C20:4 n-6	0.49 ± 0.11	0.066 (−0.294; 0.426)	**−0.013 (−0.021; −0.005)**	<0.001
C22:4 n-6	0.11 ± 0.03	0.137 (−0.191; 0.466)	**−0.014 (−0.021; −0.007)**	<0.001
C22:5 n-6	0.05 ± 0.02	0.167 (−0.171; 0.506)	**−0.013 (−0.021; −0.006)**	<0.001
n3				
C18:3 n-3	0.77 ± 0.39	−0.065 (−0.429; 0.298)	−0.001 (−0.009; 0.007)	<0.001
C20:3 n-3	0.05 ± 0.02	0.106 (−0.245; 0.458)	**−0.010 (−0.018; −0.002)**	<0.001
C20:5 n-3	0.07 ± 0.05	0.012 (−0.383; 0.408)	−0.005 (−0.014; 0.004)	<0.001
C22:5 n-3	0.15 ± 0.05	0.016 (−0.363; 0.396)	−0.007 (−0.016; 0.001)	<0.001
C22:6 n-3	0.29 ± 0.16	0.117 (−0.274; 0.508)	−0.009 (−0.017; 0.000)	<0.001
n9				
C20:3 n-9	0.02 ± 0.01	0.021 (−0.380; 0.423)	−0.003 (−0.012; 0.006)	0.018
Trans FA				
C16:1 trans	0.06 ± 0.02	−0.049 (−0.412; 0.314)	0.005 (−0.003; 0.013)	<0.001
C18:1 trans	0.31 ± 0.20	−0.296 (−0.672; 0.080)	0.004 (−0.005; 0.012)	0.001
C18:2 trans	0.10 ± 0.04	−0.010 (−0.359; 0.339)	0.002 (−0.005; 0.010)	<0.001

Influence of maternal celiac disease (CD) status and day of lactation are indicated by β estimates, and influence of country is indicated by the p-values from analysis of variance (ANOVA). Weighted effects coding was used to code the country. p-values and 95% confidence intervals were Bonferroni corrected ($n = 41$), resulting in an adjusted 99.88% confidence interval. Significant p-values and β estimates are printed in bold. SFA, saturated fatty acids; MUFA, monounsaturated fatty acids; PUFA, polyunsaturated fatty acid; IGF, insulin-like growth factor.

Table 3. Mean analyte concentrations (±SD) of late samples collected between days 101 to 163 of lactation.

	Global Mean ± SD	CD Mother β (CI: 0.06%; 99.94%)	Day of Lactation β (CI: 0.06%; 99.94%)	Country p
Hormones				
IGF-II, ng/mL	12.61 ± 3.25	0.055 (−0.393; 0.502)	−0.007 (−0.034; 0.021)	1
Insulin, mU/L	13.67 ± 9.15	−0.207 (−0.600; 0.186)	0.003 (−0.021; 0.027)	1
Adiponectin, ng/mL	17.56 ± 6.26	0.268 (−0.154; 0.689)	−0.008 (−0.034; 0.018)	1
macronutrients, g/dL				
Fat	2.4 ± 1.5	−0.035 (−0.482; 0.413)	0.000 (−0.027; 0.028)	1
Carbohydrates	6.6 ± 0.4	0.123 (−0.258; 0.504)	0.001 (−0.023; 0.024)	1
Protein	0.84 ± 0.18	0.094 (−0.339; 0.527)	−0.014 (−0.041; 0.012)	0.753
Fatty Acids, wt %				
SFA				
C8:0	0.21 ± 0.06	0.142 (−0.297; 0.580)	−0.004 (−0.031; 0.024)	1
C10:0	1.42 ± 0.33	0.317 (−0.119; 0.752)	−0.012 (−0.038; 0.015)	1
C12:0	5.76 ± 1.66	0.181 (−0.261; 0.623)	−0.007 (−0.034; 0.020)	1
C13:0	0.04 ± 0.01	0.031 (−0.386; 0.449)	0.004 (−0.022; 0.030)	**<0.001**
C14:0	6.24 ± 1.71	0.196 (−0.264; 0.656)	0.003 (−0.026; 0.031)	1
C15:0	0.33 ± 0.12	−0.097 (−0.469; 0.276)	0.010 (−0.013; 0.033)	**<0.001**
C16:0	22.33 ± 2.66	**−0.404 (−0.795; −0.012)**	−0.005 (−0.030; 0.019)	**<0.001**
C17:0	0.32 ± 0.06	−0.146 (−0.562; 0.270)	0.005 (−0.021; 0.031)	**<0.001**
C18:0	7.66 ± 1.34	−0.328 (−0.756; 0.100)	−0.002 (−0.029; 0.025)	**<0.001**
C20:0	0.24 ± 0.06	−0.197 (−0.637; 0.243)	−0.005 (−0.032; 0.022)	**0.012**
C22:0	0.09 ± 0.03	0.139 (−0.310; 0.588)	−0.002 (−0.030; 0.026)	**0.004**
C24:0	0.07 ± 0.03	0.212 (−0.240; 0.664)	−0.003 (−0.031; 0.025)	**0.024**
MUFA				
C14:1	0.23 ± 0.11	−0.114 (−0.487; 0.260)	0.005 (−0.018; 0.028)	**<0.001**
C15:1	0.08 ± 0.03	−0.010 (−0.391; 0.370)	0.009 (−0.015; 0.033)	**<0.001**
C16:1 n-7	2.14 ± 0.68	−0.033 (−0.465; 0.398)	0.003 (−0.024; 0.030)	**<0.001**
C18:1 n-9	34.87 ± 4.50	0.271 (−0.114; 0.656)	0.001 (−0.023; 0.025)	**<0.001**
C18:1 n-7	1.54 ± 0.25	0.089 (−0.365; 0.542)	−0.003 (−0.031; 0.025)	1
C20:1 n-9	0.42 ± 0.09	0.259 (−0.188; 0.706)	−0.008 (−0.036; 0.019)	1
C24:1	0.05 ± 0.02	0.304 (−0.147; 0.755)	−0.006 (−0.034; 0.022)	0.223
PUFA				
n6				
C18:2 n-6	12.93 ± 3.72	−0.106 (−0.497; 0.284)	0.005 (−0.019; 0.029)	**<0.001**
C18:3 n-6	0.15 ± 0.04	−0.243 (−0.685; 0.198)	−0.015 (−0.042; 0.012)	0.073
C20:2 n-6	0.25 ± 0.07	−0.068 (−0.456; 0.319)	−0.008 (−0.032; 0.016)	**<0.001**
C20:3 n-6	0.34 ± 0.07	−0.116 (−0.540; 0.308)	−0.025 (−0.052; 0.001)	**<0.001**
C20:4 n-6	0.43 ± 0.09	0.042 (−0.381; 0.466)	−0.016 (−0.043; 0.010)	**<0.001**
C22:4 n-6	0.09 ± 0.03	−0.015 (−0.397; 0.367)	−0.013 (−0.037; 0.011)	**<0.001**
C22:5 n-6	0.04 ± 0.02	0.064 (−0.316; 0.444)	−0.002 (−0.026; 0.022)	**<0.001**
n3				
C18:3 n-3	0.75 ± 0.39	−0.109 (−0.509; 0.291)	0.012 (−0.013; 0.037)	**<0.001**
C20:3 n-3	0.04 ± 0.01	−0.124 (−0.565; 0.317)	0.004 (−0.024; 0.031)	0.111
C20:5 n-3	0.06 ± 0.05	0.089 (−0.368; 0.546)	0.002 (−0.026; 0.031)	**0.027**
C22:5 n-3	0.14 ± 0.05	0.067 (−0.368; 0.502)	−0.001 (−0.028; 0.026)	**<0.001**
C22:6 n-3	0.24 ± 0.15	0.197 (−0.248; 0.641)	0.001 (−0.027; 0.029)	**<0.001**
n9				
C20:3 n-9	0.02 ± 0.01	−0.068 (−0.513; 0.378)	−0.012 (−0.039; 0.016)	**0.018**
Trans FA				
C16:1 trans	0.06 ± 0.02	−0.037 (−0.451; 0.378)	0.013 (−0.013; 0.038)	**<0.001**
C18:1 trans	0.30 ± 0.13	−0.229 (−0.672; 0.215)	−0.002 (−0.029; 0.026)	**0.004**
C18:2 trans	0.11 ± 0.04	−0.044 (−0.435; 0.348)	0.01 (−0.014; 0.035)	**<0.001**

Influence of maternal CD status and day of lactation are indicated by β estimates, and influence of country is indicated by the p-values from ANOVA. Weighted effects coding was used to code the country. p-values and 95% confidence intervals were Bonferroni corrected ($n = 41$), resulting in an adjusted 99.88% confidence interval. Significant p-values and β estimates are printed in bold.

Table 4. Human milk fatty acid composition found in the early samples according to the country of residence of the mothers.

	NL	It	ESP	HU	GER
SFA					
C8:0	0.29 ± 0.12 [ab]	0.22 ± 0.07 [a]	0.26 ± 0.09 [c]	0.25 ± 0.12	0.21 ± 0.06 [bc]
C10:0	1.38 ± 0.35	1.50 ± 0.34	1.37 ± 0.29	1.27 ± 0.33	1.31 ± 0.32
C12:0	5.46 ± 1.84	6.03 ± 1.73	5.23 ± 1.52	5.46 ± 1.96	5.05 ± 1.46
C13:0	0.04 ± 0.01	0.04 ± 0.01	0.04 ± 0.02 [a]	0.04 ± 0.01 [b]	0.04 ± 0.01 [ab]
C14:0	6.06 ± 1.92 [a]	6.44 ± 1.76 [b]	5.03 ± 1.42 [abc]	5.74 ± 1.73	6.33 ± 1.55 [c]
C15:0	0.31 ± 0.08 [ab]	0.35 ± 0.09 [cd]	0.25 ± 0.09 [a cef]	0.30 ± 0.11 [eg]	0.45 ± 0.12 [bdfg]
C16:0	21.54 ± 2.50 [abc]	24.09 ± 2.37 [abd]	19.79 ± 2.32 [def]	22.64 ± 2.50 [eg]	24.25 ± 2.38 [cfg]
C17:0	0.29 ± 0.05 [a]	0.33 ± 0.05	0.30 ± 0.10 [b]	0.30 ± 0.05 [c]	0.36 ± 0.06 [abc]
C18:0	7.38 ± 1.59 [a]	7.03 ± 1.03 [b]	6.97 ± 1.15 [c]	7.40 ± 1.15 [d]	8.13 ± 1.31 [abcd]
C20:0	0.30 ± 0.12 [a]	0.23 ± 0.04 [a]	0.29 ± 0.13	0.27 ± 0.12	0.27 ± 0.08
C22:0	0.12 ± 0.04 [abc]	0.08 ± 0.01	0.11 ± 0.08 [a]	0.09 ± 0.03 [b]	0.09 ± 0.03 [c]
C24:0	0.12 ± 0.05 [abc]	0.07 ± 0.02 [a]	0.09 ± 0.08	0.08 ± 0.03 [b]	0.07 ± 0.03 [c]
MUFA					
C14:1	0.25 ± 0.09 [abc]	0.24 ± 0.07 [de]	0.15 ± 0.07 [adfg]	0.20 ± 0.08 [bfh]	0.33 ± 0.12 [cegh]
C15:1	0.07 ± 0.02 [a]	0.09 ± 0.03 [bc]	0.06 ± 0.04 [bd]	0.07 ± 0.03 [ce]	0.10 ± 0.03 [ade]
C16:1n-7	2.44 ± 0.78 [a]	2.20 ± 0.42 [b]	1.77 ± 0.42 [abcc]	2.17 ± 0.55 [ce]	2.58 ± 0.82 [ce]
C18:1n-9	35.18 ± 3.54 [ab]	35.58 ± 3.83 [cd]	38.91 ± 4.33 [acef]	31.54 ± 3.07 [bdeg]	34.44 ± 3.35 [fg]
C18:1n-7	1.64 ± 0.27	1.56 ± 0.19	1.64 ± 0.21	1.59 ± 0.27	1.63 ± 0.28
C20:1n-9	0.47 ± 0.07	0.43 ± 0.06	0.47 ± 0.10	0.43 ± 0.07	0.48 ± 0.12
C24:1	0.08 ± 0.03 [abc]	0.06 ± 0.02 [a]	0.07 ± 0.04 [d]	0.06 ± 0.02 [b]	0.06 ± 0.02 [cd]
n-6 PUFA					
C18:2n-6	12.81 ± 3.21 [abc]	10.46 ± 2.71 [ade]	13.95 ± 3.94 [dfg]	16.50 ± 4.27 [befh]	10.51 ± 2.71 [cgh]
C18:3n-6	0.15 ± 0.06	0.16 ± 0.05	0.17 ± 0.09	0.17 ± 0.06	0.15 ± 0.04
C20:2n-6	0.29 ± 0.07 [a]	0.24 ± 0.07 [bc]	0.33 ± 0.10 [bd]	0.37 ± 0.08 [a ce]	0.25 ± 0.06 [de]
C20:3n-6	0.45 ± 0.09 [a]	0.44 ± 0.13	0.44 ± 0.12 [b]	0.47 ± 0.14 [c]	0.38 ± 0.10 [abc]
C20:4n-6	0.51 ± 0.12 [a]	0.45 ± 0.09 [b]	0.47 ± 0.10 [c]	0.56 ± 0.12 [bcd]	0.43 ± 0.09 [ad]
C22:4n-6	0.11 ± 0.03 [a]	0.10 ± 0.02 [b]	0.11 ± 0.08 [c]	0.14 ± 0.04 [abcd]	0.09 ± 0.02 [d]
C22:5n-6	0.05 ± 0.02 [a]	0.05 ± 0.02	0.05 ± 0.05 [b]	0.07 ± 0.02 [abc]	0.04 ± 0.01 [c]
n-3 PUFA					
C18:3n-3	1.11 ± 0.43 [abcd]	0.56 ± 0.28 [ae]	0.60 ± 0.28 [bf]	0.76 ± 0.34 [c]	0.80 ± 0.35 [def]
C20:3n-3	0.06 ± 0.02 [a]	0.03 ± 0.01 [a]	0.05 ± 0.06	0.05 ± 0.01	0.05 ± 0.02
C20:5n-3	0.09 ± 0.05 [ab]	0.05 ± 0.03 [a]	0.08 ± 0.09 [c]	0.05 ± 0.04 [bcd]	0.08 ± 0.05 [d]
C22:5n-3	0.18 ± 0.05 [abc]	0.11 ± 0.04 [ad]	0.14 ± 0.07 [b]	0.13 ± 0.04 [ce]	0.16 ± 0.06 [de]
C22:6n-3	0.31 ± 0.17 [a]	0.23 ± 0.10 [b]	0.35 ± 0.20 [bc]	0.21 ± 0.10 [a cd]	0.30 ± 0.19 [d]
n-9 PUFA					
C20:3n-9	0.03 ± 0.01 [a]	0.03 ± 0.01	0.02 ± 0.01 [ab]	0.02 ± 0.01	0.03 ± 0.01 [b]
Trans FA					
C16:1t	0.06 ± 0.02 [a]	0.06 ± 0.02	0.06 ± 0.07 [b]	0.06 ± 0.02 [c]	0.08 ± 0.03 [abc]
C18:1t	0.25 ± 0.12 [a]	0.34 ± 0.36	0.29 ± 0.21 [b]	0.46 ± 0.43 [abc]	0.31 ± 0.15 [c]
C18:2tt	0.10 ± 0.03 [a]	0.10 ± 0.04 [b]	0.09 ± 0.10 [c]	0.09 ± 0.04 [d]	0.15 ± 0.05 [abcd]

[a–h] pairs with common superscripts indicate significant country differences ($p < 0.05$ after Bonferroni adjustment.

Country effects were less pronounced for hormones (e.g., adiponectin, Figure 2B) and carbohydrates. IGF-II, protein, and total fat concentrations varied by country during the first three months, but not in the later samples (data not shown in detail).

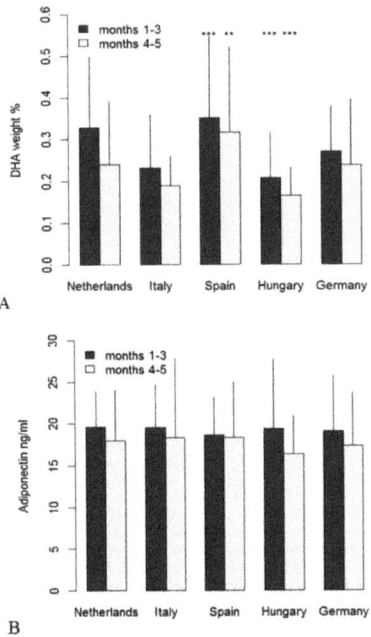

Figure 2. Mean values (+SD) of docosahexaenoic acid (DHA) weight% (**A**) and adiponectin concentration (**B**) per country in early and late milk samples; significant differences from the global means for DHA (months 1–3: 0.29%, months 4–5: 0.24%) and adiponectin (months 1–3: 19.3 ng/mL, months 4–5: 17.6 ng/mL) are indicated as ** for $p < 0.01$ and *** for $p < 0.001$.

Table 5. Human milk fatty acid composition found in the late samples according to the country of residence of the mothers.

	NL	It	ESP	HU	GER
SFA					
C8:0	0.21 ± 0.08	0.27 ± 0.17	0.21 ± 0.06	0.22 ± 0.07	0.21 ± 0.06
C10:0	1.34 ± 0.33	1.56 ± 0.34	1.40 ± 0.36	1.40 ± 0.33	1.44 ± 0.38
C12:0	5.46 ± 1.92	6.20 ± 1.77	5.79 ± 1.91	6.00 ± 1.70	5.44 ± 1.54
C13:0	0.04 ± 0.01 [ab]	0.05 ± 0.02 [ac]	0.03 ± 0.01 [cd]	0.04 ± 0.01 [e]	0.05 ± 0.01 [bde]
C14:0	6.17 ± 2.23	6.49 ± 1.50	5.77 ± 2.02	5.98 ± 1.78	6.74 ± 1.53
C15:0	0.31 ± 0.08 [ab]	0.35 ± 0.09 [cd]	0.25 ± 0.09 [ace]	0.29 ± 0.11 [f]	0.45 ± 0.12 [bdef]
C16:0	21.62 ± 2.39 [ab]	23.55 ± 2.51 [c]	19.79 ± 3.58 [acde]	22.33 ± 2.52 [df]	24.10 ± 2.19 [bef]
C17:0	0.31 ± 0.05 [a]	0.33 ± 0.06 [b]	0.28 ± 0.07 [bc]	0.31 ± 0.05 [d]	0.36 ± 0.06 [acd]
C18:0	7.63 ± 1.26	7.26 ± 0.97 [a]	6.87 ± 1.37 [bc]	7.72 ± 1.25 [b]	8.35 ± 1.46 [ac]
C20:0	0.29 ± 0.15 [ab]	0.33 ± 0.29 [cde]	0.22 ± 0.06 [ac]	0.22 ± 0.05 [bd]	0.24 ± 0.06 [e]
C22:0	0.11 ± 0.06 [abc]	0.09 ± 0.02	0.08 ± 0.03 [a]	0.08 ± 0.03 [b]	0.09 ± 0.03 [c]
C24:0	0.09 ± 0.06 [abc]	0.08 ± 0.03	0.07 ± 0.03 [a]	0.06 ± 0.02 [b]	0.06 ± 0.03 [c]
MUFA					
C14:1	0.24 ± 0.07 [ab]	0.22 ± 0.08	0.15 ± 0.07 [ac]	0.20 ± 0.09 [d]	0.33 ± 0.11 [bcd]
C15:1	0.07 ± 0.02 [a]	0.08 ± 0.03 [b]	0.05 ± 0.02 [bc]	0.07 ± 0.03 [d]	0.11 ± 0.03 [acd]
C16:1n-7	2.25 ± 0.72 [a]	1.87 ± 0.59 [b]	1.75 ± 0.51 [ac]	2.10 ± 0.53 [d]	2.49 ± 0.75 [bcd]
C18:1n-9	35.59 ± 3.57 [ab]	34.92 ± 4.21 [c]	39.58 ± 9.48 [acde]	31.31 ± 3.41 [bdf]	34.07 ± 3.19 [ef]
C18:1n-7	1.58 ± 0.26	1.45 ± 0.22	1.61 ± 0.45	1.59 ± 0.28	1.50 ± 0.25
C20:1n-9	0.48 ± 0.23	0.38 ± 0.07	0.41 ± 0.10	0.40 ± 0.09	0.44 ± 0.12
C24:1	0.06 ± 0.03	0.06 ± 0.02	0.05 ± 0.02	0.05 ± 0.01	0.05 ± 0.02
n-6 PUFA					

Table 5. Cont.

	NL	It	ESP	HU	GER
C18:2n-6	12.72 ± 2.66 [ab]	11.75 ± 3.54 [c]	12.82 ± 3.51 [de]	16.42 ± 4.38 [acdf]	10.35 ± 2.64 [bef]
C18:3n-6	0.15 ± 0.04	0.16 ± 0.04	0.15 ± 0.04	0.17 ± 0.05	0.14 ± 0.03
C20:2n-6	0.24 ± 0.05 [ab]	0.24 ± 0.05 [c]	0.26 ± 0.07 [de]	0.31 ± 0.07 [acdf]	0.20 ± 0.05 [bef]
C20:3n-6	0.33 ± 0.07	0.36 ± 0.05 [a]	0.34 ± 0.09	0.37 ± 0.10 [b]	0.30 ± 0.05 [ab]
C20:4n-6	0.42 ± 0.07 [a]	0.44 ± 0.08 [b]	0.40 ± 0.10 [c]	0.50 ± 0.11 [abcd]	0.40 ± 0.08 [d]
C22:4n-6	0.09 ± 0.03 [a]	0.10 ± 0.02 [b]	0.08 ± 0.03 [c]	0.12 ± 0.03 [abcd]	0.08 ± 0.02 [d]
C22:5n-6	0.04 ± 0.02 [a]	0.04 ± 0.01 [b]	0.04 ± 0.02 [c]	0.06 ± 0.02 [abcd]	0.04 ± 0.01 [d]
n-3 PUFA					
C18:3n-3	1.09 ± 0.43 [abc]	0.50 ± 0.28 [ad]	0.55 ± 0.23 [be]	0.72 ± 0.28 [c]	0.90 ± 0.48 [de]
C20:3n-3	0.05 ± 0.02	0.04 ± 0.02	0.04 ± 0.02	0.04 ± 0.01	0.04 ± 0.02
C20:5n-3	0.09 ± 0.11 [a]	0.05 ± 0.02	0.07 ± 0.05	0.04 ± 0.04 [ab]	0.08 ± 0.05 [b]
C22:5n-3	0.18 ± 0.09 [abc]	0.11 ± 0.03 [ad]	0.13 ± 0.06 [b]	0.12 ± 0.03 [ce]	0.16 ± 0.05 [de]
C22:6n-3	0.29 ± 0.35	0.19 ± 0.07	0.32 ± 0.20 [a]	0.18 ± 0.12 [a]	0.24 ± 0.14
n-9 PUFA					
C20:3n-9	0.02 ± 0.01 [a]	0.02 ± 0.01	0.02 ± 0.01 [ab]	0.02 ± 0.01	0.03 ± 0.01 [b]
Trans FA					
C16:1t	0.06 ± 0.02 [a]	0.07 ± 0.02	0.05 ± 0.02 [b]	0.06 ± 0.02 [c]	0.08 ± 0.02 [abc]
C18:1t	0.27 ± 0.12 [a]	0.28 ± 0.11 [b]	0.26 ± 0.13 [c]	0.42 ± 0.29 [abcd]	0.32 ± 0.12 [d]
C18:2tt	0.12 ± 0.04 [abc]	0.10 ± 0.04 [d]	0.08 ± 0.04 [ae]	0.09 ± 0.03 [bf]	0.14 ± 0.04 [cdef]

[a–h] pairs with common superscripts indicate significant country differences ($p < 0.05$ after Bonferroni adjustment).

3.2. Correlations among Human Milk Components

We focused on correlations that were consistently significant in both the early and late datasets (Table 6). Protein in milk was positively correlated with adiponectin and IGF-II levels. Milk fat content was not significantly related to adiponectin or insulin, but correlated positively with IGF-II and protein.

Table 6. Pearson correlations between the concentrations of macronutrients, hormones, and FA groups (weight%) stratified according to sample collection period.

	IGF-II	Insulin	Adip	Fat	CH	Protein	MUFA	PUFA
\multicolumn{9}{c}{Collection during the First Three Months of Lactation ($n = 319$)}								
Insulin	0.02							
Adip	0.15	−0.05						
Fat	**0.36 *****	0.09	0.06					
CH	−0.14	−0.06	−0.12	−0.16				
Protein	**0.24 *****	−0.03	**0.35 *****	**0.23 ****	−0.01			
MUFA	−0.12	**−0.20 ***	−0.1	−0.12	0.12	0.04		
PUFA	0.15	−0.04	0.02	−0.13	0.05	−0.05	**−0.33 *****	
SFA	−0.03	**0.21 ***	0.06	**0.21 ****	−0.15	0	**−0.56 *****	**−0.59 *****
\multicolumn{9}{c}{Collection during Months 4 and 5 of Lactation ($n = 250$)}								
Insulin	**0.21 ***							
Adip	**0.30 *****	0.02						
Fat	**0.51 *****	0.17	0.12					
CH	−0.11	0	−0.05	−0.13				
Protein	**0.38 *****	0.07	**0.30 *****	**0.23 ***	0.03			
MUFA	−0.1	−0.2	−0.05	−0.09	−0.04	0.02		
PUFA	0.05	−0.07	−0.05	0.12	−0.04	−0.1	**−0.34 ****	
SFA	0.06	0.25	0.09	0	0.07	0.06	**−0.68 *****	**−0.46 *****

Note: * $p < 0.05$, ** $p < 0.01$, *** $p < 0.001$ after Bonferroni correction. Significant results are given in bold. Adip = Adiponectin, CH = carbohydrates.

3.3. Intraindividual Relationships between Early and Late Milk Samples

The relationships of the milk components in early and late samples from mothers who donated two samples are summarized in Table 7. The mean difference between the days of collection was 76 days (range 11 to 109 days). Most analytes showed significant correlations ($p < 0.05$, Table 7) with the exception of carbohydrates, protein, as well as caprylic (8:0), arachidic (20:0), and nervonic acid (24:1n9). Adiponectin and IGF-II concentrations showed closer relationships between both time points than insulin. LC-PUFA and odd-chain FA showed the strongest correlations between the two sampling points (e.g., arachidonic acid (AA) and DHA in Figure 3). Significantly lower values in the 4–5 months period than in the early period for protein and most FA agree with the decreases indicated by multiple linear regression analyses during the first three months (Table 7, *t*-test). The exception is capric acid, which decreased during the early period, but was found to be higher in the later period (Table 7). Additionally, DHA showed a significant decrease from the early sample compared with the later sample.

Table 7. Comparison of the concentrations measured in the early (from lactation day 16–100) and late samples (from lactation day 101–163) from the mothers who donated two samples ($n = 202$).

	Early (M ± SD)	Late (M ± SD)	*t*-Test *p*	Correlation *r* (*p*)
	Hormones and macronutrients			
IGF-II, ng/mL	17.16 ± 5.42	12.62 ± 3.30	**<0.001**	0.303 (**0.001**)
Insulin, mU/L	12.28 ± 8.06	13.72 ± 8.74	0.531	0.480 (**<0.001**)
Adiponectin, ng/mL	19.19 ± 6.06	17.61 ± 6.45	**0.013**	0.466 (**<0.001**)
Fat, g/dL	2.22 ± 1.20	2.54 ± 1.53	0.205	0.346 (**<0.001**)
Carbohydrates, g/dL	6.54 ± 0.43	6.63 ± 0.38	0.259	0.175 (0.543)
Protein, g/dL	1.15 ± 0.22	0.85 ± 0.18	**<0.001**	0.209 (0.128)
Fatty Acids, wt %				
C8:0	0.25 ± 0.10	0.21 ± 0.06	**<0.001**	0.208 (0.123)
C10:0	1.35 ± 0.32	1.42 ± 0.34	0.305	0.508 (**<0.001**)
C12:0	5.37 ± 1.67	5.74 ± 1.66	**0.032**	0.471 (**<0.001**)
C13:0	0.04 ± 0.01	0.04 ± 0.01	1	0.460 (**<0.001**)
C14:0	5.87 ± 1.66	6.23 ± 1.78	**0.013**	0.596 (**<0.001**)
C15:0	0.34 ± 0.12	0.33 ± 0.12	1	0.676 (**<0.001**)
C16:0	22.34 ± 2.85	22.24 ± 2.72	1	0.591 (**<0.001**)
C17:0	0.32 ± 0.06	0.32 ± 0.06	1	0.504 (**<0.001**)
C18:0	7.35 ± 1.33	7.63 ± 1.35	0.97	0.410 (**<0.001**)
C20:0	0.27 ± 0.09	0.23 ± 0.06	**<0.001**	0.167 (0.995)
C22:0	0.10 ± 0.03	0.09 ± 0.03	**0.002**	0.356 (**<0.001**)
C24:0	0.09 ± 0.03	0.07 ± 0.03	**<0.001**	0.355 (**<0.001**)
C14:1	0.24 ± 0.10	0.23 ± 0.11	1	0.651 (**<0.001**)
C15:1	0.08 ± 0.03	0.08 ± 0.03	1	0.651 (**<0.001**)
C16:1 n-7	2.25 ± 0.69	2.15 ± 0.70	1	0.594 (**<0.001**)
C18:1 n-9	35.25 ± 4.53	35.23 ± 4.54	1	0.605 (**<0.001**)
C18:1 n-7	1.62 ± 0.26	1.54 ± 0.25	**0.004**	0.488 (**<0.001**)
C20:1 n-9	0.46 ± 0.07	0.42 ± 0.09	**<0.001**	0.353 (**<0.001**)
C24:1	0.06 ± 0.02	0.05 ± 0.01	**<0.001**	0.217 (0.129)
C18:2 n-6	13.02 ± 4.19	12.76 ± 3.69	1	0.628 (**<0.001**)
C18:3 n-6	0.16 ± 0.05	0.15 ± 0.04	0.393	0.660 (**<0.001**)
C20:2 n-6	0.30 ± 0.09	0.24 ± 0.06	**<0.001**	0.569 (**<0.001**)
C20:3 n-6	0.44 ± 0.11	0.33 ± 0.07	**<0.001**	0.448 (**<0.001**)
C20:4 n-6	0.49 ± 0.11	0.43 ± 0.08	**<0.001**	0.575 (**<0.001**)
C22:4 n-6	0.11 ± 0.03	0.09 ± 0.03	**<0.001**	0.557 (**<0.001**)
C22:5 n-6	0.05 ± 0.02	0.04 ± 0.02	**<0.001**	0.525 (**<0.001**)
C18:3 n-3	0.77 ± 0.38	0.75 ± 0.40	1	0.514 (**<0.001**)
C20:3 n-3	0.05 ± 0.02	0.04 ± 0.01	**<0.001**	0.388 (**<0.001**)
C20:5 n-3	0.07 ± 0.05	0.06 ± 0.04	1	0.553 (**<0.001**)
C22:5 n-3	0.15 ± 0.05	0.13 ± 0.05	0.48	0.591 (**<0.001**)
C22:6 n-3	0.28 ± 0.16	0.23 ± 0.15	**<0.001**	0.632 (**<0.001**)
C20:3 n-9	0.02 ± 0.01	0.02 ± 0.01	1	0.534 (**<0.001**)
C16:1 trans	0.07 ± 0.03	0.06 ± 0.02	1	0.417 (**<0.001**)
C18:1 trans	0.31 ± 0.20	0.30 ± 0.13	1	0.386 (**<0.001**)
C18:2 trans	0.11 ± 0.04	0.11 ± 0.04	1	0.610 (**<0.001**)

Note: Means between the two samples were compared by paired *t*-test. Correlations between the concentrations of early and late samples were calculated according to Pearson. Significant *p*-values (<0.05 after Bonferroni correction) are given in bold.

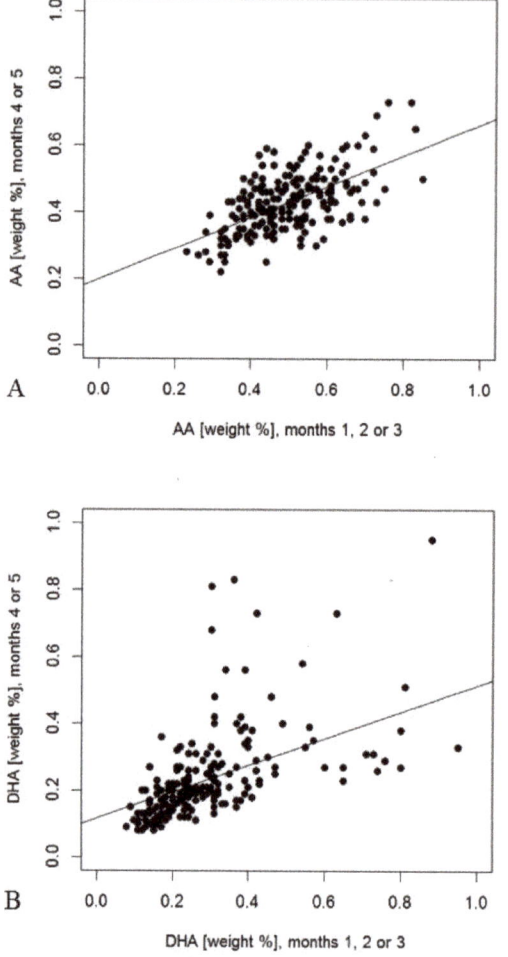

Figure 3. Weight percentages of arachidonic acid (AA) ((**A**), $r = 0.58$, $p < 0.001$) and docosahexaenoic acid (DHA) (**B**), $r = 0.63$, $p < 0.001$) measured in both early and late milk samples. Percentages were calculated for 202 mothers, who had provided samples during the first three months or during the fourth or fifth month of lactation, respectively.

4. Discussion

We observed a significant decrease of both milk protein and n-6 LC-PUFA during the first three months of lactation. Variations in the milk FA composition among the different countries were detected. Among the studied compounds, maternal CD only significantly influenced palmitic acid percentage, leading to lower values in the milk of mothers with CD.

In agreement with previous observations [31], milk carbohydrate contents remained stable over time. Human milk protein levels were higher in earlier lactation and continued to decrease beyond month 3 of lactation, similar to previous observations [7,32,33]. IGF-II and adiponectin showed a trend to decrease with time, which concurs with the assumption that IGFs and protein share common determinants or are directly associated [34]. Percentages of most saturated fatty acids SFA and MUFA (except C20:1n-9 and C24:1n-9) did not change with the duration of lactation. In contrast, LC-PUFA levels decreased. Maternal LC-PUFA stores are depleted due to the high requirements during pregnancy [35] and are further consumed during lactation, which leads to delayed recovery of DHA status after birth in breastfeeding women as compared with non-lactating women [36,37]. Diet and endogenous LC-PUFA synthesis from essential FA did not seem to compensate for the high demands. In our study population, milk fat content did not significantly increase with the day of lactation. Therefore, our results do not support the conclusion of Marangoni et al. [38], that lower LC-PUFA percentages with advancing lactation are compensated by an increase of total fat. The discrepant findings could be linked to the heterogeneity in the collection of our milk samples, while Marangoni et al. applied a defined sample collection procedure only allowing hind milk [38].

Correlations between early and late samples were low for macronutrients and only significant for milk fat, which may partially be the result of variation in conditions of sample collection. The considerable inter-individual differences of the time span between sample collections may have masked a significant intra individual correlation of protein levels. High PUFA and LC-PUFA correlation coefficients indicate a constancy of dietary habits, which define the individual PUFA and LC-PUFA levels, although levels generally decrease with time. For linoleic acid in human milk, it has been shown that about 30% is directly derived from diet and 70% is contributed by fat storage pools [39]. This also seems to apply to DHA [40] and could explain the high correlation coefficient observed for DHA. Nevertheless, single fish meals can markedly increase DHA in subsequent human milk feeds [3], although day-to-day variation is buffered by the contribution of FA from adipose tissue to milk fat [41,42]. Such single fish meals might have caused some of the observed high DHA-% and this could explain that the correlation between early and late DHA-% was not significant, as only those concentration pairs with at least one value above 0.45% were considered. A corresponding phenomenon was not seen for AA, which is assumed to be mainly contributed by endogenous synthesis from linoleic acid [43].

Our study showed that maternal CD and adherence to a GFD did not have any appreciable effect on milk macronutrients, hormones, and FA, with the exception of a lower palmitic acid percentage in the milk of CD mothers. Endogenous palmitic acid synthesis is stimulated by carbohydrate intake [44]. It is tempting to speculate that avoidance of gluten-containing grain-based foods could lead to a lower contribution of carbohydrates to energy intake, and hence lower palmitic acid synthesis. Comparisons of the diet of CD patients and healthy controls report lower carbohydrate or higher fat intake, respectively, with a GFD than in control groups in some studies, but not in all [45–47]. Previous studies have also reported lower n-6 LC-PUFA and higher or lower n-3 LC-PUFA in adult CD patients than in controls [20,21,48]. In this study population from five different countries, we could not find such differences in human milk. Patients with active CD showed higher plasma adiponectin values than healthy controls [49], but this seems not to apply for milk of mothers in remission with CD following a GFD. Olivares et al. had reported lower concentrations of protective immune mediators in milk of women with CD compared with milk from women without CD [22], but we did not identify further nutritionally important effects of CD on human milk. Therefore, breastfeeding can and should be encouraged also in women with CD and a GFD.

Total fat content was only different in the early samples between countries, whereas differences in FA composition were consistently found in both data sets. Potential differences in the lactation day at sample collection were considered in this analysis and do not explain country differences. Total lipid content in human milk is higher in hind-milk than in fore-milk [50]. As fore- or hind-milk collection was not specified in our study, differences of fat content could well be related to differences in sample collection and not reflect different dietary habits. The mode of sample collection does not affect FA composition (weight%), which remains stable during one feeding [51]. The significant variations across countries confirm previous studies that revealed that average human milk FA composition depends on the country and corresponding dietary habits [7,52,53]. While DHA levels were low in milk from Hungarian mothers, Spanish mothers showed the highest DHA levels, which corresponds with previous reports [54,55]. A contribution of at least 0.2% by DHA to total fatty acids in human milk and infant formulas has been considered important for the infant development [56,57], and even

0.3% has been recommended [58]. About 30% of the early samples and 51% of the late samples did not reach the 0.2% level, indicating that sea fish consumption or n-3 LC-PUFA supplement intake should be further encouraged in all included countries, but specifically in Hungary, increased efforts seem required.

Brenna et al. included 65 studies in their meta-analysis of human milk DHA and AA contents and found, on average, 0.32% ± 0.22% for DHA and 0.47% ± 0.13% for ARA [43]. In our study, DHA values were found to be somewhat lower (early samples 0.29% ± 0.16%, late samples 0.24% ± 015%), but AA values were very close to the reported worldwide average (0.49% ± 0.11%, 0.43% ± 0.09%, respectively). In the whole study population, and stratified according to country, the observed variation was smaller for AA than for DHA, which agrees with the findings in the meta-analysis [43] and the concept that DHA levels in milk depend mainly on dietary intake, while AA is mostly endogenously produced from linoleic acid and levels are related to the desaturase genotype [40]. Although linoleic acid status is usually not found to limit the endogenous AA synthesis [59], the finding that Hungarian samples were highest in linoleic acid and AA could suggest that a very high availability of linoleic acid supports high AA. Comparing α-linolenic and DHA percentages between the countries did not indicate associations between levels, confirming the importance of dietary DHA for appropriate milk levels.

The odd chain tridecanoic and pentadecanoic acids differed between countries and could indicate differences in dairy fat intake, as previously shown for plasma FA [60] or differences in fiber intake [61]. Protein and carbohydrate content did not show consistent differences between countries in the early and late data set. Observational studies in affluent populations and in developing countries have failed to identify dietary factors that influence milk protein or carbohydrates [62–65].

There was a significant positive correlation between human milk fat and protein. This correlation has been previously described for milk from mothers of very low birth weight infants [66] and in African mothers [67]. A joint regulation of milk protein and milk fat synthesis has been suggested based on in vitro studies [34,68]. The positive correlation of IGF-II with both protein and fat could be of interest, although regulatory effects of IGF-II in the mammary gland were proposed to be much smaller than IGF-I effects [69]. Adiponectin is significantly related to protein, but not to fat content, but models of the actions of adiponectin in the mammary gland have not been developed so far. We found no correlations between insulin and macronutrients in milk. Human milk insulin levels are related to plasma insulin levels and show comparable diurnal variations [13]. Maternal insulin levels were not available, but pre-pregnancy BMI is positively associated with human milk insulin in the present study, comparable to the results of other studies [70–72].

Strengths and Limitations

The combined analysis of macronutrients, FA, and selected hormones in our study enabled the parallel examination of day of lactation, country of residence, and CD status, which have not been studied before in large numbers of human milk samples. Some of the results have to be interpreted with caution as spot milk samples without full standardization of sampling have been studied, while 24 h collections of human milk and volume determinations would be more representative, enabling a meaningful consideration of potential effects of mixed feeding. Furthermore, interpretation is limited by partially missing information on maternal anthropometry and the ethnicity of the mothers.

5. Conclusions

Milk FA patterns depend on country of residence, which suggests significant dietary influences. In contrast, protein, fat, IGF-II, and adiponectin seem to depend on the individual metabolism. The observed relationships between protein, fat, and IGF-II could agree with an IGF involvement in the regulation of milk synthesis. As no major effects of CD on the studied human milk components were found, breastfeeding should be encouraged in women with CD and a GFD as in the general population.

Author Contributions: Conceptualization, M.L.M. and B.K.; Formal analysis, M.G. and F.F.K.; Funding acquisition, M.L.M. and B.K.; Investigation, M.G., R.A., G.C., I.R.K.-S., I.P., M.R., S.L.V., K.W., and H.D.; Methodology, H.D.; Supervision, M.L.M. and B.K.; Writing—original draft, M.G.; Writing—review & editing, C.H., F.F.K., M.L.M., R.A., G.C., I.R.K.-S., I.P., M.R., S.L.V., K.W., B.K., and H.D.

Funding: The project is supported by grants from the European Commission (FP6-2005-FOOD-4B-36383–PREVENTCD); the Azrieli Foundation; Deutsche Zöliakie Gesellschaft; Eurospital; Fondazione Celiachia; Fria Bröd; Instituto de Salud Carlos III; Spanish Society for Pediatric Gastroenterology, Hepatology, and Nutrition; Komitet Badań Naukowych (1715/B/P01/2008/34); Fundacja Nutricia (1W44/FNUT3/2013); Hungarian Scientific Research Funds (OTKA101788 and TAMOP 2.2.11/1/KONV-2012-0023); Stichting Coeliakie Onderzoek Nederland; Thermo Fisher Scientific; the European Society for Pediatric Gastroenterology, Hepatology, and Nutrition; project EarlyNutrition under grant agreement n°289346; and the European Research Council under the European Union's Seventh Framework Programme (FP/2007-2013)/ERC Advanced Grant ERC-2012-AdG-no. 322605 META-GROWTH. The funding organizations have no role in the conception, design, or conduct of the study; in the analysis or interpretation of the data; or in the writing of the manuscript or the decision to submit it for publication.

Acknowledgments: We thank Stefan Stromer (Ludwig-Maximilians-Universität München), who analysed the milk fatty acids; Yvonne Wijkhuisen, project manager of PreventCD, for support; Els Stoopman, who helped with the data management; and all the children and families who participated in this project.

Conflicts of Interest: The authors declare no conflict of interest.

References

1. Prell, C.; Koletzko, B. Breastfeeding and Complementary Feeding. *Dtsch. Arztebl. Int.* **2016**, *113*, 435–444. [CrossRef] [PubMed]
2. Van Beusekom, C.; Martini, I.A.; Rutgers, H.M.; Boersma, E.R.; Muskiet, F.A. A carbohydrate-rich diet not only leads to incorporation of medium-chain fatty acids (6:0-14:0) in milk triglycerides but also in each milk-phospholipid subclass. *Am. J. Clin. Nutr.* **1990**, *52*, 326–334. [CrossRef] [PubMed]
3. Lauritzen, L.; Jorgensen, M.H.; Hansen, H.S.; Michaelsen, K.F. Fluctuations in human milk long-chain PUFA levels in relation to dietary fish intake. *Lipids* **2002**, *37*, 237–244. [CrossRef] [PubMed]
4. Grote, V.; Verduci, E.; Scaglioni, S.; Vecchi, F.; Contarini, G.; Giovannini, M.; Koletzko, B.; Agostoni, C. Breast milk composition and infant nutrient intakes during the first 12 months of life. *Eur. J. Clin. Nutr.* **2015**, *70*, 250–256. [CrossRef] [PubMed]
5. Armand, M.; Bernard, J.Y.; Forhan, A.; Heude, B.; Charles, M.A.; EDEN Mother-Child Cohort Study Group. Maternal nutritional determinants of colostrum fatty acids in the EDEN mother-child cohort. *Clin. Nutr.* **2018**, *37*, 2127–2136. [CrossRef] [PubMed]
6. Gidrewicz, D.A.; Fenton, T.R. A systematic review and meta-analysis of the nutrient content of preterm and term breast milk. *BMC Pediatrics* **2014**, *14*, 216. [CrossRef] [PubMed]
7. Jensen, R.G. *Handbook of Milk Composition*; Academic Press: NewYork, NY, USA, 1995.
8. Bravi, F.; Wiens, F.; Decarli, A.; Dal Pont, A.; Agostoni, C.; Ferraroni, M. Impact of maternal nutrition on breast-milk composition: A systematic review. *Am. J. Clin. Nutr.* **2016**, *104*, 646–662. [CrossRef]
9. Nasser, R.; Stephen, A.M.; Goh, Y.K.; Clandinin, M.T. The effect of a controlled manipulation of maternal dietary fat intake on medium and long chain fatty acids in human breast milk in Saskatoon, Canada. *Int. Breastfeed. J.* **2010**, *5*, 3. [CrossRef]
10. Ballard, O.; Morrow, A.L. Human Milk Composition. *Pediatric Clin. N. Am.* **2013**, *60*, 49–74. [CrossRef]
11. Grosvenor, C.E.; Picciano, M.F.; Baumrucker, C.R. Hormones and growth factors in milk. *Endocr. Rev.* **1993**, *14*, 710–728. [CrossRef]
12. Newburg, D.S.; Woo, J.G.; Morrow, A.L. Characteristics and Potential Functions of Human Milk Adiponectin. *J. Pediatr.* **2010**, *11*, S41–S46. [CrossRef] [PubMed]

13. Cevreska, S.; Kovacev, V.P.; Stankovski, M.; Kalamaras, E. The presence of immunologically reactive insulin in milk of women, during the first week of lactation and its relation to changes in plasma insulin concentration. *God. Zb. Med. Fak. Skopje* **1975**, *21*, 35–41. [PubMed]
14. Woo, J.G.; Guerrero, M.L.; Altaye, M.; Ruiz-Palacios, G.M.; Martin, L.J.; Dubert-Ferrandon, A.; Newburg, D.S.; Morrow, A.L. Human milk adiponectin is associated with infant growth in two independent cohorts. *Breastfeed. Med.* **2009**, *4*, 101–109. [CrossRef] [PubMed]
15. Prentice, P.; Ong, K.K.; Schoemaker, M.H.; van Tol, E.A.; Vervoort, J.; Hughes, I.A.; Acerini, C.L.; Dunger, D.B. Breast milk nutrient content and infancy growth. *Acta Paediatr.* **2016**, *105*, 641–647. [CrossRef] [PubMed]
16. Heinig, M.J.; Nommsen, L.A.; Peerson, J.M.; Lonnerdal, B.; Dewey, K.G. Energy and protein intakes of breast-fed and formula-fed infants during the first year of life and their associationwith growth velocity: The DARLING Study. *Am. J. Clin. Nutr.* **1993**, *58*, 152–161. [CrossRef] [PubMed]
17. Eriksen, K.G.; Christensen, S.H.; Lind, M.V.; Michaelsen, K.F. Human milk composition and infant growth. *Curr. Opin. Clin. Nutr. Metab. Care* **2018**, *21*, 200–206. [CrossRef] [PubMed]
18. Green, P.H.R.; Cellier, C. Medical progress: Celiac disease. *N. Engl. J. Med.* **2007**, *357*, 1731–1743. [CrossRef]
19. Thompson, T.; Dennis, M.; Higgins, L.A.; Lee, A.R.; Sharrett, M.K. Gluten-free diet survey: Are Americans with coeliac disease consuming recommended amounts of fibre, iron, calcium and grain foods? *J. Hum. Nutr. Diet.* **2005**, *18*, 163–169. [CrossRef]
20. Van Hees, N.J.M.; Giltay, E.J.; Geleijnse, J.M.; Janssen, N.; van der Does, W. DHA Serum Levels Were Significantly Higher in Celiac Disease Patients Compared to Healthy Controls and Were Unrelated to Depression. *PLoS ONE* **2014**, *9*, e97778. [CrossRef]
21. Russo, F.; Chimienti, G.; Clemente, C.; Ferreri, C.; Orlando, A.; Riezzo, G. A possible role for ghrelin, leptin, brain-derived neurotrophic factor and docosahexaenoic acid in reducing the quality of life of coeliac disease patients following a gluten-free diet. *Eur. J. Nutr.* **2017**, *56*, 807–818. [CrossRef]
22. Olivares, M.; Albrecht, S.; De Palma, G.; Ferrer, M.D.; Castillejo, G.; Schols, H.A.; Sanz, Y. Human milk composition differs in healthy mothers and mothers with celiac disease. *Eur. J. Nutr.* **2015**, *54*, 119–128. [CrossRef] [PubMed]
23. Hogen Esch, C.E.; Rosen, A.; Auricchio, R.; Romanos, J.; Chmielewska, A.; Putter, H.; Ivarsson, A.; Szajewska, H.; Koning, F.; Wijmenga, C.; et al. The PreventCD Study design: towards new strategies for the prevention of coeliac disease. *Eur. J. Gastroenterol. Hepatol.* **2010**, *22*, 1424–1430. [CrossRef] [PubMed]
24. Vriezinga, S.L.; Auricchio, R.; Bravi, E.; Castillejo, G.; Chmielewska, A.; Crespo Escobar, P.; Kolacek, S.; Koletzko, S.; Korponay-Szabo, I.R.; Mummert, E.; et al. Randomized feeding intervention in infants at high risk for celiac disease. *N. Engl. J. Med.* **2014**, *371*, 1304–1315. [CrossRef] [PubMed]
25. Hellmuth, C.; Uhl, O.; Demmelmair, H.; Grunewald, M.; Auricchio, R.; Castillejo, G.; Korponay-Szabo, I.R.; Polanco, I.; Roca, M.; Vriezinga, S.L.; et al. The impact of human breast milk components on the infant metabolism. *PLoS ONE* **2018**, *13*, e0197713. [CrossRef] [PubMed]
26. Casadio, Y.S.; Williams, T.M.; Lai, C.T.; Olsson, S.E.; Hepworth, A.R.; Hartmann, P.E. Evaluation of a mid-infrared analyzer for the determination of the macronutrient composition of human milk. *J. Hum. Lact.* **2010**, *26*, 376–383. [CrossRef] [PubMed]
27. Polberger, S.; Lönnerdal, B. Simple and Rapid Macronutrient Analysis of Human Milk for Individualized Fortification: Basis for Improved Nutritional Management of Very-Low-Birth-Weight Infants? *J. Pediatr. Gastr. Nutr.* **1993**, *17*, 283–290. [CrossRef] [PubMed]
28. Stimming, M.; Mesch, C.M.; Kersting, M.; Kalhoff, H.; Demmelmair, H.; Koletzko, B.; Schmidt, A.; Bohm, V.; Libuda, L. Vitamin E content and estimated need in German infant and follow-on formulas with and without long-chain polyunsaturated fatty acids (LC-PUFA) enrichment. *J. Agric. Food Chem.* **2014**, *62*, 10153–10161. [CrossRef]
29. Goelz, R.; Hihn, E.; Hamprecht, K.; Dietz, K.; Jahn, G.; Poets, C.; ElmLinger, M. Effects of Different CMV-Heat-Inactivation-Methods on Growth Factors in Human Breast Milk. *Pediatr. Res.* **2009**, *65*, 458–461. [CrossRef]
30. Vittinghoff, E.; Glidden, D.; Shiboski, S.; McCulloch, C. *Regression Methods in Biostatistics: Linear, Logistic, Survival, and Repeated Measures Models*; Springer: New York, NY, USA, 2005.
31. Mitoulas, L.R.; Kent, J.C.; Cox, D.B.; Owens, R.A.; Sherriff, J.L.; Hartmann, P.E. Variation in fat, lactose and protein in human milk over 24h and throughout the first year of lactation. *Br. J. Nutr.* **2002**, *88*, 29–37. [CrossRef]

32. Shehadeh, N.; Aslih, N.; Shihab, S.; Werman, M.J.; Sheinman, R.; Shamir, R. Human milk beyond one year post-partum: Lower content of protein, calcium, and saturated very long-chain fatty acids. *J. Pediatrics* **2006**, *148*, 122–124. [CrossRef]
33. Lonnerdal, B.; Erdmann, P.; Thakkar, S.K.; Sauser, J.; Destaillats, F. Longitudinal evolution of true protein, amino acids and bioactive proteins in breast milk: A developmental perspective. *J. Nutr. Biochem.* **2017**, *41*, 1–11. [CrossRef] [PubMed]
34. Anderson, S.M.; Rudolph, M.C.; McManaman, J.L.; Neville, M.C. Secretory activation in the mammary gland: It's not just about milk protein synthesis. *Breast Cancer Res.* **2007**, *9*, 204–217. [CrossRef] [PubMed]
35. Herrera, E.; Amusquivar, E.; López-Soldado, I.; Ortega, H. Maternal lipid metabolism and placental lipid transfer. *Horm. Res. Paediatr.* **2006**, *65*, 59–64. [CrossRef] [PubMed]
36. Hornstra, G. Essential fatty acids in mothers and their neonates. *Am. J. Clin. Nutr.* **2000**, *71*, 1262s–1269s. [CrossRef] [PubMed]
37. Jorgensen, M.H.; Nielsen, P.K.; Michaelsen, K.F.; Lund, P.; Lauritzen, L. The composition of polyunsaturated fatty acids in erythrocytes of lactating mothers and their infants. *Matern. Child Nutr.* **2006**, *2*, 29–39. [CrossRef]
38. Marangoni, F.; Agostoni, C.; Lammardo, A.M.; Giovannini, M.; Galli, C.; Riva, E. Polyunsaturated fatty acid concentrations in human hindmilk are stable throughout 12-months of lactation and provide a sustained intake to the infant during exclusive breastfeeding: An Italian study. *Br. J. Nutr.* **2000**, *84*, 103–109. [PubMed]
39. Demmelmair, H.; Baumheuer, M.; Koletzko, B.; Dokoupil, K.; Kratl, G. Metabolism of U13C-labeled linoleic acid in lactating women. *J. Lipid Res.* **1998**, *39*, 1389–1396.
40. Demmelmair, H.; Koletzko, B. Lipids in human milk. *Best Pract. Res. Clin. Endocrinol. Metab.* **2018**, *32*, 57–68. [CrossRef]
41. Innis, S.M. Fatty acids and early human development. *Early Hum. Dev.* **2007**, *83*, 761–766. [CrossRef]
42. Demmelmair, H.; Sauerwald, T.; Fidler, N.; Baumheuer, M.; Koletzko, B. Polyunsaturated fatty acid metabolism during lactation. *World Rev. Nutr. Diet.* **2001**, *88*, 184–189.
43. Brenna, J.T.; Varamini, B.; Jensen, R.G.; Diersen-Schade, D.A.; Boettcher, J.A.; Arterburn, L.M. Docosahexaenoic and arachidonic acid concentrations in human breast milk worldwide. *Am. J. Clin. Nutr.* **2007**, *85*, 1457–1464. [CrossRef]
44. Hudgins, L.C.; Hellerstein, M.; Seidman, C.; Neese, R.; Diakun, J.; Hirsch, J. Human fatty acid synthesis is stimulated by a eucaloric low fat, high carbohydrate diet. *J. Clin. Investig.* **1996**, *97*, 2081–2091. [CrossRef] [PubMed]
45. Kinsey, L.; Burden, S.T.; Bannerman, E. A dietary survey to determine if patients with coeliac disease are meeting current healthy eating guidelines and how their diet compares to that of the British general population. *Eur. J. Clin. Nutr.* **2008**, *62*, 1333–1342. [CrossRef] [PubMed]
46. Melini, V.; Melini, F. Gluten-Free Diet: Gaps and Needs for a Healthier Diet. *Nutrients* **2019**, *11*, 170. [CrossRef]
47. Capristo, E.; Addolorato, G.; Mingrone, G.; De Gaetano, A.; Greco, A.V.; Tataranni, P.A.; Gasbarrini, G. Changes in body composition, substrate oxidation, and resting metabolic rate in adult celiac disease patients after a 1-y gluten-free diet treatment. *Am. J. Clin. Nutr.* **2000**, *72*, 76–81. [CrossRef] [PubMed]
48. Solakivi, T.; Kaukinen, K.; Kunnas, T.; Lehtimaki, T.; Maki, M.; Nikkari, S.T. Serum fatty acid profile in celiac disease patients before and after a gluten-free diet. *Scand. J. Gastroenterol.* **2009**, *44*, 826–830. [CrossRef] [PubMed]
49. Russo, F.; Chimienti, G.; Clemente, C.; D'Attoma, B.; Linsalata, M.; Orlando, A.; De Carne, M.; Cariola, F.; Semeraro, F.P.; Pepe, G.; et al. Adipokine profile in celiac patients: Differences in comparison with patients suffering from diarrhea-predominant IBS and healthy subjects. *Scand. J. Gastroenterol.* **2013**, *48*, 1377–1385. [CrossRef] [PubMed]
50. Mizuno, K.; Nishida, Y.; Taki, M.; Murase, M.; Mukai, Y.; Itabashi, K.; Debari, K.; Iiyama, A. Is increased fat content of hindmilk due to the size or the number of milk fat globules? *Int. Breastfeed. J.* **2009**, *4*, 7. [CrossRef] [PubMed]
51. Emery, W.B., 3rd; Canolty, N.L.; Aitchison, J.M.; Dunkley, W.L. Influence of sampling on fatty acid composition of human milk. *Am. J. Clin. Nutr.* **1978**, *31*, 1127–1130. [CrossRef]
52. Yuhas, R.; Pramuk, K.; Lien, E.L. Human milk fatty acid composition from nine countries varies most in DHA. *Lipids* **2006**, *41*, 851–858. [CrossRef]

53. Keikha, M.; Bahreynian, M.; Saleki, M.; Kelishadi, R. Macro- and Micronutrients of Human Milk Composition: Are They Related to Maternal Diet? A Comprehensive Systematic Review. *Breastfeed. Med.* **2017**, *12*, 517–527. [CrossRef] [PubMed]
54. Barreiro, R.; Diaz-Bao, M.; Cepeda, A.; Regal, P.; Fente, C.A. Fatty acid composition of breast milk in Galicia (NW Spain): A cross-country comparison. *Prostaglandins Leukot. Essent. Fatty Acids* **2018**, *135*, 102–114. [CrossRef] [PubMed]
55. Decsi, T.; Olah, S.; Molnar, S.; Burus, I. Low contribution of docosahexaenoic acid to the fatty acid composition of mature human milk in Hungary. *Adv. Exp. Med. Biol.* **2000**, *478*, 413–414. [CrossRef] [PubMed]
56. Koletzko, B.; Lien, E.; Agostoni, C.; Bohles, H.; Campoy, C.; Cetin, I.; Decsi, T.; Dudenhausen, J.W.; Dupont, C.; Forsyth, S.; et al. The roles of long-chain polyunsaturated fatty acids in pregnancy, lactation and infancy: Review of current knowledge and consensus recommendations. *J. Perinat. Med.* **2008**, *36*, 5–14. [CrossRef] [PubMed]
57. FAO. *Fats and Fatty Acids in Human Nutrition—Report of an Expert Consultation*; FAO: Rome, Italy, 2010.
58. Koletzko, B.; Boey, C.C.; Campoy, C.; Carlson, S.E.; Chang, N.; Guillermo-Tuazon, M.A.; Joshi, S.; Prell, C.; Quak, S.H.; Sjarif, D.R.; et al. Current information and Asian perspectives on long-chain polyunsaturated fatty acids in pregnancy, lactation, and infancy: Systematic review and practice recommendations from an early nutrition academy workshop. *Ann. Nutr. Metab.* **2014**, *65*, 49–80. [CrossRef] [PubMed]
59. Demmelmair, H.; MacDonald, A.; Kotzaeridou, U.; Burgard, P.; Gonzalez-Lamuno, D.; Verduci, E.; Ersoy, M.; Gokcay, G.; Alyanak, B.; Reischl, E.; et al. Determinants of Plasma Docosahexaenoic Acid Levels and Their Relationship to Neurological and Cognitive Functions in PKU Patients: A Double Blind Randomized Supplementation Study. *Nutrients* **2018**, *10*, 1944. [CrossRef] [PubMed]
60. Brevik, A.; Veierod, M.B.; Drevon, C.A.; Andersen, L.F. Evaluation of the odd fatty acids 15:0 and 17:0 in serum and adipose tissue as markers of intake of milk and dairy fat. *Eur. J. Clin. Nutr.* **2005**, *59*, 1417–1422. [CrossRef] [PubMed]
61. Weitkunat, K.; Schumann, S.; Nickel, D.; Hornemann, S.; Petzke, K.J.; Schulze, M.B.; Pfeiffer, A.F.H.; Klaus, S. Odd-chain fatty acids as a biomarker for dietary fiber intake: A novel pathway for endogenous production from propionate. *Am. J. Clin. Nutr.* **2017**, *105*, 1544–1551. [CrossRef] [PubMed]
62. Nommsen, L.A.; Lovelady, C.A.; Heinig, M.J.; Lonnerdal, B.; Dewey, K.G. Determinants of energy, protein, lipid, and lactose concentrations in human milk during the first 12 mo of lactation: The DARLING Study. *Am. J. Clin. Nutr.* **1991**, *53*, 457–465. [CrossRef]
63. Yang, T.; Zhang, Y.; Ning, Y.; You, L.; Ma, D.; Zheng, Y.; Yang, X.; Li, W.; Wang, J.; Wang, P. Breast milk macronutrient composition and the associated factors in urban Chinese mothers. *Chin. Med. J.* **2014**, *127*, 1721–1725.
64. Quinn, E.A.; Largado, F.; Power, M.; Kuzawa, C.W. Predictors of breast milk macronutrient composition in Filipino mothers. *Am. J. Hum. Biol.* **2012**, *24*, 533–540. [CrossRef] [PubMed]
65. Bzikowska-Jura, A.; Czerwonogrodzka-Senczyna, A.; Oledzka, G.; Szostak-Wegierek, D.; Weker, H.; Wesolowska, A. Maternal Nutrition and Body Composition During Breastfeeding: Association with Human Milk Composition. *Nutrients* **2018**, *10*, 1379. [CrossRef] [PubMed]
66. Weber, A.; Loui, A.; Jochum, F.; Buhrer, C.; Obladen, M. Breast milk from mothers of very low birthweight infants: Variability in fat and protein content. *Acta Paediatr.* **2001**, *90*, 772–775. [CrossRef] [PubMed]
67. Roels, O.A. Correlation between the fat and the protein content of human milk. *Nature* **1958**, *182*, 673. [CrossRef] [PubMed]
68. Qi, L.; Yan, S.; Sheng, R.; Zhao, Y.; Guo, X. Effects of Saturated Long-chain Fatty Acid on mRNA Expression of Genes Associated with Milk Fat and Protein Biosynthesis in Bovine Mammary Epithelial Cells. *Asian Australas. J. Anim. Sci.* **2014**, *27*, 414–421. [CrossRef] [PubMed]
69. Prosser, C.G. Insulin-like growth factors in milk and mammary gland. *J. Mammary Gland Biol. Neoplasia* **1996**, *1*, 297–306. [CrossRef] [PubMed]
70. Ahuja, S.; Boylan, M.; Hart, S.L.; Román-Shriver, C.; Spallholz, J.E.; Pence, B.C.; Sawyer, B.G. Glucose and Insulin Levels are Increased in Obese and Overweight Mothers' Breast-Milk. *Food Nutr. Sci.* **2011**, *2*, 201–206. [CrossRef]

71. Ley, S.H.; Hanley, A.J.; Sermer, M.; Zinman, B.; O'Connor, D.L. Associations of prenatal metabolic abnormalities with insulin and adiponectin concentrations in human milk. *Am. J. Clin. Nutr.* **2012**, *95*, 867–874. [CrossRef] [PubMed]
72. Demmelmair, H.; Koletzko, B. Variation of Metabolite and Hormone Contents in Human Milk. *Clin. Perinatol.* **2017**, *44*, 151–164. [CrossRef]

© 2019 by the authors. Licensee MDPI, Basel, Switzerland. This article is an open access article distributed under the terms and conditions of the Creative Commons Attribution (CC BY) license (http://creativecommons.org/licenses/by/4.0/).

Article

Breast Milk Content of Vitamin A and E from Early- to Mid-Lactation Is Affected by Inadequate Dietary Intake in Brazilian Adult Women

Michele R. Machado [1,2], Fernanda Kamp [1,3,*], Juliana C. Nunes [1,4], Tatiana El-Bacha [1,5,6] and Alexandre G. Torres [1,*]

1. Laboratory of Food Science and Nutritional Biochemistry, Institute of Chemistry, Federal University of Rio de Janeiro, Rio de Janeiro 21941-909, Brazil
2. School of Nutrition, Faculdade Arthur Sa Earp Neto (FMP/FASE), Petropolis, Rio de Janeiro 25680-120, Brazil
3. Biochemistry Core, Federal Institute of Education, Science and Technology of Rio de Janeiro, Rio de Janeiro 20270-021, Brazil
4. School of Nutrition, Federal University of the State of Rio de Janeiro, Rio de Janeiro 22290-250, Brazil
5. Institute of Nutrition, Federal University of Rio de Janeiro, Rio de Janeiro 21941-902, Brazil
6. Department of Physiology, Development and Neuroscience, University of Cambridge, Cambridge CB2 3EL, UK
* Correspondence: fernanda.kamp@ifrj.edu.br (F.K.); torres@iq.ufrj.br (A.G.T.); Tel.: +55-21-3938-7351 (A.G.T.)

Received: 6 July 2019; Accepted: 26 August 2019; Published: 29 August 2019

Abstract: Our aims were to investigate vitamin A and E status during lactation and the determinants of breast milk content for the appropriate nutrition of the infant in a study with nursing Brazilian women. We hypothesized that both inadequate intake and the lipoprotein distribution of vitamin A and E during lactation could have an impact on their breast milk levels from early- to mid-lactation. Nineteen adult lactating women participated in this longitudinal observational study, in which dietary records, blood and mature breast milk samples were collected for the analysis of vitamin A and E, and carotenoids in early- (2nd to 4th week) and mid-lactation (12th to 14th week). Nutrient intake was balanced by the Multiple Source Method (MSM), and the intake of vitamin A and E was inadequate in 74 and 100% of the women, respectively. However, these results were not reflected in low serum concentrations of retinol and only 37% of the volunteers were vitamin E deficient according to the blood biomarker. As lactation progressed, vitamin A and E status worsened, and this was clearly observed by the decrease in their content in breast milk. The reduced content of vitamin A and E in the breast milk was not related to their distribution in lipoproteins. Taken together, the contents of vitamin A and E in breast milk seemed to be more sensitive markers of maternal nutrition status than respective blood concentrations, and dietary assessment by the MSM in early lactation was sensitive to indicate later risks of deficiency and should support maternal dietary guidance to improve the infant's nutrition.

Keywords: retinol; α-tocopherol; inadequate intake; nutritional status; breast milk; undernourishment; dietary assessment; multiple source method

1. Introduction

Vitamin A and E are essential to newborns, and their transfer to breast milk is key for the nutrition of infants. Vitamin A is crucial for newborn development, epithelial function and protection against infections. Vitamin A deficiency affects millions of preschool-age children, especially in developing countries, and it increases the susceptibility to infection and infant mortality rate, especially before the age of 2 [1]. Vitamin A nutritional requirements are met by the intake of retinoids, such as retinyl esters, and provitamin A carotenoids, especially β-carotene and less prominently by α-carotene and β-cryptoxanthin. In addition to serving as retinoid precursors, these carotenoids are transferred to breast milk and, together with lycopene and lutein, might contribute to breastfed infant health, as they present potential bioactivity. The primary function of vitamin E is related to its activity as a potent free radical scavenger and, although overt signs of deficiency are rare, pre-term infants might be more susceptible to associated illnesses [2].

Essential nutrients, such as vitamins, are provided to the breastfed infant via transfer from the maternal circulation into the milk by the mammary gland. The contents of retinol and α-tocopherol in the breast milk are mainly determined by maternal diet and nutritional status [3,4]. However, to what extent changes in the maternal nutritional status throughout lactation might affect the transfer of vitamin A and E to the breast milk is not entirely known. Interactions between vitamin A and E seem to influence their contents in breast milk as it has been shown that postpartum retinyl-ester supplementation decreased vitamin E content in the colostrum [5]. On the other hand, the mechanisms underlying this biochemical interaction are not known and are possibly related to serum transport in lipoproteins and uptake by the mammary gland [3]. Vitamin A and E distribution in major lipoprotein fractions might fluctuate throughout lactation, which would impact their transfer to breast milk.

Our aim was to investigate vitamin A and E status, and carotenoids during lactation, and the determinants of breast milk levels for the appropriate nutrition of breastfed infants. We hypothesized that both inadequate intake and the lipoprotein distribution of vitamin A and E during lactation could have an impact on their breast milk levels from early- to mid-lactation. By taking advantage of a longitudinal study with adult nursing women in Rio de Janeiro, Brazil, we used multiple regression models to explain the impact of maternal diet and serum concentration of vitamin A and E on their breast milk content during early- and mid-lactation.

2. Materials and Methods

2.1. Experimental Design and Recruitment of Volunteers

This is a longitudinal observational study, approved by the Research Ethics Committee of the Pedro Ernesto University Hospital, Rio de Janeiro State University (CEP/HUPE-UERJ:3043/2011–CAAE:0186.0.228.228-11). Data and sample collection were conducted after signing written informed consent. The volunteers were recruited at Policlínica Américo Piquet Carneiro (UERJ) and at health facilities in the city of Rio de Janeiro. Recruitment occurred in the last month of gestation, when the first visit of the study was scheduled according to the probable delivery date. Of the 22 recruited women, three refused to participate during mid-lactation and their data were excluded. Thus, 19 pregnant women (age 20–40 years) participated in the study, exclusively or predominantly breastfeeding. They did not have any chronic or acute diseases, were not using prescribed medicines or dietary supplements, and were non-smokers and non-alcoholic who had completed single and full-term gestations. In the present study, mature breast milk was collected, because in full lactation, when mature milk is produced, tight junctions between mammary epithelial cells are fully functional, and, therefore, paracellular transport (between cells) is virtually absent. Therefore, one can expect that there are no major changes in the mammary epithelial cells transport systems during the time span assessed in this study.

2.2. Data Collection and Qualification of Lactation Practices

Demographics and medical history were obtained in early lactation (2nd to 4th week postpartum; mean 24.7 days). In mid-lactation (12th to 14th week postpartum; mean 94 days), volunteers answered a questionnaire concerning lactation practices. All women reported adopting practices consistent with exclusive or predominant breastfeeding. Total body mass (kg) and height (meters) were obtained in an anthropometric scale (Filizola, Brazil) to calculate body mass index (BMI, in kg/m^2) [6].

2.3. Assessment of Dietary Intake

Dietary intake throughout lactation was estimated using the Multiple Source Method (MSM). The MSM is a statistical method to estimate usual food intake and uses at least two different dietary inputs (e.g., dietary recalls and/or food frequency questionnaires) [7] and it has been shown to be adequate to estimate usual intake during pregnancy [8]. Using this method, 24-h dietary recalls (24hRs) were applied in early- and mid-lactation (one in each period) to evaluate possible changes in eating habits and to estimate recent intake of energy and nutrients. The nutrient composition of the 24hRs was analyzed with the NutriSurvey® software v. 2007 (Dr. Juergen Erhardt; Germany; available at www.nutrisurvey.de/nutrisurvey2007.exe) using the food composition database from the US Department of Agriculture release 22 [9] adapted for common fat sources in the Brazilian diet [10]. A semi-quantitative food frequency questionnaire (SQ-FFQ) was applied in the first lactation period to assess food sources of the nutrients of interest in the present investigation and to confirm the feasibility of the 24hR to evaluate usual dietary intake.

The prevalence of inadequate nutrient intake was estimated using Estimated Average Requirements (EARs), as set by the Institute of Medicine [11–14] as cut-off points. The prevalence of the inadequate intake of each nutrient was estimated by the proportion of individuals with intake below the EAR value. Since the EAR cut-off point method is not suitable for assessing energy adequacy, the proportion of the group with BMIs below, within, and above the desirable range was used to classify inadequate, adequate, and excessive energy intakes [11].

2.4. Collection and Processing of Biological Samples

Blood and milk samples were collected in the morning after fasting overnight, were immediately processed for the separation of plasma, serum and erythrocytes, and frozen at −80 °C until analysis. Whole blood aliquots were used for the immediate determination of hematocrit and hemoglobin concentrations. Milk samples (10 mL) were manually expressed in the morning from the breast which the baby last suckled and transferred into sterile plastic bottles. An aliquot was taken for analysis of the crematocrit. For the analysis of lactose and protein, aliquots were stored at −20 °C and for retinol, carotenoids and tocopherols analyses, at −80 °C and tubes were protected with aluminum foil.

2.5. Determination of Hematocrit and Hemoglobin, and Triglycerides and Cholesterol in Blood

Hematocrit was determined in a microhematocrit centrifuge (Hemospin, Incibrás; São Paulo, Brazil) and hemoglobin concentration was determined by the cyanmethemoglobin method by a commercial kit (BioClin; Belo Horizonte, Brazil). Triglyceride, total cholesterol and HDL-c (total serum and HDL-c fraction after LDL-c+VLDL-c precipitation) were determined by the enzymatic colorimetric method using commercial kits (BioClin, Brazil), using sodium phosphotungstate and Mg^{2+} for precipitation of LDL and VLDL particles, as described [15]. The clear supernatant, containing the HDL fraction, was used for the analysis of HDL-c (and vitamins by HPLC, described in Section 2.7), and the concentration of LDL-c+VLDL-c was determined by the difference between the concentration of HDL-c and total cholesterol. All analyses were performed in duplicate and the automatic reading was performed in triplicate.

2.6. Breast Milk Macronutrient Composition

Total fat was determined by the crematocrit method, in a microhematocrit centrifuge (Hemospin, Incibrás) [16]. Lactose was determined by a colorimetric assay [17] adapted by [18], based on the reaction with picric acid. Total protein content was determined by the Lowry method [19], after eliminating interference of fat and lactose, by precipitating the proteins with 20% trichloroacetic acid and 2.5% sodium deoxycholate, and centrifugation. The supernatant with interfering substances was discarded and precipitated milk proteins were suspended in Folin–Ciocalteu's reagent and analyzed [20]. The standard curve was prepared with casein in concentrations ranging from 20.0 to 100.0 µg/mL. All these analyses, as well as standard curves, were performed in triplicates.

2.7. Determination of Retinol, Carotenoids and Tocopherols by HPLC in Breast Milk, Whole Blood Serum, and Lipoprotein Fractions

Serum samples were prepared and extracted as described by [21], adapted from [22]. Retinol, tocopherols, and carotenoids were analyzed by HPLC in serum and in the HDL fraction, in duplicate. The concentration of these vitamins in the LDL+VLDL fraction was calculated by the difference between contents in whole serum and in the HDL fraction. Milk samples were prepared as described [21], adapted from [23], and care was taken to avoid direct light to avoid vitamin losses during sample extraction and analysis.

HPLC analysis was performed in a Shimadzu chromatographic system (Shimadzu, Kyoto, Japan), composed of a LC-20AT pump, a Shimadzu SPD-M20AV UV-Vis detector, and a CBM-20A system controller. A total of 20 µL of sample extracts and standards solutions were injected through a Rheodyne injection valve with a volumetric loop, onto a C18 reversed-phase column (Kromasil; 150 × 4.6 mm), and compounds were eluted isocratically with a ternary solvent composed of acetonitrile, tetrahydrofuran, and 15 mM methanolic ammonium acetate (65:25:10, v/v/v), at 0.9 mL/min. Each sample extract was injected twice, one run monitored at λ (nm) 450, for the analysis of carotenoids, and the other run monitored at 325 and 292 for the analysis of retinol and tocopherols, respectively. All analyses were performed in duplicate.

Identities of chromatographic peaks corresponding to the fat-soluble vitamins in samples chromatograms were determined based on standards retention times, and by exact co-elution with standards spiking in representative samples. Vitamin concentrations in samples were determined by external calibration, with linear calibration curves and direct extrapolation. Commercial standards of retinol, α-tocopherol and γ-tocopherol were from Sigma-Aldrich (São Paulo, Brazil), and carotenoid standards (purity > 95%) were isolated from foods naturally rich in these pigments, as previously described [24], by the open column method [25]. Representative chromatograms from HPLC standards are shown in Supplementary Figure S1.

2.8. Statistical Analyses

Paired *t*-tests were used to investigate differences between early- and mid-lactation on the variables studied. Variable frequency distribution was assessed by standardized coefficients of skewness and kurtosis, and those with values < −2.0 or > +2.0 were characterized as having a non-normal distribution, which were presented as median and minimum and maximum levels, in contrast to normally distributed variables that were presented as the mean ± standard deviation (SD). Non-normally distributed variables were log-transformed before running Pearson's correlation analysis, which was used to investigate associations between continuous variables. Stepwise multiple regression analyses (backward) were used to investigate the effect of independent factors on the composition of vitamins in serum and in breast milk. The criteria for the inclusion of independent variables in the multiple regression models were based on results from Pearson correlations and on data from the literature. In the final model, only significant variables that improved the model were kept (*p*-to-remove ≥ 0.05; *p*-to-persist < 0.05). The multiple regression models were further assessed by analysis of residual plots that were checked to determine if they were randomly distributed. Data analyses were performed with the Multiple Source Method (MSM) software for dietary data analysis, GraphPad Prism 7.0 (GraphPad Software, San Diego, CA, USA) and Statgraphics Centurion 18 (Statgraphics Technologies, Inc.; The Plains, VA, USA) for statistics. Values of $p < 0.05$ were considered significant.

3. Results

3.1. General Characteristics of Lactating Women

The nursing women who participated in the present study were adults (31.9 ± 5.9 years of age), primiparous and had term pregnancies (gestational age of 39.8 ± 1.1 weeks), from which 58% had vaginal delivery (Table 1). According to the BMI, 58% of women were overweight (25–30 kg/m^2 [6]) in early lactation (2nd to 4th lactation week). In mid-lactation (12th to 14th lactation week), 42% of the lactating mothers were overweight. Most of the volunteers (63%) were anemic at the beginning of the study and after 10 weeks, there was a significant 11% increase in the mean blood hemoglobin concentration and a decrease in anemia frequency to 26% in mid-lactation (Table 1).

Table 1. General characteristics of adult Brazilian lactating women in early- (2nd to 4th week) and mid-lactation (12th to 14th week) (n = 19).

Characteristics	Lactation Period		p	Recommended Value or Range [1]
	Early Lactation 2nd–4th Week	Mid-Lactation 12th–14th Week		
Body mass index (kg/m^2)	25.9 ± 4.0	25.3 ± 4.2	0.068 *	18–24.9 [6]
Hemoglobin (g/dL)	11.6 ± 1.7	12.5 ± 1.6	0.029 *	≥12.0 [26]
Hematocrit (%)	39.2 ± 3.5	37.8 ± 3.5	0.131	≥36 [26]
Triglycerides (mg/dL)	74.7 ± 21.4	57.0 ± 21.2	0.007 *	<150 [27]
Total cholesterol (mg/dL)	190.2 ± 41.9	182.2 ± 41.0	0.219	<200 [27]
LDL-c (mg/dL)	156.0 ± 44.1	146.6 ± 35.9	0.164	100–129 [27]
HDL-c (mg/dL)	34.2 ± 6.2	35.6 ± 9.4	0.601	>60 [27]

Data presented as the mean ± standard deviation; * Significant differences between lactation periods, paired t-test. [1] Shown in the same respective units as the first column.

Serum triglyceride concentration was adequate in all volunteers and reduced significantly by 24% from early- to mid-lactation. Serum HDL concentrations, on the other hand, were below the minimum value in all volunteers and 74% of the nursing women presented LDL concentrations above the desirable range, irrespective of the lactation period [27]. Regarding total cholesterol in the serum, 47 and 31% of the volunteers presented concentrations above the desirable value, in early- and mid-lactation, respectively.

3.2. Dietary Intake of Vitamin A and E Were Inadequate throughout Lactation

Dietary data assessment indicated that nutrient intake did not vary significantly between the two lactation periods, considering both recent and usual dietary consumption. Therefore, these data were presented combined, as estimated by the MSM, which assesses habitual intake considering data from two or more 24hRs and one FFQ (Table 2). The frequency of volunteers with sub-adequate nutrient intake was estimated as proposed [12,13] using EAR whenever available (Table 2). The very high prevalence of inadequate intake of vitamin A and E (Table 2) may set these lactating mothers and their offspring's health at increased risk. Although the median intake of β-carotene (Table 2) was within the range considered prudent, 47% of the volunteers presented β-carotene intake below this range. In contrast, the average intake of lycopene was lower than the level established as prudent [12], and intake was sub-adequate in 89% of the volunteers.

Table 2. Intake of energy and nutrients by Brazilian nursing women from the 2nd to the 14th week of lactation.

Energy and Nutrients	Intake (24hR-MSM)	Inadequacy (%) [3]	Nutrient Intake Adequacy	
			Reference Value (Daily Intake) [4]	Method [Ref.]
Energy (kcal)	1577 ± 164 [1]	53	—	BMI distribution [14]
Carbohydrate (en%)	51.2 ± 7.7 [1]	21	160 g [5]	EAR cut-off [14]
Protein (en%)	17.7 ± 3.1 [1]	47	1.05 g/kg [5]	EAR cut-off [14]
Total lipids (en%)	31.1 ± 0.52 [1]	26	20 to 35	AMDR [14]
Vitamin A (µg RE)	824.7 ± 21.8 [1]	74	900	EAR cut-off [13]
β-Carotene (µg)	3249 (1408–6707) [2]	—	3000–6000 [12]	—
α-Carotene (µg)	1053 (56–3712) [2]	—	—	—
Lycopene (µg)	1854 (302–6472) [2]	—	≥5000 [12]	—
Lutein + zeaxanthin (µg)	2446 (872–4873) [2]	—	—	—
Vitamin E (mg)	4.4 ± 0.9 [1]	100	16	EAR cut-off [12]

[1] Data expressed as the mean ± standard deviation; [2] Median (minimum–maximum); [3] Estimated frequency of volunteers with inadequate intake; [4] based on estimated average requirement values whenever available, and are in the same units as the first column, except stated otherwise; in the case of carotenoids (β-carotene and lycopene), the reference value was taken as the daily intake level considered prudent for maintaining health; [5] EAR is used to assess adequacy of carbohydrate and protein intake, but usual intake expressed as en% are, respectively 45–65 and 5–10 [14]. AMDR: acceptable macronutrient distribution ranges; en%: macronutrient intake as a percentage of energy intake; 24hR-MSM: two 24-h recalls were used, and data were assessed by the Multiple Source Method; BMI: body mass index; EAR: Estimated Average Requirement.

3.3. Serum Vitamin A and E Were Not Associated with Their Changes in Breast Milk Concentration throughout Lactation

Serum vitamin A was adequate in all volunteers in early- and mid-lactation (Table 3). In contrast, 68 and 84% of the volunteers were at risk of developing vitamin E deficiency in early- and mid-lactation, respectively. Although serum retinol concentration did not vary throughout lactation, retinol in breast milk showed a significant 13% decrease from early- to mid-lactation (Table 3). In early lactation, retinol concentration in the milk of all volunteers was above the limit considered adequate [1,3] for the formation of infant liver reserves. After 10 weeks of lactation, however, milk retinol concentration became inadequate in 21% of the nursing women [1]. A similar trend was observed with breast milk contents of β-carotene, lutein+zeaxanthin and α-tocopherol, which significantly decreased from early- to mid-lactation, irrespective of the serum concentrations (Table 3).

Table 3. Concentrations of fat-soluble vitamins (µmol/L) in the serum and breast milk of adult Brazilian nursing women (n = 19), in early- (2nd to 4th week) and mid- (12th to 14th week) lactation, and respective cut-off values for inadequacy [1].

Vitamin	Serum				Breast Milk			
	Lactation Period			Undernutrition Cut-Off Value	Lactation Period			Inadequacy Cut-Off Value
	2nd–4th Week	12th–14th Week	p		2nd–4th Week	12th–14th Week	p	
Retinol	1.50 ± 0.30	1.48 ± 0.31	0.913	0.7 [28]	2.3 ± 0.78	2.0 ± 0.72	0.013 *	1.05 [1,28]
β-Carotene	0.61 ± 0.07	0.59 ± 0.05	0.628	—	0.17 ± 0.02	0.14 ± 0.3	0.001 *	—
α-Carotene	0.43 ± 0.09	0.44 ± 0.08	0.720	—	0.04 ± 0.01	0.03 ± 0.01	0.304	—
Lycopene	0.20 ± 0.03	0.18 ± 0.04	0.530	—	0.04 ± 0.01	0.04 ± 0.00	1.000	—
Lutein + zeaxanthin	0.24 ± 0.04	0.23 ± 0.07	0.633	—	0.07 ± 0.02	0.06 ± 0.01	0.026 *	—
α-Tocopherol	11.1 ± 1.11	10.1 ± 1.23	0.089	11.6 [12]	1.29 ± 0.25	1.07 ± 0.13	0.001 *	7.4 [28]
γ-Tocopherol	0.76 ± 0.15	0.75 ± 0.13	0.896	—	0.38 ± 0.04	0.35 ± 0.04	0.436	—

[1] Data expressed as the mean ± standard deviation. * Significant differences between lactation periods, paired t-test.

3.4. Breast Milk Concentration of Fat-Soluble Vitamins from Early- to Mid-Lactation Is Associated with Dietary Vitamin A, Serum β-Carotene and Tocopherols

Multiple regression analysis was used to investigate the determinants of vitamin A and E and carotenoid concentration in breast milk (Table 4). Retinol concentration in breast milk was determined by the dietary intake of vitamin A, and this association was stable from early- to mid-lactation (Table 4; models 1 and 4). In contrast, the determinants of milk β-carotene changed from early- to mid-lactation (Table 4; models 2 and 5). Regarding vitamin E, serum concentration was the sole determinant of its contents in breast milk (Table 4; models 3 and 6).

Table 4. Multiple regression models of the concentrations of vitamin A and E (μmol/L) in the milk of Brazilian nursing women ($n = 19$), to assess its associations with the intake [1] and serum contents of these vitamins.

Model N	Dependent Variable Breast Milk [2]	Independent Variables Serum [A] and Diet [B]	Coefficients Value	p	Weight in the Model [3]	Adj. R^2	Error (%) [4]	p [5]
			Early-Lactation: 2nd–4th Week					
1	Retinol	Vitamin A [B]	5.33×10^{-3}	<0.0001	100%	94.9	9%	<0.0001
2	β-Carotene	β-Carotene [A]	2.25×10^{-1}	0.0045	63%	87.7	14%	<0.0001
		β-Carotene [B]	2.52×10^{-5}	0.0430	37%			
3	α-Tocopherol	α-Tocopherol [A]	1.24	<0.0001	100%	96.2	18%	<0.0001
			Mid-Lactation: 12th–14th Week					
4	Retinol	Vitamin A [B]	4.79×10^{-3}	<0.0001	100%	94.7	14%	<0.0001
5	β-Carotene	β-Carotene [A]	2.34×10^{-1}	<0.0001	100%	78.4	17%	<0.0001
6	α-Tocopherol	α-Tocopherol [A]	1.03	<0.0001	100%	97.9	12%	<0.0001

[1] Data from the average of two 24-h recalls corrected by the Multiple Source Method (Vitamin A, μg RE/day; β-Carotene, μg/day); [2] Concentrations of fat-soluble vitamins in milk; [3] Weight in the model = independent variable coefficient × variable average content in samples; [4] Relative error of estimate = (estimated absolute error × 100%)/average value of the dependent variable; [5] Model significance. Superscript letters indicate whether the vitamins included as independent variables in the model were serum concentration ([A]) or dietary intake ([B]). Variables included in the starting analysis matrices before running the backward stepwise regression analysis (p-to-remove ≥ 0.05): model 1, vitamin A intake (RE), serum retinol and fat intake; model 2, β-carotene intake and serum β-carotene; model 3, vitamin E intake and serum α-tocopherol; model 4, vitamin A intake (RE), serum retinol and fat intake; model 5, β-carotene intake and serum β-carotene; model 6, vitamin E intake and serum α-tocopherol. Adj. R^2: adjusted coefficient of determination for the fitted model.

3.5. Vitamin A and E and Carotenoids Distributed Differently in Lipoprotein Fractions during Lactation

There were no significant changes from early- to mid-lactation in the relative distribution of vitamin A and E and carotenoids in serum lipoprotein fractions (Figure 1). Retinol and γ-tocopherol were distributed equally between HDL+serum binding proteins and LDL+VLDL fractions, in both lactation periods. β-Carotene and α-carotene, lycopene and α-tocopherol were enriched in the LDL+VLDL fraction. In contrast, approximately 70% of lutein+zeaxanthin was concentrated in the HDL+binding proteins fraction, in both lactation periods.

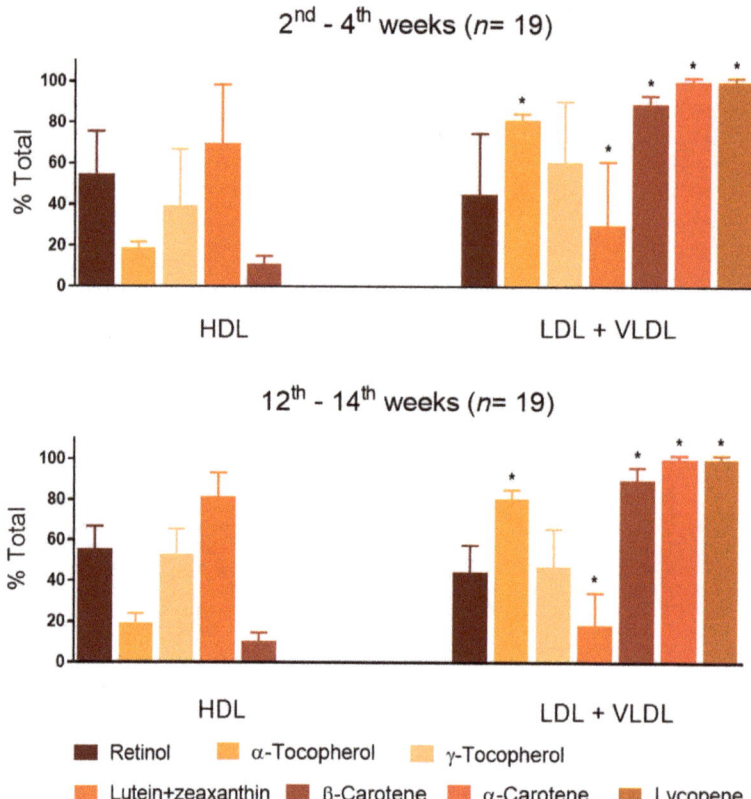

Figure 1. Distribution of fat-soluble vitamins in blood serum lipoprotein fractions in early- and mid-lactation. * Significantly different from the contents (%) in the HDL fraction (paired t-test; $p < 0.05$). Error bars represent standard deviations.

4. Discussion

Adequate breast milk contents of vitamins and minerals are crucial for infant development. In the present work, the maternal nutritional status of vitamin A and E was evaluated, and the factors associated with their content in breast milk in adult Brazilian nursing mothers. The most striking result was the fact that despite constant serum levels, the content of vitamin A and E in breast milk decreased throughout lactation, which we hypothesize as a consequence of the sustained inadequate dietary intake from early- to mid-lactation. This sub-adequate intake was insufficient to affect biochemical markers of vitamin status in serum, but negatively affected breast milk, being potentially detrimental to the lactating infant. Therefore, vitamin A and E in breast milk seem to be more sensitive markers of maternal nutrition status than their respective blood concentrations. Since the distribution of vitamin A and E in fasting serum lipoproteins did not change from early- to mid-lactation, this factor is unlikely to be related to the decrease in the transfer of fat-soluble vitamins into milk during lactation.

4.1. Dietary Intake, Nutritional Status and Milk Transfer of Vitamin A and E, and Carotenoids

Vitamin A intake was sub-adequate in 74% of the subjects [13] and was insufficient to maintain adequate breast milk levels throughout lactation. Maternal vitamin A requirement increases during lactation to fulfill the increased demand imposed by the mammary gland, and vitamin A in breast milk is essential to build up the infant's liver stores, contributing to the prevention of vitamin A deficiency when lactation ends [4,29]. A high intake of β-carotene could compensate the low intake of retinol, but the former nutrient was in the range considered prudent (3000–6000 µg/day [12]) in only 53% of the volunteers, thus it is likely that the breastfed infants were at risk of vitamin A deficiency [4,30]. Despite that, all subjects showed serum retinol levels above the cutoff value for inadequacy risk (>0.7 µmol/L [1]) throughout lactation [3].

Retinol in breast milk is also seen as a reliable marker of vitamin A nutritional status on a population basis. According to the World Health Organization, when retinol concentration in milk is below 1.1 µmol/L, the infant's body stores may be below the estimated critical amounts to supply for the increased requirements in the second half of childhood [1]. In all subjects in the present study, the breast milk concentration of retinol in early lactation was above this cutoff value. However, in mid-lactation, 21% of the nursing women were below the adequacy cutoff. Possibly, these results indicate that these nursing women presented subclinical vitamin A deficiency, which worsened throughout lactation, and that their lactating newborns were at risk of vitamin A deficiency to build up liver stores [4,29]. However, concentrations of fat-soluble vitamins are usually higher at the beginning of lactation, suffering a progressive fall as lactation progresses [3], but levels should be above the safe cut-off value throughout lactation.

The retinol concentration in breast milk in the two lactation periods evaluated was near the average values previously reported for Brazilian nursing women [5,21,30] and for Korean nursing women (2.0 to 3.0 µmol/L) [31]. As these previous works did not assess dietary intake, further discussion regarding the factors involved in the regulation of the concentration of vitamins in breast milk was not possible. Retinol content in breast milk is associated with maternal dietary fat intake, stressing the importance of dietary lipids to retinol transport and absorption [31]. It is suggested that during lactation, a large proportion of retinol from the maternal diet is transported directly into the mammary gland via blood lipoproteins, bypassing the liver [3]. Breast milk vitamin A might be more strongly affected by dietary fat intake than by serum retinol concentration, especially in groups with a low dietary intake of vitamin A and inadequate nutritional status of this vitamin [31]. Consistently, in the present study, fat intake was correlated with breast milk retinol, both in early ($r = 0.47$; $p = 0.042$) and in mid-lactation ($r = 0.46$; $p = 0.046$) and vitamin A intake was a determinant of breast milk retinol concentration in both lactation periods investigated (Table 4).

The intake of vitamin E was also sub-adequate in all volunteers [12]. Vitamin E intake might have been underestimated in the present study, because its principal dietary source was vegetable oil, and it is inherently difficult to assess cooking oil intake [32]. In Brazil, soybean oil is by far the most consumed oil and the main source of vitamin E in the subjects' diets. In contrast, it should also be considered that the possibility that vitamin E EAR (16 mg/day [12]) is overestimated, which could contribute to the high frequency of inadequacy in the present study [32]. In this sense, the recommended daily intake of vitamin E for Brazilian nursing women [32] is nearly half that recommended in North America by the Institute of Medicine [4,12]. The vitamin E status biomarker (serum α-tocopherol/cholesterol ratio; µmol/mmol) indicates that the prevalence of vitamin E deficiency (<2.25 µmol/mmol) [2,4] increased from 37 to 63% of the volunteers from early- to mid-lactation.

β-Carotene was the major carotenoid present in the serum and breast milk throughout lactation, consistent with maternal usual diets, representing over 50% of total carotenoids in milk [21,33]. Serum and breast milk β-carotene were strongly associated, as expected [34,35] and, in early-lactation, breast milk β-carotene was also associated with maternal intake, highlighting the importance of both dietary intake and blood concentration to β-carotene in breast milk [34,35]. It is noteworthy that the concentration of β-carotene and lutein+zeaxanthin in breast milk was, respectively, approximately 8- and 6-fold higher than observed in a previous study with Brazilian nursing women [21]. Compared to the values of a recent study that addressed the effect of prematurity on the content of carotenoids in the colostrum, the results of the present study were sensibly lower, particularly for lutein+zeaxanthin and lycopene concentrations, which were approximately 10-fold lower [36]. These differences might be related to the stage of lactation and also to potential differences in the habitual diets of the volunteers of both studies. The concentration of α-tocopherol in breast milk was determined by its concentration in the serum. It is important to highlight that the intake of α-tocopherol was entered as an independent variable in the multiple regression analysis matrix. However, the model did not adjust, and significance was reached only after dietary data were excluded from the analysis matrix.

4.2. Vitamin A and E, and Carotenoids Distribution among Serum Lipoprotein Fractions Was Stable from Early- to Mid-Lactation

Highly nonpolar carotenes partition more easily into lipoprotein particles that are richer in neutral lipids (VLDL+LDL), and the more polar xanthophylls partition into HDL particles [37]. Indeed, our results showed that approximately 68% of serum lutein+zeaxanthin was in the HDL fraction in both early- and mid-lactation and serum carotenes were predominant in the serum fraction of LDL+VLDL. An important quality control of the protocol used for the differential precipitation of serum lipoproteins was the analysis of HDL-c in whole serum and in the HDL fraction, after precipitation of LDL+VLDL. There was no significant difference between HDL-c concentration in whole serum and in this lipoprotein fraction, indicating that there was no co-precipitation of HDL with LDL+VLDL. Retinol was evenly distributed between HDL and non-HDL serum fractions and was unchanged from early- to mid-lactation, consistently with previous reports [3]. However, the protocol used herein to precipitate LDL+VLDL does not allow the discrimination between free retinol, transported bound to retinol binding protein (RBP), and the HDL fraction. It is worth investigating in the future whether RBP-mediated retinol transport varies during lactation. γ-Tocopherol was evenly distributed in the serum fractions of HDL and LDL+VLDL. In contrast, α-tocopherol was more concentrated in the LDL+VLDL fraction, possibly because vitamin E is transferred to HDL, and then promptly transfers into all other circulating lipoproteins [5], which possibly favors its uptake by the mammary glands and transfer into breast milk.

In summary, the distribution of fat-soluble vitamins among plasma lipoprotein fractions, HDL and LDL+VLDL, in lactating women was similar to that of non-pregnant non-lactating women [37], despite the high adipose tissue turnover that is common to exclusive lactation. In addition, this distribution pattern was stable from early- to mid-lactation in the present study. From data in Figure 1, we estimated the sample size required to assess potential differences between early- and mid-lactation in the distribution of vitamin A and E, and carotenoids among HDL and LDL+VLDL fractions (Supplementary Table S1). Accordingly, it may be speculated that for the carotenes (α- and β-carotene and lycopene) and for retinol, it is unlikely that variations in their distribution between the lipoprotein fractions would be reasonably detectable—or that if detectable, they would have a biological role. Therefore, in the case of retinol and β-carotene, it is very unlikely that the decrease in the concentrations in breast milk, until mid-lactation, is related to their distribution between HDL and LDL+VLDL serum fractions. However, changes in the distribution of the xanthophylls (Lut + Zea) and tocopherols (α and γ) in serum fractions might be detected between early- and mid-lactation with sample sizes of approximately 100 and 500 depending on the nutrient (Supplementary Table S1) and, in the case of α-tocopherol and the xanthophylls, this metabolic distribution among serum lipoproteins might have a role in the decreasing transfer to breast milk from early- to mid-lactation, if confirmed in future studies. The major limitation of this work is the limited number of subjects, that limited reaching a significance level in some analyses and the capacity to extrapolate the results found to larger population groups. However, our results indicate new potentially interesting points to be investigated in the future, which might improve nutrition guidance for lactating women, especially if at high risk of sub-adequate intake of vitamin E.

5. Conclusions

Despite constant serum levels, the content of vitamin A and E in breast milk decreased throughout lactation, imposing risks of deficiency to the breastfed infants. Subclinical vitamin A deficiency in the lactating mothers was undetectable by serum retinol concentration in early- or mid-lactation. On the other hand, dietary assessment by the MSM was fairly sensitive to detect vitamin A inadequacy and might be used in the future to identify population groups at risk and guide maternal dietary advice on food choices or eventually supplementation in order to prevent sub-adequate nutritional status in breastfed newborns. A similar trend was observed for vitamin E, because breast the milk concentration of α-tocopherol seemed largely more sensitive than its serum concentration to the sub-adequate intake of the vitamin. Future prospective clinical trials providing nutritional advice to lactating mothers with sub-adequate vitamin A and E intake and the assessment of breast milk levels and infant nutritional status of these vitamins would contribute to widen the impact of this research.

Supplementary Materials: The following are available online at http://www.mdpi.com/2072-6643/11/9/2025/s1, Figure S1: Representative chromatograms by HPLC-UV-Vis of the analysis of: (A) retinol (325 nm) and tocopherols (292 nm) standards; (B) carotenoid standards (450 nm). Refer to materials and methods for detailed methods description. Table S1: Calculated sample size necessary to disclose differences between early- and mid-lactating women in the distribution of vitamin A and E, and carotenoids in HDL and LDL+VLDL serum fractions.

Author Contributions: Conceptualization, A.G.T. and F.K.; methodology, A.G.T., F.K. and M.R.M.; formal analysis, A.G.T., J.C.N. and M.R.M.; investigation, M.R.M.; resources, A.G.T. and F.K.; data curation, A.G.T. and M.R.M.; writing—original draft preparation, M.R.M.; writing—review and editing, A.G.T. and T.E.-B.; visualization, A.G.T., F.K., J.C.N., M.R.M. and T.E.-B.; supervision, A.G.T., F.K. and J.C.N.; project administration, A.G.T.; funding acquisition, A.G.T and F.K.

Funding: This research was funded by Fundação Carlos Chagas Filho de Amparo à Pesquisa do Estado do Rio de Janeiro (FAPERJ; E-26/203.197/2015), Conselho Nacional de Desenvolvimento Científico e Tecnológico (CNPq; grant 432484/2016-7) and Coordenação de Aperfeiçoamento de Pessoal de Nível Superior (CAPES; Finance code 001), Brazil. M.R.M. was a recipient of a CNPq scholarship (grant GM). A.G.T. is a recipient of a CNPq fellowship (grant 312341/2018-0). The APC was covered partially by FAPERJ and partially by personal resources from the authors.

Acknowledgments: The authors acknowledge the contribution of the breastfeeding mothers that participated in this study, without any direct compensation. The assistance of the clinical dietitians Danúbia Incuto Silva and Vivianne Magalhães Gomes during sample and data collection is greatly appreciated. The assistance of conticom.com.br designing the graphical abstract is greatly acknowledged.

Conflicts of Interest: The authors declare no conflict of interest. The funders, public funding agencies from Brazil, had no role in the design of the study; in the collection, analyses, or interpretation of data; in the writing of the manuscript, or in the decision to publish the results.

References

1. World Health Organization (WHO). *Global Prevalence of Vitamin A Deficiency in Populations at Risk 1995–2005*; WHO: Geneva, Switzerland, 2009; pp. 9–18.
2. Ribeiro, K.D.S.; Lima, M.S.R.; Medeiros, J.F.P.; Rebouças, A.S.; Dantas, R.C.S.; Bezerra, D.S.; Osório, M.M.; Dimenstein, R. Association between maternal vitamin E status and alpha-tocopherol levels in the newborn and colostrum. *Matern. Child Nutr.* **2016**, *12*, 801–807. [CrossRef] [PubMed]
3. De Vries, J.Y.; Pundir, S.; Mckenzie, E.; Keijer, J.; Kussmann, M. Maternal circulating vitamin status and colostrum vitamin composition in healthy lactating women—A systematic approach. *Nutrients* **2018**, *28*, 687. [CrossRef] [PubMed]
4. Da Silva, A.G.C.L.; Rebouças, A.S.; Mendonça, B.M.A.; Silva, D.C.N.; Dimenstein, R.; Ribeiro, K.D.S. Relationship between the dietary intake, serum, and breast milk concentrations of vitamin A and vitamin E in a cohort of women over the course of lactation. *Matern. Child Nutr.* **2019**, *22*, e12772. [CrossRef] [PubMed]
5. Grilo, E.C.; Medeiros, W.F.; Silva, A.G.A.; Gurgel, C.S.S.; Ramalho, H.M.M.; Dimenstein, R. Maternal supplementation with a megadose of vitamin A reduces colostrum level of α-tocopherol: A randomized controlled trial. *J. Hum. Nutr. Diet.* **2016**, *29*, 652–661. [CrossRef] [PubMed]
6. World Health Organization (WHO). *Obesity: Preventing and Managing the Global Epidemic*; WHO Technical Report Series 894; WHO: Geneva, Switzerland, 2000; pp. 1–253.
7. Harttig, U.; Haubrock, J.; Knüppel, S.; Boeing, H. The MSM program: Web-based statistics package for estimating usual dietary intake using the Multiple Source Method. *Eur. J. Clin. Nutr.* **2011**, *65*, 87–91. [CrossRef] [PubMed]
8. Sartorelli, D.S.; Barbieri, P.; Perdona, G.C. Fried food intake estimated by the multiple source method is associated with gestational weight gain. *Nutr. Res.* **2014**, *34*, 667–673. [CrossRef] [PubMed]

9. U.S. Department of Agriculture, Agricultural Research Service. USDA National Nutrient Database for Standard Reference, Release 25. Available online: https://www.ars.usda.gov/northeast-area/beltsville-md-bhnrc/beltsville-human-nutrition-research-center/nutrient-data-laboratory/docs/sr25-home-page/ (accessed on 20 July 2016).
10. University of Campinas (UNICAMP)—Core Centre for Research in Food (NEPA). *Brazilian Food Composition Table, Taco*, 4th ed.; NEPA/UNICAMP: Campinas, Brazil, 2011; pp. 1–161.
11. Institute of Medicine. *Dietary Reference Intakes Research Synthesis: Workshop Summary*; National Academy Press: Washington, DC, USA, 2007; pp. 127–150.
12. Institute of Medicine. *Dietary Reference Intakes for Vitamin C, Vitamin E, Selenium and Carotenoids*; National Academy Press: Washington, DC, USA, 2000; pp. 186–283.
13. Institute of Medicine. *Dietary Reference Intakes for Vitamin A, Vitamin K, Arsenic, Boron, Chromium, Copper, Iodine, Iron, Manganese, Molybdenum, Nickel, Silicon, Vanadium, and Zinc*; National Academy Press: Washington, DC, USA, 2001; pp. 82–161.
14. Institute of Medicine. *Dietary Reference Intakes—Energy, Carbohydrate, Fiber, Fat, Fatty Acids, Cholesterol, Protein, and Amino Acids*; The National Academy Press: Washington, DC, USA, 2002; pp. 936–967.
15. Lopes-Virella, M.F.; Stone, P.; Ellis, S.; Colwell, J.A. Cholesterol determination in high-density lipoproteins separated by three different methods. *Clin. Chem.* **1977**, *23*, 882–884. [PubMed]
16. Lucas, A.; Gibbs, J.A.; Lyster, R.L.; Baum, J.D. Creamatocrit: Simple clinical technique for estimating fat concentration and energy value of human milk. *Br. Med. J.* **1978**, *1*, 1018–1020. [CrossRef]
17. Perry, N.A.; Doan, F.J. A picric acid method for the simultaneous determination of lactose and sucrose in dairy products. *J. Dairy Sci.* **1950**, *33*, 176–185. [CrossRef]
18. Costa, T.H.M.; Dorea, J.G. Concentration of fat, protein, lactose and energy in milk of mothers using hormonal contraceptives. *Ann. Trop. Pediatr.* **1992**, *12*, 203–209. [CrossRef]
19. Lowry, O.H.; Rosebrough, N.; Farr, A.L.; Randall, R.J. Protein measurement with the Folin phenol reagent. *J. Biol. Chem.* **1951**, *193*, 265–275. [PubMed]
20. Trugo, N.M.F.; Donangelo, C.M.; Koury, J.C.; Silva, M.I.; Freitas, L.A. Concentration and distribution pattern of selected micronutrients in preterm and term milk from urban Brazilian mothers during early lactation. *Eur. J. Clin. Nutr.* **1988**, *42*, 497–507. [PubMed]
21. Meneses, F.; Trugo, N.M.F. Retinol, β-carotene, and lutein+zeaxanthin in the milk of Brazilian nursing women: Associations with plasma concentrations and influences of maternal characteristics. *Nutr. Res.* **2005**, *25*, 443–451. [CrossRef]
22. Hass, D.; Keller, H.E.; Oberlin, B.; Bonfanti, R.; Schüep, W. Simultaneous determination of retinol, tocopherols, carotenes and lycopene in plasma by means of High-Performance Liquid-Chromatography on reversed phase. *Int. J. Vitam. Nutr. Res.* **1991**, *61*, 232–238.
23. Liu, Y.; Xu, M.J.; Canfield, L.M. Enzymatic hydrolysis, extraction, and quantitation of retinol and major carotenoids in mature human milk. *J. Nutr. Biochem.* **1998**, *9*, 178–183. [CrossRef]
24. Silva, L.O.; Castelo-Branco, V.N.; Carvalho, A.G.A.; Monteiro, M.C.; Perrone, D.; Torres, A.G. Ethanol extraction renders a phenolic compounds-enriched and highly stable jussara fruit (*Euterpe edulis* M.) oil. *Eur. J. Lipid Sci. Technol.* **2017**, *119*, e201700200. [CrossRef]
25. Rodriguez-Amaya, D.B.; Kimura, M. *Harvest Plus Handbook for Carotenoid Analysis*; International Food Policy Research Institute (IFPRI): Washington, DC, USA, 2004; pp. 1–57.
26. World Health Organization (WHO). *Prevention and Control of Iron Deficiency Anaemia in Women and Children*; United Nations Children's Fund (UNICEF): Geneva, Switzerland, 1999; pp. 17–30.
27. Expert Panel on Detection, Evaluation and Treatment of High Blood Cholesterol in Adults. Executive summary of the third report of the National Cholesterol Education Program (NCEP) expert panel on detection, evaluation, and treatment of high blood cholesterol in adults (Adult treatment panel III). *JAMA* **2001**, *285*, 2486–2497. [CrossRef] [PubMed]
28. Joint FAO/WHO Expert Consultation on Human Vitamin and Mineral Requirements (FAO/WHO). *Vitamin and Mineral Requirements in Human Nutrition*, 2nd ed.; WHO: Geneva, Switzerland, 2004; pp. 1–362.
29. Deminice, T.M.M.; Ferraz, I.S.; Monteiro, J.P.; Jordão, A.A.; Ambrósio, L.M.C.S.; Almeida, C.A.N. Vitamin A intake of Brazilian mothers and retinol concentrations in maternal blood, human milk, and the umbilical cord. *Nutr. Res. Pract.* **2018**, *46*, 1555–1569. [CrossRef]

30. Lira, L.Q.; De Souza, A.F.; Amâncio, A.M.; Bezerra, C.G.; Pimentel, J.B.; Moia, M.N.; Dimenstein, R. Retinol and betacarotene status in mother-infant dyads and associations between them. *Ann. Nutr. Metab.* **2018**, *72*, 50–56. [CrossRef]
31. Kim, H.; Jung, B.; Lee, B.; Kim, Y.; Jung, J.A.; Chang, N. Retinol, α-tocopherol, and selected minerals in breast milk of lactating women with full-term infants in South Korea. *Nutr. Res. Pract.* **2017**, *11*, 64–69. [CrossRef]
32. Ministry of Health, Brazil; National Sanitary Surveillance Agency (ANVISA). *Resolution RDC N° 269, Technical Regulation on Recommended Daily Intake (IDR) of Protein, Vitamins and Minerals*; ANVISA: Brasilia, DF, Brazil, 2005; pp. 1–6.
33. Garretto, D.; Kim, Y.-K.; Quadro, L.; Rhodas, R.R.; Pimentel, V.; Crnosija, N.A.; Nie, L.; Bernstein, P.; Tropper, P.; Neal-Perry, G.S. Vitamin A and β-carotene in pregnant and breastfeeding post-bariatric women in an urban population. *J. Perinat. Med.* **2019**, *47*, 183–189. [CrossRef] [PubMed]
34. Xu, X.; Zhao, X.; Berde, Y.; Low, Y.; Kuchan, M. Milk and plasma lutein and zeaxanthin concentrations in Chinese breast-feeding mother-infant dyads with healthy maternal fruit and vegetable intake. *J. Am. Coll. Nutr.* **2019**, *38*, 179–184. [CrossRef] [PubMed]
35. Zielinska, M.; Hamulka, J.; Wesolowska, A. Carotenoid content in breastmilk in the 3rd and 6th month of lactation and its associations with maternal dietary intake and anthropometric characteristics. *Nutrients* **2019**, *11*, 193. [CrossRef] [PubMed]
36. Xavier, A.A.O.; Diaz-Salido, E.; Arenilla-Velez, I.; Aguayo-Maldonado, J.; Garrido-Fernandez, J.; Fontecha, J.; Sanchez-Garcia, A.; Perez-Galvez, A. Carotenoid content in human colostrum is associated to preterm/full-term birth condition. *Nutrients* **2018**, *10*, 1654. [CrossRef] [PubMed]
37. Schweigert, F.J.; Bathe, K.; Chen, F.; Büscher, U.; Dudenhausen, J.W. Effect of the stage of lactation in humans on carotenoid levels in milk, blood plasma and plasma lipoprotein fractions. *Eur. J. Nutr.* **2004**, *43*, 39–44. [CrossRef] [PubMed]

© 2019 by the authors. Licensee MDPI, Basel, Switzerland. This article is an open access article distributed under the terms and conditions of the Creative Commons Attribution (CC BY) license (http://creativecommons.org/licenses/by/4.0/).

Article

Lipid Profile, Lipase Bioactivity, and Lipophilic Antioxidant Content in High Pressure Processed Donor Human Milk

Aleksandra Wesolowska [1,*], Joanna Brys [2,*], Olga Barbarska [1,3], Kamila Strom [3], Jolanta Szymanska-Majchrzak [4], Katarzyna Karzel [5], Emilia Pawlikowska [6], Monika A. Zielinska [7], Jadwiga Hamulka [7] and Gabriela Oledzka [3]

1. Laboratory of Human Milk and Lactation Research at Regional Human Milk Bank in Holy Family Hospital, Department of Neonatology, Faculty of Health Sciences, Medical University of Warsaw, Zwirki i Wigury Str. 63A, 02-091 Warsaw, Poland
2. Department of Chemistry, Faculty of Food Sciences, Warsaw University of Life Sciences—SGGW, Nowoursynowska St. 166, 02-787 Warsaw, Poland
3. Department of Medical Biology, Faculty of Health Sciences, Medical University of Warsaw, 14/16 Litewska St., 00-575 Warsaw, Poland
4. Department of Biochemistry, Second Faculty of Medicine, Medical University of Warsaw, Banacha 1b Str. 02-093 Warsaw, Poland
5. Faculty of Psychology, Warsaw University, Stawki 5/7, 00-183 Warsaw, Poland
6. High Pressure Physics, Polish Academy of Science, Sokolowska 29, 01-142 Warsaw, Poland
7. Department of Human Nutrition, Faculty of Human Nutrition and Consumer Sciences, Warsaw University of Life Sciences—SGGW, 159 Nowoursynowska St., 02-776 Warsaw, Poland
* Correspondence: aleksandra.wesolowska@wum.edu.pl (A.W.); joanna_brys@sggw.pl (J.B.); Tel.: +48-600-93-85-27(A.W.); +48-22-59-376-15 (J.B.)

Received: 15 July 2019; Accepted: 17 August 2019; Published: 21 August 2019

Abstract: Human milk fat plays an essential role as the source of energy and cell function regulator; therefore, the preservation of unique human milk donors' lipid composition is of fundamental importance. To compare the effects of high pressure processing (HPP) and holder pasteurization on lipidome, human milk was processed at 62.5 °C for 30 min and at five variants of HPP from 450 MPa to 600 MPa, respectively. Lipase activity was estimated with QuantiChrom™ assay. Fatty acid composition was determined with the gas chromatographic technique, and free fatty acids content by titration with 0.1 M KOH. The positional distribution of fatty acid in triacylglycerols was performed. The oxidative induction time was obtained from the pressure differential scanning calorimetry. Carotenoids in human milk were measured by liquid chromatography. Bile salt stimulated lipase was completely eliminated by holder pasteurization, decreased at 600 MPa, and remained intact at 200 + 400 MPa; 450 MPa. The fatty acid composition and structure of human milk fat triacylglycerols were unchanged. The lipids of human milk after holder pasteurization had the lowest content of free fatty acids and the shortest induction time compared with samples after HPP. HPP slightly changed the β-carotene and lycopene levels, whereas the lutein level was decreased by 40.0% up to 60.2%, compared with 15.8% after the holder pasteurization.

Keywords: donor human milk; high pressure processing; carotenoids; antioxidant capacity; lipids; bile salt stimulated lipase; preterm

1. Introduction

Lipids in human milk are not only the main source of energy, but also of bioactive components and regulatory factors, such as vitamins and polyunsaturated fatty acids. Lipids exhibit several functions in the range of biological effects connected with neurodevelopment, immunity, digestion, metabolism regulation, cell membranes communication, and signal transduction [1]. Carotenoids, which are lipid-soluble provitamins, contribute to the anti-oxidative properties of human milk, protecting preterm infants against toxic free radicals [2]. Human milk contains a high bile salt stimulated lipase (BSSL) concentration, which enables the easy digestion of mother's milk lipids even in the absence of the enzyme in premature newborns. Human milk lipids are valuable, nourishing food with high bioavailability for newborns [3]. Therefore, expressed human milk, including donor milk, should be handled with care to minimize the loss of the unique lipid composition and lipase activity [4]. Administration of donor milk within the hospital setting often requires several preparatory stages, such as pumping, freezing, thawing, and pasteurization by heat treatment (usually holder pasteurization, HoP). The potential losses in human milk lipid content could accumulate when milk passes through all these stages from donor to recipients [5]. Earlier research indicates that heat treatment does not decrease lipid content and does not alter the fat soluble vitamins A, D, and E [6]. However, it was recently shown, using mid-infrared spectroscopy, that lipids are one of the most affected components in pooled pasteurized milk compared with raw milk donated to a milk bank [7]. It was reported earlier that the total availability of lipids in donor milk is affected by the decrease of BSSL and high adhesion to containers surface [8]. In addition, human milk processing may decrease the antioxidant properties of milk, which could have nutritional and clinical implications owing to the particular susceptibility of polyunsaturated fats to peroxidation.

One of promising preservation methods that allows the unique properties of human milk to be maintained unchanged is high pressure processing (HPP). It is a non-thermal technology that is being increasingly applied in food industries worldwide. This method consists of applying hydrostatic high pressure in short-term treatment to inactivate pathogenic microorganisms and provide nutritionally intact and sensory high-quality products [9].

The aim of our study was to evaluate high pressure processing as a promising technique for the preservation the lipid profile, antioxidant properties, and lipase enzymes activity in donor milk.

2. Materials and Methods

2.1. Ethical Approval

The Bioethics Committee of Warsaw Medical University has accepted the information about conducting this non-interventional study without reservations (admission number AKBE/59/15).

2.2. Milk Sampling

Milk samples were obtained from donors between the second and sixth week of lactation from Regional Human Milk Bank in Warsaw at Holy Family Hospital. Women were given standard instructions about the best practices for expressing breast milk. Milk samples of approximately 50 mL were collected at home or in the hospital ward using an electric or manual pump, stored in the refrigerator at the temperature of 4 °C, and delivered to the human milk bank within 24 h, while maintaining refrigerated conditions.

To carry out the study, 6 pooled samples from 3–4 donors were used. Experiments were performed in duplicate and repeated three times with similar results in the following variants:

- raw milk—control
- temperature processed milk/holder pasteurization (HoP)—comparator
- high pressure processed (HPP)—experimental

The processing influence on lipase activity and the change in human milk lipids were evaluated according to these experimental designs.

Milk samples for all analyses, except for lipase activity, were centrifuged at 4400 rpm for 15 min at 4 °C (Centrifuge 5702R, Eppendorf, Darmstadt, Germany) he lipid monolayer was carefully separated, and both supernatants and lipid monolayer were frozen for later analysis. Lipase activity assay was performed on whole milk samples.

2.3. Human Milk Processing

2.3.1. High Pressure Processing (HPP)

High pressure processing was applied to human milk samples using U 4000/65 apparatus designed and produced by Unipress Equipment at the Institute of High Pressure Physics, Polish Academy of Sciences. The maximum pressure available in the apparatus was 600 MPa and the treatment chamber had a distilled water and polypropylene glycol mixture (1:1), used as pressure-transmitting fluid. The manufacturer designed the working temperature ranges of the apparatus between −10 °C and +80 °C. In our experiments, the test condition temperature was between 19 and 21 °C. Pressure was applied in the following variants: (1) 600 MPa, 10 min; (2) 100 MPa, 10 min, interval 10 min, 600 MPa, 10 min; (3) 200 MPa, 10 min, interval 10 min, 400 MPa, 10 min; (4) 200 MPa, 10 min, interval 10 min, 600 MPa, 10 min; (5) 450 MPa, 15 min. The pressure generation time was 15–25 s and the decompression time was 1–4 s.

2.3.2. Holder Pasteurization (HoP)

Human milk samples were pasteurized according to the routine procedure performed in Regional Human Milk Bank in Warsaw at Holy Family Hospital, using Sterifeed S90 ECO Pasteurise (Medicare Colgate Ltd., Devon, UK, at 62.5 °C for 30 min.

2.4. Lipase Activity

Lipase activity was determined using the lipase assay kit (QuantiChrom Lipase Assay Kit, Bioassay Systems, Hayward, CA, USA) according to the kit instructions. Whole fat human milk samples were brought to room temperature and diluted with distilled water (1:250, *v/v*). The optical density was measured by a Biotek multiple reader at room temperature.

2.5. Lipidome Profile

2.5.1. Fat Extraction

The Folch method [10] with further improvement by Boselli [11] was applied to extract fat from the studied samples.

2.5.2. The Oxidative Induction Time

Induction time (IT) was determined by the use of PDSC (pressure differential scanning calorimetry) to evaluate the samples' oxidative stability. A differential scanning calorimeter (DSC Q20 TA Instruments, New Castle, DE, USA) coupled with a high-pressure cell was used. Fat samples of 3–4 mg were weighed into an aluminium open pan and placed in the sample chamber under oxygen atmosphere with an initial pressure of 1400 kPa. The isothermal temperature for each sample was 120 °C. Obtained diagrams were analyzed using TA Universal Analysis 2000 software (TA Instruments, New Castle, DE, USA). For each sample, the output was automatically recalculated and presented as the amount of energy per one gram. The maximum oxidation induction time was determined based on the maximum rate of heat flow.

2.5.3. Acid Values and Free Fatty Acids Content

The acid value was determined by titration of fat samples with 0.1 M ethanolic potassium hydroxide solution. Free fatty acids (FFA) concentration was calculated based on acid values and the oleic acid molar mass value. Acid values were determined according to ISO method 660:2000.

2.5.4. Fatty Acid Composition

The determination of fatty acid composition was carried out by gas chromatographic (GC) analysis of fatty acid methyl esters. Methyl esters of fatty acids were prepared through transesterification with sodium methoxide according to ISO 5509:2001. A YL6100 GC chromatograph equipped with a flame ionization detector and BPX-70 capillary column of 0.25 mm i.d. x 60 m length and 0.25 μm film thickness was used. The oven temperature was programmed as follows: 60 °C for 5 min; then it was increased by 10 °C/min to 180 °C; from 180 °C to 230 °C, it was increased by 3 °C/min; and then it was kept at 230 °C for 15 min. The temperature of the injector was 225 °C, with a split ratio of 1:50 and the detector temperature of 250 °C. Nitrogen was used as the carrier gas. The results were expressed as relative percentages of each fatty acid, calculated by external normalization of the chromatographic peak area. Fatty acids were identified by comparing the relative retention times of fatty acid methyl ester (FAME) peaks with FAME chemical standard.

2.5.5. Positional Distribution of Fatty Acid in Tag

Positional distribution of fatty acid in sn-2 and sn-1,3 positions in triacylglycerols (TAG) was based on the pancreatic lipase ability to selectively hydrolyze ester bonds in sn-1,3 positions. Briefly, 20 mg of purified pancreatic lipase (porcine pancreatic lipase, crude type II), 1 mL of Tris buffer (pH 8.0), 0.25 mL of bile salts (0.05%), and 0.1 mL of calcium chloride (2.2%) were added to 50 mL centrifuge tubes and vortexed with 0.1 g of fat sample. The mixture was incubated at 40 °C in a water bath for 5 min, after which 1 mL of 6 mol/L HCl and 1 mL of diethyl ether were added, and then the mixture was centrifuged. Diethyl ether layer was collected in test tubes and evaporated under nitrogen gas to obtain 200 uL volume. A 200 uL aliquot was loaded onto a silica gel thin layer chromatography (TLC) plate with fluorescent indicator 254 nm and developed with hexane: diethyl ether acetic acid (50:50:1, v:v:v). 2-monoacylglycerol (2-MAG) band was visualised under UV light. 2-MAG band was scraped off into a screw capped test tube, extracted twice with 1 mL of diethyl ether, and centrifuged. The ether layer was collected and entirely evaporated under nitrogen, and then the sample was dissolved in n-hexane and methylated as described above.

2.6. Carotenoids Analysis

Breastmilk carotenoids' (β-carotene, lycopene, and lutein + zeaxanthin) concentration in milk samples was assessed using high-performance liquid chromatography system (HPLC). Milk samples for the analysis were prepared on the basis of the modified method published earlier [12,13]. Analysis of the studied carotenoids was carried out at a wavelength of 471 nm for lycopene, 450 nm for β-carotene, and 445 nm for lutein + zeaxanthin using the HPLC system (Japan: 2 LC-20AD pumps, CMB-20A controller system, SIL-20AC autosampler, UV/VIS SPD-20AV detector, CTD-20AC controller, Shimadzu, Kioto, Japan) using C18 Synergi Fusion-RP 80i columns (250 × 4.60 mm, Phenomenex, CA, USA). The concentration of individual carotenoids was compared to standard curves prepared with Sigma Aldrich standards (catalogue numbers: β-carotene C4582, lutein + zeaxanthin X6250, lycopene L9879). The concentration of the studied carotenoids was expressed in nmol/L. Lutein and zeaxanthin could not be completely resolved and they were summed; therefore, all references to milk lutein concentration refer to lutein + zeaxanthin.

2.7. Statistical Analysis

Relative standard deviation for the lipase, lipidome, and lipolytic antioxidant factors results was calculated, where appropriate, for all collected data, using Microsoft Excel 2012 Software (Microsoft, Redmond, Washington, DC, USA). One-way analysis of variance (ANOVA) was performed using the Statgraphics Plus, version 5.1 (Statgraphics Technologies, Inc., The Plains, VA, USA). Differences were considered to be significant at a p-value ≤ 0.05, according to Tukey's multiple range test.

3. Results

3.1. Lipase Activity

Because of the variability of lipase activity in human milk, the results from several experiments treating different milk samples were presented as a percent of activity in different samples with respect to raw milk considered 100%. As shown in Figure 1, the enzyme activity was nearly completely eliminated by HoP (2.1% residual activity). Higher lipase activity was detected in the HPP treated samples (16.5%—600 MPa, 11%—100 + 600 MPa, 13.6%—200 + 600 MPa). Almost entire lipase activity retention was observed in pressure 200 + 400 MPa, 10 min, interval 10 min; and 450 MPa, 15 min (82.2%—200 + 400 MPa, 87.3%—450 MPa).

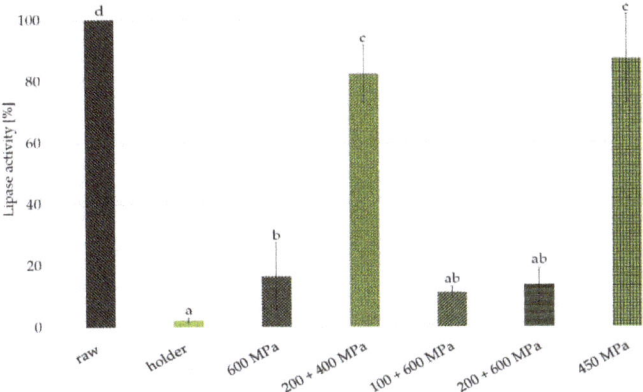

Figure 1. Lipase enzymatic activity in milk samples after holder pasteurization and variants of high pressure processing (HPP) compared with raw milk. Results are shown as a % value of raw milk with error bars representing SD. Different letters indicate that the samples are significantly different at p-value < 0.05.

3.2. FFA Content

Any type of donor milk treatment was associated with a reduction in the FFA content compared with raw milk taken as 100% (Figure 2).

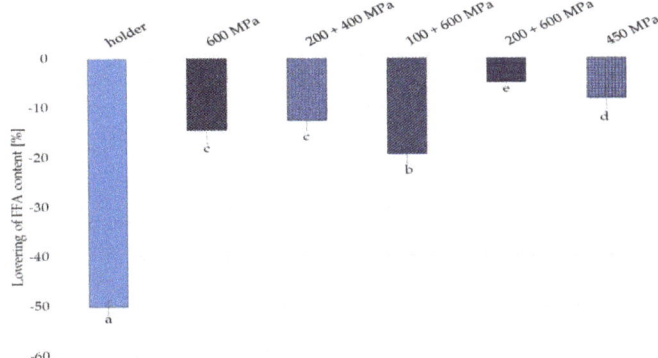

Figure 2. Decrease in free fatty acids (FFA) content compared with raw milk in fats from samples of milk after HPP and holder pasteurization. Results are shown as a % value of raw milk with error bars representing SD. Different letters indicate that the samples are significantly different at $p < 0.05$.

Holder pasteurization decreased FFA content by 50.2%. High pressure processing, depending on the variants, lowered FFA milk content by 19.5% for 100 + 600 MPa, 14.6% for 600 MPa, 12.8% for 200 + 400 MPa, 8.4% for 450 MPa, and 5.1% for 200 + 600 MPa (Figure 2).

3.3. Oxidative Stability

The PDSC measurements results of oxidative stability of sample expressed as the oxidation induction time (IT) are shown in Figure 3. Longer IT is connected with better stability. Significant differences in IT were observed especially between fat from raw and HoP milk—after heat treatment, IT was over 12 minutes shorter. The fat from samples after high pressure processing was also characterized by changed oxidative stability in comparison with the fat from raw human milk. The changes were slightly detectable in the case of 600 MPa and 200 + 600 MPa; more noticeable in 450 MPa, 15 min condition (3 minutes shorter); and about 7 min shorter induction time was observed in 200 + 400 MPa and 100 + 600 MPa (pressure 10 min, interval 10 min, pressure 10 min).

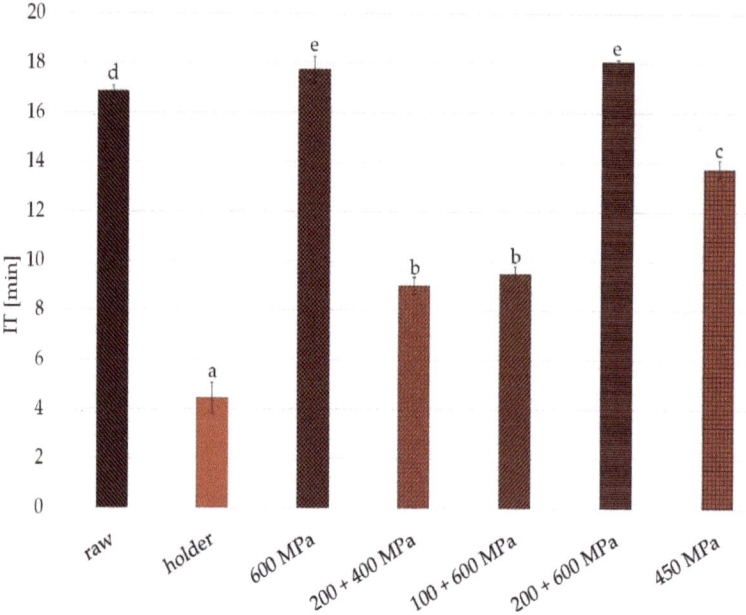

Figure 3. Oxidative stability of raw milk, in the samples after HPP and holder pasteurization. Results are shown as a % value of raw milk with error bars representing SD. Different letters indicate that the samples are significantly different at $p < 0.05$. IT, induction time.

3.4. Fatty Acid Composition

The fatty acids profile of fats from studied samples are presented in Figure 4 and Table 1. The fats from analyzed human milk contained from 10.7% to 12.2% of the polyunsaturated fatty acids (PUFA) and from 37.7% to 43.1% of the monounsaturated fatty acids (MUFA). Over 40% of all fatty acids in these fats were saturated fatty acids (SFA), of which the main representative was palmitic acid (from 23.8% to 25.1%). The main monounsaturated fatty acid present in analyzed samples of fats was oleic acid (from 34.0% to 39.4%). Polyunsaturated fatty acids found in studied samples were primarily linoleic acid (from 8.2% to 9.4%) and α-linolenic acid (1.3–1.4%). The major LCPUFA from n-6 family found in analyzed samples of fats was arachidonic acid (about 0.3–0.4%). Fats from studied human milk also contained omega-3 long chains polyunsaturated fatty acids like eicosapentaenoic acid (about 0.1%) and docosahexaenoic acid (about 0.3%). No statistically significant differences for samples after processing compared with raw milk fat were observed.

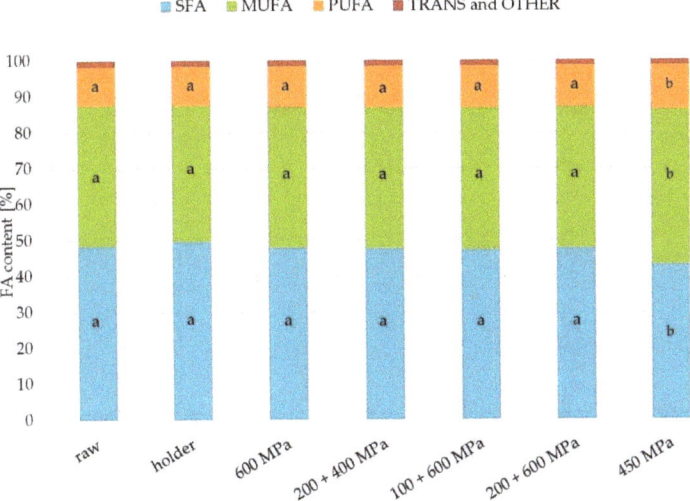

Figure 4. Content of fatty acid (SFA, saturated fatty acids; MUFA, monounsaturated fatty acids; PUFA, polyunsaturated fatty acids; TRANS, trans fatty acids) in the samples after HPP and holder pasteurization. The different lower case letters (for SFA, MUFA, and PUFA separately) indicate significantly different values ($p < 0.05$).

Table 1. Changes in fatty acid composition during high pressure processing (HPP) and holder pasteurization (mean ± SD).

Human Milk Fatty Acid Composition (%)	Raw Milk	Holder	600 MPa	200 + 400 MPa	100 + 600 MPa	200 + 600 MPa	450 MPa
Saturated fatty acids (SFA)	48.53 ± 0.94	49.77 ± 3.39	47.94 ± 0.83	47.70 ± 0.25	47.32 ± 0.08	47.72 ± 0.18	43.17 ± 0.62
C8:0	0.06 ± 0.04	0.11 ± 0.02	0.08 ± 0.04	0.04 ± 0.01	0.06 ± 0.01	0.05 ± 0.01	0.05 ± 0.00
C10:0	0.89 ± 0.09	1.07 ± 0.23	0.83 ± 0.10	0.77 ± 0.01	0.88 ± 0.02	0.85 ± 0.02	0.73 ± 0.09
C12:0	5.77 ± 0.14	6.58 ± 1.26	5.72 ± 0.36	5.28 ± 0.05	5.76 ± 0.11	5.63 ± 0.08	3.96 ± 0.51
C14:0	8.47 ± 0.40	8.89 ± 1.48	8.27 ± 0.35	8.08 ± 0.13	8.10 ± 0.07	8.15 ± 0.08	6.11 ± 0.39
C15:0	0.36 ± 0.02	0.38 ± 0.05	0.35 ± 0.02	0.35 ± 0.01	0.35 ± 0.01	0.35 ± 0.00	0.36 ± 0.01
C16:0	24.59 ± 0.49	25.13 ± 1.30	24.22 ± 0.08	24.45 ± 0.23	23.80 ± 0.01	24.14 ± 0.02	23.93 ± 0.16
C17:0	0.36 ± 0.03	0.33 ± 0.00	0.33 ± 0.01	0.34 ± 0.01	0.33 ± 0.01	0.34 ± 0.00	0.34 ± 0.01
C18:0	7.57 ± 0.03	6.94 ± 0.83	7.73 ± 0.05	7.95 ± 0.14	7.64 ± 0.09	7.81 ± 0.01	7.26 ± 0.20
C20:0	0.47 ± 0.03	0.36 ± 0.11	0.43 ± 0.01	0.46 ± 0.01	0.42 ± 0.01	0.43 ± 0.01	0.45 ± 0.01
Polyunsaturated fatty acids (PUFA)	39.07 ± 0.15	37.715 ± 2.34	39.04 ± 0.70	39.24 ± 0.42	39.46 ± 0.15	39.25 ± 0.08	43.13 ± 0.45
C20:1	0.52 ± 0.01	0.43 ± 0.13	0.52 ± 0.01	0.55 ± 0.04	0.50 ± 0.01	0.52 ± 0.01	0.72 ± 0.01
C14:1	0.25 ± 0.04	0.32 ± 0.04	0.29 ± 0.00	0.27 ± 0.01	0.28 ± 0.00	0.28 ± 0.00	0.23 ± 0.01
C15:1	0.10 ± 001	0.11 ± 0.02	0.09 ± 0.00	0.10 ± 0.01	0.09 ± 0.01	0.09 ± 0.00	0.08 ± 0.00
C16:1	2.63 ± 0.35	2.69 ± 0.25	2.44 ± 0.04	2.45 ± 0.11	2.46 ± 0.01	2.46 ± 0.01	2.49 ± 0.02
C17:1	0.20 ± 0.01	0.21 ± 0.03	0.21 ± 0.01	0.20 ± 0.01	0.23 ± 0.01	0.22 ± 0.01	0.20 ± 0.00
C18:1	35.39 ± 0.46	33.97 ± 2.50	35.49 ± 0.74	35.68 ± 0.25	35.91 ± 0.13	35.69 ± 0.09	39.41 ± 0.47
Polyunsaturated fatty acids (PUFA)	10.73 ± 1.00	11.04 ± 1.05	11.4 ± 0.12	11.53 ± 0.08	11.67 ± 0.07	11.57 ± 0.00	12.24 ± 0.30
C18:2 n-6	8.17 ± 1.15	8.75 ± 0.64	8.90 ± 0.06	8.99 ± 0.01	9.08 ± 0.04	9.05 ± 0.01	9.40 ± 0.18
C18:3 n-3	1.36 ± 0.12	1.28 ± 0.12	1.27 ± 0.01	1.30 ± 0.03	1.36 ± 0.05	1.30 ± 0.00	1.33 ± 0.06
C20:2 n-6	0.20 ± 0.01	0.18 ± 0.04	0.22 ± 0.01	0.22 ± 0.01	0.22 ± 0.01	0.22 ± 0.00	0.32 ± 0.00
C20:3 n-6	0.20 ± 0.01	0.21 ± 0.07	0.27 ± 0.03	0.25 ± 0.01	0.26 ± 0.00	0.25 ± 0.00	0.35 ± 0.01
C20:4 n-6	0.38 ± 0.01	0.34 ± 0.08	0.38 ± 0.01	0.39 ± 0.00	0.39 ± 0.00	0.38 ± 0.00	0.44 ± 0.01
C20:5 n-3	0.14 ± 0.01	0.07 ± 0.01	0.11 ± 0.04	0.11 ± 0.02	0.09 ± 0.01	0.10 ± 0.00	0.09 ± 0.03
C22:6 n-3	0.31 ± 0.01	0.22 ± 0.08	0.28 ± 001	0.27 ± 0.03	0.29 ± 0.01	0.28 ± 0.01	0.31 ± 0.00

3.5. Fatty Acid Distribution

The results regarding the percentage of fatty acids esterified at sn-2 position of TAG from fats are presented in Figure 5 and Table 2. Palmitic acid was located at sn-2 position in 64.7% TAG from raw milk fats, 66.5% fats from milk after HoP. High pressure processing causes a slight increase in the palmitic acid in sn-2 position of TAGs—to about 70% (73.6% in 600 MPa, 72.6% in 200 + 400 MPa, 70.4% 200 + 600 MPa, and 72.8% for 450 MPa). The difference was not statistically significant. Oleic acid at sn-2 was detected in 15% of TAG extracted from raw and holder pasteurized milk samples. After HPP, the percentage of oleic acid at sn-2 position of TAG diminished to 12% for 600 MPa, as well as 200 + 400 MPa and 100 + 600 MPa. In the case of 200 + 600 MPa 10 min, interval 10 min, 10 min, the percentage of the fatty acids esterified at sn-2 position of TAG was 14.1%, and almost the same for 450 MPa, 15 min. (13%).

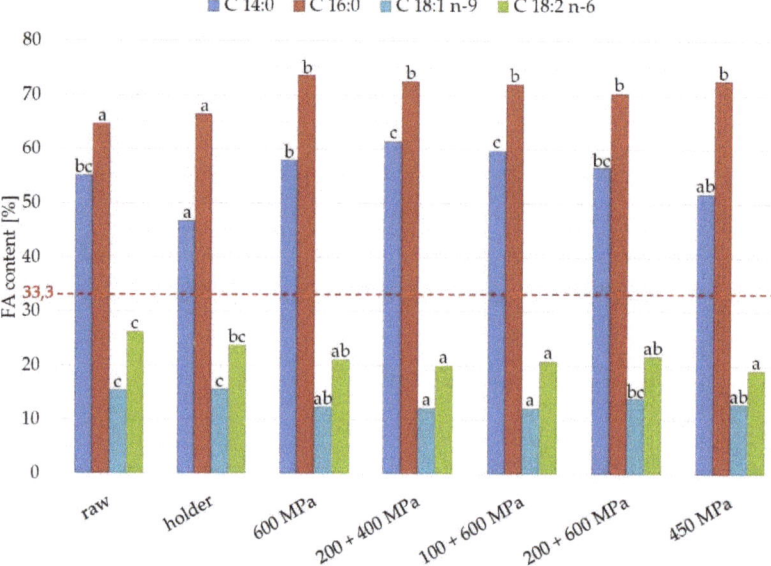

Figure 5. Fatty acid distribution in the sn-2 position of TAG from raw milk and in the samples after HPP and holder pasteurization. The different lower case letters for each fatty acid separately indicate significantly different values ($p < 0.05$). The line indicates the statistical (even) distribution of fatty acids between three TAG positions (33%).

Table 2. Changes in the fatty acid composition of sn-2 monoacylglycerols during HPP and holder pasteurization (mean ± SD; only selected fatty acids are presented).

Fatty Acid (%) in sn-2 Position of TAG	Raw Milk	Holder	600 MPa	200 + 400 MPa	100 + 600 MPa	200 + 600 MPa	450 MPa
C14:0 (myristic acid)	14.00 ± 0.71	12.50 ± 0.50	14.40 ± 0.47	14.90 ± 1.12	14.52 ± 0.62	13.90 ± 0.63	9.50 ± 0.54
C16:0 (palmitic acid)	47.70 ± 0.81	50.10 ± 1.34	53.50 ± 1.85	53.20 ± 0.59	51.50 ± 0.45	51.00 ± 1.22	52.20 ± 0.02
C18:1 (oleic acid)	16.40 ± 0.35	16.00 ± 0.18	13.40 ± 0.04	13.10 ± 1.07	13.20 ± 0.07	15.10 ± 1.65	15.40 ± 0.25
C18:2 (linoleic acid)	6.50 ± 0.28	6.20 ± 0.10	5.70 ± 0.11	5.40 ± 0.57	5.67 ± 0.28	5.90 ± 0.28	5.40 ± 0.28

3.6. Carotenoids

The studied carotenoids' concentration in raw milk was about 32.8 nmol/L for β-carotene, 85.9 nmol/L for lycopene, and 45.9 nmol/L for lutein + zeaxanthin. The content of carotenoids after processing compared with raw milk is presented in Figure 6.

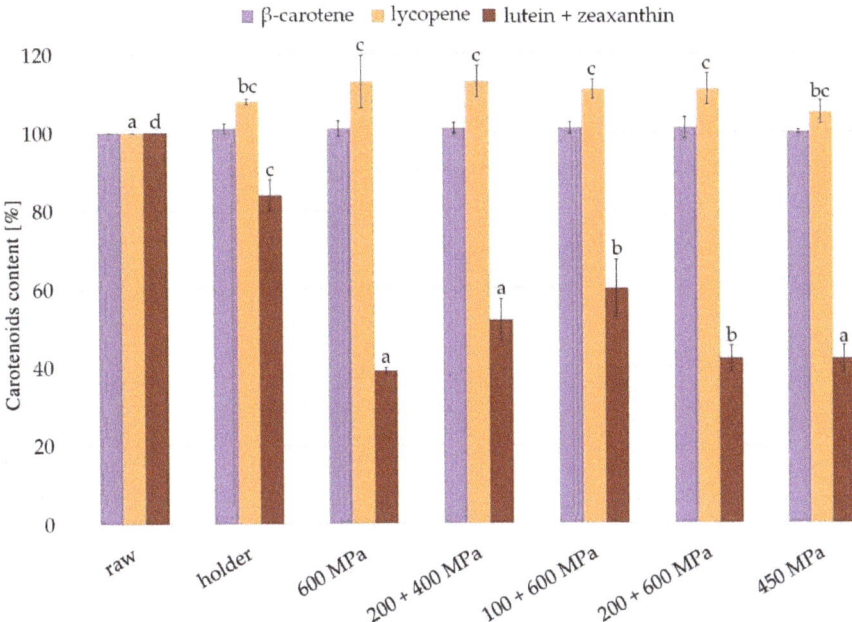

Figure 6. Content of carotenoids (β-carotene, lycopene, and lutein + zeaxanthin) in milk samples after HPP and holder pasteurization. Results are shown as a % value of raw milk with error bars representing SD. Different letters indicate that the samples are significantly different at $p < 0.05$.

β-carotene was intact after processing; therefore, the level was comparable in all milk samples. Statistically significant changes ($p \leq 0.05$) were observed in the case of lutein and zeaxanthin, as well as lycopene concentration in milk samples after processing. Pasteurization diminished the content of lutein + zeaxanthin compared with raw milk to about 84.1%, whereas HPP was so destructive as to leave 39.7% after 600 MPa, 52.9% after 200 + 400 MPa, 42.5% after 200 + 600 MPa, 60% after 100 + 600 MPa, and 42.4% after 450 MPa. In the case of lycopene, a slight content increase was observed after processing compared with raw milk—8.8% in case of pasteurization; 5.7% after 450 MPa; and 111.8% to 113.6% after other variants of high pressure processing, 200 + 400 MPa and 200 + 600 MPa, respectively.

4. Discussion

The main practical advantages of putting the baby to the breast for feeding are overall integrity of given nutrients, optimal temperature, microbiological safety, and intact bioactivity of mother's milk. Every other way of feeding involves some losses in human milk properties. Human milk is a complex body fluid with thousands of components in dynamic interplay [14–16]. For example, human milk fat breakdown by endogenous lipase is rather not noticeable when the baby is fed directly from the breast. In contrast, expressing human milk with a high lipase activity is associated with free fatty acids release from TAG, even if milk is frozen immediately. The increase of FFA levels causes pH changes that affect enzyme activity and other inherent human milk components [17–19]. In addition, unbound FFA may cause cytotoxicity in the intestinal lumen [20]. However, endogenous lipases are the prime effector of milk fats' digestion, facilitating human milk consumption in the early period of life. Therefore, it is crucial to retain enzymatic properties of human milk, especially when human milk is banked for preterms. As several previous studies proved, the activity of BSSL in donor human milk is mostly suppressed by holder pasteurization [21–23]. In contrast, high pressure processing seems to be a favorable technique for human milk preservation for this purpose [24]. Our current results concerning the residual enzymatic activity of lipase after HPP are consistent with the discovery of Demazou and co-workers, as well as the work of Pitino et al. [25,26]. All studies showed slightly diminished functional activity of this enzyme in milk after processing it in selected pressure conditions to about 80%, compared with 100% in raw milk [26]. In fact, the difference between detected enzyme activity in raw milk and milk treated in pressure of 450 MPa for 15 min and 200 + 400 MPa for 10 min with a 10 min interval was not statistically significant (Figure 1).

As human milk enzymes such as BSSL modulate the digestion to favor absorption of triglycerides hydrolysis and vitamins, human milk processing may negatively affect infants' ability to digest lipids because of the total inactivation of endogenous human milk lipases [27].

Human milk FFA content seems to be a good indicator of BSSL bioavailability in processed milk, because in the case of human milk, the FFA concentration should depend mostly on this enzyme's activity. However, available data concerning the effect of HoP on FFA concentration were inconsistent. Lepri et al. reported an 83% increase in FFA content after heat treatment [28], in contrast to several studies where no change was detected [29–32]. Only Williamson et al. report a 21% decrease in FFA concentration [33]. As shown in this study, all applied preservation techniques reduced the content of free fatty acids in fat extracted from human milk with a statistically significant difference (Figure 2). In the case of pasteurized milk, a reduction as large as 50% in the content of FFA compared with raw milk was noticed. FFA content decrease in fat from high pressure processed human milk was from 5% to 19% compared with fat from raw human milk, which is consistent with previous studies [30].

Analyzing the results, it can be observed that the differences in FFA fat content are especially significant in milk pasteurized by HoP, while after high pressure processing, the decrease in FFA content is minor. So when the enzyme was completely destroyed by heat treatment in the HoP samples, the FFA content was the lowest. This rule does not apply to high pressure samples. Among HPP variants, the largest reduction in FFA content was noticed in milk subjected to 100 + 600 MPa pressure, and lipase activity in this variant is significantly diminished. In contrast to this finding, at 200 + 600 MPa, the FFA content is almost intact while enzyme activity is severely lowered (Figure 1, Figure 2). In fact, our results of FFA and lipase activity in human milk after processing are ambiguous. The reason for this discrepancy could be a very low level of FFA in untreated human milk taken as 100%—as a result, any changes given as a percentage lead to large proportional differences [34].

Nevertheless, based on our research, we concluded that the best option seems to be 450 MPa for 15 min as the most favorable conditions for preservation of FFA concentration and lipase activity.

Oxidative stability is one of the most important factors determining the quality of foods, especially those containing high levels of unsaturated fats, including human milk. Within the fatty acid family, polyunsaturated fatty acids are highly labile molecules susceptible to oxidation, giving rise to free radicals, hydroperoxides and polymers, which might lead to loss of quality, both technological and related to their health benefits [35]. A number of methods for the assessment of oxidative stability have been developed, among which differential scanning calorimetry is one of the most frequently used. This method is very fast and convenient [36–38]. Generally, samples with longer induction times (IT) are more stable than those with a shorter induction time at the same temperature [37,39].

In our study, IT changes compared with raw milk value ranged from 5% for samples from fat after HPP in 600 MPa to 46% for HPP variant with 200–400 MPa. It appears that any kind of preservation causes changes to human milk oxidative status, but less so in selected pressure variants than in HoP (Figure 3).

Although original causes and consequences of oxidative and hydrolytic degradation processes are quite different, they seem to interact with each other and contribute to the reduction in stability of fats. Research suggests that fat FFA presence may induce oxidation as a result of a catalytic effect of carboxylic groups of FFA on the formation of free radicals. In general, the higher the level of fatty acids content, monoacylglycerols, and diacylglycerols in the fat with respect to the level of TAG, the more the oxidative stability is reduced. Pro-oxidant FFA action is thus connected with carboxylic molecular group, which accelerates the rate of hydroperoxides decomposition [35,40,41].

Analyzing the results obtained in this work, it can be concluded that the reduction in FFA content did not affect the improvement of oxidative fat stability (Figure 2, Figure 3). Pasteurized milk, despite the low content of FFA, is characterized by the shortest induction time, and thus the worst oxidative stability. Deterioration of oxidative stability in the case of pasteurized human milk could have been influenced by other factors, such as temperature, that could affect the content of antioxidants in the samples.

Human milk contains 3–5% of lipids and approximately 99% of them are TAG. The major fatty acids of TAG are palmitic acid (16:0), oleic acid (18:1), and linoleic acid (18:2 n-6), which constitute 25%, 30%, and 15%, respectively, of all fatty acids [42,43]. Therefore, fatty acid composition of human milk fat is unique, because the fat is characterized by a high content of saturated palmitic acid and also contains polyunsaturated fatty acids, which are not present in other milk fats [44–46].

Polyunsaturated fatty acid components of human milk are complex, including both C18 precursors, linoleic acid (18:2 n-6) and α-linolenic acid (18:3 n-3), as well as bioactive, very long chain polyunsaturated fatty acids (LCPUFA) of both n-6 and n-3 families [46,47].

Most of the studies on the effect of human milk processing, including the current study, have claimed that fatty acid composition was intact after holder pasteurization [21,30,47]. In addition, the results from our study proved no change in fatty acid profile in human milk after HPP, which is consistent with discoveries from other research studies on the effect of HPP on human milk [21,30]. It can be concluded that high pressure processing and holder pasteurization do not affect the composition of fatty acid (Figure 4, Table 1).

Moreover, the structure of human milk TAG is also unique, as 60–70% of palmitic acid is located at sn-2 position, and sn-1 and sn-3 positions are taken by 18:0, 18:1, and 18:2 fatty acids. This unique intramolecular structure is one of the key factors controlling the products formed by the gastric lipase in the stomach and by pancreatic or bile salt-stimulated lipases in the small intestinal absorption; therefore, it improves the efficiency of calcium absorption [48].

Considering the composition of fatty acid in sn-2 position of monoacylglycerols from the studied fats (Table 2), it can be seen that saturated fatty acids that occur mainly in this position are palmitic and myristic acid and unsaturated fatty acids—oleic and linoleic acids.

Analyzing the results regarding the percentage of fatty acids esterified at sn-2 position of TAG from the studied fats, it can be concluded that differences in the distribution of fatty acid in TAG were really small (Figure 5). There are no statistically significant differences between raw milk and HoP in the distribution of fatty acid in triacylglycerol molecules. These results are confirmed by many research studies [5,29–31,47,49]. According to scientists, no difference was found in the bioaccessibility of different fatty acids released from raw human milk and pasteurized human milk. If pasteurization affected the structure of triacylglycerols (TAG), the bioavailability of fatty acids found in these TAG would also change.

The percentage of palmitic acid and myristic acid at sn-2 of TAG in fats from all processed samples of human milk exceeded 33% (ranged from 64.7% to 73.6% for palmitic acid and from 46.7% to 61.4% for myristic acid), which means that these saturated fatty acids are located mainly in the internal position of TAG (Figure 5). This location increases the efficiency of calcium absorption in infants. Taking into account the percentage of unsaturated fatty acids in sn-2 position of TAG in studied fats, it can be stated that they are located mostly in external positions of TAG. The percentage of oleic acid at sn-2 of TAG in analyzed fats ranged from 12.2% to 15.7% and linoleic acids—from 19.2% to 26.3%, which confirms that they are located mainly in sn-1,3 positions of TAG (external position of TAG).

Carotenoids are principal non-enzymatic, lipophilic antioxidants that act as free radical scavengers. There is a substantial and growing body of research evidence showing an important role of carotenoids, especially lutein, in human development, related to their antioxidant and anti-inflammatory properties [50,51]. This is crucial for preterm infants, for whom carotenoid (mainly lutein) administration may decrease the risk of several prematurity disorders, including necrotizing enterocolitis, bronchopulmonary dysplasia, and retinopathy of prematurity related to elevated oxidative stress [52–54]. Carotenoids are one of the bioactive factors in human milk, and their amount is determined by maternal dietary intake; infant formulas are not fortified enough [51,55]. In our study, we found that all variants of HPP slightly changed the β-carotene and lycopene content (it is increased), whereas lutein + zeaxanthin content is decreased by 40.0% (at 600 MPa, 200 + 600 MPa, 450 MPa) up to 60.2% (at 100 + 600 MPa) compared with a 15.8% decrease in the holder pasteurized samples. It is the first time that the effect of high pressure on human milk with regard to β-carotene content has been evaluated. The only existing study comparing maternal breastmilk with donor milk has revealed the 18–53% lower amount of carotenoids after heat treatment, which is still higher than in infant formulas [55]. The increase in lycopene concentration after high pressure treatment was observed before in tomato products [56]. Human milk preservation and storage are not the only processes that affect carotenoid content. A study conducted by Tacken et al. also found that lutein is more sensitive to processing than other carotenoids [57]. In this case, tube feeding was related to the decreased concentration of lutein by 35% (even 42.6% with the phototherapy exposure), β-carotene by 26%. This relatively high decrease in lutein content may be caused by its high susceptibility to oxidation damage.

5. Conclusions

Human milk lipids are essential nutrition components for the growth and development of an infant. Commonly used holder pasteurization affects some of the nutritional and biological components of human milk and its lipid fractions [21,34]. Moreover, structural disintegration of lipid fraction of human milk by thermal pasteurization has been proven [7,27]. However, in the pilot clinical trial, the significance of those changes has not been proven with regards to important nutritional outcomes such as effectiveness of gastric emptying in the group of preterm infants [58,59]. Other studies revealed a decrease of fat absorption in preterms fed with donor milk [60]. We determined that high pressure processing, especially at 450 MPa for 15 min., minimizes changes in the lipidome and lipid related components such as lipase of donor human milk, with the exception of the destructive effect on lutein + zeaxanthin. As was previously shown, preterms' growth rate during their hospital stay depends on overall quality of lipidome [61]. Therefore, any methods of improving human milk lipids preservation, including high pressure processing of donor milk, could be beneficial in managing optimal infant weight gain and growth.

Author Contributions: A.W. and J.B. conceived and designed the experiments; O.B., K.S., J.S.-M., E.P., M.A.Z., and J.H. performed the experiments; J.B., A.W., O.B., and J.S.-M. analyzed the data; G.O. contributed reagents and materials; K.K. performed statistical analysis of this work and interpretation of data for the work; A.W., J.B., M.A.Z., and J.H. wrote the paper; O.B. prepared figure and tables for publication; G.O. revised manuscript critically for important intellectual content.

Funding: This work was funded by the Polish National Centre for Research and Development Social Innovation grant for project Lactotechnology as an answer to special nutritional requirements of preterm infants IS-81/NCBIR/2015.

Acknowledgments: We acknowledge Elzbieta Lodykowska from the Regional Human Milk Bank in Warsaw at Holy Family Hospital for assistance with collecting milk samples and all the mothers who participated in the study.

Conflicts of Interest: The authors declare no conflict of interest.

References

1. Koletzko, B. Human milk lipids. *Ann. Nutr. Metab.* **2016**, *69*, 27–40. [CrossRef] [PubMed]
2. Tsopmo, A. Phytochemicals in Human Milk and Their Potential Antioxidative Protection. *Antioxidants* **2018**, *7*, 32. [CrossRef] [PubMed]
3. Koletzko, B.; Rodríguez-Palmero, M.; Demmelmair, H.; Fidler, N.; Jensen, R.; Sauerwald, T.; Mis, N.F. Physiological aspects of human milk lipids. *Early Hum. Dev.* **2001**, *65*, S3–S18. [CrossRef]
4. Nessel, I.; Khashu, M.; Dyall, S.C. The effects of storage conditions on long-chain polyunsaturated fatty acids, lipid mediators, and antioxidants in donor human milk—A review. *Prostaglandins Leukot. Essent. Fatty Acids* **2019**, *149*, 8–17. [CrossRef] [PubMed]
5. Fidler, N.; Sauerwald, T.U.; Demmelmair, H.; Koletzko, B. Fat Content and Fatty Acid Composition of Fresh, Pasteurized, or Sterilized Human Milk. *Adv. Exp. Med. Biol.* **2001**, *501*, 485–495. [PubMed]
6. Van Zoeren-Grobben, D.; Schrijver, J.; Berg, H.V.D.; Berger, H.M. Human milk vitamin content after pasteurisation, storage, or tube feeding. *Arch. Dis. Child.* **1987**, *62*, 161–165. [CrossRef]
7. Cavazos-Garduño, A.; Serrano-Niño, J.; Solís-Pacheco, J.; Gutierrez-Padilla, J.; González-Reynoso, O.; García, H.; Aguilar-Uscanga, B. Effect of pasteurization, freeze-drying and spray drying on the fat globule and lipid profile of human milk. *J. Food Nutr. Res.* **2016**, *4*, 296–302.
8. Hamosh, M.; Henderson, T.R.; Ellis, L.A.; Mao, J.-I.; Hamosh, P. Digestive Enzymes in Human Milk: Stability at Suboptimal Storage Temperatures. *J. Pediatr. Gastroenterol. Nutr.* **1997**, *24*, 38–43. [CrossRef]
9. Peila, C.; Emmerik, N.E.; Giribaldi, M.; Stahl, B.; Ruitenberg, J.E.; van Elburg, R.M.; Moro, G.E.; Bertino, E.; Coscia, A.; Cavallarin, L. Human Milk Processing: A Systematic Review of Innovative Techniques to Ensure the Safety and Quality of Donor Milk. *J. Pediatr. Gastroenterol. Nutr.* **2017**, *64*, 353–361. [CrossRef]
10. Iverson, S.J.; Lang, S.L.C.; Cooper, M.H. Comparison of the bligh and dyer and folch methods for total lipid determination in a broad range of marine tissue. *Lipids* **2001**, *36*, 1283–1287. [CrossRef]
11. Boselli, E.; Velazco, V.; Caboni, M.F.; Lercker, G. Pressurized liquid extraction of lipids for the determination of oxysterols in egg-containing food. *J. Chromatogr. A* **2001**, *917*, 239–244. [CrossRef]
12. Macias, C.; Schweigert, F.J. Changes in the Concentration of Carotenoids, Vitamin A, Alpha-Tocopherol and Total Lipids in Human Milk throughout Early Lactation. *Ann. Nutr. Metab.* **2001**, *45*, 82–85. [CrossRef] [PubMed]
13. Zielinska, M.A.; Hamulka, J.; Wesolowska, A. Carotenoid Content in Breastmilk in the 3rd and 6th Month of Lactation and Its Associations with Maternal Dietary Intake and Anthropometric Characteristics. *Nutrients* **2019**, *11*, 193. [CrossRef] [PubMed]
14. Ballard, O.; Morrow, A.L. Human Milk Composition: Nutrients and Bioactive Factors. *Pediatr. Clin. N. Am.* **2013**, *60*, 49–74. [CrossRef] [PubMed]
15. Andreas, N.J.; Kampmann, B.; Le-Doare, K.M. Human breast milk: A review on its composition and bioactivity. *Early Hum. Dev.* **2015**, *91*, 629–635. [CrossRef] [PubMed]
16. Bardanzellu, F.; Fanos, V.; Strigini, F.A.L.; Artini, P.G.; Peroni, D.G. Human Breast Milk: Exploring the Linking Ring Among Emerging Components. *Front. Pediatr.* **2018**, *6*, 215. [CrossRef] [PubMed]
17. Vázquez-Román, S.; Escuder-Vieco, D.; Martín-Pelegrina, M.; Muñoz-Amat, B.; Fernández-Álvarez, L.; Brañas-García, P.; Lora-Pablos, D.; Beceiro-Mosquera, J.; Pallás-Alonso, C. Effect of refrigerated storage on the pH and bacterial content of pasteurized human donor milk. *J. Dairy Sci.* **2018**, *101*, 10714–10719. [CrossRef] [PubMed]
18. Ogundele, M.O. Effects of storage on the physicochemical and antibacterial properties of human milk. *Br. J. Biomed. Sci.* **2002**, *59*, 205–211. [CrossRef]
19. Ahrabi, A.F.; Handa, D.; Codipilly, C.N.; Shah, S.; Williams, J.E.; McGuire, M.A.; Potak, D.; Aharon, G.G.; Schanler, R.J. Effects of Extended Freezer Storage on the Integrity of Human Milk. *J. Pediatr.* **2016**, *177*, 140–143. [CrossRef]
20. Penn, A.H.; Altshuler, A.E.; Small, J.W.; Taylor, S.F.; Dobkins, K.R.; Schmid-Schönbein, G.W. Effect of Digestion and Storage of Human Milk on Free Fatty Acid Concentration and Cytotoxicity. *J. Pediatr. Gastroenterol. Nutr.* **2014**, *59*, 365–373. [CrossRef]
21. Pitino, M.A.; Alashmali, S.M.; Hopperton, K.E.; Unger, S.; Pouliot, Y.; Doyen, A.; O'Connor, D.L.; Bazinet, R.P. Oxylipin concentration, but not fatty acid composition, is altered in human donor milk pasteurised using both thermal and non-thermal techniques. *Br. J. Nutr.* **2019**, *122*, 47–55. [CrossRef] [PubMed]

22. Peila, C.; Moro, G.E.; Bertino, E.; Cavallarin, L.; Giribaldi, M.; Giuliani, F.; Cresi, F.; Coscia, A. The Effect of Holder Pasteurization on Nutrients and Biologically-Active Components in Donor Human Milk: A Review. *Nutrients* **2016**, *8*, 477. [CrossRef]
23. Ewaschuk, J.B.; Unger, S. Human milk pasteurization: Benefits and risks. *Curr. Opin. Clin. Nutr. Metab. Care* **2015**, *18*, 269–275.
24. Wesolowska, A.; Sinkiewicz-Darol, E.; Barbarska, O.; Bernatowicz-Lojko, U.; Borszewska-Kornacka, M.K.; Van Goudoever, J.B. Innovative Techniques of Processing Human Milk to Preserve Key Components. *Nutrients* **2019**, *11*, 1169. [CrossRef] [PubMed]
25. Pitino, M.A.; Unger, S.; Doyen, A.; Pouliot, Y.; Aufreiter, S.; Stone, D.; Kiss, A.; O'Connor, D.L. High Hydrostatic Pressure Processing Better Preserves the Nutrient and Bioactive Compound Composition of Human Donor Milk. *J. Nutr.* **2019**, *149*, 497–504. [CrossRef]
26. Demazeau, G.; Plumecocq, A.; Lehours, P.; Martin, P.; Couëdelo, L.; Billeaud, C. A New High Hydrostatic Pressure Process to Assure the Microbial Safety of Human Milk While Preserving the Biological Activity of Its Main Components. *Front. Public Health* **2018**, *6*, 306. [CrossRef]
27. De Oliveira, S.C.; Bourlieu, C.; Ménard, O.; Bellanger, A.; Henry, G.; Rousseau, F.; Dirson, E.; Carrière, F.; Dupont, D.; Deglaire, A. Impact of pasteurization of human milk on preterm newborn in vitro digestion: Gastrointestinal disintegration, lipolysis and proteolysis. *Food Chem.* **2016**, *211*, 171–179. [CrossRef] [PubMed]
28. Lepri, L.; Del Bubba, M.; Maggini, R.; Donzelli, G.P.; Galvan, P. Effect of pasteurization and storage on some components of pooled human milk. *J. Chromatogr. B Biomed. Sci. Appl.* **1997**, *704*, 1–10. [CrossRef]
29. Henderson, T.R.; Fay, T.N.; Hamosh, M. Effect of pasteurization on long chain polyunsaturated fatty acid levels and enzyme activities of human milk. *J. Pediatr.* **1998**, *132*, 876–878. [CrossRef]
30. Moltó-Puigmartí, C.; Permanyer, M.; Castellote, A.I.; López-Sabater, M.C.; Castellote-Bargalló, A.I. Effects of pasteurisation and high-pressure processing on vitamin C, tocopherols and fatty acids in mature human milk. *Food Chem.* **2011**, *124*, 697–702. [CrossRef]
31. Romeu-Nadal, M.; Castellote, A.; Gayà, A.; López-Sabater, M.C. Effect of pasteurisation on ascorbic acid, dehydroascorbic acid, tocopherols and fatty acids in pooled mature human milk. *Food Chem.* **2008**, *107*, 434–438. [CrossRef]
32. Borgo, L.A.; Araújo, W.M.C.; Conceição, M.H.; Resck, I.S.; Mendonça, M.A. Are fat acids of human milk impacted by pasteurization and freezing? *Nutr. Hosp.* **2015**, *31*, 1386–1393.
33. Williamson, S.; Finucane, E.; Ellis, H.; Gamsu, H.R. Effect of heat treatment of human milk on absorption of nitrogen, fat, sodium, calcium, and phosphorus by preterm infants. *Arch. Dis. Child.* **1978**, *53*, 555–563. [CrossRef]
34. Gao, C.; Miller, J.; Middleton, P.F.; Huang, Y.-C.; McPhee, A.J.; Gibson, R.A. Changes to breast milk fatty acid composition during storage, handling and processing: A systematic review. *Prostaglandins Leukot. Essent. Fat. Acids* **2019**, *146*, 1–10. [CrossRef] [PubMed]
35. Martin, D.; Reglero, G.; Señoráns, F.J. Oxidative stability of structured lipids. *Eur. Food Res. Technol.* **2010**, *231*, 635–653. [CrossRef]
36. Kowalski, B.; Gruczyńska, E.; Maciaszek, K. Kinetics of rapeseed oil oxidation by pressure differential scanning calorimetry measurements. *Eur. J. Lipid Sci. Technol.* **2000**, *102*, 337–341. [CrossRef]
37. Tan, C.P.; Man, Y.C. Recent developments in differential scanning calorimetry for assessing oxidative deterioration of vegetable oils. *Trends Food Sci. Technol.* **2002**, *13*, 312–318. [CrossRef]
38. Bryś, J.; Wirkowska, M.; Gorska, A.; Ostrowska-Ligeza, E.; Bryś, A. Application of the calorimetric and spectroscopic methods in analytical evaluation of the human milk fat substitutes. *J. Therm. Anal. Calorim.* **2014**, *118*, 841–848. [CrossRef]
39. Kowalski, B.; Tarnowska, K.; Gruczynska, E.; Bekas, W. Chemical and Enzymatic Interesterification of Beef Tallow and Rapeseed Oil Blend with Low Content of Tallow. *J. Oleo Sci.* **2004**, *53*, 479–488. [CrossRef]
40. Miyashita, K.; Takagi, T. Study on the oxidative rate and prooxidant activity of free fatty acids. *J. Am. Oil Chem. Soc.* **1986**, *63*, 1380–1384. [CrossRef]
41. Hamam, F.; Shahidi, F. Enzymatic acidolysis of an arachidonic acid single-cell oil with capric acid. *J. Am. Oil Chem. Soc.* **2004**, *81*, 887–892. [CrossRef]
42. Mu, H. Production and nutritional aspects of human milk fat substitutes. *Lipid Technol.* **2010**, *22*, 126–129. [CrossRef]

43. Visentainer, J.V.; Santos, O.O.; Maldaner, L.; Zappielo, C.; Neia, V.; Visentainer, L.; Pelissari, L.; Pizzo, J.; Rydlewski, A.; Silveira, R.; et al. Lipids and Fatty Acids in Human Milk: Benefits and Analysis. In *Biochemistry and Health Benefits of Fatty Acids*; IntechOpen: London, UK, 2018.
44. Bryś, J.; Flores, L.F.V.; Górska, A.; Wirkowska-Wojdyła, M.; Ostrowska-Ligęza, E.; Bryś, A. Use of GC and PDSC methods to characterize human milk fat substitutes obtained from lard and milk thistle oil mixtures. *J. Therm. Anal. Calorim.* **2017**, *130*, 319–327. [CrossRef]
45. Lien, E.L. The role of fatty acid composition and positional distribution in fat absorption in infants. *J. Pediatr.* **1994**, *125*, S62–S68. [CrossRef]
46. Salamon, S.; Csapó, J. Composition of the mother's milk II. Fat contents, fatty acid composition. A review. *Acta Univ. Sapientiae Aliment.* **2009**, *2*, 196–234.
47. Delgado, F.J.; Cava, R.; Delgado, J.; Ramírez, R. Tocopherols, fatty acids and cytokines content of holder pasteurised and high-pressure processed human milk. *Dairy Sci. Technol.* **2014**, *94*, 145–156. [CrossRef]
48. Nielsen, N.S.; Yang, T.; Xu, X.; Jacobsen, C. Production and oxidative stability of a human milk fat substitute produced from lard by enzyme technology in a pilot packed-bed reactor. *Food Chem.* **2006**, *94*, 53–60. [CrossRef]
49. Baack, M.L.; Norris, A.W.; Yao, J.; Colaizy, T. Long Chain Polyunsaturated Fatty Acid Levels in U.S. Donor Human Milk: Meeting the Needs of Premature Infants? *J. Perinatol.* **2012**, *32*, 598–603. [CrossRef]
50. Giampietri, M.; Lorenzoni, F.; Moscuzza, F.; Boldrini, A.; Ghirri, P. Lutein and Neurodevelopment in Preterm Infants. *Front. Mol. Neurosci.* **2016**, *10*, 4034. [CrossRef]
51. Zielińska, M.A.; Wesołowska, A.; Pawlus, B.; Hamułka, J. Health Effects of Carotenoids during Pregnancy and Lactation. *Nutrients* **2017**, *9*, 838. [CrossRef]
52. Manzoni, P.; Guardione, R.; Bonetti, P.; Priolo, C.; Maestri, A.; Mansoldo, C.; Mostert, M.; Anselmetti, G.; Sardei, D.; Bellettato, M. Lutein and zeaxanthin supplementation in preterm very low-birth-weight neonates in neonatal intensive care units: A multicenter randomized controlled trial. *Am. J. Perinatol.* **2013**, *30*, 25–32. [CrossRef]
53. Rubin, L.P.; Chan, G.M.; Barrett-Reis, B.M.; Fulton, A.B.; Hansen, R.; Ashmeade, T.; Oliver, J.; Mackey, A.; Dimmit, R.; Hartmann, E. Effect of carotenoid supplementation on plasma carotenoids, inflammation and visual development in preterm infants. *J. Perinatol.* **2012**, *32*, 418–424. [CrossRef]
54. Costa, S.; Giannantonio, C.; Romagnoli, C.; Vento, G.; Gervasoni, J.; Persichilli, S.; Zuppi, C.; Cota, F. Effects of lutein supplementation on biological antioxidant status in preterm infants: A randomized clinical trial. *J. Matern. Neonatal Med.* **2013**, *26*, 1311–1315. [CrossRef]
55. Hanson, C.; Lyden, E.; Furtado, J.; Van Ormer, M.; Anderson-Berry, A. A Comparison of Nutritional Antioxidant Content in Breast Milk, Donor Milk, and Infant Formulas. *Nutrients* **2016**, *8*, 681. [CrossRef]
56. Oey, I.; Van Der Plancken, I.; Van Loey, A.; Hendrickx, M. Does high pressure processing influence nutritional aspects of plant based food systems? *Trends Food Sci. Technol.* **2008**, *19*, 300–308. [CrossRef]
57. Tacken, K.J.M.; Vogelsang, A.; Dikkeschei, B.D.; Van Lingen, R.A.; Slootstra, J.; Van Zoeren-Grobben, D. Loss of triglycerides and carotenoids in human milk after processing. *Arch. Dis. Child. Fetal Neonatal Ed.* **2009**, *94*, 447–450. [CrossRef]
58. Bellanger, A.; Ménard, O.; Pladys, P.; Dirson, E.; Kroell, F.; De Oliveira, S.C.; Le Gouar, Y.; Dupont, D.; Deglaire, A.; Bourlieu, C. Impact of human milk pasteurization on gastric digestion in preterm infants: A randomized controlled trial. *Am. J. Clin. Nutr.* **2017**, *105*, 379–390.
59. Perrella, S.L.; Hepworth, A.R.; Gridneva, Z.; Simmer, K.N.; Hartmann, P.E.; Geddes, D.T. Gastric emptying and curding of pasteurized donor human milk and mother's own milk in preterm infants. *J. Pediatr. Gastroenterol. Nutr.* **2015**, *61*, 125–129.
60. Andersson, Y.; Sävman, K.; Bläckberg, L.; Hernell, O. Pasteurization of mother's own milk reduces fat absorption and growth in preterm infants. *Acta Paediatr.* **2007**, *96*, 1445–1449. [CrossRef]
61. Alexandre-Gouabau, M.-C.; Moyon, T.; Cariou, V.; Antignac, J.-P.; Qannari, E.M.; Croyal, M.; Soumah, M.; Guitton, Y.; David-Sochard, A.; Billard, H.; et al. Breast Milk Lipidome Is Associated with Early Growth Trajectory in Preterm Infants. *Nutrients* **2018**, *10*, 164. [CrossRef]

© 2019 by the authors. Licensee MDPI, Basel, Switzerland. This article is an open access article distributed under the terms and conditions of the Creative Commons Attribution (CC BY) license (http://creativecommons.org/licenses/by/4.0/).

Article

Quantification of Nervonic Acid in Human Milk in the First 30 Days of Lactation: Influence of Lactation Stages and Comparison with Infant Formulae

Jiahui Yu [1,2], Tinglan Yuan [1,2], Xinghe Zhang [1,2], Qingzhe Jin [1,2], Wei Wei [1,2,*] and Xingguo Wang [1,2,*]

1. State Key Lab of Food Science and Technology, Jiangnan University, Wuxi 214122, China
2. International Joint Research Laboratory for Lipid Nutrition and Safety, Collaborative Innovation Center of Food Safety and Quality Control in Jiangsu Province, School of Food Science and Technology, Jiangnan University, Wuxi 214122, China
* Correspondence: weiw@jiangnan.edu.cn (W.W.); xingguow@jiangnan.edu.cn (X.W.); Tel./Fax: +86-510-85329050 (W.W.); +86-510-85876799 (X.W.)

Received: 2 July 2019; Accepted: 12 August 2019; Published: 14 August 2019

Abstract: Nervonic acid (24:1 n-9, NA) plays a crucial role in the development of white matter, and it occurs naturally in human milk. This study aims to quantify NA in human milk at different lactation stages and compare it with the NA measured in infant formulae. With this information, optimal nutritional interventions for infants, especially newborns, can be determined. In this study, an absolute detection method that uses experimentally derived standard curves and methyl tricosanoate as the internal standard was developed to quantitively analyze NA concentration. The method was applied to the analysis of 224 human milk samples, which were collected over a period of 3–30 days postpartum from eight healthy Chinese mothers. The results show that the NA concentration was highest in colostrum (0.76 ± 0.23 mg/g fat) and significantly decreased ($p < 0.001$) in mature milk (0.20 ± 0.03 mg/g fat). During the first 10 days of lactation, the change in NA concentration was the most pronounced, decreasing by about 65%. Next, the NA contents in 181 commercial infant formulae from the Chinese market were compared. The NA content in most formulae was <16% of that found in colostrum and less than that found in mature human milk ($p < 0.05$). No significant difference ($p > 0.05$) was observed among NA content in formulae with different fat sources. Special attention was given to the variety of n-9 fatty acids in human milk during lactation, and the results indicated that interindividual variation in NA content may be primarily due to endogenous factors, with less influence from the maternal diet.

Keywords: n-9 fatty acid; nervonic acid; human milk fat; infant formula; lactational stage

1. Introduction

Human milk is generally regarded as the best source of nutrition for infants during their first six months of life [1]. Human milk contains 3%–5% fats, of which 98%–99% are triacylglycerols (TAGs), 0.26%–0.80% are phospholipids (PLs), and 0.25%–0.34% are sterols, as well as trace amounts of minor components, including monoacylglycerols (MAGs), diacylglycerols (DAGs), non-esterified fatty acids, and other substances [2]. Human milk fat is one of the most complex natural lipids with a unique fatty acids (FAs) composition. Nervonic acid (24:1 n-9, NA) is a very-long-chain monounsaturated fatty acid (>C_{20}) that naturally occurs in human milk (less than 2%) [3], but it has been rarely studied in the field of human nutrition.

NA is an essential constituent of the neuronal membrane [4] and plays a crucial role in problems such as early myelination [5], peroxisomal disorders [6,7], and undernourishment [8]. NA rapidly accumulates in the fetal brain at 32–37 weeks' gestation [9]. After birth, infants continue to obtain NA from human milk. Several publications have indicated that the NA concentration in human milk markedly decreases as lactation stages progress [10–12]. NA concentrations in colostrum (≤1 week postpartum) have been found to range from two- to six-fold higher than those in mature milk (>2 weeks and ≤16 weeks) [13,14]. The results suggested that the human brain heavily accumulates NA in the early days of life. Previous research identified that NA is an important fatty acid in the white matter and its deficiency in early development may damage the white matter [15]. The cause of the majority of preterm infants with cerebral palsy is mainly attributed to white matter diseases, such as periventricular leukomalacia [16,17].

The NA concentration in human milk is not routinely reported because NA quantitation is obscured by more abundant very-long-chain fatty acids when expressed percentage by weight of total fatty acids in gas chromatography (GC) analyses [18]. However, previous studies report that NA is detectable in human milk from different countries of the world, and a significantly higher NA content is present in colostrum compared to mature milk [11–13,19–21]. Very few data are available on the quantification of nervonic acid in human milk especially the amount in each lactation day.

The NA content in common vegetable oils is low, therefore infant formulae have much lower NA content compared with human milk [3], especially colostrum. However, relatively little attention has been paid to NA in infant formulae [22]. NA, together with docosahexaenoic acid (22:6 n-3, DHA) and arachidonic acid (20:4 n-6, AA), has a positive effect on the neural development of the neonate [23]. Nonetheless, the importance of DHA and AA in infant formulae have attracted researchers' attention [9,24,25], whereas few studies focus on the effect of NA in human milk and the comparison with infant formula which is potentially important.

This study aimed to develop an effective method for the quantification of NA concentration in human milk and systematically compare the NA concentration in human milk on different lactation days. The NA contents in infant formulae from three fat sources (cows' milk fat, goats' milk fat, and plant oil) were also studied. Special attention was given to the variation of three other n-9 FAs, including oleic acid (18:1 n-9, OA), eicosenoic acid (20:1 n-9, EiA) and erucic acid (22:1 n-9, EA), in human milk throughout lactation.

2. Materials and Methods

2.1. Standards and Chemicals

A standard mixture of 37 kinds of fatty acid methyl esters (FAMEs, from C_4 to C_{24}) was bought from Sigma-Aldrich (St. Louis, MO, USA). The standards methyl *cis*-15-tetracosenoate (24:1 n-9 FAME) and methyl tricosanoate (23:0 FAME), methanol, and *n*-hexane (HPLC grade) were obtained from J and K Scientific (Beijing, China). Other reagents were analytical grade and purchased from Sinopharm Reagent Co. Ltd. (Shanghai, China).

2.2. Human Milk Samples and Infant Formulae

Human milk samples were collected from eight healthy mothers of term infants in Wuxi, China. The study was approved by the Ethics Committee of the School of Food Science and Technology in Jiangnan University (JN No. 21212030120), and written consent was obtained from all subjects included in this study. During the lactation period from 3 to 30 days postpartum, 224 human milk samples were collected. The sample-set included colostrum (3–6 postpartum days, n = 32), transitional milk (7–14 postpartum days, n = 64), and mature milk (>15 postpartum days, n = 128). All participating mothers were non-smokers, non-medicated, and healthy. The mean age of the mothers was 28.38 ± 3.16 years, mean BMI was 21.78 ± 2.75 kg/m^2, and mean infant weight at birth was 3.35 ± 0.23 kg. The samples (5–10 mL) were collected after full expression from one breast with a breast pump between 9 and 11 am. The samples were stored at −20 °C for less than 2 h and transferred to the lab at Jiangnan University for lipid extraction.

The fatty acid composition of infant formulae is reported in a previous study [3]. A total of 181 formulae from 27 brands were collected; the tested formulae account for 75% of the Chinese infant formulae market. Three stages (infant, follow-on, and growing-up) of formulae were included. The formulae were divided into cows' milk, goats' milk, and plant oil, depending on the main fat source.

2.3. Fatty Acid Methyl Esters Preparation

Total lipids in human milk were extracted by classic Röse–Gottlieb method [26] with some modifications. Briefly, 1 mL ammonia water was added into 4 mL human milk, and then mixed in a water shaking bath at 65 °C ± 5 °C for 20 min. Then, 5 mL absolute ethanol, 10 mL absolute ether and 10 mL petroleum ether were added to extract the lipids. The samples were mixed thoroughly and stood for 2 h, and then the supernatants were collected. The lipids in the lower phase were extracted using half of the solvents as above. The solvents were removed by nitrogen and the lipids were stored in a −80 °C freezer until analysis.

The 24:1 n-9 FAME standard was prepared at a concentration of 0.5 mg/mL dissolved in *n*-hexane. The internal standard solution was prepared by weighing 10 mg of 23:0 FAME into 10 mL of *n*-hexane. Milk fat (40 mg) was suspended in 200 μL of the internal standard solution, 800 μL *n*-hexane, and 500 μL of KOH–CH$_3$OH (2 mol/L). After mixing for 2 min, the water in the solution was removed by adding the appropriate amount of anhydrous sodium sulfate. Then, the mixture was mixed thoroughly and left standing for half an hour. The supernatant was filtered by a 0.22 μm filter and analyzed by gas chromatography (GC).

2.4. GC Analysis

The samples were analyzed by an Agilent 7820A gas chromatograph (Agilent Technologies, Santa Clara, CA, USA) with a hydrogen flame ionization detector, and they were separated by a TRACE™-FAME capillary column (60 m × 0.25 mm × 0.25 μm, Thermo Fisher Scientific, Waltham, MA, USA) using nitrogen carrier gas [27]. The injection and detector temperature were set at 250 °C, and the injection volume was 2 μL. The temperature of the column was set at 60 °C for 3 min to start and then increased to 175 °C at a rate of 5 °C/min. The temperature was maintained at 175 °C for 15 min and then raised (at a rate of 2 °C/min) to 220 °C, which was maintained for 10 min. FAs were identified by comparing the retention times of the sample peaks with those of the FAME standards. The NA concentration was measured by the absolute detection method, with 23:0 FAME as an internal standard.

2.5. Quantitation of Nervonic Acid

In the first step, a series of standard solutions that contained varying concentrations of 24:1 n-9 FAME and an identical amount of 23:0 FAME were injected. The calibration line was constructed using Equation (1):

$$y = ax + b \tag{1}$$

where y is the ratio of the peak area of 24:1 n-9 FAME to that of 23:0 FAME, a is the slope of the calibration curve, x is the ratio of the weight of 24:1 n-9 FAME to that of 23:0 FAME and b is the intercept of the calibration curve.

The NA concentration in human milk fat was calculated using Equation (2), with 23:0 FAME as the internal standard.

$$C = (\frac{A_1}{A_0} - b) \times m_0 \times \frac{1}{a} \times \frac{1}{W} \times F_t \tag{2}$$

where C is the concentration of NA in human milk fat (mg/g fat), A_1 is the peak area of 24:1 n-9 FAME, A_0 is the peak area of 23:0 FAME, m_0 is the weight of the internal standard/mg, W is the weight of human milk fat in the test sample/g, and F_t is the transformation coefficient of 24:1 n-9 FAME to NA.

2.6. Validation of the Method

The following parameters (precision, recovery rate, and limits of detection and quantification) were used to validate this method based on the guidance of the United States Pharmacopeia (USP) [28]. For the precision test of the method, the same homogeneous sample was analyzed three times in one day to obtain the intra-day precision and in three different days to obtain the inter-day precision. The precision was described as peak area and retention time relative standard deviations (RSDs) [29]. The recovery of the methods was measured using three levels of concentration of standard solution that were added into three identical human milk samples, and no standard was added into the fourth sample. The rate was calculated by the detected concentration of NA standard divided by the actually added concentration. The limits of detection and quantitation were defined as three times and ten times of signal to noise ratio, respectively [30].

2.7. Statistical Analysis

All analyses of human milk were performed in duplicate. The results were expressed as means (%) ± standard deviations (SD) and were calculated using the SPSS 19.0 (SPSS, USA). Differences in NA concentration were tested by one-way analysis of variance (ANOVA) for continuous variables. Two-way ANOVA was used to evaluate the effect of fat sources and stages on the n-9 FA composition of infant formulae. Differences among all results were compared by use of Ducan's test at $p < 0.05$. Pearson's correlation test was used to determine the correlation coefficient between NA concentration and lactation days. Principal component analysis (PCA) was used to determine the differences in NA content in three lactation stages of human milk and three stages of infant formulae using SIMCA-P software version 13.5 (Demo Umetrics, Umea, Sweden).

3. Results

3.1. Nervonic Acid Concentration in Human Milk

In this study, to accurately quantify the NA concentration in human milk, a long GC program with an analysis time of 70 min was applied, and a FAME with a similar molecular structure (23:0) was added to the sample as an internal standard [31]. The gas chromatograms of NA in human milk are shown in Figure 1.

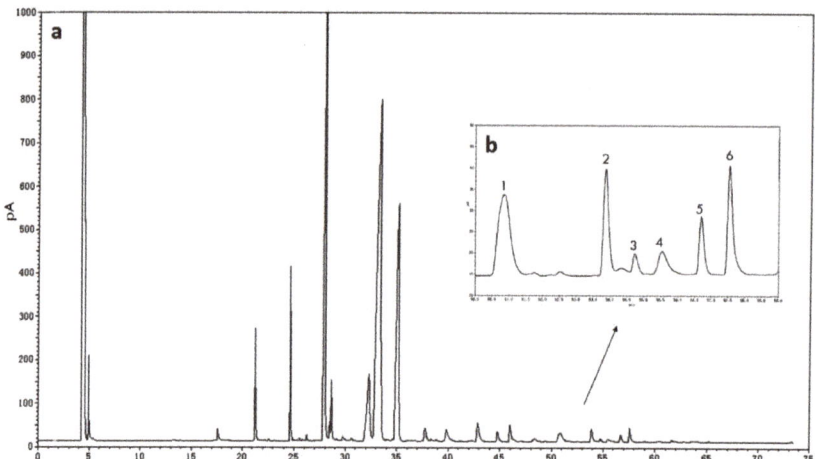

Figure 1. Gas chromatography (GC) of total fatty acids (**a**) and very-long-chain fatty acids (**b**) in one colostrum human milk. 1, tricosanoicacid (23:0); 2, docosatetraenoic acid (22:4 n-7); 3, docosapentaenoic acid (22:5 n-6); 4, nervonic acid (24:1 n-9, NA); 5, docosapentaenoic acid (22:5 n-3); 6, docosahexaenoic acid (22:6 n-3).

The calibration line of this method was defined by y = 0.733x + 0.7417 (R^2 = 0.9944) showing a good linearity [32]. The calibration lines of 24:1 n-9 and 23:0 were y = 9.47 × 10^6x (R^2 = 0.9938) and y = 9.59 × 10^6x (R^2 = 0.9965), respectively. The precision measured the dispersion degree between the tested results of the same sample under the same condition [29]. Both the intra-day and inter-day RSD% on the retention times were lower than 0.01%, and the intra-day and inter-day RSD% on the peak areas were lower than 0.02% and 0.18%, respectively. The recovery rate of the method ranged from 95% to 103%, which met the advisable international level of 80% to 115% [33]. The detection limits ranged from 1.34 µg/mL for NA to 3.25 µg/mL for 23:0 FAME and the quantitation limits ranged from 3.11 µg/mL to 5.36 µg/mL for NA and 23:0 FAME, respectively.

The NA concentrations in human milk are presented in Table S1. The NA concentration in all human milk samples decreased significantly (r = −0.822, $p < 0.001$) during the first month of lactation. The average NA concentration on day three of lactation was about five times higher than that on day 30 ($p < 0.001$), with values of 1.00 ± 0.24 mg/g and 0.18 ± 0.03 mg/g fat, respectively. No significant difference was observed in the NA concentration in human milk obtained after 15 days of lactation. The concentration of NA was highest in colostrum (0.76 ± 0.23 mg/g fat) and lowest in mature milk (0.20 ± 0.03 mg/g fat), decreasing by 82% in the first month of lactation. Figure 2 illustrates the changing trend of the mean values of NA concentration in human milk in the first month of lactation.

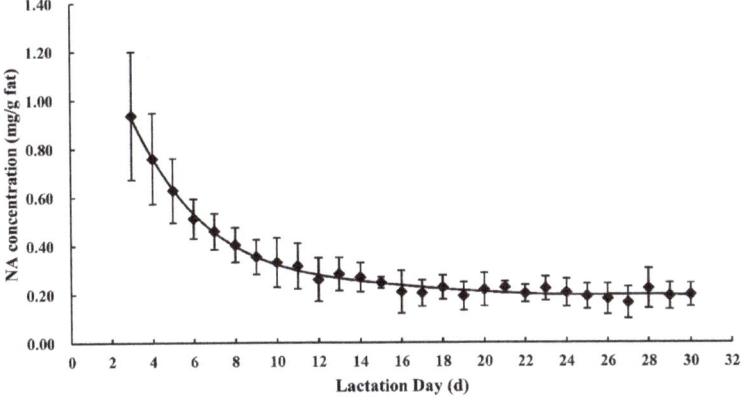

Figure 2. Nervonic acid concentrations in human milk throughout the first month of lactation.

Besides changes between lactation stages, there were significant interindividual differences in NA concentration. The differences were more pronounced from three to 15 days and marginally significant ($p > 0.05$) after day 15 of lactation. At three days postpartum, there was a two-fold variation in NA in the milk from the eight mothers, with values ranging from 0.72 to 1.44 mg/g fat. The differences between individuals decreased over time during the study period. During days 26–30 of lactation, the NA concentration varied from 0.15 mg/g fat to 0.22 mg/g fat, with a 0.07 mg/g fat difference.

3.2. Composition of n-9 Fatty Acids in Human Milk

In this study, four kinds of n-9 FAs (OA, EiA, EA, and NA) were detected in human milk samples. Changes in n-9 FA composition in human milk during the first month of lactation are shown in Figure 3. OA, whose percentage ranged from 30.23% to 33.86% of total FAs, was the predominant FA in human milk. During lactation, no significant difference was observed in OA content in human milk (Figure 3A). Contrary to the observed consistency of OA, there were significant changes ($p < 0.05$) in the percentage of EiA (from 0.17% to 0.06%) and NA (from 0.21% to 0.04%) as the milk progressed from colostrum to transitional milk and from transitional milk to mature milk (Figure 3b,d). Particularly, the NA percentage varied in the range of 0.18%–0.21% in colostrum and decreased to 0.12%–0.17% in transitional milk and 0.04%–0.10% in mature milk. The EA content decreased from 0.19% to 0.11% of the total FA content from the first to third lactation stage, but the discrepancies were less significant between transitional milk and mature milk (Figure 3c).

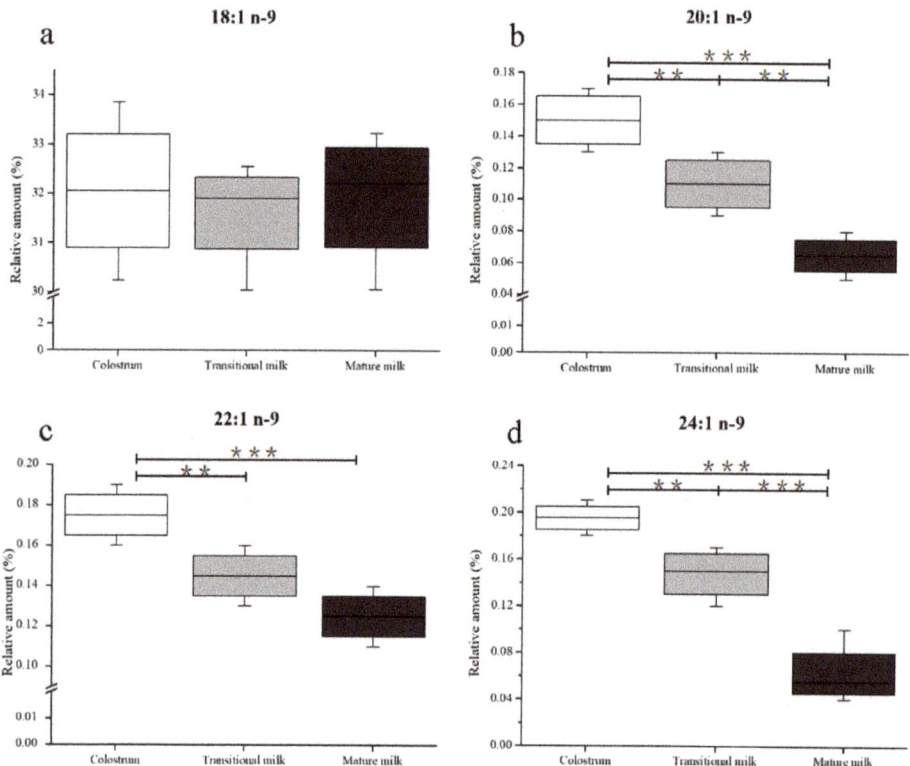

Figure 3. Changes in n-9 fatty acids including 18:1 n-9 (**a**), 20:1 n-9 (**b**), 22:1 n-9 (**c**), and 24:1 n-9 (**d**) in human milk during the first month of lactation. Significantly different from colostrum, transitional milk, and mature milk: *** $p < 0.001$; ** $p < 0.01$. No labeling indicates no significant differences.

3.3. Composition of n-9 Fatty Acids in Infant Formulae

A total of 181 infant formulae were analyzed, and 97 (53.59%) were found to contain NA. The n-9 FA composition of 97 infant formulae are presented in Table 1. The infant formulae were divided into three stages (infant formula, follow-on formula, and growing-up formula) and three sources (cows' milk formula, goats' milk formula, and plant-oil formula). There was a significant difference in fat sources of OA ($p < 0.001$), EiA ($p < 0.001$), and EA ($p < 0.01$) among different formulae. There were no significant differences in sources of NA ($p > 0.05$). The n-9 FA composition was not affected by the formulae stages or the interaction between fat sources and stages ($p > 0.05$). It is recommended that EA content in infant formulae be below 1% of the total fat content [34]. It is noteworthy that infant formulae have much lower concentrations of NA compared with human milk. The NA content of most formulae (0.02% ± 0.02%) was about <16% of that in colostrum (0.20% ± 0.04%) and less than from that found in mature milk (0.08% ± 0.04%).

Table 1. n-9 Fatty acid composition (wt%) of infant formulae.

Fatty Acids		CMF	GMF	POF	p-Value		
		(n = 30)	(n = 16)	(n = 51)	F	S	F × S
18:1 n-9	IF	32.15 ± 5.06 [a]	34.81 ± 5.66 [a,b]	40.74 ± 8.32 [b]	***	NS	NS
	FF	30.65 ± 5.95 [a]	30.15 ± 5.28 [a]	41.61 ± 6.11 [b]			
	GF	31.76 ± 5.09 [a]	29.66 ± 6.50 [a]	40.89 ± 7.31 [b]			
20:1 n-9	IF	0.23 ± 0.09	0.29 ± 0.09	0.31 ± 0.10	***	NS	NS
	FF	0.17 ± 0.10 [a]	0.24 ± 0.09 [a,b]	0.33 ± 0.09 [b]			
	GF	0.13 ± 0.04 [a]	0.21 ± 0.07 [a]	0.30 ± 0.09 [b]			
22:1 n-9	IF	0.04 ± 0.04	0.06 ± 0.04	0.04 ± 0.03	**	NS	NS
	FF	0.03 ± 0.03	0.05 ± 0.03	0.05 ± 0.03			
	GF	0.01 ± 0.01 [a]	0.04 ± 0.03 [a,b]	0.05 ± 0.04 [b]			
24:1 n-9	IF	0.03 ± 0.13	0.03 ± 0.02	0.03 ± 0.02	NS	NS	NS
	FF	0.02 ± 0.00 [a]	0.02 ± 0.01 [a]	0.03 ± 0.01 [b]			
	GF	0.02 ± 0.01	0.02 ± 0.01	0.03 ± 0.01			

CMF, cows' milk formula; GMF, goats' milk formula; POF, plant-oil formula; IF, infant formula; FF, follow-on formula; GF, growing-up formula. F, fat source; S, stage. Different superscript lowercase letters indicate significant differences ($p < 0.05$) with a row. *** $p < 0.001$; ** $p < 0.01$. NS, $p > 0.05$.

4. Discussion

NA occurs naturally in human milk; however, it has not been routinely reported in the global FA profile of human milk [18]. The main reason for its frequent omission is that the concentration of NA in human milk is low, and its retention time is similar to several polyunsaturated fatty acids (PUFAs). These factors make the quantitation of NA very difficult. In this study, 23:0 FAME was used as an internal standard to quantify NA concentration in human milk fat. In general, the validation of the method for the quantification of NA in human milk indicated that the method parameters' were well within internationally accepted limits.

The NA concentration significantly decreased ($p < 0.05$) with the progression of lactation days. Furthermore, the NA level decreased markedly during the first 10 days then decreased slowly after 15 days. The fat contents of human milk increased during lactation (Table S2). As shown in Table 2, the NA contents in human milk from mothers in different geographical regions and at different lactation stages were summarized. Most of the studies show a significantly higher NA content in colostrum compared to mature milk [11–13,19,20]. This study indicates the concentration of NA in colostrum was three times higher than that in mature milk, which is similar to Hua et al. [19] and Rueda et al. [20]. The results of the NA concentration in human milk during the three stages (colostrum, transitional milk, and mature milk), were similar to those found by Xiang [35]. Other publications show ten times higher NA content in colostrum compared to mature milk [11,12]. However, there are also a few publications that report the NA concentrations decrease from 0.35 to 0.27 mg/g of human milk, with no significant difference between colostrum and mature milk [36].

There are some discrepancies between studies, as shown in Table 2. The percentage of NA in human milk from individuals in Wuxi ranged from 0.06% to 0.20%. This result is marginally different from the values from several previous studies, which have reported values ranging from 0.19% to 0.99%. The difference could be due to analytical methods, as well as individuals from different locations and the lactation day. Further studies on the absolute detection of NA concentration from different countries and the correlation with maternal diet are needed.

Table 2. Mean value of nervonic acid (wt%) in human milk from individuals in different regions.

Regions	Colostrum	Transitional Milk	Mature Milk	References
Wuxi, China	0.20 (3–6 d)	0.15 (7–14 d)	0.06 (15–30 d)	This study
Taiwan, China	0.99 (1–6 d)	-	0.28 (2 m)	Hua et al. [19]
Switzerland	0.39 (1 week)	0.13 (2 weeks)	0.07 (3–8 weeks)	Thakkar et al. [13]
Beijing, China	0.54 (4 d)	-	0.25 (30 d)	Zhao et al. [14]
Thailand	-	-	0.06	Golfetto et al. [37]
Korean	-	0.27	-	
Bangladeshi	-	-	0.20	Yakes et al. [38]
Northern Sudanese	0.19	0.15	0.02	Nyuar et al. [11]
Wenzhou, China	0.45	-	-	Peng et al. [39]
Changzhou, China	0.25	-	-	
Shanghai, China	-	0.08	0.05	Jing et al. [40]
Guangzhou, China	-	0.06	0.06	
Nanchang, China	-	0.12	0.11	
Harbin, China	-	0.06	0.04	
Hohhot, China	-	0.21	0.19	
Granada, Spain	0.28	0.08	0.07	Sala-Vila A et al. [21]
Congolese	-	-	0.04	Rocquelin et al. [41]
Panama	0.32	0.16	0.10	Rueda et al. [20]
Spain	0.24	0.17	0.10	
Dominica	-	-	0.05	Beusekoma et al. [42]
Saint Lucia	0.41	0.11	0.04	Boersma et al. [12]
Belize	-	-	0.06	Cheristien et al. [43]
Dominica	-	-	0.02	

Dietary habits differ among regions and can cause internal physiological differences [44]. Jing and co-workers determined the NA content of transitional milk and mature milk from five regions (Shanghai, Guangzhou, Nanchang, Harbin, and Hohhot) of China, and the content in Hohhot was highest, regardless of whether the sample was transitional milk or mature milk [40]. Diets containing NA were fed to lactating rats, and the NA in the diet influenced the NA concentration in the animal's milk [45]. Although it is rare for NA to be provided by a typical diet, earlier studies have reported that dietary OA and EA influence the nervonic acid content in humans [46]. Lactating women are able to synthesize NA by carbon chain elongation with enzyme catalysts [47]. The large variation in NA concentration in the milk of human mothers from different regions may reflect differences in the mothers' diets [48].

There are differences in the diets of lactating mothers around the world that can account for the NA concentration differences between individuals. Additionally, the individual variation may also be related to maternal age [49], and Body Mass Index (BMI) [50]. However, in this study and previous studies, the NA concentration in all samples had a downward trend throughout the entire lactation period. This consistency, despite regional differences, indicates that the decline in NA content in human milk is more likely to be influenced by an endogenous factor [51] than a dietary habit.

The amounts of DHA and AA in human milk, similarly to NA, had a decreasing trend over the lactation period (Table S3). This has been attributed to the genes that encode delta (5)- and delta (6)-desaturases and their effect on the proportions of these FAs in human milk [52]. The DHA and AA precursors linolenic acid (18:3 n-3, ALA) and linoleic acid (18:2 n-6, LA) increase in parallel [53]. The hypothetical pathways of n-9 FAs metabolism are shown in Figure 4. NA is synthesized from OA by elongation. The OA content measured in this study (30.23%–33.86%) was in accordance with Qi's results (30.66%–32.71%) [27], but inconsistent with Weiss's study (43.96%–48.21%) [53]. OA content was stable in this study, which was contrary to the study by Lopez-Lopez et al., who claimed that OA content decreased as lactation progressed [54]. Additionally, the percentage of EiA, EA, and NA decreased with the progression of lactation, which agreed with the results obtained by Nyuar et al. [11]. In general, the amount of OA was stable over time, and the other precursors, EiA and EA, decreased during the same period, which differs from the precursors of DHA or AA. The mechanism by which genes regulate the metabolism of n-9 FAs remains unclear and needs further study. In human milk, OA is the richest FA and serves as a structural component. It can be assumed that the OA content might be sufficient to synthesize NA and other n-9 FAs.

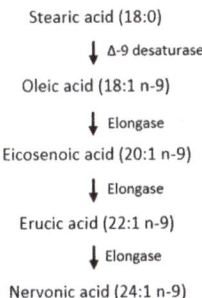

Figure 4. Hypothetical pathway for the synthesis of nervonic acid in infants.

It has been demonstrated that infants have the capacity to convert ALA and LA to DHA and AA, respectively, and that elongases and desaturases have activity in the first week after birth [55]. However, the ability of infants to synthesis FAs is weak, and the amount of DHA formed from ALA is inadequate to support brain development [56]. For the optimal development of an infant's brain, a sufficient amount of NA, DHA, and AA is necessary for newborns.

Human milk is a rich source of NA compared with commercial infant formulae. In the 118 collected infant formulae samples, only 53.59% contained NA. The PCA revealed differences of NA content between human milk (including colostrum, transitional milk, and mature milk) and infant formulae (Figure 5). A significant distinction of NA content was observed between human milk and infant formula ($p < 0.05$). NA content in the three-stages (colostrum, transitional milk, and mature milk) of human milk differs obviously ($p < 0.05$). However, there is no significant difference of NA content in the three types (infant, follow-on, growing-up) of infant formula. Most NA-containing formulae had NA concentrations that were <16% of that measured in colostrum. NA content in formulae needs to be enhanced urgently for term newborns and especially for preterm newborns. It has been reported that the NA concentration in preterm milk is seven-fold higher compared to that in milk from mothers with full-term babies at a similar stage of lactation [9].

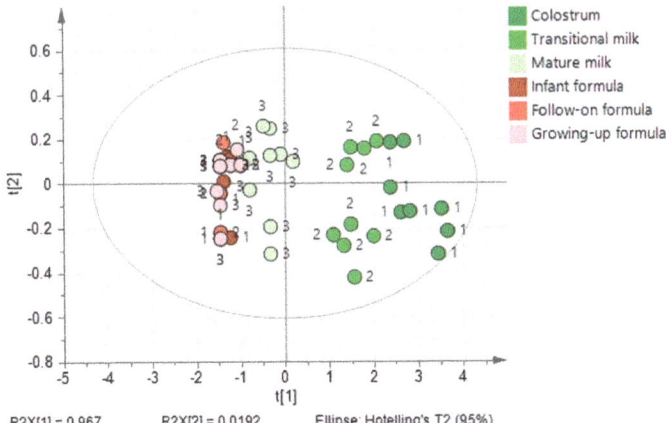

Figure 5. Principal component analysis of stage difference in the NA content in human milk (1, colostrum; 2, transitional milk; 3, mature milk) and infant formulae (1, infant formula; 2, follow-on formula; 3, growing-up formula).

The NA level is considered to reflect brain maturation, and its accumulation in the brain is a sign of the onset of myelinogenesis [57]. Newborns who cannot be breastfed fail to get sufficient levels of NA from infant formulae. Long-term NA deficiency will hamper the development of the nervous system and cause visual impairment [58]. Supplying dietary lipids to stimulate the synthesis of n-9 FAs has been advised to support myelination in newborns [23]. Thus, infant formulae need to enhance their NA content to match the NA content in human milk. This applies to formulae for full-term infants, and it is especially the case for preterm infants.

NA content in formulae was not related to fat sources or stages. However, the differences observed between other n-9 FAs in infant formulae were particularly influenced by fat sources. The current formula fat sources (generally used cow milk fat, goat milk fat, and vegetable oils) contain no or trace amount of NA. NA has been found in some plant species [59] for example *Malania oleifera* [60] and *Lunaria annua* [61]. However, oils of these plants are rarely developed and used in infant formulae because of their high EA content. It has been demonstrated that the heart may be damaged by EA, and EA is undesirable for human consumption [62]. It is necessary to explore the effect of NA as well as different forms (FA, TAG or PLs) for the optimal brain development of infants.

5. Conclusions

The NA concentration was at its highest at the onset of lactation, with an overall decrease of 82% throughout the 3–30 days postpartum study period. The concentration dropped at a faster rate in the first 10 days and then flattened out. Thus, it is crucial for infants, especially preterm infants, to get enough NA within the first ten days of lactation. The NA concentration differs between breastfeeding mothers, in part because of the regions in which they live. Further studies are needed to clarify the NA biosynthetic system in infants. Most of the 97 infant formulae in which NA was detected had less than 16% of the NA content in colostrum. There were significant differences between the fat sources of OA, EiA, and EA in the formulae, but this was not the case for NA. The effect of the stages or the interaction between sources and stages on infant formulae FA composition was not obvious. Because of the low level of NA found in infant formulae, it is suggested that the NA content in infant formulae for newborns be increased. Further work is also warranted to identify the best form of NA for infants.

Supplementary Materials: The following are available online at http://www.mdpi.com/2072-6643/11/8/1892/s1, Table S1: Nervonic acid (mg/g fat) concentration in human milk during the first month of lactation, Table S2: Total fat content (mg/mL) in human milk during the first month of lactation; Table S3: Total fatty acid composition (%, wt) in human milk during the lactation days.

Author Contributions: Data curation, X.Z.; formal analysis, Q.J.; methodology, T.Y.; project administration, W.W.; supervision, X.W.; writing—original draft, J.Y.

Funding: This research was funded by the National Natural Science Foundation of China, grant number 31701558. The APC was funded by the Young Elite Scientists Sponsorship Program by the China Association for Science and Technology (CAST), grant number 2017QNRC001.

Acknowledgments: This work was supported by the National Natural Science Foundation of China (grant number 31701558) and the Young Elite Scientists Sponsorship Program by CAST (grant number 2017QNRC001). The authors particularly thank L. Ding from Wuxi Wanyi Infant Care and mothers who volunteered to contribute human milk.

Conflicts of Interest: The authors declare no competing financial interest.

References

1. Cheong, L.Z.; Jiang, C.Y.; He, X.Q.; Song, S.; Lai, O.M. Lipid Profiling, Particle Size Determination, and in Vitro Simulated Gastrointestinal Lipolysis of Mature Human Milk and Infant Formula. *J. Agric. Food Chem.* **2018**, *66*, 12042–12050. [CrossRef] [PubMed]
2. Wei, W.; Jin, Q.; Wang, X. Human milk fat substitutes: Past achievements and current trends. *Prog. Lipid Res.* **2019**, *74*, 69–86. [CrossRef] [PubMed]

3. Sun, C.; Zou, X.; Yao, Y.; Jin, J.; Xia, Y.; Huang, J.; Jin, Q.; Wang, X. Evaluation of fatty acid composition in commercial infant formulas on the Chinese market: A comparative study based on fat source and stage. *Int. Dairy J.* **2016**, *63*, 42–51. [CrossRef]
4. Amminger, G.P.; Schaefer, M.R.; Klier, C.M.; Slavik, J.M.; Holzer, I.; Holub, M.; Goldstone, S.; Whitford, T.J.; McGorry, P.D.; Berk, M. Decreased nervonic acid levels in erythrocyte membranes predict psychosis in help-seeking ultra-high-risk individuals. *Mol. Psychiatry* **2012**, *17*, 1150–1152. [CrossRef] [PubMed]
5. Kinney, H.C.; Brody, B.A.; Kloman, A.S.; Gilles, F.H. Sequence of central nervous system myelination in human infancy. II. Patterns of Myelination in Autopsied Infants. *J. Neuropathol. Exp. Neurol.* **1988**, *47*, 217–234. [CrossRef] [PubMed]
6. Moser, A.B.; Jones, D.S.; Raymond, G.V.; Moser, H.W. Plasma and red blood cell fatty acids in peroxisomal disorders. *Neurochem. Res.* **1999**, *24*, 187–197. [CrossRef] [PubMed]
7. Moser, A.B.; Kreiter, N.; Bezman, L.; Lu, S.; Raymond, G.V.; Naidu, S.; Moser, H.W. Plasma very long chain fatty acids in 3000 peroxisome disease patients and 29,000 controls. *Ann. Neurol.* **1999**, *45*, 100–110. [CrossRef]
8. Yeh, Y.Y. Long chain fatty acid deficits in brain myelin sphingolipids of undernourished rat pups. *Lipids* **1988**, *23*, 1114–1118. [CrossRef]
9. Ntoumani, E.; Strandvik, B.; Sabel, K.G. Nervonic acid is much lower in donor milk than in milk from mothers delivering premature infants-Of neglected importance? *Prostaglandins Leukot. Essent. Fat. Acids* **2013**, *89*, 241–244. [CrossRef]
10. Wu, K.; Gao, R.; Tian, F.; Mao, Y.; Wang, B.; Zhou, L.; Shen, L.; Guan, Y.; Cai, M. Fatty acid positional distribution (sn-2 fatty acids) and phospholipid composition in Chinese breast milk from colostrum to mature stage. *Br. J. Nutr.* **2018**, *121*, 65–73. [CrossRef]
11. Nyuar, K.B.; Min, Y.; Ghebremeskel, K.; Khalil, A.K.H.; Elbashir, M.I.; Cawford, M.A. Milk of northern Sudanese mothers whose traditional diet is high in carbohydrate contains low docosahexaenoic acid. *Acta Pãediatr.* **2010**, *99*, 1824. [CrossRef]
12. Boersma, E.R.; Offringa, P.J.; Muskiet, F.A.; Chase, W.M.; Simmons, I.J. Vitamin E, lipid fractions, and fatty acid composition of colostrum, transitional milk, and mature milk: An international comparative study. *Am. J. Clin. Nutr.* **1991**, *53*, 1197–1204. [CrossRef]
13. Thakkar, S.K.; De Castro, C.A.; Beauport, L.; Tolsa, J.F.; Fumeaux, C.J.F.; Affolter, M.; Giuffrida, F. Temporal Progression of Fatty Acids in Preterm and Term Human Milk of Mothers from Switzerland. *Nutrients* **2019**, *11*, 12. [CrossRef]
14. Zhao, P.; Zhang, S.W.; Liu, L.; Pang, X.Y.; Yang, Y.; Lu, J.; Lv, J.P. Differences in the Triacylglycerol and Fatty Acid Compositions of Human Colostrum and Mature Milk. *J. Agric. Food Chem.* **2018**, *66*, 4571–4579. [CrossRef]
15. Jamieson, E.C.; Farquharson, J.; Logan, R.W.; Howatson, A.G.; Patrick, W.J.A.; Weaver, L.T.; Cockburn, F. Infant cerebellar gray and white matter fatty acids in relation to age and diet. *Lipids* **1999**, *34*, 1065–1071. [CrossRef]
16. Nagae, L.M.; Hoon, A.H.; Stashinko, E.; Lin, D.; Zhang, W.; Levey, E.; Wakana, S.; Jiang, H.; Leite, C.C.; Lucato, L.T. Diffusion tensor imaging in children with periventricular leukomalacia: Variability of injuries to white matter tracts. *Ajnr Am. J. Neuroradiol.* **2007**, *28*, 1213. [CrossRef]
17. De Vries, L.S.; Van Haastert, I.L.C.; Rademaker, K.J.; Corine, K.; Floris, G. Ultrasound abnormalities preceding cerebral palsy in high-risk preterm infants. *J. Pediatr.* **2004**, *144*, 815–820. [CrossRef]
18. Liu, Z.; Rochfort, S.; Cocks, B. Milk lipidomics: What we know and what we don't. *Prog. Lipid Res.* **2018**, *71*, 70–85. [CrossRef]
19. Hua, M.C.; Su, H.M.; Kuo, M.L.; Chen, C.C.; Yao, T.C.; Tsai, M.H.; Liao, S.L.; Lai, S.H.; Chiu, C.Y.; Su, K.W.; et al. Association of maternal allergy with human milk soluble CD14 and fatty acids, and early childhood atopic dermatitis. *Pediatr. Allergy Immunol.* **2019**, *30*, 204–213. [CrossRef]
20. Rueda, R.; Ramírez, M.; Garcíasalmerón, J.L.; Maldonado, J.; Gil, A. Gestational Age and Origin of Human Milk Influence Total Lipid and Fatty Acid Contents. *Ann. Nutr. Metab.* **1998**, *42*, 12–22. [CrossRef]
21. Salavila, A.; Castellote, A.I.; Rodriguezpalmero, M.; Campoy, C.; Lópezsabater, M.C. Lipid composition in human breast milk from Granada (Spain): Changes during lactation. *Nutrition* **2005**, *21*, 467–473. [CrossRef]

22. Zou, L.; Pande, G.; Akoh, C.C. Infant Formula Fat Analogs and Human Milk Fat: New Focus on Infant Developmental Needs. In *Annual Review of Food Science and Technology*; Doyle, M.P., Klaenhammer, T.R., Eds.; Annual Reviews: Palo Alto, CA, USA, 2016; Volume 7, pp. 139–165.
23. Billeaud, C.; Boue-Vaysse, C.; Couedelo, L.; Steenhout, P.; Jaeger, J.; Cruz-Hernandez, C.; Ameye, L.; Rigo, J.; Picaud, J.C.; Saliba, E.; et al. Effects on Fatty Acid Metabolism of a New Powdered Human Milk Fortifier Containing Medium-Chain Triacylglycerols and Docosahexaenoic Acid in Preterm Infants. *Nutrients* **2018**, *10*, 690. [CrossRef]
24. Dingess, K.; Valentine, C.; Davidson, B.; Peng, Y.; Guerrero, M.; Ruiz-Palacios, G.; Brenna, J.; McMahon, R.; Morrow, A. Docosahexaenoic acid, nervonic acid and iso-20 (BCFA) concentrations in human milk from the Global Exploration of Human Milk Project. *FASEB J.* **2014**, *28*, 623.15.
25. Sala-Vila, A.; Castellote, A.I.; Campoy, C.M.; Rodriguez-Palmero, M. The source of long-chain PUFA in formula supplements does not affect the fatty acid composition of plasma lipids in full-term infants. *J. Nutr.* **2004**, *134*, 868–873. [CrossRef]
26. Jensen, R.G. *The Lipids of Human Milk*; CRC Press: Boca Raton, FL, USA, 1989.
27. Qi, C.; Sun, J.; Xia, Y.; Yu, R.Q.; Wei, W.; Xiang, J.Y.; Jin, Q.Z.; Xiao, H.; Wang, X.G. Fatty Acid Profile and the sn-2 Position Distribution in Triacylglycerols of Breast Milk during Different Lactation Stages. *J. Agric. Food Chem.* **2018**, *66*, 3118–3126. [CrossRef]
28. The United States Pharmacopeia. Validation of Compendial Procedures. In *The United States Pharmacopeia, the National. Formulary, Chapter 1225*; Stanford University: Stanford, CA, USA, 2009.
29. Quintela, M.; Báguena, J.; Gotor, G.; Blanco, M.J.; Broto, F. Estimation of the uncertainty associated with the results based on the validation of chromatographic analysis procedures: Application to the determination of chlorides by high performance liquid chromatography and of fatty acids by high resolution gas chro. *J. Chromatogr. A* **2012**, *1223*, 107–117. [CrossRef]
30. Taverniers, I.; De Loose, M.; Van Bockstaele, E. Trends in quality in the analytical laboratory. II. Analytical method validation and quality assurance. *Trac Trends Anal. Chem.* **2004**, *23*, 535–552. [CrossRef]
31. IOFI Working Group on Methods of Analysis. Guidelines for the quantitative gas chromatography of volatile flavouring substances, from the Working Group on Methods of Analysis of the International Organization of the Flavor Industry (IOFI). *Flavour Fragr. J.* **2011**, *26*, 297–299.
32. Gravador, R.S.; Harrison, S.M.; Monahan, F.J.; Gkarane, V.; Farmer, L.J.; Brunton, N.P. Validation of a Rapid Microwave-Assisted Extraction Method and GC-FID Quantification of Total Branched Chain Fatty Acids in Lamb Subcutaneous Adipose Tissue. *J. Food Sci.* **2019**, *84*, 80–85. [CrossRef]
33. AOAC. *AOAC Peer-Verified Methods Program, Manual on Policies and Procedures*; The Association of Official Analytical Chemists: Arlington, VA, USA, 1998.
34. Koletzko, B.; Baker, S.; Cleghorn, G.; Neto, U.F.; Gopalan, S.; Hernell, O.; Hock, Q.S.; Jirapinyo, P.; Lonnerdal, B.; Pencharz, P.; et al. Global standard for the composition of infant formula: Recommendations of an ESPGHAN Coordinated International Expert Group. *J. Pediatr. Gastroenterol. Nutr.* **2005**, *41*, 584–599. [CrossRef]
35. Xiang, M.; Harbige, L.S.; Zetterstrā, M.R. Long-chain polyunsaturated fatty acids in Chinese and Swedish mothers: Diet, breast milk and infant growth. *Acta Pædiatr.* **2005**, *94*, 1543–1549. [CrossRef]
36. Rydlewski, A.A.; Silva, P.D.; Manin, L.P.; Tavares, C.B.G.; Paula, M.G.; Figueiredo, I.L.; Neia, V.; Santos, O.O.; Visentainer, J.V. Lipid Profile Determination by Direct Infusion ESI-MS and Fatty Acid Composition by GC-FID in Human Milk Pools by Folch and Creamatocrit Methods. *J. Braz. Chem. Soc.* **2019**, *30*, 1063–1073. [CrossRef]
37. Golfetto, I.; Mcgready, R.; Ghebremeskel, K.; Min, Y.; Dubowitz, L.; Nosten, F.; Drury, P.; Simpson, J.A.; Arunjerdja, R.; Crawford, M.A. Fatty acid composition of milk of refugee Karen and urban Korean mothers. Is the level of DHA in breast milk of Western women compromised by high intake of saturated fat and linoleic acid? *Nutr. Health* **2007**, *18*, 319. [CrossRef]
38. Yakes, E.A.; Arsenault, J.E.; Munirul, I.M.; Hossain, M.B.; Ahmed, T.; Bruce, G.J.; Gillies, L.A.; Shafiqur, R.A.; Drake, C.; Jamil, K.M. Intakes and breast-milk concentrations of essential fatty acids are low among Bangladeshi women with 24–48-month-old children. *Br. J. Nutr.* **2011**, *105*, 1660–1670. [CrossRef]
39. Peng, Y.; Zhou, T.; Wang, Q.; Liu, P.; Zhang, T.; Zetterström, R.; Strandvik, B. Fatty acid composition of diet, cord blood and breast milk in Chinese mothers with different dietary habits. *Prostagland. Leukot. Essent. Fat. Acids* **2009**, *81*, 325–330. [CrossRef]

40. Li, J.; Fan, Y.W.; Zhang, Z.W.; Yu, H.; An, Y.; Kramer, J.K.G.; Deng, Z.Y. Evaluating the trans fatty acid, CLA, PUFA and erucic acid diversity in human milk from five regions in China. *Lipids* **2009**, *44*, 257. [CrossRef]
41. Rocquelin, G.; Tapsoba, S.; Dop, M.C.; Mbemba, F.; Traissac, P.; Martinprével, Y. Lipid content and essential fatty acid (EFA) composition of mature Congolese breast milk are influenced by mothers' nutritional status: Impact on infants' EFA supply. *Eur. J. Clin. Nutr.* **1998**, *52*, 164–171. [CrossRef]
42. van Beusekom, C.M.; Nijeboer, H.J.; Van der Veere, C.N.; Luteyn, A.J.; Offringa, P.J.; Muskiet, F.A.; Boersma, E.R. Indicators of long chain polyunsaturated fatty acid status of exclusively breastfed infants at delivery and after 20–22 days. *Early Hum. Dev.* **1993**, *32*, 207. [CrossRef]
43. Van, B.C.; Martini, I.A.; Rutgers, H.M.; Boersma, E.R.; Muskiet, F.A. A carbohydrate-rich diet not only leads to incorporation of medium-chain fatty acids (6:0–14:0) in milk triglycerides but also in each milk-phospholipid subclass. *Am. J. Clin. Nutr.* **1990**, *52*, 326.
44. Nishimura, R.Y.; Barbieiri, P.; Castro, G.S.; Jordão, A.A., Jr.; Perdoná, G.S.; Sartorelli, D.S. Dietary polyunsaturated fatty acid intake during late pregnancy affects fatty acid composition of mature breast milk. *Nutrition* **2014**, *30*, 685. [CrossRef]
45. Bettger, W.J.; DiMichelle-Ranalli, E.; Dillingham, B.; Blackadar, C.B. Nervonic acid is transferred from the maternal diet to milk and tissues of suckling rat pups. *J. Nutr. Biochem.* **2003**, *14*, 160–165. [CrossRef]
46. Bettger, W.J.; Blackadar, C.B. Dietary very long chain fatty acids directly influence the ratio of tetracosenoic (24:1) to tetracosanoic (24:0) acids of sphingomyelin in rat liver. *Lipids* **1997**, *32*, 51–55. [CrossRef]
47. Fulco, A.J.; Mead, J.F. The biosynthesis of lignoceric, cerebronic, and nervonic acids. *J. Biol. Chem.* **1961**, *236*, 2416–2420.
48. Lauritzen, L.; Jorgensen, M.H.; Hansen, H.S.; Michaelsen, K.F. Fluctuations in human milk long-chain PUFA levels in relation to dietary fish intake. *Lipids* **2002**, *37*, 237–244. [CrossRef]
49. Argov-Argaman, N.; Mandel, D.; Lubetzky, R.; Kedem, M.H.; Cohen, B.C.; Berkovitz, Z.; Reifen, R. Human milk fatty acids composition is affected by maternal age. *J. Matern. Fetal Neonatal Med.* **2017**, *30*, 34–37. [CrossRef]
50. Marín, M.C.; Sanjurjo, A.; Rodrigo, M.A.; de Alaniz, M.J. Long-chain polyunsaturated fatty acids in breast milk in La Plata, Argentina: Relationship with maternal nutritional status. *Prostagland. Leukot. Essent. Fat. Acids* **2005**, *73*, 355–360. [CrossRef]
51. Muskiet, F.A.; Hutter, N.H.; Martini, I.A.; Jonxis, J.H.; Offringa, P.J.; Boersma, E.R. Comparison of the fatty acid composition of human milk from mothers in Tanzania, Curacao and Surinam. *Hum. Nutr. Clin. Nutr.* **1987**, *41*, 149–159.
52. Molto-Puigmarti, C.; Plat, J.; Mensink, R.P.; Muller, A.; Jansen, E.; Zeegers, M.P.; Thijs, C. FADS1 FADS2 gene variants modify the association between fish intake and the docosahexaenoic acid proportions in human milk. *Am. J. Clin. Nutr.* **2010**, *91*, 1368–1376. [CrossRef]
53. Weiss, G.A.; Troxler, H.; Klinke, G.; Rogler, D.; Braegger, C.; Hersberger, M. High levels of anti-inflammatory and pro-resolving lipid mediators lipoxins and resolvins and declining docosahexaenoic acid levels in human milk during the first month of lactation. *Lipids Health Dis.* **2013**, *12*, 89. [CrossRef]
54. Lópezlópez, A.; Lópezsabater, M.C.; Campoyfolgoso, C.; Riverourgell, M.; Castellotebargalló, A.I. Fatty acid and sn-2 fatty acid composition in human milk from Granada (Spain) and in infant formulas. *Eur. J. Clin. Nutr.* **2002**, *56*, 1242–1254. [CrossRef]
55. Salem, N.; Wegher, B.; Mena, P.; Uauy, R. Arachidonic and docosahexaenoic acids are biosynthesized from their 18-carbon precursors in human infants. *Proc. Natl. Acad. Sci. USA* **1996**, *93*, 49–54. [CrossRef]
56. Gibson, R.A.; Makrides, M.; Neumann, M.A.; Simmer, K.; Mantzioris, E.; James, M.J. Ratios of linoleic acid to alpha-linolenic acid in formulas for term infants. *J. Pediatr.* **1994**, *125*, 48–55. [CrossRef]
57. Martinez, M. Tissue levels of polyunsaturated fatty acids during early human development. *J. Pediatr.* **1992**, *120*, S129. [CrossRef]
58. Uzman, L.L.; Rumley, M.K. Changes in the composition of the developing mouse brain during early myelination. *J. Neurochem.* **2010**, *3*, 170–184. [CrossRef]
59. Fan, Y.; Meng, H.-M.; Hu, G.-R.; Li, F. Biosynthesis of nervonic acid and perspectives for its production by microalgae and other microorganisms. *Appl. Microbiol. Biotechnol.* **2018**, *102*, 3027–3035. [CrossRef]
60. Wang, X.Y.; Fan, J.S.; Wang, S.Q. Development situation and outlook of nervonic acid plants in China. *China Oils Fats* **2006**, *31*, 69–71.

61. Yiming, G.; Elzbieta, M.; Tammy, F.; Vesna, K.; Brost, J.M.; Michael, G.; Barton, D.L.; Taylor, D.C. Increase in nervonic acid content in transformed yeast and transgenic plants by introduction of a *Lunaria annua* L. 3-ketoacyl-CoA synthase (KCS) gene. *Plant Mol. Biol.* **2009**, *69*, 565–575.
62. Das, S.; Roscoe, T.J.; Delseny, M.; Srivastava, P.S.; Lakshmikumaran, M. Cloning and molecular characterization of the Fatty Acid Elongase 1 (FAE 1) gene from high and low erucic acid lines of *Brassica campestris* and *Brassica oleracea*. *Plant Sci.* **2002**, *162*, 245–250. [CrossRef]

© 2019 by the authors. Licensee MDPI, Basel, Switzerland. This article is an open access article distributed under the terms and conditions of the Creative Commons Attribution (CC BY) license (http://creativecommons.org/licenses/by/4.0/).

Article

Longitudinal Changes of Mineral Concentrations in Preterm and Term Human Milk from Lactating Swiss Women

Magalie Sabatier [1,*], Clara L. Garcia-Rodenas [1], Carlos A. De Castro [2], Peter Kastenmayer [1], Mario Vigo [1], Stéphane Dubascoux [1], Daniel Andrey [1], Marine Nicolas [1], Janique Richoz Payot [1], Valentine Bordier [1], Sagar K. Thakkar [2], Lydie Beauport [3], Jean-François Tolsa [3], Céline J. Fischer Fumeaux [3] and Michael Affolter [1]

1. Nestlé Research, Société des Produits Nestlé SA, 1000 Lausanne, Switzerland
2. Nestle Research Singapore Hub, Société des Produits Nestlé SA, 29 Quality Road, Singapore 618802, Singapore
3. Clinic of Neonatology, Department Woman Mother Child, University Hospital of Lausanne, 1011 Lausanne, Switzerland
* Correspondence: magalie.sabatier@rdls.nestle.com; Tel.: +41-21-784-83-42

Received: 15 July 2019; Accepted: 6 August 2019; Published: 9 August 2019

Abstract: An adequate mineral supply to preterm infants is essential for normal growth and development. This study aimed to compare the mineral contents of human milk (HM) from healthy mothers of preterm (28–32 weeks) and full term (>37 weeks) infants. Samples were collected weekly for eight weeks for the term group ($n = 34$) and, biweekly up to 16 weeks for the preterm group ($n = 27$). Iron, zinc, selenium, copper, iodine, calcium, magnesium, phosphorus, potassium, and sodium were quantitatively analyzed by Inductively Coupled Plasma-Mass Spectrometry. The mineral contents of both HM showed parallel compositional changes over the period of lactation, with occasional significant differences when compared at the same postpartum age. However, when the comparisons were performed at an equivalent postmenstrual age, preterm HM contained less zinc and copper from week 39 to 48 ($p < 0.002$) and less selenium from week 39 to 44 ($p < 0.002$) than term HM. This translates into ranges of differences (min–max) of 53% to 78%, 30% to 72%, and 11% to 33% lower for zinc, copper, and selenium, respectively. These data provide comprehensive information on the temporal changes of ten minerals in preterm HM and may help to increase the accuracy of the mineral fortification of milk for preterm consumption.

Keywords: human milk; term; preterm; calcium; phosphorus; magnesium; zinc; iron; copper; iodine; selenium; potassium; sodium

1. Introduction

An estimated 15 million babies are born preterm (i.e., born before 37 weeks of gestation) worldwide every year, with the incidence ranging from 5% to 18% of births according to the area [1]. Very preterm infants (i.e., born before 32 weeks of gestation) represent a highly vulnerable population with complications that can lead to growth failure, lifetime disability including learning difficulties, visual and hearing impairments, or even death. This population requires a high level of medical care, including nutritional support, to offset the risk of nutritional deficiencies and associated adverse health outcomes. Indeed, term infants are generally born with adequate stores of minerals, the accrual of which occurs during the third trimester of pregnancy. A consequence of being born before term is a high risk of mineral shortfalls due to the limited stores, but also to the rapid postnatal growth, immature gastrointestinal tracts, high endogenous losses, and variable intakes [2]. Due to their essentiality, minerals need to be provided at a level to meet the in utero accretion rates to favor catch-up growth and to not compromise long-term neuronal and cognitive development [3]. The essential minerals or trace elements include iron, zinc, copper, iodine, selenium, calcium, phosphorus, and magnesium, but also potassium and sodium (non-exhaustive list). Among other physiological roles in humans, iron, zinc, copper, and iodine are essential for optimal brain development [4]; calcium, phosphorus, and magnesium are required for adequate bone health [5]; selenium, zinc, and iodine play a crucial role in thyroid hormone metabolism [2]; and potassium and sodium are key electrolytes involved in water fluxes and enzymatic functions [6]. Breastfeeding is the gold standard source of nutrition for term infants and is also highly recommended for preterm infants, as it is protective against several complications of preterm birth [7,8]. However, because of the risk of inadequate nutrient intakes, the fortification of mothers' own milk or donor human milk (HM) is generally prescribed for very preterm infants to enhance the protein, energy, and micronutrients intake [7–9]. Fortification is often recommended to be continued until a postconceptional age of 40 weeks, but possibly until about 52 weeks if infants are discharged with a suboptimal weight for a postconceptional age [3]. Nevertheless, there is no consensus on nutrition after discharge from the neonatal intensive care unit (NICU) [10]. However, it may be assumed that to support adequate growth and development, breastfed preterm infants should have, at the minimum, a similar mineral intake to breastfed term infants of an equivalent developmental stage [11].

Optimal HM mineral fortification during an NICU stay and, potentially, after discharge, requires a precise understanding of HM mineral content and its potential sources of variability. Although decades of research have been allocated to HM mineral concentrations, most of the research has been performed on HM of mothers delivering at term, and for some elements, there are a limited number of reports on preterm HM. When available, comparisons between term HM and preterm HM have been performed at an equivalent postpartum age (i.e., lactation stage), but, to our knowledge, there is no evaluation of differences at the same postmenstrual age (i.e., infant developmental stage). As concentrations of several minerals and trace elements in milk change over lactation [12], the assumption was that the comparison by postmenstrual age could reveal unexplored differences between both groups of HM in the period that surrounds preterm infant discharge from NICU. Therefore, the objectives of the present study were (i) to precisely describe the concentration of zinc, copper, iodine, selenium, iron, calcium, phosphorus, magnesium, potassium, and sodium in HM from mothers of preterm (28–32 weeks of gestation) infants over a period of 16 weeks; (ii) to compare preterm HM mineral content with that of milk from mothers of full term (>37 weeks of gestation) infants at equivalent infant postpartum and postmenstrual ages.

2. Materials and Methods

2.1. Subjects

The aim of the study was to characterize the nutritional composition of preterm and term HM. The study design (monocentric, prospective cohort) and characteristics of the subjects were already reported elsewhere [13], with the results on longitudinal changes for other nutrients (i.e., macronutrients [14], proteins [13], lipids [15], and human milk oligosaccharide [16]). In brief, the study was conducted at the University Hospital in Lausanne (CHUV), Switzerland, between October 2013 and July 2014. The study included women older than 18 years of age and intending to breastfeed their infants for a minimum of 4 months. The preterm group included infants with a gestational age from 28 0/7 to 32 6/7 weeks and the full-term group referred to infants with a gestational age of 37 0/7 to 41 6/7 weeks. The exclusion criteria included any counter-indication to breastfeeding, the mother having been diagnosed with diabetes (type I or II) before pregnancy, alcohol or drugs consumption during pregnancy, and/or mothers having insufficient French language skills to follow the study guidelines.

Mothers who participated in the study were followed until postpartum week 16 for the preterm group, and week 8 for the term group, or until lactation discontinuation (whatever came first), by a dedicated research nurse who was qualified as a lactation consultant (International Board of Lactation Consultant, IBCLC).

The study protocol was established following the Declaration of Helsinki's guidelines and was approved by the local Ethical Committee (Commission cantonale d'éthique de la recherche sur l'être humain du Canton de Vaud, Switzerland; Protocol 69/13, clinical study 11.39.NRC) on April 9, 2013. All the subjects participating in the study signed an inform consent before the enrollment. Registration of this trial was completed at ClinicalTrials.gov under NCT02052245.

2.2. Data Collection

Neonatal demographic and delivery data were prospectively collected and recorded in electronic case report forms. The following infant characteristics were collected: single or multiple gestation, mode of delivery, sex, weight, length, head circumference, and gestational age at birth. Birth weight was monitored with an electronic scale accurate to the nearest 5 g, crown-heel length with a height gauge, and head circumference with a tape. An electronic scale was used to measure the maternal weight at delivery, whereas the maternal age, height, and weight before pregnancy were self-reported and completed during face-to-face interviews between mothers and the research nurse, and were verified by one of the two referent pediatricians (LB, CJFF).

2.3. Human Milk Sampling and Processing

A schematic representation of the HM sampling and processing has already been published previously [16]. Preterm HM samples were collected every 7 days ± 1 day during the first 8 weeks and then every 14 days ± 1 day during the following 8 weeks, which yielded a total of 12 samples during the 16 weeks. For term HM, a total of 8 samples were collected every 7 days ± 1 day for 8 weeks.

Full HM expression from a single breast was performed between 6 am and noon (i.e., corresponding to first morning expression) using an electric breast pump (Symphony®, Medela, 6340 Baar, Switzerland). After full milk collection, the milk was homogenized. A maximum volume of 10 mL milk was reserved for analysis. Lower volumes were collected for the two first time points in the preterm group (i.e., max 5 mL). The remaining HM volumes were provided to the infant for feeding. Samples of HM for analysis were transferred by the mothers into 15 mL polypropylene tubes (Falcon™, Fisher Scientific, Reinach, Switzerland), previously labelled with the subject number and collection information, and stored at −18 °C in the home freezer until being transferred to the hospital. Samples were temporarily kept at −80 °C at the hospital before shipment to the Nestlé Research Centre (Lausanne, Switzerland). To avoid multiple thawing/freezing cycles, HM samples were thawed once for splitting into 15 aliquots before storage at −80 °C until analysis.

2.4. Mineral Quantification

Quantification of minerals was realized using Inductively Coupled Plasma Mass Spectrometry (ICP-MS, Nexion 300-D, PerkinElmer, SCIEX, Norwalk, CT, USA). For sodium, magnesium, phosphorous, potassium, calcium, iron, copper, zinc, and selenium, 0.7 mL of human breast milk was transferred into perfluoroalkoxy alkane vessels and mineralized in a MARS XPress microwave digestion system (CEM® Corp., Matthews, NC, USA) using HNO_3/H_2O_2. Mineralized samples were transferred to PE tubes, diluted with Milli Q water and germanium and tellurium were added as internal standards. Quantification was realized by ICP-MS using helium or CH_4 as a collision or reaction gas. For iodine, 1 mL of human breast milk was treated with 1 mL tetramethylammonium hydroxide (25%) and 4.5 mL of Milli Q water in a drying oven at 90 °C for 3 h. Sample solutions were diluted to 15 mL and centrifuged at 1730 g for 15 min. Quantification was realized using ICP-MS in normal mode using germanium as the internal standard. Certified Reference Materials (CRM) were added to all analytical series to control the quality of the quantification.

2.5. Statistics

Statistical analyses were done using SAS (version 9.3, SAS Institute Inc., Cary, NC, USA, 2013) and R (version 3.2.1, R Foundation for Statistical Computing, Vienna, Austria). The longitudinal changes of mineral contents were compared in preterm HM and term HM at equivalent infant (1) postpartum ages and (2) postmenstrual ages. Mixed linear models were used for both comparisons to estimate the differences between the groups. The models used age (either postpartum or postmenstrual), term/preterm status, interaction between age and term/preterm status, sex of infant, and mode of delivery as fixed effects as follows: concentration = visit × term_status + child_sex + delivery + twin_type. There was only one case of mixed sex twins out of six "sets". Since the milk samples analyses were conducted according to the mothers and not the infants, the statistical models used the mother as an individual and the sex of the infant was chosen as the sex of the first child (for twins). This entails that one mother is represented as having a male infant where indeed she has both (because of twins). Within-subject variability was accounted for by declaring the subject ID as a random effect. Contrast estimates of the model were calculated by comparing preterm and term HM groups at each time point. No imputation method was applied for missing data (both in between visits and loss to follow up) as the method used is adapted to handle incomplete data. A conventional two-sided test with a 5% error rate was used without adjusting for multiplicity. Mixed linear models were produced using the package nlme and contrast estimates were made with contrast package. R graphs were created using the package ggplot2.

3. Results

3.1. General Characteristics of the Subjects

The subjects' characteristics were reported elsewhere [13,16] and are summarized in Table 1. In brief, 27 mothers having delivered 33 preterm infants and 34 mothers having delivered 34 term infants were included. In the respective groups, two (7.4% for preterm) and six (17.4% for term) mothers dropped out the study. In total, 473 HM samples, 257 from preterm and 216 from full-term infant mothers, were available for mineral analyses. Except for the rate of Caesarean delivery being higher in the preterm mothers than in the term ones, none of the baseline characteristics were significantly different between the two groups of mothers. Regarding the infants' characteristics, thirty-six percent of the preterm births were twins. As expected, the weight, height, and head circumference were statistically significantly lower in the preterm group than in the term group, but not the sex distribution.

Table 1. General characteristics of the two groups of mothers and infants.

Study Population	Preterm	Term	p-Value *
Mothers (drop outs)	N = 27 (2)	N = 34 (2)	
Age (years)	32.4 ± 5.6	31.2 ± 4.2	0.3173
BMI before pregnancy (Kg/m^2)	22.8 ± 3.3	23.2 ± 4.9	0.6990
BMI at child birth (Kg/m^2)	25.8 ± 3.7	26.9 ± 4.7	0.3141
Delivery type: caesarean (%)	63	23.5	0.0019
Infants	N = 33	N = 34	
Gestational age at birth (weeks)	30.8 ± 1.4	39.5 ± 1.0	<0.0001
Males (%)	54.5	52.9	0.8952
Twins (%)	36.4	0.0	0.0001
Height (cm)	40.4 ± 3.2	49.4 ± 1.7	<0.0001
Weight (g)	1421.4 ± 372.8	3277.6 ± 353.6	<0.0001
Head circumference (cm)	27.8 ± 2.1	34.4 ± 1.5	<0.0001

* t-test and Fisher test of proportions were used for the comparison of continuous and discrete variables, respectively.

3.2. Mineral Concentration and Longitudinal Change in Preterm and Term Human Milk at the Same Postpartum Age

A summary of the results i.e., mean ± SD and median (min-max) for each mineral, is provided in Table 2 along with the range of values reported in the literature. The most concentrated element in both HMs was found to be potassium, followed by, in order of decreasing concentrations, calcium, sodium, phosphorus, magnesium, zinc, iron, copper, iodine, and selenium. The longitudinal concentrations (mean ± SD) of minerals in preterm and term HM per visit at equivalent postpartum ages are reported in Supplementary Table S1. As depicted in Figure 1, in preterm and term HM, the changes in concentration over the 4-month period of sample collection converge into three main patterns. First, the mineral concentrations were observed to increase during the first week postpartum and then to decline during the lactation, like for phosphorus and copper. Secondly, the concentrations were highest in the first week and then either decreased quickly (i.e., zinc, potassium), or gradually (i.e., calcium, selenium, sodium) over the lactation, or remained stable in mature milk (i.e., iodine, iron). Finally, for magnesium, the concentrations remained fairly stable during the course of lactation.

Table 2. Summary of mineral concentrations (i.e., mean ± SD and median (min-max)) in preterm (from sample collection from postpartum week 1 to 16) and term human milk (as collected over 8 weeks postpartum). Retrieved published ranges of mineral concentration in human milk * are also provided for comparison.

Minerals	Statistic	Preterm Milk	Term Milk	Literature Range
(mg/L)				
Potassium	Mean	578 ± 107	575 ± 92	515$ [17], 688$ [18]
	Median	569 (209–907)	562 (308–908)	-
Calcium	Mean	281 ± 51	286 ± 47	-
	Median	282 (145–459)	287 (136–433)	252 (84–462)# [19]
Sodium	Mean	205 ± 177	235 ± 237	135–371# [20]
	Median	160 (54–1577)	170 (84–1969)	-
Phosphorus	Mean	145 ± 32	148 ± 30	-
	Median	145 (48–225)	146 (45–257)	143 (17–278)# [19]
Magnesium	Mean	37 ± 9	32 ± 7	-
	Median	36 (12–70)	33 (17–53)	31 (15–64)# [21]
Zinc	Mean	2.4 ± 1.7	3.2 ± 1.9	2.2–2.5; 2.9–3.9; 1.7–5.3# [22]
	Median	2.1 (0.2–15.5)	2.7 (0.4–11.9)	-
Iron	Mean	0.36 ± 0.23	0.44 ± 0.26	-
	Median	0.32 (0.13–2.78)	0.36 (0.13–1.8)	0.47 (0.04–1.92)# [23]
Copper	Mean	0.36 ± 0.16	0.44 ± 0.15	-
	Median	0.35 (0.06–1.1)	0.42 (0.13–1.0)	0.33 (0.03–2.19)# [23]
(µg/L)				
Iodine	Mean	92 ± 67	87 ± 41	15–150# [20]
	Median	76 (2–422)	76 (18–228)	62 (5.4–2170)# [24]
Selenium	Mean	14.3 ± 4.7	15.0 ± 4.2	11$ [25]–28$ [26]
	Median	13 (4–45)	14 (6–34)	-

* Published median (min-max) value where preferred for reporting. # indicates that values were calculated from term human milk (HM); $ corresponds to data calculated from preterm HM.

No statistical difference was found in mineral concentrations between preterm and term HM at the same postpartum age, except at a few time points. The phosphorus concentration was higher in preterm HM than in term HM at week 1 ($p = 0.024$). The same direction of difference was observed for magnesium at week 4 ($p = 0.027$), week 6 ($p = 0.0357$), and week 7 ($p = 0.047$). Inversely, the sodium and copper concentrations were lower in preterm HM than in term HM at week 2 ($p = 0.032$ and $p = 0.024$, respectively). The calcium to phosphorus mass ratios were found to range between 1.57 and 2.32 in preterm HM, and between 1.84 and 2.22 in term HM, with a statistically significant difference between groups found at week 1 ($p = 0.042$; preterm HM < term HM) and at week 2 ($p = 0.018$, preterm HM < term HM). The zinc to copper molar ratio ranged between 20.6 (at week 1) and 5.2 in preterm HM, and between 29.8 (at week 1) and 5.6 in term HM, with no significant difference between the two HMs.

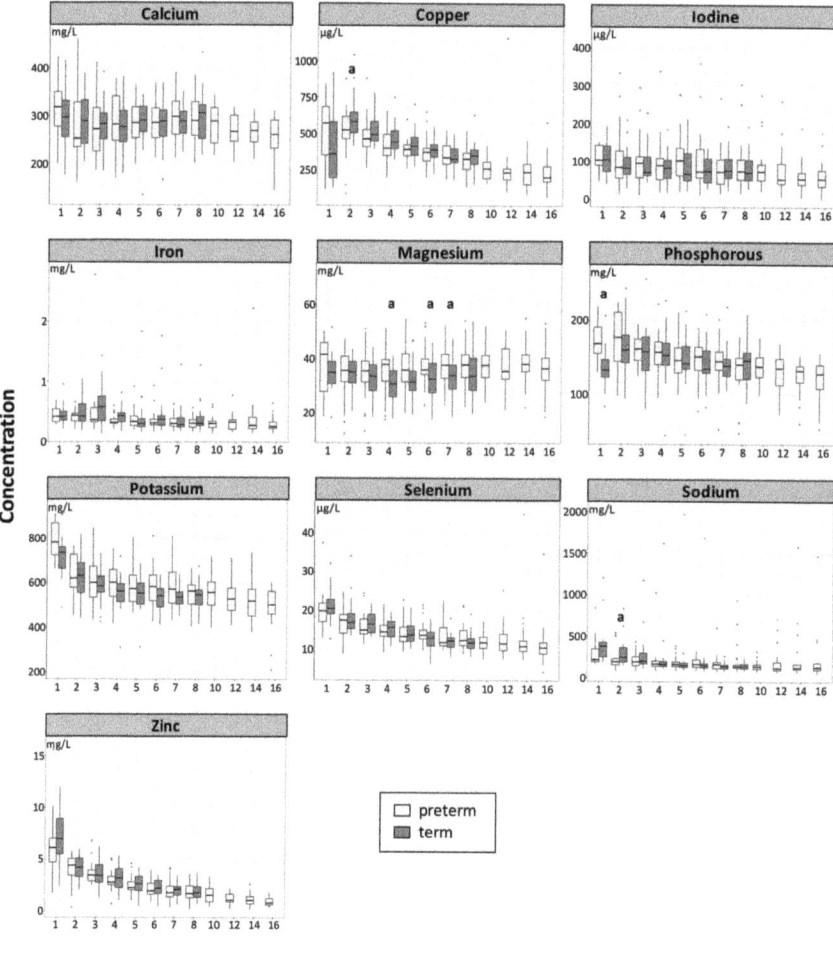

Figure 1. Longitudinal changes of mineral concentration (means) in term (dark grey) and preterm (light grey) human milk (HM) at an equivalent postpartum age. The letter (a) indicates if differences between term and preterm are significant $p < 0.05$. Box plots represent medians with 25th and 75th percentile, min-max range and outliers.

No significant differences were found when comparing mineral concentrations by sex, mode of delivery, or twin/single delivery.

3.3. Longitudinal Changes of Mineral Concentration in Term and Preterm Human Milk at an Equivalent Postmenstrual Age

When the HM from mothers of preterm and full term infants was compared at an equivalent postmenstrual age, the main significant differences were observed for zinc, copper, selenium, and magnesium. As depicted in Figure 2, zinc and copper concentrations in term HM were significantly higher than in preterm HM from week 39 to 48 of postmenstrual age ($p < 0.002$). Similar observations were made for selenium, with significant differences being found at week 39 to 44 ($p < 0.002$). In comparison to term HM, preterm HM contained 53% to 78% (average 62%), 30% to 72% (average 44%), and 11% to 33% (average 24%) less zinc, copper, and selenium, respectively.

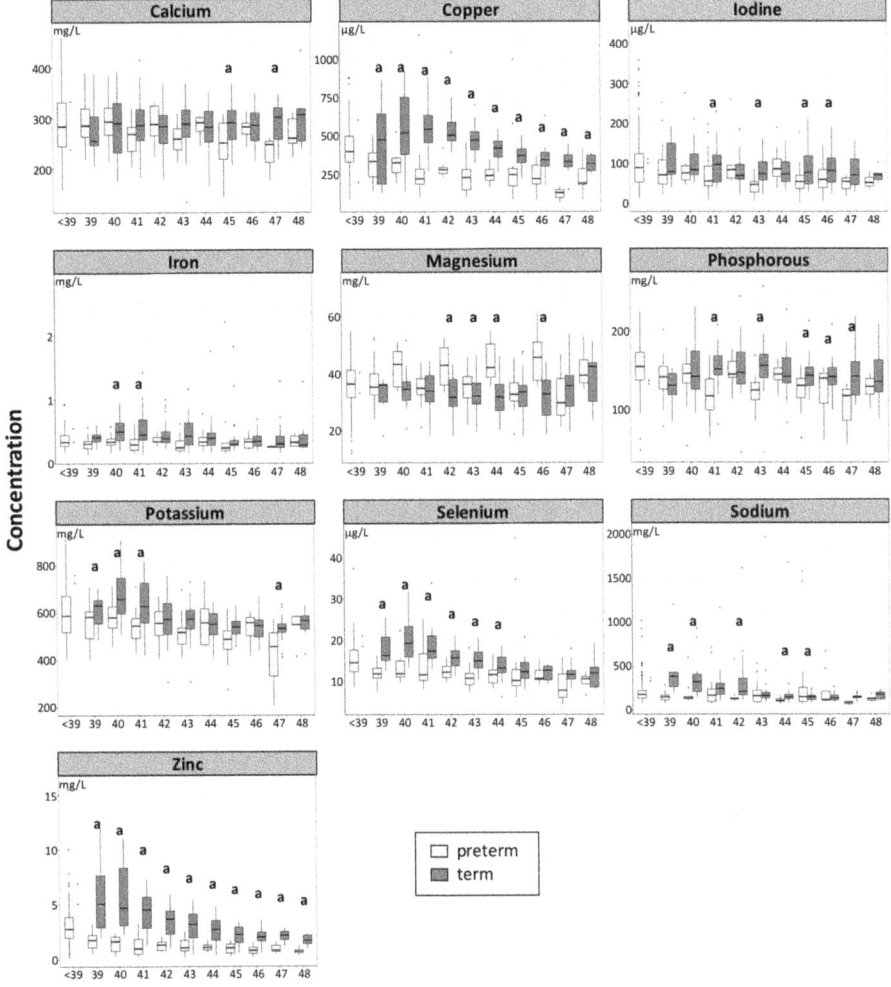

Figure 2. Longitudinal changes of mineral concentration (means) in term (dark grey) and preterm (light grey) human milk (HM) at an equivalent postmenstrual age. The letter (a) indicates if differences between term and preterm are significant $p < 0.05$. Box plots represent medians with 25th and 75th percentile, min-max range and outliers.

Magnesium is the only element for which the concentration in preterm HM was found to be statistically significantly higher than in term HM from weeks 42 to 44 and week 46 ($p < 0.05$). The differences between preterm HM and term HM per postmenstrual age ranged from minus 17% to plus 39% (average + 12%).

Statistically significant differences were also found for sodium, potassium, phosphorus, iron, iodine, and calcium at scattered postmenstrual weeks, with higher values in favor of term HM. The sodium concentration in preterm HM was significantly lower at weeks 39, 40, and 42, and weeks 44 to 45 ($p < 0.01$). The potassium concentration was significantly lower at week 39 to 41 ($p < 0.006$); for phosphorus, at week 41, 43, 45, and 47 ($p < 0.05$); for iodine, at week 41, 43, 45, and 46; for iron, at week 40 and 41 ($p < 0.002$); and for calcium, at week 45 and 47 ($p < 0.05$). The longitudinal concentrations (mean ± SD) of minerals in preterm and term HM per visit at equivalent postmenstrual ages are reported in Supplementary Table S2.

4. Discussion

We have reported a comprehensive longitudinal overview of the content of 10 minerals in 500 HM samples from women delivering before term or at term. The study design enabled the comparison of preterm and term HM mineral contents at the same postpartum weeks (i.e., same lactation stages) and at equivalent postmenstrual weeks (i.e., equivalent infant developmental stages).

4.1. Mineral Concentration and Longitudinal Change in Preterm and Term Human Milk at the Same Postpartum Age

To our knowledge, the results of our study represent the first set of mineral analyses from breast milk in apparently healthy Swiss mothers of preterm and term infants. As illustrated in Table 2, the mean and median concentration determined for each individual element in preterm HM over the period of sample collection was found to be in the range of retrieved published values for HM [17–28]. These ranges can be wide and this variability is mainly due to the stage of lactation and to the intra and intersubject variability, including the circadian rhythm [12,29]. The presented result is in line with previous research performed in HM from a well-nourished population which showed, for several minerals, that their diet, status, or supplementation have a limited or no impact on their milk concentration. That is, for instance, the case for iron, copper, zinc, and magnesium [12], and to some extent, for phosphorus, potassium, and sodium [30–32]. Inversely, the calcium, iodine, and selenium concentration in HM can be impacted by the diet of the mother, and therefore, except for calcium, by the use of dietary supplements [12]. In the present study, the dietary intakes of the mother were not recorded, and no recent survey on either dietary intakes or mineral status in Swiss pregnant or lactating women was retrieved for these three elements. Nevertheless, data on iodine intake in the general Swiss population was classified as sufficient by the Iodine Global Network (http://www.ign.org/scorecard.htm). In our study, this is reflected by an iodine median concentration in milk above the published median value of 62 µg/L [24]. A similar observation is made for calcium, despite the results of a Swiss cross-sectional study showing that in 28% of subjects following an omnivore diet, calcium intake does not reach the corresponding nutritional recommendation [33]. No recent data on the prevalence of intake inadequacy or deficiency was retrieved for selenium, but the rather low levels in milk observed in our study may indicate a need for a higher selenium intake in Swiss lactating women, as underlined in a review published earlier for the Swiss population [34]. This may help to increase the selenium milk content. In any case, even if dietary intakes of the mother do not influence their milk mineral concentration, they should give particular attention to the mineral density of their diet to match their own increased requirements if adaptation of the mechanism of absorption is not enough to avoid putting themselves at risk of mineral deficiencies.

The comparison of concentration at the same postpartum age shows no statistically significant difference between preterm HM and term HM for six minerals, i.e., potassium, zinc, iodine, selenium, iron, and calcium. These data are in agreement with previous published results for potassium [17,18,28,35,36], zinc [22,37,38], iodine [24,39], and selenium [26,40]. Few papers have reported higher levels of iron in preterm HM during the early lactation period [28,41,42], but some others [43,44] and a more recent one [45], failed to observe any differences. In HM, 2% to 9% of iron is linked to lactoferrin [45] and our iron content results correlate with the results of lactoferrin analyzed from the same set of samples [13]. Only minor and scattered differences between preterm HM and term HM have previously been reported for calcium. A recent systematic review and meta-analysis performed with 11 studies showed no differences from day 1 to week 12 postpartum, except at week 7 to 9, where a statistically significant difference of 15% was reported (preterm HM > term HM) [46]. Calcium along with phosphorus are the two main minerals required for the multifaceted mechanisms of bone metabolism [47]. Therefore, the meta-analysis was also performed for phosphorus from eight independent studies, and showed that the phosphorus concentration in preterm HM was not statistically different than in term HM, except between week 3 and 4 and week 5 to 6, with reported differences of 14% and 16%, respectively (preterm HM < term HM) [46]. In our study, a significant difference was only found at week 1 for this mineral in the opposite direction (preterm HM > term HM). We also found a higher concentration in preterm HM than in term HM for magnesium and only at three time points during the transitional phase (i.e., week 1 to 4). These differences were below 20%. In the literature, only one study out of 16 reported a similar difference [21,48]. In addition, we found lower concentrations of copper and sodium in preterm HM than in term HM, but only at postpartum week 2. Like for iron, inconsistent results were retrieved for copper in the literature [23]. Preterm HM copper concentrations were reported to be significantly lower [42,44], higher [41], or no different from the concentrations of term HM [44,49,50]. Except for one study [17], the sodium concentration in preterm HM is consistently reported as higher in preterm HM than in term HM, at the very early stage of lactation [18,51]. Our result goes in an opposite direction, but only at week 2. These discrepancies are most probably due to the difference in the timing of sample collection and comparison that, in most studies, started at day 1 postpartum versus day 7 in the present report. Other analytical reasons may be the differences in the number of samples, the collection method (e.g., foremilk vs. hindmilk as compared to full-breast expression in our study), and the storage procedures [45]. Beside this reading, although it is well-established that the levels of minerals in preterm HM are not sufficient to meet the needs of the preterm infant, it is not unusual to read that when higher concentrations are found in preterm HM than in term HM, it could be due to a physiological adaption to meet the increased requirement of the preterm infants [28,41,42,51]. This explanation goes along with the hypothesis of a reduced blood flow to the mammary gland and thus milk volume, and the immaturity of the gland (i.e., incomplete differentiation of epithelial cells, and the absence of tight junctions between the cells) [52]. However, some authors reported no difference in milk volumes produced by both preterm and term mothers, stressing that understanding of the physiological processes responsible for the micronutrient composition in human milk still requires investigation [17]. Some authors also report that it is likely that these discrepancies are due to differences in hormonal balance and metabolic regulation due to the shorter gestational period [17,53,54].

Overall, in agreement with previous reports, the present results show parallel compositional changes in the mineral content of milk from mothers of preterm and term infants throughout the 8 postpartum weeks of common sample collection. While breastfeeding is the gold standard for term and preterm feeding, for the latter, it is prescribed that human milk should be fortified with nutrients in short supply [7–9]. Despite the punctual differences described above, when comparing the median of values for each individual element to the recommendations by the ESPGHAN [9], it appeared that, except for iodine and potassium, none of the concentrations reached the minimum recommended level [9] (results presented in Supplementary Figure S1). This emphasizes the need for fortifying with only those minerals whose content in HM is lower than the recommended levels, in order to avoid overloads of those minerals that are present in HM at sufficient levels.

4.2. Longitudinal Changes of Mineral Concentration in Term and Preterm Human Milk at an Equivalent Postmenstrual Age

Fortification of human milk is recommended up to preterm infant discharge from the neonatal care unit, to address the mismatch between its content and the nutritional recommendations for optimal growth and development. However, there is a lack of consensus about the need to fortify human milk after discharge (usually occurring at around term corrected age) [10]. Considering that the references for an optimal growth of preterm infants after they reach the term corrected age are calibrated against the standards of infants born at term, we assumed that the preterm infant nutritional requirements during this period would be at least the same as those of their term counterparts. Therefore, the mineral contents of preterm HM were compared to those of term HM at the same postmenstrual ages (i.e., from postmenstrual weeks 39 to 48). This comparison showed highly significant differences for zinc, copper, and selenium, with a lower concentration in preterm HM than in term HM. Zinc and copper contribute to many cellular and molecular processes and therefore are essential for growth, the immune response, and cognitive function [4]. The serum zinc concentration in term and preterm infants is reported to be high at birth and then to progressively decrease, which was correlated with normal growth [55]. Selenium is an important factor for optimal function of the antioxidant systems and its serum concentration has been showed to increase in healthy term breastfed infants after birth [56]. Therefore, the observed differences of concentrations of these three trace elements in preterm HM and term HM at the same postmenstrual age may be physiologically relevant and require further investigation.

4.3. Strengths and Limitations of the Study

This study has some limitations. First, the samples were collected at only one time of the day. As the fluctuation of mineral concentration in milk has been previously observed over the day [29], one may argue that the data are not representative of the whole day concentration. Nevertheless, the sampling time was standardized and it allowed us to perform a comparison of the HM of the two infant populations at a postpartum and postmenstrual age. Finally, the number of subjects was limited in both groups, reducing the power of the study and limiting the number of association analyses.

5. Conclusions

This study represents the first set of simultaneous analyses of ten minerals in human term and preterm milk comprehensively covering the period of 2 months postpartum. It confirms previous observations of concentrations and temporal changes, as well as the global similarity in the composition of term and preterm milk when compared at the same lactation stage. Interestingly, zinc, copper, and selenium contents were found to be consistently lower in preterm milk than in term milk when compared at an equivalent infant developmental stage. The relevance of the observed differences to the health, growth, and development of preterm infants, as well as the impact on fortification practices, remain to be further investigated.

Supplementary Materials: The following are available online at http://www.mdpi.com/2072-6643/11/8/1855/s1: Table S1: Concentration of human milk minerals and trace elements in term or preterm milk at different weeks postpartum; Table S2: Concentration of human milk minerals and trace elements in term or preterm milk at equivalent postmenstrual ages; Figure S1: Longitudinal changes of mineral concentration (means) in term (dark grey) and preterm (light grey) human milk (HM) at an equivalent postmenstrual age, with dotted lines representing the range of recommended intakes (converted in mg/L, using measured energy density of preterm HM), as provided by ESPGHAN [3].

Author Contributions: Conceptualization: C.L.G.-R., M.A., L.B., J.-F.T., and C.J.F.F.; investigation: L.B., J.-F.T., and C.J.F.F.; analysis: ICP-MS: P.K., M.V., S.D., D.A., M.N., and J.R.P.; statistics: C.A.D.C.; results interpretation: M.S., C.L.G.R., M.A., and V.B.; writing—original draft: M.S. and M.A.; writing—review and editing: M.S., C.A.D.C., M.A., C.L.G.-R., S.D.; L.B., J.-F.T., S.K.T., and C.J.F.F.; visualization: C.A.D.C. and M.A.; supervision: C.L.G.-R., M.A., and C.J.F.F.; project administration: M.A.; funding acquisition: C.L.G.-R. and M.A.

Funding: This research was funded by Nestlé Research, Lausanne (Société des Produits Nestlé SA).

Acknowledgments: Sincere thanks to all mothers who enthusiastically participated in this study; to Nassima Grari, the study lactation nurse who admirably cared about all mothers during the full study duration; and to Céline Romagny and Emilie Darcillon, who managed the organizational and operational part of the clinical study and the data management.

Conflicts of Interest: All authors, except L.B., J.-F.T., and C.J.F.F., are Nestlé employees.

References

1. March of Dimes; PMNCH; Save the Children; WHO. *Born Too Soon: The Global Action Report on Preterm Birth*; World Health Organization: Geneva, Switzerland, 2012.
2. Finch, C.W. Review of trace mineral requirements for preterm infants: What are the current recommendations for clinical practice? *Nutr. Clin. Pract.* **2015**, *30*, 44–58. [CrossRef] [PubMed]
3. Aggett, P.J.; Agostoni, C.; Axelsson, I.; De Curtis, M.; Goulet, O.; Hernell, O.; Koletzko, B.; Lafeber, H.N.; Michaelsen, K.F.; Puntis, J.W.; et al. Feeding preterm infants after hospital discharge: A commentary by the espghan committee on nutrition. *J. Pediatr. Gastroenterol. Nutr.* **2006**, *42*, 596–603. [CrossRef] [PubMed]
4. Ramel, S.E.; Georgieff, M.K. Preterm nutrition and the brain. *World Rev. Nutr. Diet.* **2014**, *110*, 190–200. [PubMed]
5. Mimouni, F.B.; Mandel, D.; Lubetzky, R.; Senterre, T. Calcium, phosphorus, magnesium and vitamin d requirements of the preterm infant. *World Rev. Nutr. Diet.* **2014**, *110*, 140–151. [PubMed]
6. Fusch, C.; Jochum, F. Water, sodium, potassium and chloride. *World Rev. Nutr. Diet.* **2014**, *110*, 99–120. [PubMed]
7. Patel, P.; Bhatia, J. Human milk: The prefered firts food for premature infants. *J. Hum. Nutr. Food Sci.* **2016**, *4*.
8. Tudehope, D.I. Human milk and the nutritional needs of preterm infants. *J. Pediatr.* **2013**, *162*, S17–S25. [CrossRef]
9. Agostoni, C.; Buonocore, G.; Carnielli, V.P.; De Curtis, M.; Darmaun, D.; Decsi, T.; Domellof, M.; Embleton, N.D.; Fusch, C.; Genzel-Boroviczeny, O.; et al. Enteral nutrient supply for preterm infants: Commentary from the european society of paediatric gastroenterology, hepatology and nutrition committee on nutrition. *J. Pediatr. Gastroenterol. Nutr.* **2010**, *50*, 85–91. [CrossRef]
10. Arslanoglu, S.; Boquien, C.Y.; King, C.; Lamireau, D.; Tonetto, P.; Barnett, D.; Bertino, E.; Gaya, A.; Gebauer, C.; Grovslien, A.; et al. Fortification of human milk for preterm infants: Update and recommendations of the european milk bank association (emba) working group on human milk fortification. *Front. Pediatr.* **2019**, *7*, 76. [CrossRef]
11. van de Lagemaat, M.; Rotteveel, J.; van Weissenbruch, M.M.; Lafeber, H.N. Increased gain in bone mineral content of preterm infants fed an isocaloric, protein-, and mineral-enriched postdischarge formula. *Eur. J. Nutr.* **2013**, *52*, 1781–1785. [CrossRef]
12. Dror, D.K.; Allen, L.H. Overview of nutrients in human milk. *Adv. Nutr.* **2018**, *9*, 278S–294S. [CrossRef] [PubMed]
13. Garcia-Rodenas, C.L.; De Castro, C.A.; Jenni, R.; Thakkar, S.K.; Beauport, L.; Tolsa, J.F.; Fischer-Fumeaux, C.J.; Affolter, M. Temporal changes of major protein concentrations in preterm and term human milk. A prospective cohort study. *Clin. Nutr.* **2018**. [CrossRef] [PubMed]
14. Fischer Fumeaux, C.J.; Garcia-Rodenas, C.L.; De Castro, C.A.; Courtet-Compondu, M.C.; Thakkar, S.K.; Beauport, L.; Tolsa, J.F.; Affolter, M. Longitudinal analysis of macronutrient composition in preterm and term human milk: A prospective cohort study. *Nutrients* **2019**, *11*, 1525. [CrossRef] [PubMed]
15. Thakkar, S.K.; De Castro, C.A.; Beauport, L.; Tolsa, J.F.; Fischer Fumeaux, C.J.; Affolter, M.; Giuffrida, F. Temporal progression of fatty acids in preterm and term human milk of mothers from switzerland. *Nutrients* **2019**, *11*, 112. [CrossRef] [PubMed]
16. Austin, S.; De Castro, C.A.; Sprenger, N.; Binia, A.; Affolter, M.; Garcia-Rodenas, C.L.; Beauport, L. Human milk oligosaccharides in the milk of mothers delivering term versus preterm infants. *Nutrients* **2019**, *11*, 1282. [CrossRef] [PubMed]
17. Bauer, J.; Gerss, J. Longitudinal analysis of macronutrients and minerals in human milk produced by mothers of preterm infants. *Clin. Nutr.* **2011**, *30*, 215–220. [CrossRef] [PubMed]
18. Gross, S.J.; David, R.J.; Bauman, L.; Tomarelli, R.M. Nutritional composition of milk produced by mothers delivering preterm. *J. Pediatr.* **1980**, *96*, 641–644. [CrossRef]

19. Dorea, J.G. Breast milk calcium and phosphorus concentrations. *Nutrition* **2000**, *16*, 146–147. [CrossRef]
20. Salamon, S.; Csapó, J. Composition of mother's milk iii. Macroand microelements contents. A review. *Acta Univ. Sapiente. Alimentaria* **2009**, *2*, 235–275.
21. Dorea, J.G. Magnesium in human milk. *J. Am. Coll. Nutr.* **2000**, *19*, 210–219. [CrossRef]
22. Dorea, J.G. Zinc and copper concentrations in breastmilk. *Indian Pediatr.* **2012**, *49*, 592. [CrossRef] [PubMed]
23. Dorea, J.G. Iron and copper in human milk. *Nutrition* **2000**, *16*, 209–220. [CrossRef]
24. Dorea, J.G. Iodine nutrition and breast feeding. *J. Trace Elem. Med. Biol.* **2002**, *16*, 207–220. [CrossRef]
25. Kim, S.Y.; Park, J.H.; Kim, E.A.; Lee-Kim, Y.C. Longitudinal study on trace mineral compositions (selenium, zinc, copper, manganese) in korean human preterm milk. *J. Korean Med. Sci.* **2012**, *27*, 532–536. [CrossRef] [PubMed]
26. Ellis, L.; Picciano, M.F.; Smith, A.M.; Hamosh, M.; Mehta, N.R. The impact of gestational length on human milk selenium concentration and glutathione peroxidase activity. *Pediatr. Res.* **1990**, *27*, 32–35. [CrossRef] [PubMed]
27. Dutta, S.; Saini, S.; Prasad, R. Changes in preterm human milk composition with particular reference to introduction of mixed feeding. *Indian Pediatr.* **2014**, *51*, 997–999. [CrossRef]
28. Lemons, J.A.; Moye, L.; Hall, D.; Simmons, M. Differences in the composition of preterm and term human milk during early lactation. *Pediatr. Res.* **1982**, *16*, 113–117. [CrossRef]
29. Picciano, M.F.; Guthrie, H.A. Copper, iron, and zinc contents of mature human milk. *Am. J. Clin. Nutr.* **1976**, *29*, 242–254. [CrossRef]
30. Zhao, A.; Ning, Y.; Zhang, Y.; Yang, X.; Wang, J.; Li, W.; Wang, P. Mineral compositions in breast milk of healthy chinese lactating women in urban areas and its associated factors. *Chin. Med. J. (Engl.)* **2014**, *127*, 2643–2648.
31. Ereman, R.R.; Lonnerdal, B.; Dewey, K.G. Maternal sodium intake does not affect postprandial sodium concentrations in human milk. *J. Nutr.* **1987**, *117*, 1154–1157. [CrossRef]
32. Finley, D.A.; Lonnerdal, B.; Dewey, K.G.; Grivetti, L.E. Inorganic constituents of breast milk from vegetarian and nonvegetarian women: Relationships with each other and with organic constituents. *J. Nutr.* **1985**, *115*, 772–781. [CrossRef] [PubMed]
33. Schupbach, R.; Wegmuller, R.; Berguerand, C.; Bui, M.; Herter-Aeberli, I. Micronutrient status and intake in omnivores, vegetarians and vegans in switzerland. *Eur. J. Nutr.* **2017**, *56*, 283–293. [CrossRef] [PubMed]
34. Berger, M.M. Oligoéléments en suisse et en europe. *Rev. Med. Suisse* **2012**, *8*, 2078–2084. [PubMed]
35. Atkinson, S.A.; Radde, I.C.; Chance, G.W.; Bryan, M.H.; Anderson, G.H. Macro-mineral content of milk obtained during early lactation from mothers of premature infants. *Early Hum. Dev.* **1980**, *4*, 5–14. [CrossRef]
36. Grumach, A.S.; Jeronimo, S.E.; Hage, M.; Carneiro-Sampaio, M.M. Nutritional factors in milk from brazilian mothers delivering small for gestational age neonates. *Rev. Saude Publica* **1993**, *27*, 455–462. [CrossRef] [PubMed]
37. De Figueiredo, C.S.; Palhares, D.B.; Melnikov, P.; Moura, A.J.; dos Santos, S.C. Zinc and copper concentrations in human preterm milk. *Biol. Trace Elem. Res.* **2010**, *136*, 1–7. [CrossRef]
38. Fernandez-Menendez, S.; Fernandez-Sanchez, M.L.; Fernandez-Colomer, B.; de la Flor St Remy, R.R.; Cotallo, G.D.; Freire, A.S.; Braz, B.F.; Santelli, R.E.; Sanz-Medel, A. Total zinc quantification by inductively coupled plasma-mass spectrometry and its speciation by size exclusion chromatography-inductively coupled plasma-mass spectrometry in human milk and commercial formulas: Importance in infant nutrition. *J. Chromatogr. A.* **2016**, *1428*, 246–254. [CrossRef] [PubMed]
39. Ares, S.; Quero, J.; Duran, S.; Presas, M.J.; Herruzo, R.; Morreale de Escobar, G. Iodine content of infant formulas and iodine intake of premature babies: High risk of iodine deficiency. *Arch Dis. Child. Fetal Neonatal Ed.* **1994**, *71*, F184–F191. [CrossRef]
40. Hunt, C.D.; Friel, J.K.; Johnson, L.K. Boron concentrations in milk from mothers of full-term and premature infants. *Am. J. Clin. Nutr.* **2004**, *80*, 1327–1333. [CrossRef]
41. Atinmo, T.; Omololu, A. Trace element content of breastmilk from mothers of preterm infants in nigeria. *Early Hum. Dev.* **1982**, *6*, 309–313. [CrossRef]
42. Perrone, L.; Di Palma, L.; Di Toro, R.; Gialanella, G.; Moro, R. Interaction of trace elements in a longitudinal study of human milk from full-term and preterm mothers. *Biol. Trace Elem. Res.* **1994**, *41*, 321–330. [CrossRef]

43. Trugo, N.M.; Donangelo, C.M.; Koury, J.C.; Silva, M.I.; Freitas, L.A. Concentration and distribution pattern of selected micronutrients in preterm and term milk from urban brazilian mothers during early lactation. *Eur. J. Clin. Nutr.* **1988**, *42*, 497–507.
44. Mendelson, R.A.; Anderson, G.H.; Bryan, M.H. Zinc, copper and iron content of milk from mothers of preterm and full-term infants. *Early Hum. Dev.* **1982**, *6*, 145–151. [CrossRef]
45. Fernández-Sánchez, M.L.; De la Flor St. Remy, R.; González Iglesias, H.; López-Sastre, J.B.; Fernández-Colomer, B.; Pérez-Solís, D.; Sanz-Medel, A. Iron content and its speciation in human milk from mothers of preterm and full-term infants at early stages of lactation: A comparison with commercial infant milk formulas. *Microchem. J.* **2012**, *105*, 108–114.
46. Gidrewicz, D.A.; Fenton, T.R. A systematic review and meta-analysis of the nutrient content of preterm and term breast milk. *BMC Pediatr.* **2014**, *14*, 216. [CrossRef]
47. Rigo, J.; Pieltain, C.; Salle, B.; Senterre, J. Enteral calcium, phosphate and vitamin d requirements and bone mineralization in preterm infants. *Acta Paediatr.* **2007**, *96*, 969–974. [CrossRef]
48. Butte, N.F.; Garza, C.; Johnson, C.A.; Smith, E.O.; Nichols, B.L. Longitudinal changes in milk composition of mothers delivering preterm and term infants. *Early Hum. Dev.* **1984**, *9*, 153–162. [CrossRef]
49. Aquilio, E.; Spagnoli, R.; Seri, S.; Bottone, G.; Spennati, G. Trace element content in human milk during lactation of preterm newborns. *Biol. Trace Elem. Res.* **1996**, *51*, 63–70. [CrossRef]
50. Sann, L.; Bienvenu, F.; Lahet, C.; Bienvenu, J.; Bethenod, M. Comparison of the composition of breast milk from mothers of term and preterm infants. *Acta Paediatr. Scand.* **1981**, *70*, 115–116. [CrossRef]
51. Koo, W.W.; Gupta, J.M. Breast milk sodium. *Arch Dis. Child.* **1982**, *57*, 500–502. [CrossRef]
52. Anderson, G.H. The effect of prematurity on milk composition and its physiological basis. *Fed. Proc.* **1984**, *43*, 2438–2442.
53. Chatterton, R.T., Jr.; Hill, P.D.; Aldag, J.C.; Hodges, K.R.; Belknap, S.M.; Zinaman, M.J. Relation of plasma oxytocin and prolactin concentrations to milk production in mothers of preterm infants: Influence of stress. *J. Clin. Endocrinol. Metab.* **2000**, *85*, 3661–3668. [CrossRef] [PubMed]
54. Hill, P.D.; Aldag, J.C.; Demirtas, H.; Naeem, V.; Parker, N.P.; Zinaman, M.J.; Chatterton, R.T., Jr. Association of serum prolactin and oxytocin with milk production in mothers of preterm and term infants. *Biol. Res. Nurs.* **2009**, *10*, 340–349. [CrossRef] [PubMed]
55. Ackland, M.L.; Michalczyk, A.A. Zinc and infant nutrition. *Arch Biochem. Biophys.* **2016**, *611*, 51–57. [CrossRef] [PubMed]
56. Tindell, R.; Tipple, T. Selenium: Implications for outcomes in extremely preterm infants. *J. Perinatol.* **2018**, *38*, 197–202. [CrossRef] [PubMed]

© 2019 by the authors. Licensee MDPI, Basel, Switzerland. This article is an open access article distributed under the terms and conditions of the Creative Commons Attribution (CC BY) license (http://creativecommons.org/licenses/by/4.0/).

Communication

Hormones in Breast Milk and Effect on Infants' Growth: A Systematic Review

Alessandra Mazzocchi [1], Maria Lorella Giannì [1,2], Daniela Morniroli [1,2], Ludovica Leone [1], Paola Roggero [1,2], Carlo Agostoni [1,3,*], Valentina De Cosmi [3] and Fabio Mosca [1,2]

1. Department of Clinical Sciences and Community Health, University of Milan, 20122 Milan, Italy
2. Fondazione IRCCS Ca' Granda Ospedale Maggiore Policlinico, NICU, 20122 Milan, Italy
3. Fondazione IRCCS Ca' Granda Ospedale Maggiore Policlinico, Pediatric Intermediate Care Unit, 20122 Milan, Italy
* Correspondence: carlo.agostoni@unimi.it; Tel.: +39-02-55-032-497; Fax: +39-02-55-030-226

Received: 19 July 2019; Accepted: 7 August 2019; Published: 9 August 2019

Abstract: Breast milk is characterized by a dynamic and complex composition which includes hormones and other bioactive components that could influence infant growth, development, and optimize health. Among the several beneficial effects associated with prolonged breastfeeding, a 13% decrease in the risk of overweight and obesity has been reported. Recent research has focused on breast milk hormones contributing to the appetite and energy balance regulation and adiposity. Accordingly, we conducted a literature systematic review with the aim to provide an update on the effect of leptin, ghrelin, Insulin Growth Factor 1, adiponectin, and insulin on infants' and children's growth and body composition. The revised literature reveals contrasting findings concerning the potential role of all these hormones on modeling growth and fat mass apposition and health outcomes later in life. Further studies are needed to gain further insight into the specific role of these bioactive components in metabolic pathways related to body composition. This could help gain a further insight on infants' growth, both in physiological and pathological settings.

Keywords: hormones; adipokines; breast milk; growth; body composition; term infant

1. Introduction

Breast milk (BM) is the only species-specific human food tailored to meet infants' needs through the adaptation of its micro and macronutrient content and bioactive components, depending upon the mother–infant dyad characteristics (i.e., gestational age at birth, birth weight, etc.) [1,2].

Furthermore, the dynamic composition of BM also allows mother–infant signaling over lactation [3,4]. As a result, the infant is guided in the developmental and physiological processes through breastfeeding, whose crucial role in promoting healthy growth and optimal cognitive development is widely acknowledged [2,5]. Breastfeeding is beneficial through an epigenetic effect that is demonstrated to be dose-dependent [6]. However, although our understanding of the bioactive, non-nutritional compounds of BM and their dynamic changes over time has improved, the complexity of BM composition and the synergistic mechanisms responsible for its life-long biological effects have not yet been unraveled [7]. Among the several beneficial health effects associated with prolonged breastfeeding, a 13% decrease in the risk of overweight and obesity development has been reported [8]. In light of the well-known different growth trajectory and body composition development that characterize breastfed infants in comparison to formula-fed ones [9,10], gaining further insight into the factors that contribute to infants' appetite regulation and metabolic programming in the long run is, therefore, crucial for developing adequate strategies aimed at obesity prevention [11,12]. Accordingly, recent research has focused on BM hormones involved in the regulation of appetite and energy balance and adiposity, with the aim of identifying the factors that modulate their concentration throughout lactation and elucidating their potential biological relevance within this context [13,14]. We conducted a literature review with the aim of providing an update on the effects of BM leptin, ghrelin, Insulin Growth Factor 1 (IGF-1), adiponectin, and insulin found within BM on infants' and children's growth and body composition.

2. Methods

We performed a review using PUBMED, searching for all trails published in English from 2009 up to July 2019. The following key words were used: "Leptin" OR "Ghrelin" OR "IGF-1" OR "Adiponectin" OR "Insulin" OR "Adipokines" AND "BM" OR "donor milk" OR "banked milk" AND "weight gain" OR "body composition" OR "obesity" OR "adiposity" OR "growth" AND "infant" OR "children". In this review, observational studies examining the effects of these BM hormones on infants' and children's anthropometry and growth were selected. A total of 344 articles were initially identified and, among them, 30 studies were retrieved and evaluated for inclusion as relevant studies for the analysis by reviewing the abstract, and, when necessary, the full text. A manual bibliographic cross-referencing was also performed. We also reviewed the reference lists of relevant studies in order to detect other relevant primary sources.

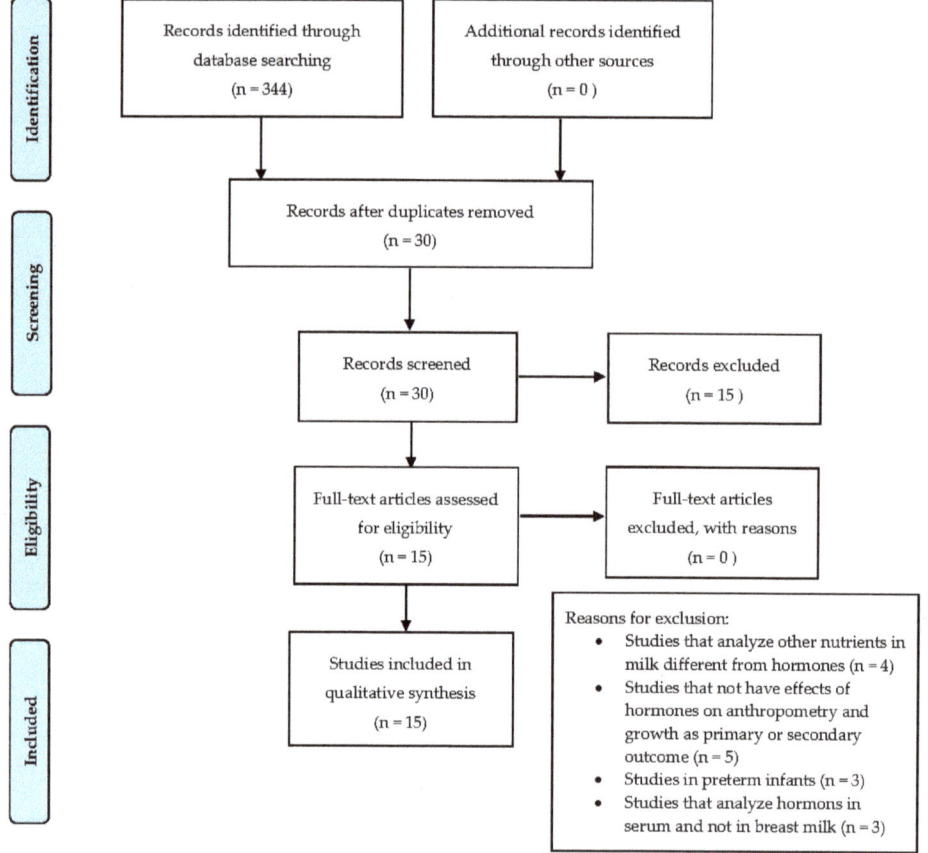

Figure 1. PRISMA diagram of search strategy.

Articles were considered as relevant and included in the analysis if (a) they reported the effect of milk hormones on infant growth; (b) included weight gain, body composition, obesity, adiposity as outcome measures; and (c) enrolled human participants. The selection procedure used for identifying and including the studies is summarized in Figure 1. Finally, we included 15 observational studies for the review. Three researchers independently searched, screened, and identified studies and abstracted and tabulated data. Discrepancies were addressed and sorted out by discussion.

3. Results

Young et al. [15] performed a prospective study of 41 healthy, term, breastfed infants, with the aim of assessing components in BM that are related with infant growth and may affect fat-free mass (FFM) or fat mass (FM) deposition thoroughout the first four months of life. A significant association between the trajectory of weight for length Z-scores (WLZ) and BM insulin was detected. However, the effect differed by maternal body mass index (BMI), being the association negative among infants of normal weight mothers. In the multivariable model, mean BM insulin quintiles were significantly associated with infant fat gain rate, suggesting the role of this hormone in regulating fat-mass accretion. On the contrary, an inverse association among BM adiponectin content and rate of fat gain was found.

Kon et al. [16] assessed 103 mother–infant pairs during the first three months of lactation and reported higher BM concentration of IGF-1 in the mothers of infants with high weight gain than in the mothers of those with low and normal weight gain at all study points. BM leptin and ghrelin content tended to show a similar behavior at two and three months of lactation and at one and two months of lactation, respectively. These results suggest that both leptin and ghrelin contribute to infants' weight gain. On the contrary, in another prospective study, Yis et al. [17] found that neither BM ghrelin nor BM leptin was correlated with anthropometric data in breastfed infants. They only observed positive correlations when they analyzed hormones sampled in infants' serum, in particular between ghrelin and triceps skinfold thickness (TSF), and leptin and weight, TSF and weight gain at three months of age. Gridneva and colleagues [18] showed a different effect of BM leptin and adiponectin content on infant FFM and FM deposition through the first 12 months of life: Specifically, they saw an inverse association between the daily intake of adiponectin and the infant fat-free mass but a positive relationship with infant fat mass. At one year, higher intake of BM leptin was associated with higher fat mass deposition. Interestingly, contrasting data were found in this study by Fields et al. [19], where authors observed that milk leptin concentration reduces through lactation and it was higher in obese mothers; furthermore, this hormone shows a negative association with infant length, FM percentage, total FM, and trunk fat at six months of age. In a previous pilot study on 19 exclusively breastfeeding mother–infant dyads [20], Fields observed that greater BM leptin was associated with lower BMI-for-age z-score (BMIZ), and higher content of BM insulin was associated with lower infant weight, BMIZ, and FFM. All these results concerning leptin suggest that this hormone may have a role both with regard to FM and FFM deposition. The relationship between BM adiponectin, leptin, and insulin with infant body composition was assessed in 430 mother–infant dyads taking part into the Canadian Healthy Infant Longitudinal Development (CHILD) Study [21]. Data showed an inverse correlation between BM leptin and insulin content and infant WLZ and BMIZ at four months of age, whereas no significant association was found between BM adiponectin concentrations and infant body composition. Meyer et al. [22] longitudinally investigated the correlation among BM leptin and adiponectin with body composition in children aged three to five years. The authors previously reported a positive association between BM adiponectin concentration and FM and weight gain in infants in the first two years of age. However, the findings of this analysis failed to demonstrate a significant relationship of BM leptin and adiponectin levels, assessed at six weeks and four months after delivery, with later body composition characteristics. Brunner and colleagues [23] investigated the association of leptin and adiponectin in BM with infant weight gain and body composition using skinfold thickness assessment in the first 24 months of life. Although BM leptin assessed at four months was negatively associated with concurrent infant weight, BMI, and FFM, no relationship with infant growth and body composition at later ages was reported. On the contrary, there was a tendency for BM adiponectin to be negatively associated with infant growth parameters in the first four months of age, but afterwards, it was positively associated with infant weight gain and FM up to the first 24 months of life. Mohamad et al. [24], assessed the relationship of maternal serum and BM adipokines with infant adiposity development. A higher BM adiponectin concentration at two months after delivery was associated with reduced infant body weight, BMIZ, and abdominal circumference at two months of age. After adjusting for confounders, BM adiponectin at two months postpartum was the only factor that remained independently associated with infant adiposity at two months of age. This result indicates that BM adiponectin may be protective on infant adiposity development in the early postnatal period. Cesur et al. [25] investigated the relationship between ghrelin and adiponectin content in BM and serum samples of 25 mother–infant pairs with the anthropometry of newborn infants at one and four months of life. BM ghrelin content was significantly higher than the infant and maternal serum ghrelin at both study points. Moreover, the level of the fourth month BM ghrelin level positively correlated with infants' weight gain throughout the study. Larsson et al. [26] aimed to identify the determinants of early growth by comparing two groups of exclusively breastfed infants, one with excessive weight gain and the second one with a normal growth pattern during the first five months. The lack of difference in the BM intake at the age of five months between the groups suggests that the high weight gain could be driven by factors related to milk composition. Accordingly, the leptin content was significantly decreased in the BM of the group characterized by high weight gain, thus stimulating appetite and milk intake in this group of infants. Anyhow, it has to be considered that, in this study, measurement of BM was carried out only at five months of age and not earlier, when growth velocity was considerably higher. Ucar and colleagues [27] investigated whether leptin may exert an effect on satiety by comparing the formilk and hindmilk leptin concentrations in 18 lactating women. However, no differences were found between the leptin levels of BM samples assessed during the two different phases of feeding at breast. Additionally, no association between leptin concentrations in both BM and maternal plasma and infants' body weight, BMI, TSF, and left upper arm circumference measurements was found. Concerning the role of adiponectin in BM, a longitudinal observational study of two cohorts of 45 mother–infant pairs was carried out to determine an association between milk adiponectin and infant growth. No significant difference was detected among the two cohorts at the beginning of the study. However, higher BM adiponectin was associated, through the first six months, with lower infant adiposity—in particular, with weight-for-age z-score (WAZ) and WLZ, but not length for age z-score (LAZ), after adjusting for confounders and irrespective of the cohort [28]. Similarly, a negative association between adiponectin and infant growth (i.e., weight-for-height z-score (WHZ) and head circumference) in two cohorts of infants from 48 healthy mothers and from 48 mothers with gestational diabetes mellitus was found [12]. No associations between BM insulin, leptin, and ghrelin concentrations and infant WHZ and head circumference resulted [12]. In Table 1 the main findings of the studies meeting the inclusion criteria are shown.

Table 1. Observational studies on hormones' dosage in breast milk and infants' growth and anthropometry.

Study	Sample Size	Growth and Anthropometric Outcomes	Hormones in Breast Milk	Major Findings
[12]	96 BF infants	Infant body weight (kg), length (cm), WLZ, head circumference (cm).	Adiponectin, leptin, insulin and ghrelin	Adiponectin inversely associated with WLZ and head circumference ($p \leq 0.003$). Association between adiponectin and insulin and head circumference ($p \leq 0.007$ and $p \leq 0.049$, respectively).
[15]	41 BF infants	WLZ, % body fat from skinfolds from 0 to 4 mo.	Leptin, adiponectin, ghrelin, insulin	Negative association between HM insulin and WLZ trajectory in infants of normal weight mothers ($p = 0.028$)
[16]	103 BF infants	Weight gain (g/mo).	Leptin, adiponectin, ghrelin, IGF-1	Correlation between breast milk IGF-1 and infant weight gain ($r = 0.294$, $p = 0.043$). Higher ghrelin levels at 1 and 2 mo and higher leptin levels at 2 and 3 mo of lactation ($p < 0.05$) in infants with high weight gain (>1000 g/mo).
[17]	24 BF infants	Infant's body weight (g), length (cm), triceps skinfold thickness (mm), postnatal weight gain (g) at 3 and 6 mo of age.	Ghrelin, leptin	No correlation between breast-milk ghrelin or breast-milk leptin with anthropometric data.
[18]	20 BF infants	Infant's body weight (g), length (cm), BMI (kg/m^2), ultrasound skinfolds, bioimpedance spectroscopy and FMI (FM (kg)/length (m)^2), FFMI (FFM (kg)/length (m)^2).	Adiponectin, leptin	Higher intake of adiponectin associated with lower infant FFM ($p = 0.005$) and FFM index ($p = 0.009$) and higher FM ($p < 0.001$), FM index (FMI; $p < 0.001$), and %FM ($p < 0.001$). Higher intake of leptin associated with larger increases in infant adiposity (2-12 month): FM, $p = 0.0006$; %FM, $p = 0.0004$.
[19]	37 BF infants	Infant's body weight (g), length (cm), FM (g and %), FFM (g), trunk fat mass (g).	Insulin, leptin	Inverse association between leptin levels at 1 mo and infant length ($p = 0.0257$), FM % ($p = 0.0223$), FM (g) ($p = 0.0226$), and trunk fat mass ($p = 0.0111$) at 6 mo.

Table 1. *Cont.*

Study	Sample Size	Growth and Anthropometric Outcomes	Hormones in Breast Milk	Major Findings
[20]	19 BF infants	Infant's body weight (g), length (cm), WLZ, BMIZ, FM (g and %), FFM (g), trunk fat mass (g) at 1 mo of age	Leptin, insulin	Leptin associated with lower BMIZ ($r = -0.54$, $p = 0.03$). Higher concentrations of insulin associated with lower infant weight, relative weight, and FFM ($r = -0.49$–0.58, $p < 0.06$).
[21]	430 BF infants	Infant's body weight (g), length (cm), WFL, BMIZ at 4 mo and 1 yrs of age.	Adiponectin, leptin, insulin	Higher leptin associated with lower infant WLZ at 4 mo ($\beta - 0.67$, 95% confidence interval (CI): -1.17, -0.17 for highest vs lowest quintile) and 1 yrs ($\beta - 0.58$, 95% CI: -1.02, -0.14). Insulin showed a U-shaped association, with intermediate concentrations predicting the lowest infant WLZ at 4 mo ($\beta - 0.51$, 95% CI: -0.87, -0.15 for third vs lowest quintile) and 1 yrs ($\beta - 0.35$, 95% CI: -0.66, -0.04). Adiponectin not associated with infant body composition.
[22]	147 BF infants	Infant's body weight (kg), BMI percentiles, sum of four skinfolds (mm), FM (kg and %), FFM (kg) at 3, 4, and 5 yrs of age.	Adiponectin, leptin	No association between leptin or total adiponectin levels assessed at 6 weeks post-delivery with children's body weight, BMI percentiles, sum of four skinfolds, measurements, FM (kg and %), or FFM (kg).
[23]	188 BF infants	The relationship of BM leptin and adiponectin with infant weight gain and body composition up to the age of 2 yrs.	Adiponectin, leptin	Milk leptin at 4 mo negatively associated with infant weight ([95%CI]: -604.96 g [-1166.19; -43.72], $p = 0.037$] and FFM (-400.95 g [-777.64; -24.25], $p = 0.039$) at the age of 4 mo. Adiponectin tended to be negatively associated with infant FFM ($p = 0.015$) and weight ($p = 0.054$) in the first 4 mo, but afterwards was positively related to weight gain ($p = 0.027$) and the sum of skinfolds ($p = 0.047$) up to 2 years.

Table 1. *Cont.*

Study	Sample Size	Growth and Anthropometric Outcomes	Hormones in Breast Milk	Major Findings
[24]	155 BF infants	Infant's body weight (kg) and BMIZ, abdominal circumference (cm).	Leptin, adiponectin	The higher level of adiponectin at 2 mo postpartum associated with reduced infant body weight ($\beta = -0.54$ $p = 0.003$), BMIZ ($\beta = -0.79$, $p = 0.008$) and abdominal circumference at 2 mo of age ($\beta = -2.34$, $p = 0.003$). No association between adiponectin at birth and 2 mo with infant adiposity at 6 and 12 mo of age. An increased maternal ALR was related to reduced infant BMIZ at birth.
[25]	25 BF infants	Infant's body weight (g), length (cm), BMI (kg/m^2).	Ghrelin, adiponectin	Positive correlation between the level of the 4th mo ghrelin level and infants' weight gain ($r = 0.51$, $p = 0.025$).
[26]	30 BF infants	Infant's body weight (g), length (cm), WAZ, LAZ, BAZ, TSFZ, SSFZ.	Leptin, adiponectin	A 40% reduction of median leptin content at 5 mo in the high weight gain group ($p = 0.045$). At 5 mo, no significant associations between milk concentrations of hormones and infants' WAZ, BAZ or LAZ, or energy and hormones and infant's anthropometry (WAZ, BAZ, or LAZ) or change in these z-scores from birth to the 5 mo visit (all $p > 0.11$).
[27]	18 BF infants	Infant's body weight (kg), BMI (kg/m^2), triceps skinfold thickness (mm), left upper arm circumference (mm).	Leptin	No correlation between Log leptin concentrations and infants' body weight, BMI, triceps skinfold thickness, and left upper arm circumference measurements ($p > 0.05$).
[22]	322 BF infants	Infant's body weight (kg), length (cm), WAZ, BMI (kg/m^2), LAZ, WLZ.	Adiponectin	During the first 6 months, higher adiponectin associated with lower infant WAZ ($\beta = 0.20 \pm$ standard error (SE) 0.04, $p = 0.0001$) and WLZ ($\beta = 0.29 \pm 0.08$, $p = 0.0002$). Adiponectin not associated with infant length.

Mo = months; yrs = years; BF = breastfed; FM = fat mass; FFM = fat-free mass; FMI = FM index; FFMI = FFM index; BMI = body mass index; WAZ = weight for age z-score; WLZ = weight-for-length/height z-score; LAZ = length-for-age z-score; BMIZ = BMI z-score; TSFZ = triceps skinfold-for-age z-score; SSFZ = subscapular skinfold-for-age; ALR = maternal adiponectin to leptin ratio; β = beta regression coefficient.

4. Discussion

In this systematic review, we explored updated evidences on five hormones (leptin, ghrelin, IGF-1, adiponectin, insulin) described in BM that are also related with hunger, fat deposition, and adipose tissue metabolism in all stages of life. Leptin is an hormone synthetized by adipose cells and mucosal cells in the small intestine; it regulates energy homeostasis by interacting with its receptors in the hypothalamus, inhibiting hunger [29]. As a result, fat storage in adipocytes is decreased [30]. In newborns, Chaoimh et al. [31] found a negative relationship between cord leptin content and weight gain, demonstrating that a low weight gain is accompanied by a reduced fat mass deposition in the early postnatal period. Accordingly, BM leptin has been reported to be inversely associated with infant global adiposity and trunk fat at six months of age [19]. Moreover, a higher content of BM leptin was associated with lower WLZ and BMIZ in infants [20,21]. However, this effect appears to be temporary, as indicated by the lack of effect of leptin levels in BM on body composition in the first years of age [22,23]. Interestingly, despite this well-known anti-adiposity effect, BM leptin content has also been positively correlated with higher weight gain in infants in the early postnatal period, and increased adiposity as far as 12 months of lactation [16,18]. Consistently with these results, a possible relationship of BM leptin with body fat stores and infant adiposity has been suggested by Savino and colleagues, who determined leptin values in serum and in the mothers' BM. The authors found positive correlations between BM leptin and infant serum leptin, and between infant serum leptin and both infants' BMI and weight [32]. Yis et al. also found a positive correlation with leptin serum level in infants, but failed to demonstrate a significant correlation with BM leptin levels [17]. These inconsistent results may pose a doubt on BM leptin's hormonal role during lactation, which could differ from what is known to be its primary role in human body, as hypothesized by Ucar and colleagues who failed to demonstrate a role played by leptin BM levels as a satiety factor, and also showed no correlation with infants' weight and adiposity [27]. Opposite to leptin, ghrelin is considered to be the "hunger hormone". It is a neuropetide secreted by specific gastrointestinal cells when the stomach is empty, that acts on the hypothalamus to stimulate appetite, increase gastric acid secretion, and improve gastrointestinal motility [33]. A different relationship between the serum content of hormones involved in satiety and appetite regulation, including ghrelin, has been found by Vásquez-Garibay et al. [34]. Specifically, the authors demonstrated a more significant association in breastfeeding mother–infant pairs than in formula-fed ones. Moreover, Cesur et al. found that ghrelin in BM is significantly higher than serum ghrelin, both in the infants' and the mothers' samples through first months of lactation [25]. However, although there is agreement regarding the role of breastfeeding in modulating the mechanisms underlying satiety and appetite regulation, studies available in the literature on BM ghrelin and infants' weight gain have shown inconsistent results. While Kon and colleagues [16] reported a positive correlation between BM ghrelin in the first two months of lactation with higher weight gain in infants, the results were not confirmed by Yis et al., despite again finding a positive correlation with infants' serum ghrelin levels and TSF [17]. Equally to ghrelin, adiponectin is a peptide with hormonal functions that stimulate hunger, acting on the hypothalamus. It is produced specifically by the adipose tissue, and its serum levels are inversely correlated with the FM percentage of the body in the adult population [35]. Its levels in BM and association with infants' FM and growth have been explored by many studies included in this review, with inconsistent results. Young et al. inversely correlated BM adiponectin levels with FM gain in healthy term infants, consistent with its hormonal function [12], whereas Gridneva and colleagues found a correlation with a higher adiposity in the first year of life [18]. Similar results, both with a positive or negative correlation, were obtained in all other studies explored, whereas the largest cohort CHILD study failed to find any association between adiponectin BM levels and infants' body composition [21]. IGF-1 is part of countless metabolic pathways of cells' signaling with their environment, often referred as IGF "axis" that plays a primary role in cell proliferation and inhibition of cell death (apoptosis) in both physiological and pathological states [36]. IGF-1 expression is also required for achieving maximal growth in developing organisms [37]. In this review, we found only one study that explored IGF-1 BM levels and infants' growth. Kon et al. reported a higher BM level of IGF-1 in those mothers whose infants showed a higher weight gain [16]. Insulin is a peptide that acts as a hormone, produced by specific endocrine cells located in the pancreatic tissue. It is considered to be the most important anabolic hormone and promotes cellular intake glucose in muscle and adipose tissue [38]. According to these functions, its levels in BM have been demonstrated to positively correlate with infant WLZ trajectory in the first months of life [12]. However, similarly to other hormones listed above, these findings were not confirmed by other studies considered in this review. However, limitations should be noted when interpreting the results. First of all, more longitudinal investigations are needed, especially during the early lactation period, to clarify the effects of breast milk hormones on growth and regulation of nutrition in infancy and childhood. The use of different outcomes to measure the child's growth was a further limitation of the review. The duration of the studies was until different ages and in some studies, it was not possible to arrange clinical visits before the age of five to six months. Finally, infant growth was evaluated using anthropometric measurements: More accurate measurements of body composition will help accurately evaluate the mediation effect of hormones in BM.

5. Conclusions

The inconsistent findings of the studies considered in this review could be explained with the concomitant action of countless factors interfering with infants' body composition during the first years of life. While some of these, including basic subject characteristics, could be identified and controlled, other factors such as synergetic or antagonist functions of many other bioactive components of BM, that could play a role in modeling growth and FM apposition, are difficult—if not impossible—to take into account. Although further studies are needed to determine the specific role of hormones and adipokines in BM, undoubtedly, these bioactive components participate in metabolic pathways related to body composition. The understanding of these pathways could help gain further insight on infants' growth, both in physiological and pathological settings.

Author Contributions: A.M., V.D.C., M.L.G., and D.M. drafted the manuscript; proof read, ordered the references, made determinations and gave technical support in data interpretation, and final revision of the draft. C.A. and F.M. critically reviewed the paper. All the others authors contributed significantly to the paper and agreed to the manuscript in its current form.

Funding: This research received no external funding.

Acknowledgments: Supported by a contribution from the Italian Ministry of Health (IRCCS grant).

Conflicts of Interest: The authors declare no conflict of interest.

References

1. Kramer, M.S.; Guo, T.; Platt, R.W.; Shapiro, S.; Collet, J.P.; Chalmers, B.; Hodnett, E.; Sevkovskaya, Z.; Dzikovich, I.; Vanilovich, I. PROBIT Study Group. Breastfeeding and infant growth: Biology or bias? *Pediatrics* **2002**, *110*, 343–347. [CrossRef] [PubMed]
2. Mosca, F.; Giannì, M.L. Human milk: Composition and health benefits. *Pediatr. Med. Chir.* **2017**. [CrossRef] [PubMed]
3. Sharp, J.A.; Modepalli, V.; Enjapoori, A.K.; Bisana, S.; Abud, H.E.; Lefevre, C.; Nicholas, K.R. Bioactive Functions of Milk Proteins: A Comparative Genomics Approach. *J. Mammary Gland Biol. Neoplasia* **2014**, *19*, 289–302. [CrossRef] [PubMed]
4. Andreas, N.J.; Kampmann, B.; Mehring Le-Doare, K. Human breast milk: A review on its composition and bioactivity. *Early Hum. Dev.* **2015**, *91*, 629–635. [CrossRef] [PubMed]
5. Verduci, E.; Banderali, G.; Barberi, S.; Radaelli, G.; Lops, A.; Betti, F.; Riva, E.; Giovannini, M. Epigenetic effects of human breast milk. *Nutrients* **2014**, *6*, 1711–1724. [CrossRef] [PubMed]
6. Pauwels, S.; Symons, L.; Vanautgaerden, E.L.; Ghosh, M.; Duca, R.C.; Bekaert, B.; Freson, K.; Huybrechts, I.; Langie, S.A.S.; Koppen, G.; et al. The Influence of the Duration of Breastfeeding on the Infant's Metabolic Epigenome. *Nutrients* **2019**, *11*, 1408. [CrossRef] [PubMed]

7. Fields, D.A.; Schneider, C.R.; Pavela, G. A narrative review of the associations between six bioactive components in breast milk and infant adiposity. *Obesity* **2016**, *24*, 1213–1221. [CrossRef] [PubMed]
8. Victora, C.G.; Bahl, R.; Barros, A.J.; França, G.V.; Horton, S.; Krasevec, J.; Murch, S.; Sankar, M.J.; Walker, N.; Rollins, N.C. Lancet Breastfeeding Series Group. Breastfeeding in the 21st century: Epidemiology, mechanisms, and lifelong effect. *Lancet* **2016**, *387*, 475–490. [CrossRef]
9. Lind, M.V.; Larnkjær, A.; Mølgaard, C.; Michaelsen, K.F. Breastfeeding, Breast Milk Composition, and Growth Outcomes. In *Recent Research in Nutrition and Growth*; Karger Publishers: Basel, Switzerland, 2018; Volume 89, pp. 63–77.
10. Kugananthan, S.; Gridneva, Z.; Lai, C.T.; Hepworth, A.R.; Mark, P.J.; Kakulas, F.; Geddes, D.T. Associations between Maternal Body Composition and Appetite Hormones and Macronutrients in Human Milk. *Nutrients* **2017**, *9*, 252. [CrossRef] [PubMed]
11. Rito, A.I.; Buoncristiano, M.; Spinelli, A.; Salanave, B.; Kunešová, M.; Hejgaard, T.; Solano, M.G.; Fijałkowska, A.; Sturua, L.; Hyska, J.; et al. Association between Characteristics at Birth, Breastfeeding and Obesity in 22 Countries: The WHO European Childhood Obesity Surveillance Initiative—COSI 2015/2017. *Obes. Facts* **2019**, *12*, 226–243. [CrossRef]
12. Yu, X.; Rong, S.S.; Sun, X.; Ding, G.; Wan, W.; Zou, L.; Wu, S.; Li, M.; Wang, D. Associations of breast milk adiponectin, leptin, insulin and ghrelin with maternal characteristics and early infant growth: A longitudinal study. *Br. J. Nutr.* **2018**, *120*, 1380–1387. [CrossRef] [PubMed]
13. Neville, M.C.; Anderson, S.M.; McManaman, J.L.; Badger, T.M.; Bunik, M.; Contractor, N.; Crume, T.; Dabelea, D.; Donovan, S.M.; Forman, N.; et al. Lactation and neonatal nutrition: Defining and refining the critical questions. *J. Mammary Gland Biol. Neoplasia* **2012**, *17*, 167–188. [CrossRef] [PubMed]
14. Badillo-Suárez, P.A.; Rodríguez-Cruz, M.; Nieves-Morales, X. Impact of Metabolic Hormones Secreted in Human Breast Milk on Nutritional Programming in Childhood Obesity. *J. Mammary Gland Biol. Neoplasia* **2017**, *22*, 171–191. [CrossRef] [PubMed]
15. Young, B.E.; Levek, C.; Reynolds, R.M.; Rudolph, M.C.; MacLean, P.; Hernandez, T.L.; Friedman, J.E.; Krebs, N.F. Bioactive components in human milk are differentially associated with rates of lean and fat mass deposition in infants of mothers with normal vs. elevated BMI. *Pediatric Obes.* **2018**, *13*, 598–606. [CrossRef] [PubMed]
16. Kon, I.Y.; Shilina, N.M.; Gmoshinskaya, M.V.; Ivanushkina, T.A. The study of breast milk IGF-1, leptin, ghrelin and adiponectin levels as possible reasons of high weight gain in breast-fed infants. *Ann. Nutr. Metab.* **2014**, *65*, 317–323. [CrossRef] [PubMed]
17. Yiş, U.; Oztürk, Y.; Sişman, A.R.; Uysal, S.; Soylu, O.B.; Büyükgebiz, B. The relation of serum ghrelin, leptin and insulin levels to the growth patterns and feeding characteristics in breast-fed versus formula-fed infants. *Turk. J. Pediatrics* **2010**, *52*, 35–41.
18. Gridneva, Z.; Kugananthan, S.; Rea, A.; Lai, C.T.; Ward, L.C.; Murray, K.; Hartmann, P.E.; Geddes, D.T. Human Milk Adiponectin and Leptin and Infant Body Composition over the First 12 Months of Lactation. *Nutrients* **2018**, *10*, 1125. [CrossRef] [PubMed]
19. Fields, D.A.; George, B.; Williams, M.; Whitaker, K.; Allison, D.B.; Teague, A.; Demerath, E.W. Associations between human breast milk hormones and adipocytokines and infant growth and body composition in the first 6 months of life. *Pediatric Obes.* **2017**, *12*, 78–85. [CrossRef] [PubMed]
20. Fields, D.A.; Demerath, E.W. Relationship of insulin, glucose, leptin, IL-6 and TNF-α in human breast milk with infant growth and body composition. *Pediatric Obes.* **2012**, *7*, 304–312. [CrossRef]
21. Chan, D.; Goruk, S.; Becker, A.B.; Subbarao, P.; Mandhane, P.J.; Turvey, S.E.; Lefebvre, D.; Sears, M.R.; Field, C.J.; Azad, M.B. Adiponectin, leptin and insulin in breast milk: Associations with maternal characteristics and infant body composition in the first year of life. *Int. J. Obes.* **2018**, *42*, 36–43. [CrossRef]
22. Meyer, D.M.; Brei, C.; Stecher, L.; Much, D.; Brunner, S.; Hauner, H. The relationship between breast milk leptin and adiponectin with child body composition from 3 to 5 years: A follow-up study. *Pediatric Obes.* **2017**, *12*, 125–129. [CrossRef] [PubMed]
23. Brunner, S.; Schmid, D.; Zang, K.; Much, D.; Knoeferl, B.; Kratzsch, J.; Amann-Gassner, U.; Bader, B.L.; Hauner, H. Breast milk leptin and adiponectin in relation to infant body composition up to 2 years. *Pediatric Obes.* **2015**, *10*, 67–73. [CrossRef] [PubMed]

24. Mohamad, M.; Lim, P.; Wang, Y.; Soo, K.; Mohamed, H. Maternal Serum and Breast Milk Adiponectin: The Association with Infant Adiposity Development. *Int. J. Environ. Res. Public Health* **2018**, *15*, 1250. [CrossRef] [PubMed]
25. Cesur, G.; Ozguner, F.; Yilmaz, N.; Dundar, B. The relationship between ghrelin and adiponectin levels in breast milk and infant serum and growth of infants during early postnatal life. *J. Physiol. Sci.* **2012**, *62*, 185–190. [CrossRef] [PubMed]
26. Larsson, M.; Lind, M.; Larnkjær, A.; Due, A.; Blom, I.; Wells, J.; Michaelsen, K. Excessive weight gain followed by catch-down in exclusively breastfed infants: An exploratory study. *Nutrients* **2018**, *10*, 1290. [CrossRef] [PubMed]
27. Ucar, B.; Kırel, B.; Bör, Ö.; Kılıç, F.S.; Doğruel, N.; Aydoğdu, S.D.; Tekin, N. Breast milk leptin concentrations in initial and terminal milk samples: Relationships to maternal and infant plasma leptin concentrations, adiposity, serum glucose, insulin, lipid and lipoprotein levels. *J. Pediatric Endocrinol. Metab.* **2000**, *13*, 149–156. [CrossRef]
28. Woo, J.G.; Guerrero, M.L.; Altaye, M.; Ruiz-Palacios, G.M.; Martin, L.J.; Dubert-Ferrandon, A.; Newburg, D.S.; Morrow, A.L. Human milk adiponectin is associated with infant growth in two independent cohorts. *Breastfeed. Med.* **2009**, *4*, 101–109. [CrossRef] [PubMed]
29. Flier, J.S.; Maratos-Flier, E. Leptin's Physiologic Role: Does the Emperor of Energy Balance Have No Clothes? *Cell Metab.* **2017**, *26*, 24–26. [CrossRef] [PubMed]
30. Brennan, A.M.; Mantzoros, C.S. Drug Insight: The role of leptin in human physiology and pathophysiology—Emerging clinical applications. *Nat. Clin. Pract. Endocrinol. Metab.* **2006**, *2*, 318–327. [CrossRef]
31. Chaoimh, C.N.; Murray, D.M.; Kenny, L.C.; Irvine, A.D.; Hourihane, J.O.; Kiely, M. Cord blood leptin and gains in body weight and fat mass during infancy. *Eur. J. Endocrinol.* **2016**, *175*, 403–410. [CrossRef]
32. Savino, F.; Sardo, A.; Rossi, L.; Benetti, S.; Savino, A.; Silvestro, L. Mother and Infant Body Mass Index, Breast Milk Leptin and Their Serum Leptin Values. *Nutrients* **2016**, *8*, 383. [CrossRef] [PubMed]
33. Schwartz, M.W.; Woods, S.C.; Porte, D., Jr.; Seeley, R.J.; Baskin, D.G. Central nervous system control of food intake. *Nature* **2000**, *404*, 661–671. [CrossRef] [PubMed]
34. Vásquez-Garibay, E.M.; Larrosa-Haro, A.; Guzmán-Mercado, E.; Muñoz-Esparza, N.; García-Arellano, S.; Muñoz-Valle, F.; Romero-Velarde, E. Serum concentration of appetite-regulating hormones of mother-infant dyad according to the type of feeding. *Food Sci. Nutr.* **2019**, *7*, 869–874. [CrossRef] [PubMed]
35. Ukkola, O.; Santaniemi, M. Adiponectin: A link between excess adiposity and associated comorbidities? *J. Mol. Med.* **2002**, *80*, 696–702. [CrossRef] [PubMed]
36. Nemet, D.; Cooper, D.M. Exercise, diet, and chilhood obesity: The GH-IGF-I connection. *J. Pediatric Endocrinol. Metab.* **2002**, *15*, 751–757. [CrossRef]
37. De Jong, M.; Cranendonk, A.; Twisk, J.W.; van Weissenbruch, M.M. IGF-I and relation to growth in infancy and early childhood in very-low-birth-weight infants and term born infants. *PLoS ONE* **2017**, *12*, e0171650. [CrossRef] [PubMed]
38. Dimitriadis, G.; Mitrou, P.; Lambadiari, V.; Maratou, E.; Raptis, S.A. Insulin effects in muscle and adipose tissue. *Diabetes Res. Clin. Pract.* **2011**, *93*, S52–S59. [CrossRef]

© 2019 by the authors. Licensee MDPI, Basel, Switzerland. This article is an open access article distributed under the terms and conditions of the Creative Commons Attribution (CC BY) license (http://creativecommons.org/licenses/by/4.0/).

Article

Validation and Application of Biocrates Absolute*IDQ*® p180 Targeted Metabolomics Kit Using Human Milk

Daniela Hampel [1,2,*], Setareh Shahab-Ferdows [1], Muttaquina Hossain [3], M. Munirul Islam [3], Tahmeed Ahmed [3] and Lindsay H. Allen [1,2]

1. USDA/ARS Western Human Nutrition Research Center, 430 West Health Sciences Drive, Davis, CA 95616, USA
2. Department of Nutrition, University of California, One Shields Ave, Davis, CA 95616, USA
3. Nutrition and Clinical Services Division, International Centre for Diarrhoeal Disease Research, 68 Shaheed Tajuddin Ahmed Sarani, Mohakhali, Dhaka 1212, Bangladesh
* Correspondence: daniela.hampel@usda.gov or dhampel@ucdavis.edu; Tel.: +1-530-752-9540

Received: 22 June 2019; Accepted: 17 July 2019; Published: 26 July 2019

Abstract: Human-milk-targeted metabolomics analysis offers novel insights into milk composition and relationships with maternal and infant phenotypes and nutritional status. The Biocrates Absolute*IDQ*® p180 kit, targeting 40 acylcarnitines, 42 amino acids/biogenic amines, 91 phospholipids, 15 sphingolipids, and sum of hexoses, was evaluated for human milk using the AB Sciex 5500 QTRAP mass-spectrometer in liquid chromatography-tandem mass-spectrometry (LC-MS/MS) and flow-injection analysis (FIA) mode. Milk (<6 months lactation) from (A) Bangladeshi apparently healthy mothers (body mass index (BMI) > 18.5; n = 12) and (B) Bangladeshi mothers of stunted infants (height-for-age Z (HAZ)-score <−2; n = 13) was analyzed. Overall, 123 of the possible 188 metabolites were detected in milk. New internal standards and adjusted calibrator levels were used for improved precision and concentration ranges for milk metabolites. Recoveries ranged between 43% and 120% (coefficient of variation (CV): 2.4%–24.1%, 6 replicates). Milk consumed by stunted infants vs. that from mothers with BMI > 18.5 was lower in 6 amino acids/biogenic amines but higher in isovalerylcarnitine, two phospholipids, and one sphingomyelin ($p < 0.05$ for all). Associations between milk metabolites differed between groups. The Absolute*IDQ*® p180 kit is a rapid analysis tool suitable for human milk analysis and reduces analytical bias by allowing the same technique for different specimens. More research is needed to examine milk metabolite relationships with maternal and infant phenotypes.

Keywords: human milk; targeted metabolomics; amino acids; lipid metabolites; LC-MS; flow injection analysis

1. Introduction

The postnatal period is a critical stage for infant's physiology, accompanied by rapid changes in brain development, and in metabolic, immunologic, intestinal, and physiological systems [1]. Human milk undergoes dynamic changes in composition during lactation to provide all nutrients and other bioactive substrates for optimal infant growth and development, and exclusive breastfeeding (EBF) is recommended during the first 6 months of life [2]. Even beyond the EBF period, it remains a vital dietary source of nutrients; however, the mechanisms behind its ability to drive and affect infant metabolism, health, development, and long-term outcomes are not fully understood. The recent resurgent interest in human milk composition triggered the development of state-of-the-art methods for analyzing micronutrients [3–10], peptides and oligosaccharides (HMOs) [11,12], the human milk microbiome [13,14], and milk lipids [15,16]. Within this scope, metabolomics provides novel insight into the complexity of the milk metabolome and its influence on infant metabolism and physiology via intestinal enzymes [17,18].

The effects of human milk bioactives on early infant physiology and development are diverse; effects may reach the insulin signaling cascade, the hepatic mitochondrial system, and possibly the mechanistic target of rapamycin (mTOR) signaling pathway, vital to cell growth, protein and lipid synthesis, or lipid and adiposity accumulation and adipogenesis [1,19–22]. On the other hand, associations of human milk metabolites were also found with maternal phenotype and diet [22,23]. The observed relationships of milk metabolites and maternal and infant status and phenotypes demand further investigation and analyzing the human milk metabolome as the interface between mother and infant can provide key information for maternal and infant health. Moreover, milk metabolites are not only derived from maternal blood but also from *de novo* synthesis in the mammary gland [23,24]. Consequently, obtaining the metabolic profile of the mothers as well as their milk may offer important insights into the contributions of the maternal metabolic state to the milk metabolome compared to the mammary-gland-derived compounds. Using the same analytical technique on different matrices reduces method bias and produces more robust results.

The Absolute*IDQ*® p180 kit (https://www.biocrates.com/images/p180_Folder_HP_v01-2018.pdf) from Biocrates Life Science AG covers multiple compound classes (acylcarnitines, amino acids, biogenic amines, sphingolipids, phospholipids, and sum of hexoses) which are involved in various central metabolic processes, such as immune regulation, fatty acid oxidation, membrane composition, cell cycle control, insulin resistance, or nutritional status. While this assay has been validated for various species and matrices, the human milk matrix has not yet been evaluated. Therefore, the purpose of this study was to assess application of this kit for the analysis of human milk metabolites. The validated assay was then used in a feasibility plate to analyze milk samples from (A) apparently healthy Bangladeshi mothers ("BMI > 18.5 mothers") (BMI—body mass index), < 6 months lactation) and (B) mothers of stunted infants ("stunted infants", height-for-age Z (HAZ)-score <−2) as proof of concept.

2. Materials and Methods

2.1. Chemicals and Materials

All reagents, internal and calibration standards, quality controls, test mix, and a patented 96-well filter plate required for the Absolute*IDQ*®p180 analysis were included in the kit or provided by Biocrates Life Science AG (Innsbruck, Austria).

2.2. AbsoluteIDQ® p180 Assay and Sample Preparation

The Absolute*IDQ*® p180 kit is a fully automated assay based on phenylisothiocyanate (PITC) derivatization of the target analytes in bodily fluids (e.g., plasma, serum, urine) using internal standards for quantitation. Amino acids and biogenic amines are determined in LC-MS mode, acylcarnitines, phospholipids (lyso-phosphatidylcholines with acyl residue at CXX:X, phosphatidylcholine with diacyl residue sum CXX:X (PC aa), and phosphatidylcholine with acyl-alkyl residue sum CXX:X (PC ae)), sphingomyelins, and the sum of hexoses are analyzed in flow injection analysis (FIA). Human milk sample preparation was carried out according to the manufacturer's protocol. Briefly, 2 to 10 µL of human milk was transferred to the upper 96-well plate and dried under a nitrogen stream. Thereafter, 50 µL of a 5% PITC solution was added to derivatize amino acids and biogenic amines. After incubation, the filter spots were dried again before the metabolites were extracted using 5 mM ammonium acetate in methanol (300 µL) into the lower 96-well plate for analysis after further dilution using the MS running solvent A. Quantification was carried out using internal standards and a calibration curve (Cal 1 to Cal 7). The experimental metabolomics measurement technique is described in detail by patents EP1897014B1 and EP1875401B1 [25,26]. See Supplementary Materials (Table S1) for the full list of metabolites and their abbreviations.

2.3. LC-MS

The LC-MS/MS system was comprised of an ACQUITY UPLC-system (Waters, Milford, MA, USA) coupled to a QTRAP 5500 mass spectrometer (AB Sciex, Redwood City, CA, USA) in electrospray ionization (ESI) mode. Amino acids and biogenic amines were analyzed via LC-MS in positive mode. Two microliters of the sample extract were injected onto an ACQUITY UPLC BEH C18 column, 2.1 × 7.5 mm, 1.7 µm protected by an ACQUITY BEH C18, 1.7 µm VanGuard pre-column (Waters, Milford, MA, USA) at 50 °C using a 7.3 min solvent gradient employing 0.2% formic acid in water (solvent A) and 0.2% formic acid in acetonitrile (solvent B).

Twenty microliters of the sample extract were used in the flow injection analysis (FIA) in positive mode to capture acylcarnitines, glycerophospholipids, and sphingolipids, while hexoses were monitored in a subsequent run in negative mode. All FIA injections were carried out using the Biocrates MS Running Solvent. Additional LC and MS settings for LC-MS and FIA mode are described in Table 1. All metabolites were identified and quantified using isotopically-labeled internal standards and multiple reaction monitoring (MRM) as optimized and provided by Biocrates Life Sciences AG (Innsbruck, Austria).

Table 1. LC and MS parameters [1].

Instrument	Parameter	LC-MS	FIA (MS Only)
5500 QTRAP MS	CUR	45	20
	IS	5500	5500
	TEM	500	200
	GS1	60	40
	GS2	70	50
	CAD	9	9
	EP	10	10
	CXP	15	15
ACQUITY UPLC	Time [min]	Flow rate [mL/min]	Solvent A [%]
LC-MS	0	0.8	100
	0.45	0.8	100
	3.3	0.8	85
	5.9	0.8	30
	6.05	0.8	0
	6.2	0.9	0
	6.42	0.9	0
	6.52	0.8	0
	6.7	0.8	100
	7.3	0.8	100
FIA	0	0.03	
	1.6	0.03	Biocrates Solvent I MS running buffer in isocratic mode
	2.4	0.2	
	2.8	0.2	
	3.0	0.03	

[1] CUR—curtain gas; IS—ion spray voltage (V);—temperature (°C); GS1/GS2—ion source gas 1 and 2 (psi); CAD—CAD gas (psi); EP—entrance potential (V); CXP—collision cell exit potential (V); LC-MS—liquid chromatography—mass spectrometry; FIA—flow injection analysis.

2.4. Method Validation

Pooled milk was used to evaluate matrix effects and metabolite recovery. The samples were diluted in phosphate buffered saline solution (PBS) 1:2 and 1:5 to assess ion suppression due to the matrix. A 3-level analyte recovery was carried out using Cal 2, 3, and 5 in half, normal, and double concentrations, covering up to 230-fold of the endogenous milk concentrations. All experiments were conducted in replicates of six. Individual milk samples (n = 25) from Bangladeshi women were prepared according to the manufacturer's protocol and analyzed at 3 different sample volumes (2, 5, and 10 µL).

Analyte recovery was calculated using the following equation:

$$\text{Recovery [\%]} = (C_{measured} - C_{endogenous}) \times 100/C_{added} \quad (1)$$

where $C_{measured}$ is the measured concentrations of the spiked sample, $C_{endogenous}$ the measured concentration of the non-spiked sample, and C_{added} the theoretically added concentration by spiking.

Relative recovery in the diluted samples was calculated as follows:

$$\text{Relative recovery [\%]} = C_{X/dF} \times dF \times 100/C_X \quad (2)$$

where $C_{X/dF}$ is the concentration measured in the diluted sample, dF the dilution factor (2 or 5), and C_X the concentrations measured in the non-diluted milk samples.

2.5. Human Milk Samples

To determine recovery rates and evaluate matrix effects, a human milk pool, obtained from milk samples provided by apparently healthy women in the Vancouver, BC, Canada area, was used. Individual milk samples, available in the laboratory but not collected for the purpose of this validation, were analyzed in the feasibility plate. These convenient samples were collected during the first 6 months of lactation from apparently healthy Bangladeshi lactating women ("BMI > 18.5 mothers", body mass index (BMI) > 18.5, n = 12) whose infants' anthropometry was unknown, and Bangladeshi mothers with stunted infants ("stunted infants", HAZ-score < −2, n = 13), to try to obtain a wide range of metabolite concentrations in the milk. All milk samples were shipped from the United States Department of Agriculture – Agricultural Research Service (USDA/ARS) Western Human Nutrition Research Center in Davis, CA, USA on dry ice to Biocrates Life Science AG in Innsbruck, Austria for testing. Upon arrival, samples were stored at −80 °C until analysis.

2.6. Statistical Analysis

Raw data was computed in MetIDQ^{TM} version Carbon (Biocrates Life Science AG, Innsbruck, Austria). Mean, standard deviation, and coefficient of variation (CV) for the validation were calculated in Excel 2016 (Microsoft Corporation, Redmond, WA, USA). R statistical software (3.5.2, R Foundation for Statistical Computing) was used for statistical analysis and visualization of the results for the feasibility study. The Wilcoxon rank sum test was used to explore differences in analyte concentrations based on group assignment (BMI > 18.5 mothers vs. stunted infants), and metabolite associations were assessed using Spearman's rank correlation. p-values < 0.05 were considered significant.

3. Results

3.1. Validation Using Pooled Human Milk

3.1.1. LC-MS (Amino Acids and Biogenic Amines)

Since the development of the assay, new internal standards became available for lysine, acetyl-ornithine, alpha-aminoadipic acid, cis-4-OH-proline, histamine, kynurenine, nitro-tyrosine, phenylethylamine (PEA), symmetric dimethylarginine (SDMA), and trans-4-OH-proline. Using these new internal standards, mean CVs (range) were reduced from 11.3% (3.6%–26.6%) to 6.6% (2.3%–11.5%; p = 0.017, Student's paired t-test). The quantification of glutamate and taurine required additional calibrators (Cal 8 and 9), which was also true for putrescine (Cal 8). Saturation effects were observed for proline, tryptophan, and valine at Cal 5 and Cal 6 levels. The sensitivity of the 5500 QTRAP MS-system allowed the extension to calibrators (Cal 0.5 and 0.25) for most of the metabolites and was needed for asparagine, methionine, ornithine, tryptophan, asymmetric dimethylarginine (ADMA), kynurenine, methionine-sulfoxide, sarcosine, and SDMA. However, alanine, spermidine, spermine, and ADMA could not be accurately measured at the Cal 0.25 level.

All 21 amino acids and 12 biogenic amines were detectable in the pooled human milk sample. Overall recovery across three spiking levels ranged between 79%–106% (CV: 4.5%–19.2%) for amino acids, and 81%–108% (CV: 2.4%–11.0%) for biogenic amines, as shown in Figure 1a and Supplementary Table S2. Compared to the non-diluted milk samples, the diluted pooled milk revealed a relative concentration range for amino acids of (1:2/1:5 dilution) 70%–100%/67%–104% (CV: 2.4%–15.7%/2.7%–11.0%) and 87%–119%/80%–102% (CV: 0.8%–13.5%/10.8%–14.6%) for biogenic amines, as shown in Figure 1b and Supplementary Table S3. Ornithine relative concentration was in good agreement with the non-diluted milk samples for the 1:2 dilutions (96.7%) but revealed a higher variation of CV = 25.3%. However, some of the amino acids and biogenic amines, including ornithine, were only observed in very low concentrations in the non-diluted pooled milk samples.

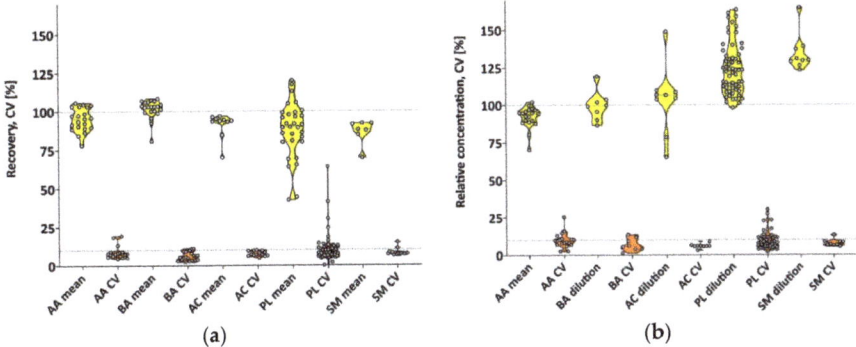

Figure 1. (a) Violin plots of mean milk metabolite recovery and coefficients of variation (CVs) determined by 3-level spiking experiments (n = 6); (b) relative milk metabolite recovery and CVs of the diluted milk samples. AA—amino acids; BA—biogenic amines; AC—acylcarnitines; PL—phospholipids; SM—sphingomyelins.

3.1.2. FIA (Acylcarnitines, Phospholipids, Sphingomyelins, Hexoses)

Sample extracts were diluted 1:20 for analysis using the 5500 QTRAP mass spectrometer. A relatively low recovery (CV) for PC aa C32:1 (42.6% (41.3%)) and to a lesser extent for PC aa C42:1 (44.4% (15.0%)) were observed. Overall, 89 (11 acylcarnitines, 63 phospholipids, 15 sphingomyelins) out of 146 FIA-analytes were quantifiable, with five additional metabolites detectable above (LOD) with no measurable carry-over effects. Overall recovery across the three spiking levels ranged between 70%–97% (CV: 5.8%–9.6%; acylcarnitines), 70%–92% (CV: 6.4%–8.6%; sphingomyelins), and 91.4% (CV: 7.1%) for the sum of hexoses. Excluding the above mentioned phospholipids, the recovery ranges between 64%–120% (CV: 3.9%–15.9%). PC ae C30:1 showed a good recovery of 85.7% but high variation (CV: 24.1%), as shown in Figure 1a and Supplementary Table S4.

Compared to the concentrations measured in the non-diluted pooled milk samples, acylcarnitines revealed relative concentration ranges (1:2/1:5 dilution) of 70%–134%/60%–164% (CV: 3.0%–7.0%/6.7%–10.7%), sphingomyelins ranged between 110%–167%/135%–228% (CV: 4.1%–13.0%/6.1%–20.5%), and hexoses 111%/143% (CV: 6.0%/10.5%). Phospholipids relative concentrations spanned between 97%–158%/95%–173% (CV: 1.8%–19.3%/2.3%–28.5%). Only five of the metabolites had a CV > 20% at the higher dilution. PC aa C32:1 and PC ae C30:1 experienced higher variations (CVs: 27% to 36.3%). PC aa C 38:1 was not detectable at a higher dilution and showed extreme variations at the lower dilution level with a mean relative concentration of 157% and CV = 55.7%, as shown in Figure 1b and Supplementary Table S5.

3.2. Feasibility Plate

Across all analyzed milk samples (BMI > 18.5 mothers and stunted infants), all 21 amino acids, 15 sphingomyelins, and the sum of hexoses were quantifiable. In addition, 10 biogenic amines, 11 acylcarnitines, and 54 phospholipids were observed, as shown in Supplementary Table S6.

Milk from BMI > 18.5 mothers had significantly higher concentrations of the amino acids and biogenic amines citrulline, glutamate, glycine, phenylalanine, serine, and the biogenic amine sarcosine (all $p < 0.045$), and lower concentrations of isovalerylcarnitine, phosphatidylcholine with diacyl residue sum C36:6 (PC aa), phosphatidylcholine with acyl-alkyl residue sum C30:2 (PC ae), and sphingomyelin with acyl residue sum C22:3 (all $p < 0.048$), as shown in Table 2, when compared to milk fed to stunted infants.

Table 2. Median concentrations and interquartile range (IQR) for milk metabolites affected by group (BMI > 18.5 mothers vs. stunted infants).

Metabolite [1]	BMI > 18.5 Mothers	Stunted Infants	p-Value [2]
	[µmol/L]		
C 5	0.13 (0.12, 0.14)	0.19 (0.17, 0.37)	0.011
PC aa C36:6	0.028 (0.025, 0.028)	0.031 (0.028, 0.035)	0.044
PC ae C30:2	0.011 (0.01, 0.011)	0.012 (0.012, 0.013)	0.047
SM C22:3	0.017 (0.014, 0.019)	0.022 (0.018, 0.037)	0.014
Citrulline	18.5 (12.2, 22.3)	10.6 (8.7, 16.3)	0.021
Glutamate	1825 (1477, 1977)	1181 (976, 1472)	0.002
Glycine	116 (103, 147)	96.0 (79.5, 104)	0.017
Phenylalanine	12.5 (11.7, 16.1)	11.2 (10.3, 12.0)	0.012
Serine	142 (94.2, 184)	85.0 (79.4, 108)	0.026
Sarcosine	0.61 (0.52, 0.72)	0.46 (0.38, 0.51)	0.044

[1] C 5—isovalerylcarnitine/2-methylbutyrylcarnitine/valerylcarnitine; PC aa C36:6—phosphatidylcholine with diacyl residue sum C36:6; PC ae C30:2—phosphatidylcholine with acyl-alkyl residue sum C30:2; SM C22:3—sphingomyelin with acyl residue sum C22:3. [2] p-values by Wilcoxon rank sum test.

While milk metabolites were significantly associated ($p < 0.05$) in both groups, the correlation profile differed considerably. These significant relationships were found not only among metabolites that were significantly different in concentration between the two groups but also among metabolites that had similar concentrations in all milk samples across groups, as shown in Figure 2.

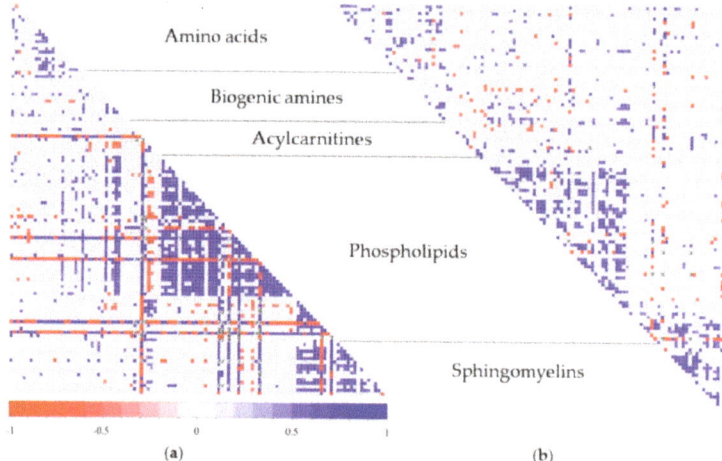

Figure 2. Spearman rank correlation map of milk metabolites in (**a**) BMI > 18.5 mothers and (**b**) stunted infants groups. Only significant correlations are shown ($p < 0.05$).

4. Discussion

4.1. Validation Using Pooled Human Milk

LC-MS: The new internal standards for amino acids and biogenic amines, and the extension of the calibrator curve allowed a more accurate analysis of the milk metabolites. While proline, tryptophan, and valine revealed saturation effects at mid calibrator levels, the measured milk concentrations of these amino acids, however, were well within the linear calibrator curve and the high level calibrators were not needed. While the low range of the calibrator curve was also extended to Cal 0.5 and 0.25, alanine, spermidine, spermine, and ADMA were not quantifiable at the lowest calibrator. Given that these metabolites were present in milk above Cal 0.5, the low accuracy at Cal 0.25 concentrations was not problematic. Some of the amino acids and biogenic amines, including ornithine, were found already in very low concentrations in the non-diluted pooled milk sample; consequently, their concentrations in the diluted samples were around or below the limit of detection (LOD) and reliable data could not be obtained. For metabolites within the calibrator curve, the diluted samples were mostly within 80%–120%, indicating some but not serious effects for quantification due to the matrix. The lower ion spray voltage (IS) intensity found for milk samples when compared to plasma further confirmed effects due to the milk matrix. To reduce matrix interferences due to early eluting matrix components (e.g., lactose), results for creatinine, the first eluting metabolite in LC-MS mode could be sacrificed to maintain MS-integrity and accuracy over time.

FIA: Phospholipids showed the greatest variation in metabolite recovery, and also represented the largest compound class of metabolites analyzed. Some of the variation could be explained by low abundance of metabolites at the edges of the calibrator range, or standard additions possibly to levels of saturation and therefore loss of linearity and accuracy. However, selectivity issues by milk-specific interferences cannot be excluded. No extended calibrator curve was necessary in FIA mode. The relative concentration of the analytes in the diluted samples compared to the non-diluted samples of generally above 100%, and often above 150% at higher dilutions, suggested some ion suppression due to the matrix which can be overcome by sample dilution. However, no metabolite was affected by the matrix to a degree that would not allow quantitation. However, given the abundance of lactose in the sample (~7%), a greater dilution would be preferable to reduce contamination of the MS by matrix constituents in FIA mode.

We found all detectable amino acids, sphingomyelins, and the sum of hexoses in human milk, but only 25% of the acylcarnitines, 50% of the biogenic amines, and 68% of the phospholipids. A recent study on plasma metabolites of healthy men and women (18–80 years) using the AbsoluteIDQ® p180 kit reported fewer of the amino acids, biogenic amines, acylcarnitines, and sphingomyelins in plasma, but more phospholipids [27], illustrating the variations in metabolite profile depending on specimen analyzed. Moreover, that study reported that the metabolic profile of plasma and urine can predict sex with >90% accuracy, as well as age (based on gender), and within females the menopausal status could be predicted with >80% accuracy. Other studies showed that differences in plasma metabolite profiles could be a useful diagnostic tool for Alzheimer's disease and cognitive impairment, or type 2 diabetes [28,29], further indicating the importance and relevance of the information than can be derived from the milk metabolome to the phenotype and status [27].

4.2. Feasibility Plate

Differences by Group Assignment

Overall, the metabolites detected in the feasibility plate using milk from Bangladeshi women were similar to those in the pooled human milk sample. The observed higher concentrations in milk from BMI > 18.5 mothers for selected amino amides and biogenic amines and lower concentrations of phospholipids, acylcarnitines, and sphingomyelins are consistent with alterations found in the serum metabolome of stunted compared to non-stunted children (1 to 5 years of age) in Malawi by Semba et al., using the same AbsoluteIDQ® p180 kit [30]. Stunting can be caused by intrauterine growth retardation, inadequate nutrition for optimal infant growth and development, or repeated infections early in life, and usually originates in utero [31]. Thus, the reported low circulating amino acids and other metabolome alterations in the blood of stunted children in Malawi may stem from challenges in utero and early infancy, including possible effects derived from the lower supply of amino acids in human milk as reported here. This hypothesis is supported by the fact that human growth is regulated by the mechanistic target of rapamycin complex C1 (mTORC1), a major growth regulating pathway, which suppresses protein and lipid synthesis and cellular growth when intake of specific amino acids is insufficient [30,32]. However, since we do not have the full information about maternal and infant characteristics, including exact stage of lactation for each sample, we cannot exclude that additional maternal or infant factors possibly affect the milk metabolome. Nonetheless, correlation patterns changed in milk fed to stunted infants, indicating alterations of the milk metabolome and its pattern could be related to the infant phenotype. Since the alterations in the metabolome correlation profile were not limited to metabolites showing significant differences in concentrations, these relationships may be also affected by maternal or infant status. However, our sample size was very small, the anthropometric status of infants in the group of BMI > 18.5 mothers was unknown, and further research is needed to confirm and explore the results found in this feasibility study.

5. Conclusions

This is the first study to validate the commercially available AbsoluteIDQ® p180 assay for targeted metabolomics in the human milk matrix. A greater range of calibrators is needed to accommodate the wide range of milk metabolite concentrations; new available internal standards improved accuracy of the results. While the primary purpose of this study was the validation of the AbsoluteIDQ® p180 assay for the human milk matrix, and it was not our intention to draw conclusions about group differences in metabolites from the feasibility plate, the observed differences in concentrations and metabolome profiles in the two groups warrant further investigation of the mother–infant relationship and the importance of milk metabolome for infant optimal growth and development, which is now possible using identical targeted metabolomics assays for blood and milk samples.

Supplementary Materials: The following are available online at http://www.mdpi.com/2072-6643/11/8/1733/s1, Table S1: List and abbreviations of metabolites, Table S2: Metabolite concentrations (mean ± SD) and recovery rates (CV) of pooled milk in LC-MS mode, Table S3: Relative recovery (mean ± SD) of milk metabolites in diluted pooled human milk based on non-diluted samples in LC-MS mode, Table S4: Metabolite concentrations (mean ± SD) and recovery rates (CV) of pooled milk in FIA mode, Table S5: Relative recovery (mean ± SD) of milk metabolites in diluted pooled human milk based on non-diluted samples in FIA mode, Table S6: Median concentrations and interquartile ranges (IQR) of milk metabolites in the BMI > 18.5 mothers and the stunted infants groups.

Author Contributions: Conceptualization and Study Design and Validation, D.H., S.S.-F., and L.H.A.; Field Study: M.H., M.M.I., and T.A.; Formal Analysis, D.H.; Writing—Original Draft Preparation, D.H.; Writing—Review and Editing, D.H. and L.H.A.; Funding Acquisition, L.H.A. All authors read and approved the final version of the manuscript.

Funding: This research was funded by the Bill & Melinda Gates Foundation (OPP1148405 & OPP1164613), and USDA-Agricultural Research Service intramural Project 5306-51000-004-00D.

Acknowledgments: The authors thank Denise Sonntag and Cornelia Roehring at Biocrates Life Science AG for support in data interpretation, and Excel Que for statistical analysis. USDA is an equal opportunity employer and provider.

Conflicts of Interest: The authors declare no conflict of interest.

References

1. Slupsky, C. Metabolomics in Human Milk Research. In *Human Milk: Composition, Clinical Benefits and Future Opportunities*; Donovan, S.M., German, B.J., Lonnerdal, B., Lucas, A., Eds.; Karger AG: Basel, Switzerland, 2019; Volume 90, pp. 179–190.
2. World Health Organization. *The Optimal Duration of Exclusive Breastfeeding*; Report of an Expert Consultation; WHO: Geneva, Switzerland, 2001.
3. Hampel, D.; York, E.R.; Allen, L.H. Ultra-performance liquid chromatography tandem mass-spectrometry (UPLC-MS/MS) for the rapid, simultaneous analysis of thiamin, riboflavin, flavin adenine dinucleotide, nicotinamide and pyridoxal in human milk. *J. Chromatogr. B Anal. Technol. Biomed. Life Sci.* **2012**, *903*, 7–13. [CrossRef] [PubMed]
4. Hampel, D.; Shahab-Ferdows, S.; Domek, J.M.; Siddiqua, T.; Raqib, R.; Allen, L.H. Competitive chemiluminescent enzyme immunoassay for vitamin B12 analysis in human milk. *Food Chem.* **2014**, *153*, 60–65. [CrossRef] [PubMed]
5. Hampel, D.; Shahab-Ferdows, S.; Adair, L.S.; Bentley, M.E.; Flax, V.L.; Jamieson, D.J.; Ellington, S.R.; Tegha, G.; Chasela, C.S.; Kamwendo, D.; et al. Thiamin and Riboflavin in Human Milk: Effects of Lipid-Based Nutrient Supplementation and Stage of Lactation on Vitamer Secretion and Contributions to Total Vitamin Content. *PLoS ONE* **2016**, *11*, e0149479. [CrossRef] [PubMed]
6. Redeuil, K.; Bénet, S.; Affolter, M.; Thakkar, S.; Campos-Giménez, E. A Novel Methodology for the Quantification of B-Vitamers in Breast Milk. *J. Anal. Bioanal. Tech.* **2017**, *8*. [CrossRef]
7. Ren, X.N.; Yin, S.A.; Yang, Z.Y.; Yang, X.G.; Bing, S.; Ren, Y.P.; Zhang, J. Application of UPLC-MS/MS Method for Analyzing B-vitamins in Human Milk. *Biomed. Environ. Sci.* **2015**, *28*, 738–750. [PubMed]
8. Lildballe, D.L.; Hardlei, T.F.; Allen, L.H.; Nexo, E. High concentrations of haptocorrin interfere with routine measurement of cobalamins in human serum and milk. A problem and its solution. *Clin. Chem. Lab. Med.* **2009**, *47*, 182–187. [CrossRef] [PubMed]
9. Hampel, D.; Shahab-Ferdows, S.; Gertz, E.; Flax, V.L.; Adair, L.S.; Bentley, M.E.; Jamieson, D.J.; Tegha, G.; Chasela, C.S.; Kamwendo, D.; et al. The effects of a lipid-based nutrient supplement and antiretroviral therapy in a randomized controlled trial on iron, copper, and zinc in milk from HIV-infected Malawian mothers and associations with maternal and infant biomarkers. *Matern. Child. Nutr.* **2017**, *14*. [CrossRef]
10. Redeuil, K.; Vulcano, J.; Prencipe, F.P.; Bénet, S.; Campos-Giménez, E.; Meschiari, M. First quantification of nicotinamide riboside with B3 vitamers and coenzymes secreted in human milk by liquid chromatography-tandem-mass spectrometry. *J. Chrom. B Analyt. Technol. Biomed. Life Sci.* **2019**, *1110–1111*, 74–80. [CrossRef]
11. Huang, J.; Kailemia, M.J.; Goonatilleke, E.; Parker, E.A.; Hong, Q.; Sabia, R.; Smilowitz, J.T.; German, J.B.; Lebrilla, C.B. Quantitation of human milk proteins and their glycoforms using multiple reaction monitoring (MRM). *Anal Bioanal. Chem.* **2017**, *409*, 589–606. [CrossRef]
12. Wu, S.; Tao, N.; German, J.B.; Grimm, R.; Lebrilla, C.B. Development of an annotated library of neutral human milk oligosaccharides. *J. Proteome Res.* **2010**, *9*, 4138–4151. [CrossRef]
13. Fernández, L.; Langa, S.; Martín, V.; Maldonado, A.; Jiménez, E.; Martín, R.; Rodríguez, J.M. The human milk microbiota: Origin and potential roles in health and disease. *Pharmacol. Res.* **2013**, *69*, 1–10. [CrossRef] [PubMed]
14. McGuire, M.K.; McGuire, M.A. Got bacteria? The astounding, yet not-so-surprising, microbiome of human milk. *Cur. Opin. Biotechnol.* **2017**, *44*, 63–68. [CrossRef] [PubMed]
15. Sokol, E.; Ulven, T.; Færgeman, N.J.; Ejsing, C.S. Comprehensive and quantitative profiling of lipid species in human milk, cow milk and a phospholipid-enriched milk formula by GC and MS/MSALL. *Eur. J. Lipid Sci. Technol.* **2015**, *117*, 751–759. [CrossRef] [PubMed]
16. Sinanoglou, V.J.; Cavouras, D.; Boutsikou, T.; Briana, D.D.; Lantzouraki, D.Z.; Paliatsiou, S.; Volaki, P.; Bratakos, S.; Malamitsi-Puchner, A.; Zoumpoulakis, P. Factors affecting human colostrum fatty acid profile: A case study. *PLoS ONE* **2017**, *12*, e0175817. [CrossRef] [PubMed]
17. Marincola, F.C.; Dessì, A.; Corbu, S.; Reali, A.; Fanos, V. Clinical impact of human breast milk metabolomics. *Clin. Chim. Acta* **2015**, *451*, 103–106. [CrossRef] [PubMed]

18. Gay, M.; Koleva, P.; Slupsky, C.; Toit, E.; Eggesbo, M.; Johnson, C.; Wegienka, G.; Shimojo, N.; Campbell, D.; Prescott, S.; et al. Worldwide Variation in Human Milk Metabolome: Indicators of Breast Physiology and Maternal Lifestyle? *Nutrients* **2018**, *10*, 1151. [CrossRef] [PubMed]
19. Melnik, B. Milk—A nutrient system of mammalian evolution promoting mTORC1-dependent translation. *Int. J. Mol. Sci.* **2015**, *16*, 17048–17087. [CrossRef] [PubMed]
20. Russell, H.; Taylor, P.M.; Hundal, H.S. Amino acid transporters: Roles in amino acid sensing and signalling in animal cells. *Biochem. J.* **2003**, *373*, 1–18.
21. Moran-Ramos, S.; Ocampo-Medina, E.; Gutierrez-Aguilar, R.; Macías-Kauffer, L.; Villamil-Ramírez, H.; López-Contreras, B.E.; León-Mimila, P.; Vega-Badillo, J.; Gutierrez-Vidal, R.; Villarruel-Vazquez, R.; et al. An amino acid signature associated with obesity predicts 2-year risk of hypertriglyceridemia in school-age children. *Sci. Rep.* **2017**, *7*, 5607. [CrossRef]
22. Isganaitis, E.; Venditti, S.; Matthews, T.J.; Lerin, C.; Demerath, E.W.; Fields, D.A. Maternal obesity and the human milk metabolome: Associations with infant body composition and postnatal weight gain. *Am. J. Clin. Nutr.* **2019**, *110*, 111–120. [CrossRef]
23. Smilowitz, J.T.; O'Sullivan, A.; Barile, D.; German, J.B.; Lönnerdal, B.; Slupsky, C.M. The human milk metabolome reveals diverse oligosaccharide profiles. *J. Nutr.* **2013**, *143*, 1709–1718. [CrossRef] [PubMed]
24. Anderson, S.M.; Rudolph, M.C.; McManaman, J.L.; Neville, M.C. Key stages in mammary gland development. Secretory activation in the mammary gland: It's not just about milk protein synthesis! *Breast Cancer Res.* **2007**, *9*, 204. [CrossRef] [PubMed]
25. Ramsay, S.L.; Stoeggl, W.M.; Weinberger, K.M.; Graber, A.; Guggenbichler, W. Apparatus and method for analyzing a metabolite profile. Google Patents US8265877 B2, 11 September 2012.
26. Ramsay, S.L.; Guggenbichler, W.; Weinberger, K.M.; Graber, A.; Stoeggl, W.M. Device for Quantitative Analysis of a Drug or Metabolite Profile. Google Patents US8116983 B2, 14 February 2012.
27. Rist, M.J.; Roth, A.; Frommherz, L.; Weinert, C.H.; Krüger, R.; Merz, B.; Bunzel, D.; Mack, C.; Egert, B.; Bub, A.; et al. Metabolite patterns predicting sex and age in participants of the Karlsruhe Metabolomics and Nutrition (KarMeN) study. *PLoS ONE* **2017**, *12*, e0183228. [CrossRef] [PubMed]
28. Klavins, K.; Koal, T.; Dallmann, G.; Marksteiner, J.; Kemmler, G.; Humpel, C. The ratio of phosphatidylcholines to lysophosphatidylcholines in plasma differentiates healthy controls from patients with Alzheimer's disease and mild cognitive impairment. *Alzheimers Dement* **2015**, *1*, 295–302. [CrossRef] [PubMed]
29. Floegel, A.; Stefan, N.; Yu, Z.; Mühlenbruch, K.; Drogan, D.; Joost, H.G.; Fritsche, A.; Häring, H.U.; de Angelis, M.H.; Peters, A.; et al. Identification of serum metabolites associated with risk of type 2 diabetes using a targeted metabolomic approach. *Diabetes* **2013**, *62*, 639–648. [CrossRef] [PubMed]
30. Semba, R.D.; Shardell, M.; Ashour, F.A.S.; Moaddel, R.; Trehan, I.; Maleta, K.M.; Ordiz, M.I.; Kraemer, K.; Khadeer, M.A.; Ferrucci, L.; et al. Child stunting is associated with low circulating essential amino acids. *EBioMedicine* **2016**, *6*, 246–252. [CrossRef] [PubMed]
31. Dewey, K.G.; Begum, K. Long-term consequences of stunting in early life. *Matern. Child. Nutr.* **2011**, *7*, 5–18. [CrossRef] [PubMed]
32. Laplante, M.; Sabatini, D.M. mTOR signaling in growth control and disease. *Cell* **2012**, *149*, 274–293. [CrossRef] [PubMed]

© 2019 by the authors. Licensee MDPI, Basel, Switzerland. This article is an open access article distributed under the terms and conditions of the Creative Commons Attribution (CC BY) license (http://creativecommons.org/licenses/by/4.0/).

Article

The Effect of High-Dose Postpartum Maternal Vitamin D Supplementation Alone Compared with Maternal Plus Infant Vitamin D Supplementation in Breastfeeding Infants in a High-Risk Population. A Randomized Controlled Trial

Adekunle Dawodu [1,*], Khalil M. Salameh [2], Najah S. Al-Janahi [3], Abdulbari Bener [4] and Naser Elkum [5]

1. Department of Pediatrics, College of Medicine, University of Cincinnati, Cincinnati, OH 45267, USA
2. Division of Pediatrics, Al-Wakra Hospital, Hamad Medical Corporation, Doha, Qatar
3. Department of Obstetrics and Gynecology, Women's Hospital, Hamad Medical Corporation, Doha, Qatar
4. Department of Biostatistics and Medical Informatics, Cerrahpasa Faculty of Medicine, Istanbul University Cerrahpasa and Istanbul Medipol University, 34098 Cerrahpasa-Istanbul, Turkey
5. Sidra Medicine, Doha, Qatar
* Correspondence: adekunle_dawodu@yahoo.com or dawodua@ucmail.uc.edu; Tel.: +15136971546

Received: 10 May 2019; Accepted: 12 July 2019; Published: 17 July 2019

Abstract: In view of continuing reports of high prevalence of severe vitamin D deficiency and low rate of infant vitamin D supplementation, an alternative strategy for prevention of vitamin D deficiency in infants warrants further study. The aim of this randomized controlled trial among 95 exclusively breastfeeding mother–infant pairs with high prevalence of vitamin D deficiency was to compare the effect of six-month post-partum vitamin D_3 maternal supplementation of 6000 IU/day alone with maternal supplementation of 600 IU/day plus infant supplementation of 400 IU/day on the vitamin D status of breastfeeding infants in Doha, Qatar. Serum calcium, parathyroid hormone, maternal urine calcium/creatinine ratio and breast milk vitamin D content were measured. At baseline, the mean serum 25-hydroxyvitamin D (25(OH)D) of mothers on 6000 IU and 600 IU (35.1 vs. 35.7 nmol/L) and in their infants (31.9 vs. 29.6) respectively were low but similar. At the end of the six month supplementation, mothers on 6000 IU achieved higher serum 25(OH)D mean ± SD of 98 ± 35 nmol/L than 52 ± 20 nmol/L in mothers on 600 IU ($p < 0.0001$). Of mothers on 6000 IU, 96% achieved adequate serum 25(OH)D ($\geq$50 nmol/L) compared with 52%in mothers on 600 IU ($p < 0.0001$). Infants of mothers on 600 IU and also supplemented with 400 IU vitamin D_3 had slightly higher serum 25(OH)D than infants of mothers on 6000 IU alone (109 vs. 92 nmol/L, $p = 0.03$); however, similar percentage of infants in both groups achieved adequate serum 25(OH)D $\geq$50 nmol/L (91% vs. 89%, $p = 0.75$). Mothers on 6000 IU vitamin D_3/day also had higher human milk vitamin D content. Safety measurements, including serum calcium and urine calcium/creatinine ratios in the mother and serum calcium levels in the infants were similar in both groups. Maternal 6000 IU/day vitamin D_3 supplementation alone safely optimizes maternal vitamin D status, improves milk vitamin D to maintain adequate infant serum 25(OH)D. It thus provides an alternative option to prevent the burden of vitamin D deficiency in exclusively breastfeeding infants in high-risk populations and warrants further study of the effective dose.

Keywords: vitamin D deficiency; supplementation; breastfeeding; mothers; infants

1. Introduction

Chronic vitamin D deficiency resulting in rickets remains a serious childhood public health problem worldwide, particularly in breastfeeding infants who lack sun exposure and vitamin D supplementation in Asia, parts of Africa and the Middle East, including Qatar, and among the immigrant population from the above countries to Europe, Australia and New Zealand [1,2]. Although vitamin D is effective in preventing vitamin D deficiency, recent reports confirmed that nutritional rickets due to vitamin D deficiency is common and the prevalence is significantly higher in Middle Eastern countries [3], some parts of Asia [4] and among children of immigrant and minority populations than the Western population in prospective and cross-section studies [5]. Of 540 Qatari children of less than five years of age attending a primary health care clinic, 24% were found to have nutritional rickets [3]. A national survey of vitamin D deficiency in Mongolian children <5 years found that 42% of the children had serum 25-hydroxyvitamin D [25(OH)D] < 23 nmol/L and 50% of those with serum 25(OH)D < 23 nmol/L had clinical rickets [4]. Despite reports of high prevalence of nutritional rickets, especially in high risk populations worldwide [1–5], and the different recommendations from professional bodies and expert advisory groups on preventive measures [1,6], the prevalence of vitamin D deficiency associated with increased risk of rickets continues to be a significant public health problem, especially in breastfed infants [2,7–9] due to low vitamin D intake from the breast milk, low vitamin D supplementation and lack of sunlight exposure. The prevalence and magnitude of vitamin D deficiency depends on the definition used in the reported studies. The Institute of Medicine defines vitamin D deficiency as serum 25(OH)D < 30 nmol/L [9] while the Endocrine Society defines serum 25(OH)D < 50 nmol/L as vitamin D deficiency [10]. Some studies report the prevalence of 25(OH)D < 25 nmol/L as cutoff for vitamin D deficiency in unsupplemented breastfed infants with 43% in New Delhi, India [11], and 58% in a cohort of Arab breastfed infants in Doha, Qatar [12]. Furthermore, the prevalence of serum 25(OH)D < 30 nmol/L associated with increased risk of nutritional rickets [1] range from 13% in Cincinnati, Ohio [13], and 22% in Mexico City [13]. In a review of global vitamin D status based on articles published in PubMed/Medline, the prevalence of vitamin D deficiency as defined in the present study (serum 25(OH)D < 50 nmol/L) [10] in infants range from 24%, 40%, 46% to as high as 93%, 96% and 99% in Argentina, Australia, Black American (US), in Iran, Kuwait and India respectively [14].

Besides rickets, vitamin D deficiency or suboptimal vitamin D intake in infancy has also been shown to be associated with increased risk of lower respiratory tract infections in infancy and childhood [15,16] and vitamin D deficiency should, therefore, be regarded as a significant public health problem in children and prevention, particularly in the breastfed infants, requires heightened attention. The natural sources of vitamin D after birth in breastfeeding infants are previous transplacental transfer, human milk and sunlight exposure. The infant vitamin D stores at birth are dependent on maternal vitamin D status during pregnancy [17]. There are many documented reports of high prevalence of vitamin D deficiency during pregnancy worldwide associated with lack of sunlight exposure and inadequate vitamin D corrective supplements [8] that will predispose their infants to low vitamin D stores at birth [17]. The few studies that have also reported vitamin D status of breastfeeding mothers, indicate high prevalence of vitamin D deficiency (serum 25(OH)D < 50 nmol/L) ranging from 15% in Cincinnati [13], 51% in Shanghai [13], and 62% in Mexico City [13] which would theoretically be associated with low milk vitamin D intake for the nursing infant.

Since human milk contains low vitamin D (20–70 IU/L), it would be insufficient to meet the recommended daily intake of 400 IU of vitamin D for infants [6,9] when infant sun exposure is limited and the mothers are vitamin D deficient. Therefore, the current recommendation from professional body in the U.S. [6], and experts' opinion [1,9,10] is that all infants receive direct oral supplementation of 400 IU vitamin D/day to maintain adequate vitamin D status to prevent vitamin D deficiency. Compliance with vitamin D supplementation in breastfeeding infants has been reported to be very low and recent studies from the US found vitamin D supplementation rate of 5%–19% in fully breastfed infants during the first six months of life [18,19]. In addition, vitamin D supplementation of breastfeeding infants does not address the concomitant high prevalence of vitamin D deficiency in their mothers. Therefore, ensuring maternal vitamin D sufficiency has been suggested as part of the strategy to prevent vitamin D deficiency in breastfed infants and their mothers [8,20]. This led to a pilot study of high-dose maternal vitamin D supplementation alone to prevent vitamin D deficiency in breastfeeding mother and her nursing infant.

Because of increasing reports of rickets [21] and the high prevalence of vitamin D deficiency in breastfeeding infants [6,8,21] and the new data in adults on the safety of up to 10,000 IU/day of vitamin D supplementation [22], a larger study from Charleston, South Carolina in the US, compared the effect of daily high-dose maternal vitamin D supplementation alone of 6400 IU with maternal supplementation of 400 IU/day and direct infant supplementation of 300 IU/day for six months on vitamin D status of breastfeeding infants [23]. The authors found significantly increased maternal serum 25(OH)D and breast milk vitamin D content in the maternal high-dose group [23]. The mean serum 25(OH)D in the nursing infants following high-dose maternal supplementation alone were similar to those in infants following maternal plus infant supplementation. There were no safety concerns related to vitamin D supplementation in mothers and infants as measured by serum calcium levels and urine calcium–creatinine ratio [23]. In a more recent large National Institute of Health-funded randomized controlled trial the same group from Charleston, South Carolina, compared the effectiveness of maternal vitamin D_3 supplementation of 6400 IU/day alone with maternal plus infant supplementation with 400 IU/day in 334 exclusively breastfeeding mother–infant pairs. The authors confirmed that maternal supplementation alone with 6400 IU/day significantly increased maternal vitamin D status to meet the requirement of the nursing infant without safety issues [24]. The vitamin D content of the breast milk was not measured in this later study to assesses the contribution of milk vitamin D. In view of the reported high prevalence of vitamin D deficiency in Arab mother–infant breastfeeding dyads [8], reported low rate of infant vitamin D supplementation, and the possible effect of baseline vitamin D status on the response to supplementation dose [25,26], the present randomized controlled trial was conducted to compare the effect of high-dose maternal vitamin D supplementation alone with combined maternal and direct infant vitamin D supplementation on vitamin D status of breastfeeding infant in this high-risk population. To our knowledge, this was the first study to further examine and compare with previous landmark US studies the effect of daily maternal vitamin D supplementation alone on maternal and breast milk vitamin D status as part of a strategy to reduce the burden of vitamin D deficiency in breastfeeding mother–infant pairs in a population with endemic vitamin D deficiency.

1.1. Primary Aim

The effect of 6000 IU/day maternal vitamin D_3 supplementation alone was compared with maternal supplementation of 600 IU/day plus direct infant supplementation of 400 IU/day vitamin D_3 on the serum 25(OH)D levels of breastfeeding mothers and their infants including the percentage of mothers and infants that achieved a priori criteria of adequate serum 25(OH)D levels of ≥50 nmol/L [9].

1.2. Secondary Aim

The effect of maternal high-dose (6000 IU/day) vitamin D_3 and 600 IU/day maternal vitamin D_3 supplementation on human milk vitamin D content was also evaluated.

The hypothesis was that 6000 IU/day maternal vitamin D_3 supplementation alone would optimize vitamin D status of exclusively breastfeeding mother and maintain vitamin D status of the nursing infant at equivalent level to that of an infant on direct oral vitamin D_3 supplementation of 400 IU/day plus maternal 600 IU/day vitamin D_3 supplementation.

2. Materials and Methods

2.1. Trial Design, Setting and Participants

This was a randomized, controlled, double-blind trial of the effect of 6000 IU/day maternal vitamin D_3 supplementation alone versus maternal vitamin D_3 supplementation of 600 IU/day plus direct infant vitamin D_3 supplementation of 400 IU/day on the vitamin D status of breastfeeding infants in a sunny environment of Doha, Qatar. Arab breastfeeding mothers in Doha and Al-Wakra, Qatar, who delivered at term (37–42 weeks) at Al-Wakra Hospital, Hamad Medical Corporation (HMC) and planned to fully breastfeed their babies for the first 4–6 months postpartum were eligible for the study and were enrolled within four weeks of delivery and followed up for six months (during the period of August 2013–May 2016).

The study was approved by Hamad Medical Corporation and Weill Cornell Medical College Qatar Joint Institutional Review Board (JIRB No. 13-00036), and Cincinnati Children's Hospital Medical Center Institutional Review Board (IRB) (study No. 2013–4909). Subjects were included in the study if they met the following criteria: (1) Arab women who delivered at term and presented for routine follow-up within 4 weeks after delivery, (2) were self-reported to be in good health, (3) agreed to blood and milk collection at enrollment and follow up to blood draw from the infant for study investigations, (4) planned to fully breastfeed for at least 4–6 months and (5) would be available for follow up visits. Exclusion criteria were mothers with pre-existing calcium disorders, active thyroid disease, Type 1 diabetes or liver diseases, which are likely to affect vitamin D status of the mothers and those of the infants. Mothers of infants with major or multiple congenital anomalies were also excluded.

2.2. Initial Visit and Baseline Data

Each mother completed questionnaires on socio-demographic and health status. These included maternal age, nationality, and educational level and occupation of herself and the father of the infant. Pregnancy and delivery information, and infant dietary and neonatal history including growth parameters were recorded. Maternal and infant sunlight exposure behaviors was based on usual outdoor clothing and included body surface area (BSA) exposed to sunlight, duration of sun exposure outdoor (h/week) and sun index score (%BSA x h/week of sun exposure), which correlate with serum vitamin D status in adults and infants [18,27]. The season in which the blood samples were drawn were defined as "hot season" (April–September) and "cool season" (October–March). Baseline biochemical parameters including serum vitamin D status and calcium homeostasis were assessed as a function of season and sunlight exposure characteristics at first visit. The data provided information with which to compare the results derived from different supplementation groups and at other different time points.

Maternal weight and height were recorded to determine the body mass index (BMI), (weight (kg)/height (meter squared)). BMI was not included as criterion for exclusion because of the high prevalence of overweight and obesity in the Arab population. It was taken into account in the analysis as possible confounding variable.

2.3. Interventions

Vitamin D Supplementation, Randomization, Blinding

The mother's vitamin D tablets of 6000 IU or 600 IU were of similar color and taste and were manufactured and supplied by Tischon Corp (Salisbury, Maryland) as in our previous randomized controlled trial (RCT) [25] and each mother received a 100-day supply for three months visit and was repeated at the fourth visit. Biotics Research Corporation (Rosenberg, Texas, USA) manufactured and supplied infant vitamin D_3 drops and the placebo administered daily to each infant. The medication met FDA guidelines and has been used in breastfeeding infants in the USA [24]. The vitamin D drops contained 0 IU (placebo) and 400 IU that were similar in taste, appearance and smell. Mothers were instructed to administer one drop daily to their nursing infants. The investigators, patients, and health care providers were blinded to treatment.

The onsite physician-investigators, assisted by research nurse coordinators, enrolled eligible patients after obtaining informed consent. The consented mother–infant pairs were randomly assigned to compare two treatment regimens of vitamin D_3 supplementations: (a) high-dose maternal supplementation with 6000 IU/day of vitamin D_3 and the infant received placebo and (b) maternal supplementation with 600 IU/day of vitamin D_3 plus the infant receiving 400 IU/day of vitamin D_3 orally. The literature suggests that 1000 IU vitamin D supplementation would increase milk vitamin D by 80 IU/L [23,28]; therefore, high-dose 6000 IU vitamin D_3 maternal supplementation was chosen to improve maternal vitamin D status and increase milk vitamin D to a level that could meet the current need of the nursing infant. The current recommended lactating mother vitamin D intake of 600 IU and 400 IU infant intake [9] were chosen as control. The sampling procedure was designed by a statistician to achieve a seasonally balanced study population so that equal number of each group were enrolled during each of the two major seasons. A random assignment conducted as a stratified block design was computer generated to ensure equal number of mothers were randomly assigned to each of the two treatment groups monthly. Mothers on 6000 IU/day were provided with infant vitamin D drops with 0 IU vitamin D_3, and mothers on 600 IU/day were provided with infant vitamin D drops with 400 IU vitamin D_3. A secretary not involved in the project kept a list of randomization code.

2.4. Follow Up of Subjects

Research nurses assisted in the screening, enrollment, data collection and the follow up of mothers and the infants. The research coordinators completed the questionnaires and schedule appointments for blood draw for vitamin D and calcium homeostatic parameters, anthropometric measurement, pill count, as well as urine and breast milk collection. A computer generated calendar served as a reminder to contact the patients prior to their appointment. Each month, the study data manager and the research nurses generated electronic report on patient recruitment and retention, and every effort was made to reschedule patient for missed appointment.

2.5. Measurement of Outcome Variables

Maternal serum 25(OH)D, parathyroid hormone (PTH) and calcium were monitored at visit 1(enrollment) within 4 weeks postpartum, visit 4 (4 months postpartum after 3 months of vitamin D supplementation), and visit seven (seven months postpartum after six months of vitamin D supplementation). Infant serum 25(OH)D, PTH, and Ca, were also monitored at visits 1, 4 and 7. Serum calcium and maternal urine Ca/Cr ratio detects any possible episodes of hypercalcemia and hypercalciuria. The PTH versus vitamin D status relationship was included to compare response to vitamin D supplementation. Maternal urine pregnancy tests were performed monthly. If the mother was found to be pregnant, she would be informed and exit the study because of the uncertainty of the effect of giving high-dose maternal medication to the mother during the first trimester of pregnancy.

2.6. Laboratory Methods

Maternal and infant serum 25(OH)D and intact PTH. Total serum 25(OH)D levels were measured in HMC chemical laboratory by direct competitive chemiluminescence immunoassay on DiaSorin liaison platform (DiaSorin Liaison, Saluggia, VC, Italy). Serum intact PTH levels were measured also by chemiluminescence immunoassay (Unicel DxL 600, Beckman Coulter, Inc, CA, USA) as previously reported [12]. Vitamin D deficiency was defined a priori as serum 25(OH)D levels < 50 nmol/L [10] and vitamin D adequacy as serum 25(OH)D 50 nmol/L or greater [9] for this study.

Maternal and infant calcium measurement. The serum concentrations of calcium and urinary calcium and creatinine were measured in HMC Clinical Laboratory using standard analytic methods.

Breast milk vitamin D content. A 25 ml aliquot of full breast milk expression was collected at visits 1, 4 and 7 and frozen. The milk vitamin D content was measured by the use of LC-MS/MS Mass spectrometry techniques. Solvents were removed under a vacuum and samples were purified with the use of 2 chemically different HPLC systems. The final quantitation of the vitamin D and 25(OH)D was achieved by LC_MS/MS. The day-to-day inter-assay and intra-assay CVs for quantification were ≤ 12% [29]. The milk vitamin D contents were converted to antirachitic activity (ARA) using accepted conversion methods [29,30].

2.7. Safety Outcome

Each mother was monitored monthly for hypercalciuria. Vitamin D metabolites, serum calcium and urinary calcium/creatinine (Ca/Cr) ratio were monitored closely, and the results checked to detect any values in cautionary or higher than the predetermined parameters. Vitamin D_3 supplementation was stopped if maternal serum calcium was >2.75 mmol/L and urine Ca/Cr ratio > 1 mmol/mmol [31]. Serum 25(OH)D > 95 ng/ml was defined as upper limit of serum 25(OH)D [10]. In recent studies with high-dose vitamin D supplementation of lactating women there were no evidence of toxicity as shown by either hypercalcemia or hypercalciuria [23,24]. Nonetheless, to ensure safety, Data Safety Monitoring Board (including chemical pathologist, endocrinologist, and a pediatrician) monitored reports of adverse events and compliance.

2.8. Sample Size Calculation

The primary outcome was infant serum 25(OH)D concentration at the seventh month postpartum and the secondary outcome was percentage of infants that achieved adequate vitamin D status defined a priori as serum 25(OH)D 50 nmol/L or greater. Based upon the data from published study [23], we calculated a percentage of infants in each group with serum 25(OH)D levels of at least 50 nmol/L (20 ng/ml) as successful outcome. Using these percentages as probability of a successful outcome, we estimated that the sample size of 160 mother–infant pairs will provide 90% power to show statistical non-inferiority (defined as <5% difference in success rate) in infants whose mothers alone were supplemented with vitamin D compared with infants whose mothers were supplemented with 600 IU and the infants also received oral supplementation and allowing for unexpected attrition rate which resulted in a total of 190 mother–infant pairs with 95 mother–infant pairs per group. This sample size also allowed us to compare maternal serum and milk vitamin D levels in between the groups over time using appropriate statistical analysis including linear mixed model.

2.9. Statistical Analysis

All patient demographic and outcome data were entered into a Red Cap database. The primary variables collected were maternal and infant serum 25(OH)D concentrations at visits 1, 4 and 7. Univariate statistics were generated to examine their distribution as well as missing values and potential outliers. We investigated the distribution of measured variables across groups to observe if there were any differential distribution problems. Kolmogorov Smirnov test, histogram, Q-Q plot, and box plot were used to control for normality. If variables were not normally distributed, they were presented as medians and differences examined by non-parametric tests. Student-t test was used to ascertain the significance of differences between mean values of two continuous variables and confirmed by non-parametric Mann–Whitney test. Spearman's rank correlation coefficient was used to evaluate the strength of concordance between variables. A series of simple linear regressions were conducted to examine the contribution of each of the potential confounder or covariate variables on the outcome. Variables found to be significant in this analysis were included in the final model. The model was a general linear mixed model that incorporates the repeated nature of the data as well as potential confounder variables. Specifically, we used PROC and GENMOD with an identity link function and generalized estimating equation approach to model the relationship between vitamin D and the dose group while accounting for the potential confounding variables. We compared as secondary outcome the percentage of breastfeeding infants in the two groups that achieved serum 25(OH)D levels 50 nmol/L or greater at each time point in the study using Chi-squared test.

As secondary aim, analysis of variance was conducted to test the difference in terms of milk vitamin D content between the two maternal supplementation dose groups at the different time points. This was followed up with pairwise tests of the groups using Tukey's post hoc tests to adjust for the potential multiple testing and control for other factors that may be associated with changes in vitamin

D levels. Statistical analyses were performed using SAS 9.3 (SAS Institute Inc., Carry, NC, USA). The level $p < 0.05$ was considered as the cut-off value for significance.

3. Results

Of the 420 mothers that were assessed for eligibility, 190 consented to participate and were randomized with IRB acceptance to two treatment groups: maternal 6000 IU/day group plus infant placebo (0 IU/day vitamin D) versus maternal 600 IU/day group plus direct infant supplementation with 400 IU/day vitamin D_3. Sixty-two (32%) of the mothers did not continue participation after randomization without any specific explanation but their baseline data were included in the analysis. One hundred twenty-eight mothers (67%) continued active participation through visit 4 while 104 (55%) completed the study to visit 7 for analysis (Figure 1).

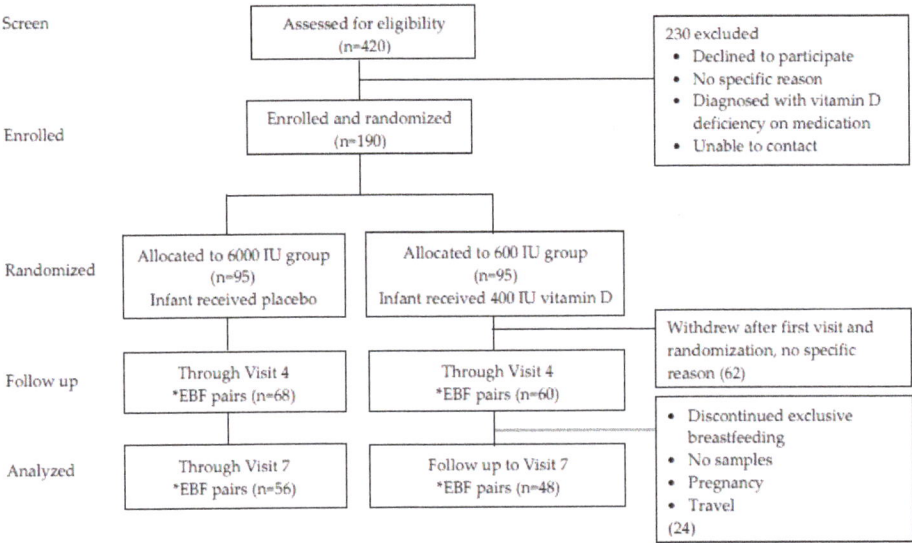

Figure 1. Flow chart of the participants throughout the study. *EBF—Exclusive Breastfeeding.

The mean age of the mothers was 29.6 years, weight was 76 kg, and BMI 29.3. There were no significant differences in all the baseline characteristics including vitamin D status of the mothers and infants between the two groups and breast milk vitamin D content (Tables 1 and 2). Of the mothers and infants with available data, 83% of the mothers were vitamin D deficient (serum 25(OH)D < 50 nmol/L) and 21% had serum 25(OH)D < 25 nmol/L, which is below serum 25(OH)D considered to be associated with osteomalacia [9]. Similarly, 82% of the infants were vitamin D deficient and 47% had serum 25(OH)D < 25 nmol/L, values which are associated with rickets [9]. The mean percentage of maternal body surface area (BSA) exposure outdoors of 13% with 0.32 hr/wk of sun exposure and mean sun index score of 6.6 were low. Although the infants BSA exposure outdoors (26%) was higher than in the mothers, the duration of sun exposure, 0.13 hr/wk, was lower. All the results indicated low maternal and infant sunlight exposure. On univariate analysis, maternal 25(OH)D correlated with BSA (rs = 0.2, $p = 0.005$) and infant 25(OH)D correlated with maternal BSA (rs = 0.19, $p = 0.009$) (Spearman's correlation).

Table 1. Baseline (Visit 1) characteristics and vitamin D status of exclusively breastfeeding mothers by maternal supplementation group.

Variables	N++	6000 IU Group (n = 95) Mean ± SD	N++	600 IU Group (n = 95) Mean ± SD	p-value
Age (years)	95	29.7 ± 5.0	95	29.5 ± 4.6	0.87
Weight (Kg)	95	77.4 ± 16.4	95	75.4 ± 16.4	0.40
Body Mass Index	95	29.4 ± 5.6	95	29.1 ± 6.0	0.33
Education	95		95		
• None/elementary		4		3	0.76
• High school		17		14	
• College/University		74		78	
Subjective health score	95	7.6		7.6	0.95
Season at enrollment (hot) %	95	50.5	95	51.6	0.88
Sun exposure behavior	95		95		
• % BSA exposure outdoors		11.7 ± 7.2		13.6 ± 12.1	0.19
• Sun exposure (h/week)		0.27 ± 0.88		0.38 ± 1.55	0.56
• Sun index score (% BSA x sun exposure h/week)		4.5 ± 15.9		8.7 ± 34.9	0.29
Serum Ca (mmol/L)	95	2.36 ± 0.08	94	2.34 ± 0.07	0.11
Urine CA/Cr ratio (mmol/mmol)	90	0.17 ± 0.15	90	0.14 ± 0.12	0.24
Serum 25(OH)D (nmol/L)	94	35.1 ± 16.3	94	35.7 ± 13.6	0.76
Serum PTH (pg/ml)	94	46.4 ± 25.7	93	50.4 ± 25.2	0.29
Breast milk vitamin D (IU/L)	92	25.5 ± 72.0	90	17.4 ± 42.4	0.36

++ N = number of observations. No significant differences between the two groups. PTH = parathyroid hormone.

Table 2. Baseline (visit 1) characteristics and vitamin D status of exclusively breastfeeding infants by maternal supplementation group.

Variables	N++	6000 IU Group (n = 95) Mean ± SD	N++	600 IU Group (n = 95) Mean ± SD	p-value
Weight (g)	95	3404 ± 509	95	3338 ± 448	0.34
Length (cm)	95	50.8 ± 2.9	95	50.8 ± 2.1	0.86
Head circumference (cm)	95	34.8 ± 2.2	95	34.7 ± 1.4	0.58
Sun exposure behavior					
• % BSA exposure outdoors	95	25.6 ± 13.2	95	26.3 ± 16.8	0.75
• Sun exposure (h/week)	95	0.15 ± 0.5	95	0.11 ± 0.39	0.55
• Sun index score (% BSA x sun exposure h/week)	95	5.1 ± 18.1	95	3.95 ± 14.7	0.63
Serum Ca (mmol/L)	95	2.70 ± 0.08	94	2.69 ± 0.09	0.79
Serum 25(OH)D (nmol/L)	93	31.9 ± 21.7	94	29.6 ± 16.1	0.41
Serum PTH (pg/ml)	91	30.4 ± 21.4	91	31.1 ± 21.8	0.84

++ N = number of observations. No significant differences were observed between the two groups.

3.1. Follow Up

3.1.1. Primary Outcome

The primary outcomes of the study were the serum 25(OH)D concentrations in both the mothers and the infants. Following intervention, maternal and infant serum 25(OH)D increased in both groups from baseline. The mean 25(OH)D concentrations were higher at visit 4 (88 nmol/L vs. 51 nmol/L, $p < 0.0001$) and at visit 7 (98 nmol/L vs. 51 nmol/L, $p < 0.0001$) in mothers on 6000 IU/day than those on 600 IU/day (Table 3). There was a higher serum 25(OH)D level in infants on direct vitamin D supplementation of 400 IU/day plus maternal supplementation of 600 IU/day than in infants of mothers on 6000 IU/day alone without infant supplementation ($p < 0.001$) at visit 4. At visit 7, serum 25(OH)D was slightly higher in infants on direct 400 IU/day vitamin D plus maternal 600 IU/day vitamin D supplementation compared with infants whose mothers alone were supplemented with 6000 IU/day vitamin D_3 ($p < 0.03$) (Table 3).

Table 3. Comparison of serum 25(OH)D concentrations in exclusively breastfeeding mothers and infants by group and visit.

Variables	Visit	N++	6000 IU Group (n = 95) Mean ± SD	N++	600 IU Group (n = 95) Mean ± SD	p-value
Maternal Serum 25(OH)D nmol/L	1	94	35.1 ± 16.3	94	35.7 ± 13.6	0.76
	4	68	88.3 ± 32.2	60	51.4 ± 15.7	<0.0001 *
	7	56	98.2 ± 36.5	48	51.7 ± 19.8	<0.0001 *
Infant Serum 25(OH)D nmol/L	1	93	31.9 ± 21.7	94	29.6 ± 16.1	0.41
	4	67	81.4 ± 26.5	60	105.5 ± 50.4	0.001 *
	7	55	92.2 ± 35.5	47	109.1 ± 43.3	0.03 *

++ N = number of observations. * Maternal serum 25(OH)D (nmol/l) and infant serum 25(OH)D (nmol/L) were significantly different between the two groups at visit 4 and visit 7 (Two-sided T-tests).

3.1.2. Secondary Outcome

The percentage of mothers and infants that achieved serum 25(OH)D ≥ 50 nmol/L, considered adequate by the Institute of Medicine based on the needs of 97.5% of the healthy population for bone health [9], were similar in the two groups at baseline; 18% in the 6000 IU group vs. 16% in the 600 IU group ($p = 0.70$). However, by visit 7, there was a significantly higher percentage of mothers in 6000 IU group with adequate vitamin D status than mothers in 600 IU group (96% vs. 52%, $p < 0.0001$) (Table 4).

Table 4. Categories of maternal and infant serum 25(OH)D status by visit and group.

Variables	Visit	6000 IU Group [n/N (%)]	600 IU Group [n/N (%)]	p-value
Maternal Serum 25(OH)D ≥ 50 nmol/L	1	17/94 (18%)	15/94 (16%)	0.70
	4	64/68 (94%)	30/60 (50%)	<0.0001 *
	7	54/56 (96%)	25/48 (52%)	<0.0001 *
Infant Serum 25(OH)D ≥ 50 nmol/L	1	18/93 (19%)	15/94 (16%)	0.54
	4	59/67 (88%)	53/60 (88%)	0.96
	7	49/55 (89%)	43/47 (91%)	0.75

n = number of observations with serum 25(OH)D value ≥ 50 nmol/L within the group and N is number of observations in each group. (%) is the percentage of serum 25(OH)D ≥ 50 nmol/L within each group. * Significant difference between the two groups of mothers at visits 4 and 7 (Chi-squared test).

For the infants, the percentage of infants with adequate vitamin D status (serum 25(OH)D ≥ 50 nmol/L) in the two groups were similar at all the visits and at visit 7, the end of study, (91% in infants on direct plus maternal supplementation vs. 89% in infants whose mothers alone were supplemented, $p = 0.75$) (Table 4).

3.1.3. Serum PTH Findings

Serum PTH were evaluated to assess the PTH and vitamin D response to vitamin D supplementation during the intervention. At visit 7, the higher PTH in mothers in 600 IU group was associated with significantly lower serum 25(OH)D than those of mothers in 6000 IU group (51.3 vs. 39.7 pg/mL) ($p < 0.003$) (Table 5). There were no differences in the serum PTH levels of infants in the two groups at any of the visit times (Table 5).

Table 5. Comparison of maternal and infant serum calcium, PTH and maternal urine Ca/Cr ratio and infant serum calcium and PTH by group and visit.

Variables	Visit	N++	6000 IU Group (n = 95) Mean ± SD	N++	600 IU Group (n = 95) Mean ± SD	p-value
Maternal						
Serum Ca (mmol/L)	1	95	3.36 ± 0.08	94	2.33 ± 0.07	0.11
	4	68	2.35 ± 0.08	60	2.33 ± 0.08	0.26
	7	58	2.34 ± 0.08	47	2.34 ± 0.09	0.86
Serum PTH (pg/ml)	1	94	46.4 ± 25.7	93	50.4 ± 25.1	0.29
	4	68	40.1 ± 22.3	61	47.0 ± 20.1	0.06
	7	58	39.7 ± 18.1	48	51.3 ± 25.6	0.003 *
Urine Ca/Cr ratio (mmol/mmol)	1	90	0.16 ± 0.15	90	0.14 ± 12.0	0.24
	4	66	0.24 ± 0.27	60	0.20 ± 0.15	0.24
	7	56	0.24 ± 0.20	49	0.19 ± 0.13	0.13
Infant						
Serum Ca (mmol/L)	1	95	2.69 ± 0.08	94	2.69 ± 0.09	0.31
	4	67	2.62 ± 0.09	60	2.64 ± 0.09	0.31
	7	57	2.57 ± 0.09	48	2.54 ± 0.09	0.15
Serum PTH (pg/ml)	1	91	30.4 ± 21.4	91	31.1 ± 21.8	0.84
	4	64	20.7 ± 12.0	59	21.3 ± 13.0	0.78
	7	57	26.7 ± 28.0	47	26.7 ± 13.2	0.98

++ N = number of observations. * The maternal serum PTH significantly different between the two groups (Two-sided T-tests).

3.1.4. Breast Milk Vitamin D Content

The breast milk vitamin D content in mothers in the two groups was evaluated to compare the effect of maternal vitamin D_3 supplementation of 6000 IU/day and 600 IU/day on milk vitamin D supply to their nursing infants. There was a very significant interaction between the milk vitamin D during the intervention (Figure 2). The mean vitamin D content at baseline were low in both groups of mothers (25 IU/L in 6000 IU group vs. 17.4 IU/L in 600 IU group, $p = 0.36$) (Table 1) and 57% of 182 lactating mothers had values below detection of the assay. After six months of vitamin D supplementation (visit 7), the mothers in 6000 IU group achieved higher mean vitamin D milk content of 202 ± 190 compared with 26.2 ± 38 in mothers in 600 IU group ($p < 0.0001$). The median values of milk vitamin D are compared in Table 6 using nonparametric tests since the data was not normally distributed.

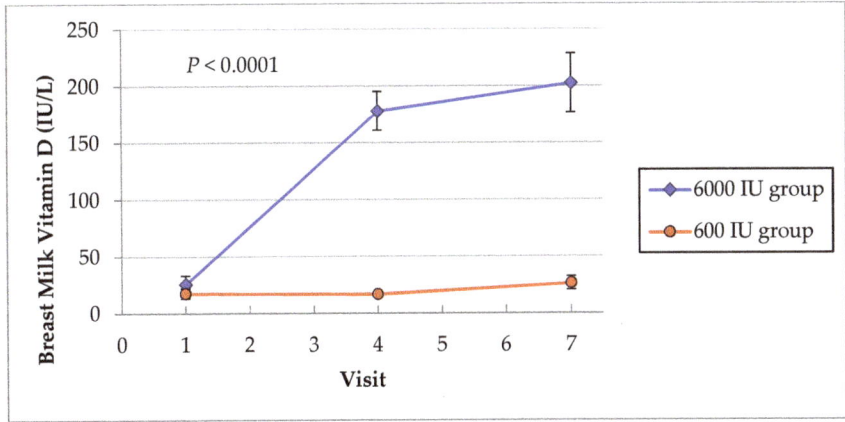

Figure 2. Breast milk vitamin D showed significant interactions between the groups. Mothers in 6000 IU group had substantial higher mean vD milk content of 202 IU/L compared with 26 IU/L in mothers in the 600 IU group at visit 7 ($p < 0.0001$).

Table 6. Comparison of breast milk vitamin D between groups at Visit 1, Visit 4, Visit 7.

Variables	Visit	N++	6000 IU Group ($n = 95$) Median (Range)	N++	600 IU Group ($n = 95$) Median (Range)	p-value
Breast Milk Vitamin D (IU/L)	1	92	8.1 (8.1, 532.0)	90	8.1 (8.1, 379.9)	0.7657
	4	68	185.9 (8.1, 644.0)	60	12.1 (8.1, 70.8)	<0.0001 *
	7	54	143.7 (8.1, 852.5)	41	14.3 (8.1, 203.1)	<0.0001 *

++ N = number of observations. * The median milk vitamin D content were significantly different between the two groups at visits 4 and 7 (Wilcoxon Mann-Whitney Test).

3.1.5. Safety Assessment

Total serum calcium levels were similar at baseline and at visits 4 and 7 in both groups of mothers. Similarly, the mean urine maternal calcium/creatinine ratios were similar at visits 1, 4 and 7 (Table 5) and there were no safety concerns. Serum calcium levels in the two groups of infants were also similar at visits 1, 4 and 7 (Table 5). The Data Safety Monitoring Board (DSMB) monitored the progress of the study and did not identify serious adverse events that warranted stopping the study.

3.1.6. Health Outcomes

A. The mothers on 6000 IU (group 1) had a self-reported health status that was not significantly different than those on 600 IU (group 2) at visit 4 ($p = 0.95$) and at visit 7 ($p = 0.83$).

B. The health status reported for infants of mother in group 1 was not significantly different than those of mothers in group 2 at visit 4 (Fisher's Exact test $p = 0.83$) and at visit 7 ($p = 1.0$).

C. Infant growth parameters, weight, length and head circumference, were also similar at baseline and on follow up between the two groups at visits 4 and 7 (Table 7).

Table 7. Comparison of Infant Birth Weight (g), Length (cm), and Head Circumference (cm) by Visit and Group.

Variables	Visit	N++	6000 IU Group (n = 95) Mean ± SD	N++	600 IU Group (n = 95) Mean ± SD	p-value
Weight (g)	1	95	3404.7 ± 509.3	95	3338.6 ± 448.4	0.34
Length (cm)		95	50.8 ± 2.9	95	50.9 ± 2.1	0.86
Head circumference (cm)		95	34.8 ± 2.2	95	34.7 ± 1.4	0.58
Weight (g)	4	68	7250.9 ± 849.3	62	7021.2 ± 783	0.11
Length (cm)		68	64.3 ± 2.5	62	64.1 ± 2.4	0.72
Head circumference (cm)		68	41.9 ± 1.5	62	41.6 ± 1.1	0.20
Weight (g)	7	58	8917.0 ± 1085.7	49	8789.0 ± 915.3	0.50
Length (cm)		58	70.7 ± 2.8	49	70.8 ± 3.1	0.84
Head circumference (cm)		58	44.8 ± 1.5	49	44.6 ± 1.0	0.32

++ N = number of observations. There were no significant differences between the two groups.

4. Discussion

The high prevalence of vitamin D deficiency in the present study as early as one month postpartum justifies the need for a modified strategy for prevention of vitamin D deficiency in mother–infant dyads in Arab breastfeeding population possibly starting from pregnancy. The severity of vitamin D deficiency suggests that vitamin D deficiency in mothers and infants is an unrecognized public health problem [12] which is detrimental to mothers and their infants.

Based on the definition of vitamin D deficiency, four out of five breastfed infants in this study had vitamin D deficiency at the first month of life, 47% had very low serum 25(OH)D level of <25 nmol/L, and over half (57%) had serum 25(OH)D lower than 30 nmol/L which is consistent with increased risk of rickets and lower respiratory infection without vitamin D supplementation or sunlight exposure [9,15,16]. The findings support the current reports that vitamin D deficiency in infants continues to be a significant global public health problem in low and middle resource countries worldwide [2,7] and warrants heightened attention. The explanations for the high prevalence of vitamin D deficiency in the exclusively breastfeeding infants in this study include lack of sunlight exposure shown by low mean sun index score of 4.52 compared with a mean sun index score of 68 in a cohort of breastfeeding infants in Cincinnati in summer [18], lack of vitamin D supplementation soon after birth due to "delay policy" of the Health Ministry of the country to supplement breastfeeding infants with vitamin D from age 2–12 months, and maternal vitamin D deficiency. Sunshine deprivation is common among mothers and infants in many Middle Eastern countries for cultural reasons and vitamin D supplementation rate of infants is low as in high income countries [18,19]. Since changes in sunlight exposure behavior may be difficult culturally, an alternative approach with high-dose maternal vitamin D supplementation alone was evaluated to optimize maternal vitamin D and milk vitamin D status to improve vitamin D nutrition of the nursing infant. Under research condition, the currently recommended direct supplementation of infants with 400 IU/day plus maternal supplementation of 600 IU achieved a slightly higher mean serum 25(OH)D of 109 nmol/L after 6 months of supplementation (visit 7) than 92 nmol/L in infants whose mothers alone were supplemented with 6000 IU/day and the infant received no supplementation ($p < 0.03$). Nonetheless, the percentage of infants in the two groups that achieved adequate serum 25(OH)D level of ≥50 nmol/L were similar (91% in infants on direct 400 IU/day plus maternal supplementation of 600 IU/day vs. 89% in infants of mothers in 6000 IU group and the infant received no supplementation ($p = 0.75$)) (Table 4). Furthermore, there were no differences in serum PTH response to vitamin D supplementation in the two groups at any of the visit times providing additional evidence of similarity in vitamin D status in response to vitamin D supplementation in both groups of infants (Table 5). Therefore, in the present study, maternal supplementation alone with 6000 IU/day vitamin D_3 achieved adequate infant vitamin D status, similar to direct infant supplementation of 400 IU/day plus maternal supplementation of 600 IU/day vitamin D_3.

A recent large randomized controlled trial indicated that serum 25(OH)D of infants who had direct vitamin D supplementation with 400 IU/day plus maternal supplementation of 400 IU/day and infants of mothers on 6400 IU/day supplementation alone and the infant received no supplementation were similar (108 vs. 98 nmol/L) [24] and our findings in infants are similar to theirs. It is also of interest that mean serum 25(OH)D of 90 nmol/L had been reported among unsupplemented infants of indigenous populations in East Africa who do not lack sunlight exposure [32], supporting a concept that "normal infant vitamin D concentration" could be naturally higher than is currently understood. The finding also suggests that vitamin D status could be "naturally higher" depending on sunshine and limited sun exposure in unsupplemented infants and be a significant contributor to infant vitamin D status.

Addressing maternal findings, the magnitude of vitamin D deficiency in the mothers was similar to those reported from India [11] but higher than 62% in breastfeeding mothers in Mexico City and 52% in mothers from Shanghai or 11% in a cohort of breastfeeding mothers in Cincinnati, USA [13,18]. In addition, the serum 25(OH)D levels of 35 nmol/L in the mothers were low compared with the baseline values of 82–90 nmol/L reported in breastfeeding mothers in a recent study from South Carolina, US [24]. Over one third of the mothers (36%) at baseline had serum 25(OH)D < 30 nmol/L which is consistent with increased risk of osteomalacia and non-skeletal disorders in adults [9,10] and is detrimental to the mothers, and vitamin D nutrition of their infants as referred to in earlier reports [23,24]. After 6 months of vitamin D supplementation, 6000 IU/day vitamin D_3 supplementation was significantly more effective in optimizing maternal vitamin D status than 600 IU/day. Serum 25(OH)D increased by 62 nmol/L between visit 1 and visit 7 in mothers in 6000 IU group compared with only 17.5 nmol/L in mothers in 600 IU group ($p < 0.0001$). In addition, a higher proportion of mothers (96%) in 6000 IU group compared with 52% of mothers in 600 IU group achieved adequate vitamin D status at visit 7 defined a priori as serum 25(OH)D ≥ 50 nmol/L ($p < 0.0001$). PTH levels in mothers in 600 IU group was higher than in mothers in 6000 IU group probably due to significantly lower vitamin D status in 600 IU group following vitamin D supplementation (Table 3). In the randomized controlled trial from the U.S. [24], maternal supplementation alone with 6400 IU/day vitamin D_3 also was more effective in elevating maternal serum 25(OH)D after 6 months of vitamin D supplementation than maternal supplementation with 400 IU/day. In that study there was a decline of 10.5 nmol/L in the 400 IU group but a 52 nmol/L increase after 6 months supplementation in the 6400 IU group ($p < 0.0001$). The increment in the serum 25(OH)D after six months of 6400 IU vitamin D supplementation is similar to the findings in this study. Therefore, the high-dose supplementation is more effective in optimizing maternal vitamin D status and the low dose is inadequate supplementation for high-risk, sunshine deprived population. It is of interest that at the end of the intervention 2% of mothers in the 6400 group and 13% in the 400 group would be considered vitamin D deficient (<50 nmol/L) in the US study [24]. In contrast, 4% of mothers in the 6000 group in the present study and almost half (48%) of those in the 600 group were considered vitamin D deficient. Part of the explanation for the difference in response to supplementation could be related to differences in baseline vitamin D status between mothers in this and the other study which has been shown to affect response to supplementation [25,26]. The mean 25(OH)D level of 98 nmol/L at the end of the study in 6000 IU group is still lower than mean 25(OH)D of 135 nmol/L reported at delivery among unsupplemented traditional female Tanzanian population in East Africa where mothers had no significant sunlight restriction as in breastfeeding mothers in Qatar [32]. Maintaining serum 25(OH)D ≥ 75 nmol/L has been associated with reduced risk of non-skeletal health disorders [10] and may be of theoretical benefit to a higher number of mothers in the 6000 IU group. If the objective of supplementation during lactation is to optimize vitamin D status in mother–infant pairs, then effective high-dose maternal supplementation still needs to be identified especially in high-risk population.

Monitoring of serum calcium, Ca/Cr ratio and serum 25(OH)D in our study did not show any significant difference between the two groups nor any serious adverse events that warrants stopping the study. Two infants had serum 25(OH)D of 94 ng/ml, below the predetermined upper limit of 25(OH)D without any other biochemical or clinical evidence of vitamin D toxicity. Safety parameters in two recent studies from South Carolina, US [23,24], which included serum calcium, and urine Ca/Cr ratios were similar in the two groups of mothers and infants, and revealed no evidence of vitamin D toxicity.

Although it is known that low human milk vitamin D content (range 5–80 IU/L) [24,33] contributes to the low vitamin D nutrition in unsupplemented breastfeeding infants who lack sun exposure, recent studies indicate that maternal vitamin D supplementation of 1000 IU/day would be expected to increase milk vitamin D content by 80 IU/L to improve vitamin D intake of nursing infants [23,28]. In the present study, baseline milk vitamin D levels were within the lower range of previously reported levels but 57% of the mothers in the two groups had undetectable vitamin D content and the milk provided no vitamin D for the nursing infant. After six months of vitamin D supplementation, the mean milk vitamin D level at visit 7 rose to a higher mean value of 202 IU/L in the 6000 IU group (Figure 2) or a median value of 143.7 IU/L (Table 6) than in the 600 group ($p < 0.0001$). The efficacy of the high-dose supplementation of 6000 IU to meet the current recommended infant vitamin D intake was lower than predicted. The lower efficacy may also be related to the effect of limited maternal sun exposure on breast milk vitamin D content as shown in another previous study [23]. It, however, supports our hypothesis that mothers in 6000 IU group would improve milk vitamin D, which would benefit her infant while mothers on 600 IU/day would only have marginal or lack milk vitamin D for their nursing infants. The mean vitamin D content of 202 IU/L in mothers in 6000 IU group was within the lower range of recommended vitamin D intake of 200–400 IU/day for nursing infants [24], while the milk vitamin D content of mothers in 600 IU group was very insufficient without direct vitamin D supplementation to meet the recommended daily intake of nursing infants, as shown in another study [23] in which milk vitamin D was measured. A recent large RCT which compared vitamin D status in infants of the mother who received 6400 IU/day alone with direct infant supplementation of 400 IU plus maternal supplementation of 400 IU/day also found similar vitamin D levels in the two groups of infants but the contribution of milk vitamin D was not compared because vitamin D content in the milk was not measured [24]. To the best of our knowledge, this would be the first study of the effect of high-dose maternal vitamin D supplementation on serum and milk vitamin D status as part of an alternative strategy to prevent vitamin D deficiency in infants in an Arab population with severe magnitude of vitamin D deficiency, and which provided comparison with results of previous landmark US studies [23,24]. Putting all the results together, 6000 IU/day maternal vitamin D supplementation alone appeared safe, optimizes maternal vitamin D status, increased milk vitamin D intake for nursing infants, and achieved similar adequate vitamin D status as in infants on direct 400 IU/day plus maternal 600 IU/day vitamin D supplementation.

There were no differences in the reported health status of the mothers and infants, and in the growth parameters of the infants between the two groups on follow up.

The limitations of our study include moderately adequate sample size, but a high dropout rate among the mothers mostly due to lack of compliance with exclusive breastfeeding and withdrawal from follow up with no specific reasons. Because the study is conducted in Arab breastfeeding mother and infant dyads, we only enrolled Arab mothers for this specific study. The results may be modified in other high-risk populations because of differences in sunlight exposure behavior and dose of vitamin D supplementation.

The present study provided additional valuable data to inform future research in other high-risk populations with similar population characteristics on the effective dose of daily high-dose maternal supplementation alone as an alternative option to reduce global burden of vitamin D deficiency in breastfeeding infants [24]. If successful, the strategy could provide data for health care providers to show that breast milk alone could be a source of vitamin D for breastfeeding infant with appropriate effective maternal high-dose vitamin D supplementation alone (6000–6400 IU or more). This option to prevent vitamin D deficiency in the breastfeeding mother–infant dyads would warrant public health education of health care professionals and providers. It is of interest that a recent study [34] found maternal preference for taking medication themselves as opposed to giving the vitamin D supplement to their infants. The authors suggested that maternal supplementation alone may improve vitamin D intake of the nursing infant in an environment where infant supplementation rate is low.

5. Conclusions

Maternal vitamin D_3 supplementation alone with 6000 IU/day achieved similar adequate vitamin D status as in infants on direct vitamin D_3 supplementation of 400 IU/day plus maternal 600 IU/day supplementation, and safely optimizes maternal vitamin D status with increase in milk vitamin D content. It supports high-dose maternal vitamin D supplementation alone as a possible alternative strategy to reduce the burden of global vitamin D deficiency in breastfeeding infants and their mothers and warrants more studies on appropriate effective dose especially in high-risk populations.

Author Contributions: Conceptualization, A.D., K.M.S., N.S.A.-J. and A.B.; formal analysis, N.E.; funding acquisition, A.D. and A.B.; investigation, A.D. and K.M.S.; methodology, A.D., K.M.S. and A.B.; project administration, A.D. and K.M.S.; software, N.E.; supervision, A.D. and K.M.S.; validation, N.E.; writing—original draft, A.D.; writing—review and editing, A.D., K.M.S., N.S.A.-J., A.B. and N.E.

Funding: This research was fully funded by Qatar Research Fund (NPRP6-1151-3-275) in Doha, Qatar.

Acknowledgments: Adriana Reedy and Bin Zhang provided data editing and statistical support; Ghadeer Mustafa provided supervision of data entry; Maram Fakhory and Heba Ibrahim for data collection; Zenica Ramos for project administrative matters; and Donna Bridges for administrative support. Bruce Hollis and Carol Wagner gave expert advice during the implementation of the project.

Conflicts of Interest: The authors declare no conflicts of interest. The funder had no role in the design of the study; in the collection, analysis or interpretation of data; in the writing of the manuscript; or in the decision to publish the results.

References

1. Munns, C.F.; Shaw, N.; Kiely, M.; Specker, B.L.; Thacher, T.D.; Ozono, K.; Michigami, T.; Tiosano, D.; Mughal, M.Z.; Makitie, O.; et al. Global consensus recommendations on prevention and management of nutritional rickets. *J. Clin. Endocrinol. Metab.* **2016**, *101*, 394–415. [CrossRef] [PubMed]
2. Creo, A.L.; Thacher, T.D.; Pettifor, J.M.; Strand, M.A.; Fischer, P.R. Nutritional rickets around the world: An update. *Paediatr. Int. Child Health* **2017**, *37*, 84–98. [CrossRef] [PubMed]
3. Bener, A.; Hoffmann, G.F. Nutritional rickets among children in a sun rich country. *Int. J. Pediatr. Endocrinol.* **2010**, *2010*, 410502. [CrossRef] [PubMed]
4. Uush, T. Prevalence of classic signs and symptoms of rickets and vitamin D deficiency in Mongolian children and women. *J. Steroid Biochem. Mol. Biol.* **2013**, *136*, 207–210. [CrossRef] [PubMed]
5. Wheeler, B.J.; Dickson, N.P.; Houghton, L.A.; Ward, L.M.; Taylor, B.J. Incidence and characteristics of vitamin D deficiency rickets in New Zealand children: A New Zealand Paediatric Surveillance Unit study. *Aust. N. Z. J. Public Health* **2015**, *39*, 380–383. [CrossRef]
6. Wagner, C.L.; Greer, F.R. Prevention of rickets and vitamin D deficiency in infants, children, and adolescents. *Pediatrics* **2008**, *122*, 1142–1152. [CrossRef]
7. Roth, D.E.; Abrams, S.A.; Aloia, J.; Bergeron, G.; Bourassa, M.W.; Brown, K.H.; Calvo, M.S.; Cashman, K.D.; Combs, G.; De-Regil, L.M.; et al. Global prevalence and disease burden of vitamin D deficiency: A roadmap for action in low- and middle-income countries. *Ann. N. Y. Acad. Sci.* **2018**, *1430*, 44–79. [CrossRef]
8. Dawodu, A.; Wagner, C.L. Prevention of vitamin D deficiency in mothers and infants worldwide—A paradigm shift. *Paediatr. Int. Child Health* **2012**, *32*, 3–13. [CrossRef]
9. Institute of Medicine. *Dietary Reference Intakes for Calcium and Vitamin D*; The National Academies Press: Washington, DC, USA, 2011.
10. Holick, M.F.; Binkley, N.C.; Bischoff-Ferrari, H.A.; Gordon, C.M.; Hanley, D.A.; Heaney, R.P.; Murad, M.H.; Weaver, C.M. Evaluation, treatment, and prevention of vitamin D deficiency: An Endocrine Society clinical practice guideline. *J. Clin. Endocrinol. Metab.* **2011**, *96*, 1911–1930. [CrossRef]

11. Seth, A.; Marwaha, R.K.; Singla, B.; Aneja, S.; Mehrotra, P.; Sastry, A.; Khurana, M.L.; Mani, K.; Sharma, B.; Tandon, N. Vitamin D nutritional status of exclusively breast fed infants and their mothers. *J. Pediatr. Endocrinol. Metab.* **2009**, *22*, 241–246. [CrossRef]
12. Salameh, K.; Al-Janahi, N.S.; Reedy, A.M.; Dawodu, A. Prevalence and risk factors for low vitamin D status among breastfeeding mother-infant dyads in an environment with abundant sunshine. *Int. J. Womens Health* **2016**, *8*, 529–535. [CrossRef] [PubMed]
13. Dawodu, A.; Davidson, B.; Woo, J.G.; Peng, Y.M.; Ruiz-Palacios, G.M.; de Lourdes Guerrero, M.; Morrow, A.L. Sun exposure and vitamin D supplementation in relation to vitamin D status of breastfeeding mothers and infants in the global exploration of human milk study. *Nutrients* **2015**, *7*, 1081–1093. [CrossRef] [PubMed]
14. Palacios, C.; Gonzalez, L. Is vitamin D deficiency a major global public health problem? *J. Steroid Biochem. Mol. Biol.* **2014**, *144PA*, 138–145. [CrossRef] [PubMed]
15. Wayse, V.; Yousafzai, A.; Mogale, K.; Filteau, S. Association of subclinical vitamin D deficiency with severe acute lower respiratory infection in Indian children under 5 y. *Eur. J. Clin. Nutr.* **2004**, *58*, 563–567. [CrossRef] [PubMed]
16. Roth, D.E.; Shah, R.; Black, R.E.; Baqui, A.H. Vitamin D status and acute lower respiratory infection in early childhood in Sylhet, Bangladesh. *Acta Paediatr.* **2010**, *99*, 389–393. [CrossRef]
17. Hollis, B.W.; Wagner, C.L. Assessment of dietary vitamin D requirements during pregnancy and lactation. *Am. J. Clin. Nutr.* **2004**, *79*, 717–726.
18. Dawodu, A.; Zalla, L.; Woo, J.G.; Herbers, P.M.; Davidson, B.S.; Heubi, J.E.; Morrow, A.L. Heightened attention to supplementation is needed to improve the vitamin D status of breastfeeding mothers and infants when sunshine exposure is restricted. *Matern. Child Nutr.* **2012**. [CrossRef]
19. Perrine, C.G.; Sharma, A.J.; Jefferds, M.E.; Serdula, M.K.; Scanlon, K.S. Adherence to vitamin D recommendations among US infants. *Pediatrics* **2010**, *125*, 627–632. [CrossRef]
20. Hollis, B.W.; Wagner, C.L. Vitamin D requirements during lactation: High-dose maternal supplementation as therapy to prevent hypovitaminosis D for both the mother and the nursing infant. *Am. J. Clin. Nutr.* **2004**, *80*, 1752S–1758S. [CrossRef]
21. Thacher, T.D.; Fischer, P.R.; Strand, M.A.; Pettifor, J.M. Nutritional rickets around the world: Causes and future directions. *Ann. Trop. Paediatr.* **2006**, *26*, 1–16. [CrossRef]
22. Heaney, R.P.; Davies, K.M.; Chen, T.C.; Holick, M.F.; Barger-Lux, M.J. Human serum 25-hydroxycholecalciferol response to extended oral dosing with cholecalciferol. *Am. J. Clin. Nutr.* **2003**, *77*, 204–210. [CrossRef] [PubMed]
23. Wagner, C.L.; Hulsey, T.C.; Fanning, D.; Ebeling, M.; Hollis, B.W. High-dose vitamin D3 supplementation in a cohort of breastfeeding mothers and their infants: A 6-month follow-up pilot study. *Breastfeed Med* **2006**, *1*, 59–70. [CrossRef] [PubMed]
24. Hollis, B.W.; Wagner, C.L.; Howard, C.R.; Ebeling, M.; Shary, J.R.; Smith, P.G.; Taylor, S.N.; Morella, K.; Lawrence, R.A.; Hulsey, T.C. Maternal Versus infant vitamin D supplementation during lactation: A randomized controlled trial. *Pediatrics* **2015**, *136*, 625–634. [CrossRef] [PubMed]
25. Dawodu, A.; Saadi, H.F.; Bekdache, G.; Javed, Y.; Altaye, M.; Hollis, B.W. Randomized controlled trial (RCT) of vitamin D supplementation in pregnancy in a population with endemic vitamin D deficiency. *J. Clin. Endocrinol. Metab.* **2013**, *98*, 2337–2346. [CrossRef] [PubMed]
26. Heaney, R.P. Vitamin D–baseline status and effective dose. *N. Engl. J. Med.* **2012**, *367*, 77–78. [CrossRef] [PubMed]
27. Barger-Lux, M.J.; Heaney, R.P. Effects of above average summer sun exposure on serum 25-hydroxyvitamin D and calcium absorption. *J. Clin. Endocrinol. Metab.* **2002**, *87*, 4952–4956. [CrossRef] [PubMed]
28. Greer, F.R.; Hollis, B.W.; Napoli, J.L. High concentrations of vitamin D2 in human milk associated with pharmacologic doses of vitamin D2. *J. Pediatrics* **1984**, *105*, 61–64. [CrossRef]
29. Wall, C.R.; Stewart, A.W.; Camargo, C.A., Jr.; Scragg, R.; Mitchell, E.A.; Ekeroma, A.; Crane, J.; Milne, T.; Rowden, J.; Horst, R.; et al. Vitamin D activity of breast milk in women randomly assigned to vitamin D3 supplementation during pregnancy. *Am. J. Clin. Nutr.* **2016**, *103*, 382–388. [CrossRef]
30. Hollis, B.W.; Pittard, W.B., 3rd; Reinhardt, T.A. Relationships among vitamin D, 25-hydroxyvitamin D, and vitamin D-binding protein concentrations in the plasma and milk of human subjects. *J. Clin. Endocrinol. Metab.* **1986**, *62*, 41–44. [CrossRef]

31. Vieth, R.; Chan, P.C.; MacFarlane, G.D. Efficacy and safety of vitamin D3 intake exceeding the lowest observed adverse effect level. *Am. J. Clin. Nutr.* **2001**, *73*, 288–294. [CrossRef]
32. Luxwolda, M.F.; Kuipers, R.S.; Kema, I.P.; Van der Veer, E.; Dijck-Brouwer, D.A.; Muskiet, F.A. Vitamin D status indicators in indigenous populations in East Africa. *Eur. J. Nutr* **2013**, *52*, 1115–1125. [CrossRef] [PubMed]
33. Dawodu, A.; Tsang, R.C. Maternal vitamin D status: Effect on milk vitamin D content and vitamin D status of breastfeeding infants. *Adv. Nutr.* **2012**, *3*, 353–361. [CrossRef] [PubMed]
34. Umaretiya, P.J.; Oberhelman, S.S.; Cozine, E.W.; Maxson, J.A.; Quigg, S.M.; Thacher, T.D. Maternal preferences for vitamin D supplementation in breastfed infants. *Ann. Fam. Med.* **2017**, *15*, 68–70. [CrossRef] [PubMed]

 © 2019 by the authors. Licensee MDPI, Basel, Switzerland. This article is an open access article distributed under the terms and conditions of the Creative Commons Attribution (CC BY) license (http://creativecommons.org/licenses/by/4.0/).

Article

The Concentration of Omega-3 Fatty Acids in Human Milk Is Related to Their Habitual but Not Current Intake

Agnieszka Bzikowska-Jura [1], Aneta Czerwonogrodzka-Senczyna [1], Edyta Jasińska-Melon [2], Hanna Mojska [2], Gabriela Olędzka [3], Aleksandra Wesołowska [4,*] and Dorota Szostak-Węgierek [1]

[1] Department of Clinical Dietetics, Faculty of Health Sciences, Medical University of Warsaw, E Ciolka Str. 27, 01-445 Warsaw, Poland
[2] Department of Metabolomics Food and Nutrition Institute, 61/63 Powsińska Str., 02-903 Warsaw, Poland
[3] Department of Medical Biology, Faculty of Health Sciences, Medical University of Warsaw, Litewska Str. 14/16, 00-575 Warsaw, Poland
[4] Laboratory of Human Milk and Lactation Research at Regional Human Milk Bank in Holy Family Hospital, Faculty of Health Sciences, Department of Neonatology, Medical University of Warsaw, Zwirki i Wigury Str. 63A, 02-091 Warsaw, Poland
* Correspondence: aleksandra.wesolowska@wum.edu.pl; Tel.: +48-22-317-9343

Received: 11 June 2019; Accepted: 9 July 2019; Published: 12 July 2019

Abstract: This study determined fatty acid (FA) concentrations in maternal milk and investigated the association between omega-3 fatty acid levels and their maternal current dietary intake (based on three-day dietary records) and habitual dietary intake (based on intake frequency of food products). Tested material comprised 32 samples of human milk, coming from exclusively breastfeeding women during their first month of lactation. Milk fatty acids were analyzed as fatty acid methyl ester (FAME) by gas chromatography using a Hewlett-Packard 6890 gas chromatograph with MS detector 5972A. We did not observe any correlation between current dietary intake of omega-3 FAs and their concentrations in human milk. However, we observed that the habitual intake of fatty fish affected omega-3 FA concentrations in human milk. Kendall's rank correlation coefficients were 0.25 ($p = 0.049$) for DHA, 0.27 ($p = 0.03$) for EPA, and 0.28 ($p = 0.02$) for ALA. Beef consumption was negatively correlated with DHA concentrations in human milk (r = −0.25; $p = 0.046$). These findings suggest that current omega-3 FA intake does not translate directly into their concentration in human milk. On the contrary, their habitual intake seems to markedly influence their milk concentration.

Keywords: human milk; omega-3 fatty acids; docosahexaenoic acid; eicosapentaenoic acid; α-linolenic acid; dietary intake; food frequency questionnaire

1. Introduction

Human milk is universally recognized as the optimal food for infants. Many studies have shown the role of fat in human milk as the main source of energy, selected fatty acids (FAs), crucial fat-soluble vitamins, and key nutrients for the infant development [1–3]. Among FAs, polyunsaturated fatty acids (PUFAs) are of principal importance. The two major classes of PUFAs are those of omega-3 and omega-6 families. Omega-3 fatty acids have a carbon–carbon double bond located in the third position from the methyl end of the chain. There are several different omega-3 FAs, but the majority of human milk research focuses on three: docosahexaenoic acid (DHA), α-linolenic acid (ALA), and eicosapentaenoic acid (EPA). ALA contains 18 carbon atoms, whereas EPA and DHA are considered "long-chain" (LC) omega-3 FAs, because EPA contains 20 carbons and DHA contains 22 [4]. Omega-3 FAs, mainly DHA, are important components of retinal photoreceptors and brain cell membranes. Therefore, DHA is essential for infant visual and cognitive development [5–7]. The European Food Safety Authority (EFSA) recommends 100 mg/day as the adequate intake of DHA for infants [8].

Fatty acids in human milk may originate either from the maternal dietary FAs, from FAs released from maternal adipose tissue, or from *de novo* synthesis in maternal tissues [5]. The human fatty acid desaturase can form only carbon–carbon double bonds located in the ninth position from the methyl end of a fatty acid. ALA may be endogenously converted to EPA and then to DHA. Therefore, ALA is considered an essential fatty acid, which means that it must be obtained from the diet [9,10]. However, the results of the studies show that the ability to convert ALA to DHA in humans is low, as less than 10% of ALA is converted to DHA [11]. For that reason, DHA from the maternal diet is a much more efficient source of DHA for neural tissue than an equivalent amount of ALA [10,12]. Therefore, consuming EPA and DHA directly from food and/or supplements is the only practical way to increase the levels of these fatty acids in the body.

Many national health authorities [13,14], including the Polish Society for Pediatric Gastroenterology, Hepatology, and Nutrition [15] recommend that maternal intake of DHA should be at least 200 mg per day. Women can meet the recommendation by consuming one to two portions of fatty fish (e.g., salmon, sardines) per week (equivalent of 150–300 g). Although the maternal intake of DHA is crucial for infant brain and retina development, studies carried out in the United States [16], Canada [17,18], and Europe [19] have reported that breastfeeding women do not meet dietary recommendations. This probably results from low fatty fish consumption [20], and is partly related to concerns of methylmercury fish contamination [21], as well as low DHA supplements use [16–18].

The tissue levels of FAs in a woman during lactation are directly related to her reserve capacity and the metabolic utilization of fatty acids (synthesis, oxidation and transport). Hence, the maternal diet and metabolism of FAs of women during lactation seem to be the most important factors affecting DHA concentration in human milk. Human milk FA composition changes continuously as dietary FAs are rapidly transported from chylomicrons into human milk with a peak between six and 12 h after dietary DHA intake [21,22].

To investigate the relationship between maternal diet and human milk composition, several dietary assessment methods have been developed and evaluated. The most common are food frequency questionnaires (FFQs) and multiple-day food records. Since food records do not rely on memory, they have been used as a reference method to validate other dietary assessment methods. On the other hand, day-to-day variations and seasonal variations in food consumption may decrease their objectivity. Furthermore, individuals are not always able to recall all the foods consumed or the specific components of the food (especially when dining out), and have difficulty in determining accurate portion sizes and typically underreport dietary intake. This is in contrast to FFQs, which often overestimate the intake of energy and nutrients [23]. Nonetheless, the FFQ has been suggested as an optimal tool in estimating dietary intake of omega-3 fatty acids as it evaluates long-term diet rather than food records [24]. Most of the FFQs available in the literature were designed to assess a wide range of nutrients; however, they were not appropriate for dietary assessment focusing specifically on fatty acids. Given that omega-3 FAs are contained in a particular range of foods, we used a tailored omega-3 FA FFQ. Serra-Majem et al. [25] suggested that its validity is comparable to the whole diet-based FFQs. (0.42–0.52 versus 0.19–0.54). We hypothesize that the concentration of omega-3 fatty acids in human milk is related to their habitual but not current intake; for this reason, in this study, we aimed to determine FA concentrations in maternal milk and assess the association between omega-3 fatty acids levels and their maternal dietary intake evaluated with two methods: dietary intake based on the three-day dietary record, and intake frequency of food products (FFQ, or food frequency questionnaire).

2. Materials and Methods

2.1. Subjects and Study Session Design

The Ethics Committee of the Medical University of Warsaw (KB/172/115) approved the study protocol, and all the participating women signed informed consents. A convenience sample of exclusively breastfeeding women (n = 32) was recruited from the Holy Family Hospital in Warsaw. Participants were enrolled during their first month of lactation (weeks two to four). Inclusion criteria comprised: age ≥18 years, singleton pregnancy, and full-term delivery (gestational age ≥37 weeks). Exclusion criteria were as follows: pre-existing chronic or gestational diseases, smoking during pregnancy and/or breastfeeding, low birth weight of the newborn, and low milk supply. The survey consisted of two parts. Firstly, we collected data about socio-demographic and other maternal characteristics, such as: age, education level, material status, pre-pregnancy anthropometric parameters (weight and height), and total weight gain during pregnancy. Then, we collected dietary information involving three-day dietary record and intake frequency of food products. During the study session, the actual body weight and height of every mother were measured using a Seca 799 measurement station and column scales (±0.1 kg/cm; Seca, Chino, CA, USA). The pre-pregnancy and actual body mass index (BMI) was calculated as the ratio of the body weight to the height squared (kg/m^2). Interpretation of these results followed the international classification proposed by the World Health Organization (WHO): below 18.5 kg/m^2, underweight; 18.5–24.9 kg/m^2, normal weight; 25.0–29.9 kg/m^2, overweight; 30.0 kg/m^2 and above, obese [26].

2.2. Human Milk Collection

Twenty-four-hour human milk samples (n = 32) were collected by women at home after they had been given detailed instructions on taking, storing, and transporting samples to the Holy Family Hospital in Warsaw. Foremilk and hindmilk samples were collected from all the participants from four time periods (06:00–12:00, 12:00–18:00, 18:00–24:00, and 24:00–06:00) to minimize possible circadian influences on the milk fatty acid composition. The term foremilk refers to the milk at the beginning of a feeding, and hindmilk refers to milk at the end of a feeding, which has a higher fat content than the milk at the beginning of that particular feeding. A total of 5 to 10 mL of foremilk and hindmilk samples were obtained from the breast(s) from which the infant was fed. Samples were collected into pre-labeled polypropylene containers provided to each woman. Participants were instructed to store milk in the refrigerator (~4 °C) during the 24-h collection process. Then, milk samples were stored at −20 °C for later analysis.

2.3. Lipid Concentration and Fatty Acid Analysis of Human Milk

Tested material comprised 32 samples of human milk. The lipid concentration in human milk was analyzed using the Miris human milk analyzer (HMA) (Miris, Uppsala, Sweden) with a validated protocol, as discussed in a previous study [27]. Collected milk samples for fatty acids analysis were immediately frozen in plastic test tubes at a temperature of −20 °C and delivered to the Department of Metabolomics (Food and Nutrition Institute, Warsaw, Poland) into thermic bags. Samples were stored at −80 °C until analysis. Frozen samples were thawed only once at room temperature, without light. After thawing, samples were shaken at room temperature (3–5 min) to obtain a homogeneous mixture. Aliquots were extracted and analyzed for fatty acid (FA) composition and content.

2.3.1. Fat Extraction

Milk samples were extracted from 1 mL of sample with chloroform-methanol (2:1) (Avantor Performance Materials S.A., Warsaw, Poland) containing 0.02% butyl-hydroxytoluene (2,6-tert-butyl-4-methylphenol, BHT, ≥99.0%, GC, powder) (Sigma-Aldrich CHEMIE GmbH, CA, USA) as an antioxidant, according to Folch et al. [28].

2.3.2. Gas Chromatography-Mass Spectrometry Analysis

Milk fatty acids were analyzed as fatty acid methyl ester (FAME) by gas chromatography using a Hewlett-Packard 6890 gas chromatograph with MS detector 5972 A. The methylation procedure was as follows: organic extracts were evaporated at 40 °C in a gentle nitrogen stream, and then were saponified with 0.5 mL of potassium hydroxide in methanol (0.5 N) (Avantor Performance Materials S.A., Poland) for 10 min at 80 °C in an electric multiblock heater and subsequently methylated with 1 mL of hydrochloric acid in methanol (3 N) (Sigma-Aldrich CHEMIE GmbH, USA) for 15 min at 85 ± 2 °C. After cooling to room temperature, fatty acid methyl esters were extracted with 1 mL of isooctane (2,2,4 trimethylpentane) (Avantor Performance Materials S.A., Poland). One microliter of the sample was injected into the GC column. The GC-MS analysis has been used with a split injector (1:100 ratio), injector and detector temperatures of −250 °C, and carrier gas helium (20 mL/s; the pressure of 43.4 psi). The chromatography oven was programmed to 175 °C for 40 min; thereafter, it was increased by 5 °C per min until the temperature reached 220 °C, and was held at this temperature for 16 min. FAMEs separations were performed on a CP Sil 88 fused silica capillary column (100 m × 0.25 mm i.d., film thickness: 0.20 μm; Agilent J & W GC Columns, CA, USA). Peak identification was verified by comparison with authentic standards (Supelco FAME Mix 37 Component; Sigma-Aldrich, CA, USA) and by mass spectrometry. The obtained results were expressed as a percentage by weight (% wt/wt) of all the fatty acids detected with a chain length between eight and 24 carbon atoms. The method was validated and accredited by the Polish Centre of Accreditation (accreditation certificate AB 690). Quality control was also implemented by the use of certified reference materials: BCR-163 (Beef-Pork FAT blend; ABP cat. 3; 8 g; Sigma-Aldrich, CA, USA).

2.4. Fatty Acids Dietary Intake and Intake Frequency of Food Products

The assessment of women's fatty acids intake was based on a three-day dietary record. Mothers were asked to note each food and dietary supplement they had consumed in the tree consecutive days prior to the human milk sampling day. No dietary recommendation was given before the study; participants were allowed to consume self-chosen diets. To verify the sizes of declared food portions, we used the "Album of Photographs of Food Products and Dishes" developed by the National Food and Nutrition Institute [29]. Fatty acids dietary intake was calculated using Dieta 5.0 nutritional software (National Food and Nutrition Institute, Warsaw, Poland). Additionally, the habitual intake of fatty acids was assessed using a FFQ containing 19 items. The FFQ provided information about the consumption frequency of DHA sources in the last three months, such as fish, seafood, meat (poultry, turkey, pork, beef), and eggs. We also collected information about the consumption frequency of other fatty acids sources, including vegetable oils (e.g., canola oil, olive oil, linseed oil, coconut oil), butter, milk and dairy products, nuts, and seeds. The response options were arranged in five categories, from "never", "less than once a week", "once or twice a week", "more than twice a week but not every day", to "every day".

2.5. Statistical Analysis

Statistical analyses were performed using Statistica 12PL, Tulusa, USA and IBM Statistics 21, New York, NY, USA. A *p*-value below 0.05 was adopted as statistically significant. Variables distributions were evaluated with a Shapiro–Wilk test and descriptive statistics. Data were presented as means and standard deviations as well as medians and interquartile ranges. Correlations between the intake of fatty acids and fatty acids concentrations in human milk were estimated with Pearson's r correlation coefficient. Correlations between omega-3 fatty acids (DHA, EPA, ALA) concentrations in human milk, and food consumption frequency were estimated with Kendall's tau correlation coefficients.

3. Results

3.1. Maternal Characteristics

The mean maternal age was 30.9 ± 6.5 years and most of them were primiparous (75%; n = 24). Detailed anthropometric data are shown in Table 1. Before pregnancy and during the first month postpartum, none of the participants was classified as being underweight (BMI <18.5 kg/m^2). In both periods of time, most of them (n = 23, 72%) had normal body mass, and 28% (n = 9) were classified as being overweight or obese. All the participants declared high university education and high material status. We do not observed statistically significant differences between pre-pregnancy and postpartum BMI values (*t*-test was 1.13; $p = 0.26$).

Table 1. Characteristics of the mothers.

Characteristic	Mean ± SD	Range
Age (years)	30.9 ± 6.5	27–44
Height (cm)	1.66 ± 0.1	1.54–1.8
Pre-pregnancy weight (kg)	62.2 ± 11.8	44–90
Pre-pregnancy body mass index (kg/m^2)	22.6 ± 3.4	18.6–30.9
Weight gain during pregnancy (kg)	15.1 ± 4.8	7–30
Weight at first month postpartum (kg)	65.5 ± 13.2	45.6–95
Body mass index at first month postpartum (kg/m^2)	23.6 ± 3.8	18.5–32.1

3.2. Fatty Acids Concentrations in Human Milk

The fatty acids profile of human milk is shown in Table 2. The relative proportion of saturated, monounsaturated, and polyunsaturated fatty acids was 41.9 ± 4.9%, 39.6 ± 3.1%, and 15.1 ± 3.4%, respectively. The predominant fatty acids in human milk were oleic acid (35.4 ± 3.1%), palmitic acid (19.7 ± 2.5%), and linoleic acid (11.1 ± 2.6%). No significant correlation was found between DHA concentrations, palmitic (r = −0.24; $p = 0.2$) and oleic (r = 0.13; $p = 0.48$) FAs; however, we found correlation with linoleic acid (r = 0.44; $p = 0.013$). Also, a significant negative correlation was found between DHA concentration and the omega-6:omega-3 ratio in human milk (r = −0.45; $p = 0.012$). When only the milk of mothers supplementing their diet with DHA were considered (n = 22), the mean concentration of DHA was 0.78% of total fatty acids. The difference between the concentration of DHA in supplementing and not supplementing mothers was not statistically significant (r = 0.29; $p = 0.37$).

Table 2. Fatty acids composition (%) of human milk [1].

Fatty Acids	Mean ± SD	Median (Interquartile Range)
Saturated fatty acids (SFA)	41.9 ± 4.9	42.3 (38.0–45.7)
C4:0 (butanoic acid)	0.0 ± 0.0	0.0 (0.0–0.0)
C6:0 (caproic acid)	0.0 ± 0.0	0.0 (0.0–0.0)
C8:0 (caprylic acid)	0.1 ± 0.0	0.1 (0.1–0.1)
C10:0 (capric acid)	1.1 ± 0.3	1.1 (1.0–1.4)
C12:0 (lauric acid)	3.5 ± 1.1	3.3 (2.5–4.1)
C13:0 (tridecanoic acid)	0.1 ± 0.0	0.1 (0.0–0.1)
C14:0 (myristic acid)	9.5 ± 2.5	9.3 (7.9–11.1)
C15:0 (pentadecanoic acid)	0.8 ± 0.2	0.7 (0.7–0.8)
C16:0 (palmitic acid)	19.7 ± 2.5	19.9 (17.4–22.0)
C17:0 (margaric acid)	0.5 ± 0.1	0.5 (0.4–0.6)
C18:0 (stearic acid)	6.4 ± 1.5	6.0 (5.4–7.3)
C20:0 (arachidic acid)	0.2 ± 0.1	0.2 (0.1–0.2)
C21:0 (henicosanoic acid)	0.0 ± 0.0	0.0 (0.0–0.0)
C22:0 (behenic acid)	0.1 ± 0.0	0.1 (0.1–0.1)
C23:0 (tetracosanoic acid)	0.0 ± 0.0	0.0 (0.0–0.0)

Table 2. Cont.

Fatty Acids	Mean ± SD	Median (Interquartile Range)
Monounsaturated fatty acids (MUFA)	39.6 ± 3.1	39.0 (38.0–42.0)
C14:1 (myristoleic acid)	0.2 ± 0.1	0.2 (0.2–0.3)
C15:1 (pentadecenoic acid)	0.0 ± 0.0	0.0 (0.0–0.0)
C16:1 trans	0.4 ± 0.1	0.4 (0.3–0.4)
C16:1 cis	2.6 ± 0.5	2.6 (2.3–2.9)
C17:1 (heptadecenoic acid)	0.2 ± 0.0	0.2 (0.2–0.3)
C18:1 cis (oleic acid)	35.4 ± 3.1	34.9 (33.6–37.6)
C18:1 trans (vaccenic acid)	1.2 ± 0.5	1.2 (0.8–1.5)
C20:1 (gadoleic acid)	0.8 ± 0.2	0.7 (0.7–0.9)
C22:1 (erucic acid)	0.2 ± 0.1	0.1 (0.1–0.2)
C24:1 (lignoceric acid)	0.2 ± 0.0	0.2 (0.1–0.2)
Polyunsaturated fatty acids (PUFA)	15.1 ± 3.4	15.3 (12.7–16.8)
n-3 polyunsaturated	2.7 ± 0.9	2.6 (2.1–3.1)
C18:3 (α-linolenic acid, ALA)	1.5 ± 0.6	1.4 (1.0–1.8)
C20:3 (eicosatrienoic acid)	0.1 ± 0.0	0.1 (0.0–0.1)
C20:5 (eicosapentaenoic acid, EPA)	0.2 ± 0.1	0.2 (0.2–0.3)
C22:6 (docosahexaenoic acid, DHA)	0.7 ± 0.3	0.7 (0.5–1.0)
n-6 polyunsaturated	12.1 ± 2.7	12.1 (10.4–13.4)
C18:2 (linoleic acid, LA)	11.1 ± 2.6	11.1 (9.5–12.3)
C18:3 (γ-linoleic acid)	0.1 ± 0.0	0.1 (0.1–0.1)
C20:3 (dihomo-γ-linoleic acid)	0.3 ± 0.1	0.3 (0.2–0.4)
C20:4 (arachidonic acid, ARA)	0.5 ± 0.1	0.5 (0.4–0.6)
Ratio		
n-6:n-3	4.6 ± 1.0	4.8 (4.1–5.1)
DHA:LA [2]	0.1 ± 0.0	0.1 (0.0–0.1)
ARA [3]:DHA [4]	0.9 ± 0.4	0.7 (0.5–1.1)
LA:ALA [5]	8.1 ± 2.4	7.8 (6.4–9.6)
Total fat concentration [6]	3.49 ± 1.0	3.5 (3.0–4.2)

[1] Data are presented as the relative proportion of each fatty acid (% of total fatty acids). [2] LA linoleic acid; [3] ARA arachidonic acid; [4] DHA docosahexaenoic acid; [5] ALA α-linolenic acid. [6] Total fat concentration is presented as grams per 100 mL.

3.3. Fatty Acids Dietary Intake and Its Association with Concentration in Human Milk

The fatty acids dietary intake is shown in Table 3. Mean energy intake was 1752 kcal ± 228.3 kcal, which was lower than the recommended level (EER, estimated energy requirement = 2555 kcal per day). The risk of deficient energy intake was observed in 100% of the participants. According to the Polish nutritional standards, the recommended intake for ALA is 0.5% of total energy, which in our participants it corresponded to 0.97 g per day. The mean dietary intake of ALA (1.5 ± 0.8% of total fatty acids) was higher than recommended levels; nevertheless, 22% of participants did not meet the recommendation. When only dietary sources were considered, the mean intake of DHA (243 mg ± 333.5) (Table 4) reached the Polish [15] and European [13,14] recommendation of 200 mg of DHA daily, whereas among 59% of the women, we observed insufficient DHA intake. Including taken supplements, the percentage of deficient DHA intake decreased, and was 16%. The majority of the participants (69%; n = 22) reported taking DHA supplements, and 10% (n = 3) of women declared taking supplements containing DHA and EPA.

Table 5 presents correlation coefficients (Pearson's r) between human milk fatty acids concentrations and maternal fatty acids dietary intake, as well as maternal dietary intake together with supplementation. We did not observe any statistically significant correlation between these factors ($p > 0.05$).

Table 3. Fatty acids content in mothers' diet.

Fatty Acids	Mean ± SD (g)	Median (Interquartile Range) (g)
Saturated fatty acids (SFA)	23.9 ± 10.3	20.9 (7.3–69.5)
C4:0 (butanoic acid)	0.4 ± 0.3	0.5 (0–1.1)
C6:0 (caproic acid)	0.3 ± 0.2	0.3 (0–0.8)
C8:0 (caprylic acid)	0.2 ± 0.2	0.2 (0–1.0)
C10:0 (capric acid)	0.5 ± 0.3	0.5 (0–1.6)
C12:0 (lauric acid)	0.9 ± 0.9	0.7 (0.2–5.5)
C14:0 (myristic acid)	2.8 ± 1.4	2.7 (0.7–7.9)
C15:0 (pentadecanoic acid)	0.3 ± 0.2	0.3 (0–0.9)
C16:0 (palmitic acid)	12.6 ± 5.2	11.6 (4.5–35.0)
C17:0 (margaric acid)	0.2 ± 0.1	0.2 (0–0.7)
C18:0 (stearic acid)	5.2 ± 2.8	5.1 (1.2–18.3)
C20:0 (arachidic acid)	0.1 ± 0.1	0.1 (0–0.3)
Monounsaturated fatty acids (MUFA)	24.7 ± 8.9	24.8 (13.2–55.0)
C14:1 (myristoleic acid)	0.2 ± 0.2	0.2 (0–0.8)
C15:1 (pentadecenoic acid)	0.1 ± 0.1	0.1 (0–0.2)
C16:1	1.3 ± 0.4	1.2 (0.6–2.4)
C17:1 (heptadecenoic acid)	0.1 ± 0.1	0.1 (0–0.5)
C18:1 (oleic and vaccenic acids)	22.1 ± 8.3	21.7 (11.5–50.3)
C20:1 (gadoleic acid)	0.3 ± 0.2	0.2 (0.1–0.9)
C22:1 (erucic acid)	0.3 ± 0.3	0.1 (0–1.1)
Polyunsaturated fatty acids (PUFA)	10.7 ± 4.0 13.5 ± 14.2 [s]	11.2 (4.9–22.9) 11.3 (4.9–88.4) [s]
C18:2 (linoleic acid, LA)	8.5 ± 3.4	8.5 (4.1–21.1)
C18:3 (α-linolenic acid, ALA)	1.5 ± 0.8	1.4 (0.5–4.3)
C20:5 (eicosapentaenoic acid, EPA)	0.1 ± 0.2 0.1 ± 0.2 [s]	0 (0–0.5) 0.1 (0–0.5) [s]
C22:6 (docosahexaenoic acid, DHA)	0.2 ± 0.3 0.6 ± 0.6 [s]	0.1 (0–1.2) 0.4 (0–1.8) [s]

[s] Diet + supplementation.

Table 4. Estimated daily intake of EPA and DHA from food, supplement, and food + supplement in lactating women.

	EPA (mg)	DHA (mg)
Food	104.4 ± 152.0 (18, 0–190)	243.3 ± 333.5 (50, 37–411)
Supplement [1]	29.7 ± 43.6 (0, 0–43)	370.6 ± 465.1 (250, 8–549)
Food + supplement	134.1 ± 153.9 (85, 1–215)	614.0 ± 574.6 (354, 202–905)

Data are presented as means ± SD (median, interquartile range). [1] Participants who did not take supplements or those who took supplements that did not contain EPA/DHA were considered as 0 mg supplement on EPA and DHA.

Table 5. Correlations between human milk fatty acids and fatty acids in the mother's diet and supplementation.

Concentration in Human Milk	Dietary Intake, Sole or Together with Supplementation					
	SFA	MUFA	PUFA	ALA	EPA	DHA
SFA [1]	0.26	0.16	−0.20	−0.13	−0.18 −0.19 [s]	−0.10 −0.16 [s]
MUFA [2]	−0.14	−0.04	0.14	0.26	0.21	0.19
PUFA [3]	−0.20	−0.14	0.20	0.01	0.08	−0.01
ALA [4]	−0.19	−0.09	0.04	0.32	0.20	−0.06
EPA [5]	−0.16	0.05	0.05	−0.11	0.20 0.17 [s]	−0.02 0.17 [s]
DHA [6]	−0.24	−0.18	−0.04	−0.26	0.16 0.23 [s]	0.04 0.24 [s]

Data are presented as Pearson's r coefficients.; [1] SFA, Saturated fatty acids; [2] MUFA, Monounsaturated fatty acids; [3] PUFA, Polyunsaturated fatty acids; [4] ALA, α-linolenic acid; [5] EPA, eicosapentaenoic acid; [6] DHA, docosahexaenoic acid; [s] Pearson's r coefficients diet + supplementation.

3.4. Association between Intake Frequency of Food Products and DHA Concentrations in Human Milk

Table 6 presents Kendall's rank correlation coefficients between the intake frequency of food products and DHA concentrations in human milk. According to the FFQ, almost half of the participants (~47%) declared fatty fish consumption once or twice a week. On the other hand, almost 43% of the women consumed fish less than once a week or never. Based on the three-day dietary record, the most frequently consumed fatty fish species were salmon and mackerel, which were reported by seven (22%) and six women (19%), respectively. Butter and milk were the most frequently used foods, which were consumed by approximately ~44% and ~38% of participants, respectively.

Table 6. Intake frequency (%) of food products and correlations with DHA concentrations in human milk [1].

Food	Never	Less than Once a Week	Once or Twice a Week	More than Twice a Week but Not Every Day	Every Day	Correlation [1] with Concentrations in Human Milk		
						DHA [2]	EPA [3]	ALA [4]
Fatty fish (e.g., salmon, herring)	12.50	31.25	46.88	9.38	0.00	0.25 *	0.27 *	0.28 *
Lean fish (e.g., cod, sole)	21.88	31.25	43.75	3.13	0.00	0.14	0.08	0.21
Seafood	31.25	43.75	25.17	0.00	0.00	0.21	0.13	0.19
Poultry and turkey	3.13	3.13	43.75	40.63	9.38	0.09	0.13	0.08
Pork	6.25	40.63	37.50	15.63	0.00	0.02	0.29 *	0.18
Beef	9.38	50.00	34.38	6.25	0.00	−0.25 *	−0.14	0.11
Meat products (e.g., sausages, sliced meats)	6.25	12.50	37.50	18.75	25.00	0.24	−0.11	−0.10
Eggs	15.63	6.25	34.38	34.38	9.38	−0.14	−0.06	−0.00
Milk	31.25	15.63	9.38	6.25	37.50	0.02	0.05	0.26 *
Fermented dairy products	37.50	28.13	21.88	12.50	0.00	0.17	0.10	0.29 *
Cheese	25.00	28.13	15.63	18.75	12.50	−0.02	−0.02	0.00
Cottage cheese	21.88	12.50	43.75	15.63	6.25	0.22	0.06	0.17
Milk desserts	46.88	28.13	9.38	15.63	0.00	0.11	−0.12	−0.18
Butter	15.63	9.38	12.50	18.75	43.75	−0.14	−0.11	0.10
Canola oil	18.75	28.13	21.88	25.00	6.25	0.09	−0.0	0.13
Olive oil	12.50	18.75	25.00	43.75	0.00	0.08	0.02	0.03
Linseed oil	62.50	15.63	9.38	9.38	3.13	0.01	0.10	0.30 *
Coconut oil	56.25	25.00	15.63	3.13	0.00	−0.12	−0.07	0.29 *
Nuts and seeds	6.25	18.75	21.88	28.13	25.00	−0.03	0.02	0.01

[1] Data are presented as Kendall's rank correlation coefficients. * $p < 0.05$. [2] DHA, docosahexaenoic acid; [3] EPA, eicosapentaenoic acid; [4] ALA, α-linolenic acid.

We found a significant positive correlation between fatty fish consumption and all the omega-3 fatty acids concentrations in human milk. Kendall's rank correlation coefficients were 0.25 ($p = 0.049$) for DHA, 0.27 ($p = 0.03$) for EPA, and 0.28 ($p = 0.02$) for ALA. The ALA content in maternal milk was also positively correlated with the intake frequency of linseed oil (r = 0.30; $p = 0.01$), coconut oil (r = 0.29; $p = 0.02$), milk (r = 0.26; $p = 0.04$), and fermented dairy products (r = 0.29; $p = 0.02$), whereas EPA concentration was positively correlated with intake frequency of pork (r = 0.29; $p = 0.02$). Beef consumption, by contrast, was negatively correlated with DHA concentration in human milk (r = −0.25; $p = 0.046$). No other significant correlations between intake frequency of food products and DHA concentration in human milk were found.

The major food contributor of total omega-3 FAs intake was fatty fish (45%) and lean fish (17%) (Figure 1). Seafood (mainly shrimps) (10%), poultry products (8%), and meat products (6%) also made significant contributions to the estimated intake of omega-3 FAs. Within seafood and fish categories, salmon was found to be the main source of all omega-3 FAs, DHA (67%), EPA (53%), and ALA (59%).

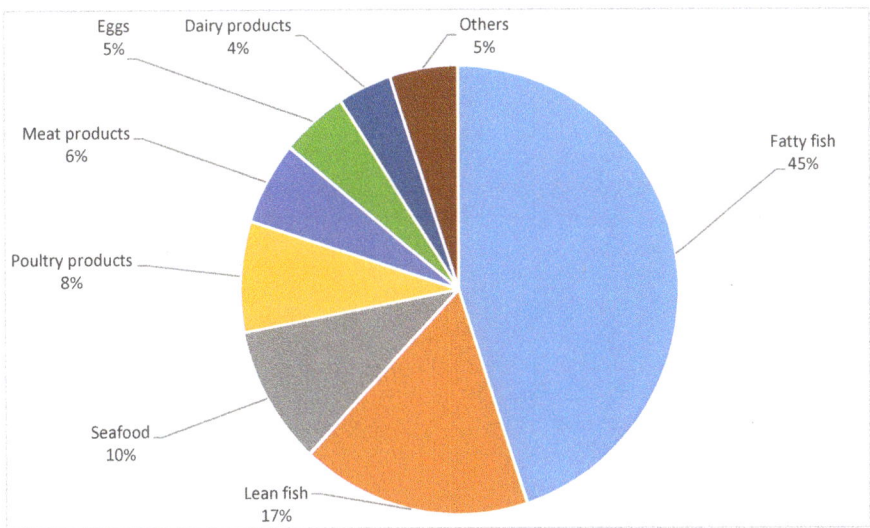

Figure 1. The relative contribution of food groups to total omega-3 FAs intake from the three-day dietary record.

Food items were divided into groups based on the Dieta 5.0 nutritional software database. Food items that contributed to less than 1% of total omega-3 FAs intake were categorized as "others". This included fats and vegetable oils, mixed dishes, and sauces.

4. Discussion

Our study has revealed three primary findings concerning the studied population. First, the mother's dietary intake of omega-3 FAs met Polish and European standards. Secondly, milk DHA concentration averaged about 0.70% of total fatty acids, which was twofold higher than the worldwide average (WWA) [30]. The third finding of this study is that there were no correlations between dietary intake of omega-3 FAs measured with three-day dietary records and their concentrations in human milk, whereas the intake frequency of food products—mainly fatty fish—was positively correlated with the concentration of ALA, EPA, and DHA in human milk.

In our study, the average maternal dietary DHA and ALA intakes (including taken supplements) were 613 ± 575 mg and 152 ± 0.81 mg, respectively. For DHA, this value was threefold higher than Polish [15] and European [13,14] recommendations (200 mg DHA per day). Our finding is not consistent with previous studies conducted in the United States [31,32], Canada [17,33], and Europe [34], which reported that the majority of breastfeeding women in the Alberta Pregnancy Outcomes and Nutrition (APrON) cohort were not meeting any of the various authorities' omega-3 FAs recommendations [35]. Similarly, in the Latvian study (Latvia and Poland are situated in the same geographic region, both countries have access to the Baltic Sea), DHA daily intake was lower than recommended, and was only 136 ± 26 mg [36]. It is widely reported that supplementation increases mother's milk DHA concentrations [37,38], and our results confirmed these findings. DHA concentration in the milk of mothers who supplemented their diet with DHA was higher than in the whole group, and was 0.78% of total fatty acids. However, we did not observe statistically significant differences between DHA concentrations in the milk of mothers who supplemented and did not supplement their diet with DHA ($r = 0.29$; $p = 0.37$). When only dietary sources were considered, the insufficient intake of DHA was observed among 59% of participants, whereas including taken supplements, the percentage decreased to 16%.

Based on the three-day dietary record, we observed that the largest contributor of omega-3 FAs intake was fatty fish, lean fish, and seafood, which was consumed once or twice a week by 47%, 44%, and 25% of participants, respectively. Other foods significantly contributing to total omega-3 FAs intake were poultry, red meat (pork and beef), and eggs. These findings were consistent with other studies [18,19], which reported that seafood, fish, poultry, meat, and eggs were the primary dietary sources of total omega-3 FAs.

The mean concentration of DHA in human milk in our study (0.7 ± 0.3% of total fatty acids) was higher than those reported in the meta-analysis based on 106 studies published between 1986–2006 year [30]. The authors found that the WWA DHA concentration in human milk was 0.32 ± 0.22% with a wide range of 0.06% to 1.4% of total fatty acids. We observed that our result was twofold compared to the data from Mediterranean countries, such as Italy (0.28–0.35%) [39,40], Spain (0.31–0.38) [41,42], and also Nordic countries such as Iceland (0.30%) [43] and Denmark (0.35%) [44]. Our mean DHA concentration in human milk was also higher than that reported by Jackson and Harris [45]. They suggested that the target of DHA level in human milk should be 0.3% of total fatty acids, emphasizing however that the optimal DHA level remains to be established. An approximate DHA concentration in human milk may be achieved when DHA intake exceeds 200 mg. Brenna and Lapillonne [14] have developed a regression analysis equation to calculate human milk DHA concentration depending on dietary DHA intake, based on the data from Gibbson et al. [46]:

human milk DHA concentration (% of total fatty acids) = 0.72 × maternal DHA intake (g per day) + 0.20.

Inserting our mean maternal DHA intake into this equation, we obtain a value of 0.63%, which is similar to the mean DHA concentration in human milk from our analysis (0.7%). In our study, the daily DHA intake was significantly higher than recommended, which may explain that the target DHA level in human milk was reached.

In our study, the mean concentration of ALA in human milk was 1.5 ± 0.8%, which was higher than those reported by Presa-Owens et al. (0.79%) [47] and Kotelzko et al. (0.81%) [48]. Some oils, such as soybean oil and canola oil, and legumes such as soybeans contain large amounts of ALA. Our findings suggest that women in our study consume these types of products to a greater extent than those in other Western countries [49]. However, in our study, the ratio of LA to ALA in maternal milk was higher than recommended (2.8:1) [50], and was 8:1. As LA and ALA compete for the same key enzymes in long-chain polyunsaturated fatty acids (LCPUFAs) biosynthesis, and ALA has a higher affinity than LA for δ-6-desaturase, the dietary ratio of LA:ALA is more relevant for proper LCPUFAs production than the intake of each fatty acid individually [49]. Furthermore, there is evidence that the DHA concentration in maternal milk has a positive correlation with the intelligence quotient of the child, whereas LA concentration has a negative correlation. It indicates that the FA profile during the lactation period is of particular significance [51–53]. In our study, the ratio of DHA:LA (0.1:1) was higher than the mean ratio from 28 countries [54]. Lassek and Gaulin [54] reported that the DHA:LA ratio at the level of ~0.04:1–0.07:1 is linked to higher cognitive test score results. It is worth noting that in our study, the majority of women (94%, n = 30) had the DHA:LA ratio in their milk of at least 0.04:1. Further, their mean ratio of omega-6:omega-3 FAs (4.6:1) was three times lower than the value reported by Silva et al. (14:1) [55], and nearly two times lower than that in the Latvian study (7:1) [36]. Silva et al. [52] and Nishimura et al. [56] suggested that a diet low in fish and high in vegetable oils, mainly soybean oil, facilitated a higher omega-6:omega-3 ratio (>10:1). However, we did not observe a significant correlation between fish or vegetable oils consumption and the omega-6:omega-3 ratio in human milk. The EFSA recommended a 4:1 proportion for dietary omega-6:omega-3 FAs [8]. It has been suggested that a very high omega-6:omega-3 ratio (~15:1) may promote cardiovascular diseases and cancers [57].

We found a significant positive correlation (r = 0.25; p = 0.049) between habitual fatty fish consumption and DHA concentration in human milk, which is consistent with results from other studies [36,49,56,58]. A Danish observational study reported a higher proportion of DHA concentration in the milk of mothers who consumed fish than in those who did not consume fish (0.63% compared with 0.41%; p = 0.018) and in women who consumed fatty fish compared with those who did not consume fish (0.73% compared with 0.41%; $p < 0.01$) [59]. Contrary to these findings, Juber et al. [59] reported that intake of fish did not correlate with mother's milk DHA levels. The results may depend on fish breeding (farmed or wild) and the fish species that are most commonly consumed in a specific region. For instance, when consuming 100 g of wild salmon, the total DHA intake is lower (0.31% of total fatty acids) compared to eating the same quantity of farmed salmon (0.88% of total fatty acids). It is caused by the higher total fat content in farmed salmon (12.3 g/100 g) than in wild salmon (2.07 g/100 g) [60]. In our study, the most commonly consumed fish were salmon and mackerel, but we did not have information about breeding, so the dietary intake of DHA based on three-day dietary record may be overestimated on underestimated in some cases. Using Dieta 5.0 nutritional software, we based our analysis on its nutritional database, which did not have distinctions between wild and farmed fish. However, Qiunn and Kuawa [61] concluded that an additional portion of fish (no information about breeding) per week led to a 0.014% increase of DHA concentration in human milk. We also observed that the DHA levels in human milk were negatively correlated ($r = -0.25$; $p = 0.046$) with beef consumption, the fat of which consists mainly of saturated fatty acids (SFA) (~52.3 g/100 g) and contains low amounts of PUFA (3.8 g/100 g) [62].

We noted that similarly to DHA, the ALA and EPA concentrations in human milk were also significantly positively correlated with the frequency of fatty fish consumption ($r = 0.27$; $p = 0.02$ for ALA and $r = 0.28$; $p = 0.03$ for EPA). ALA concentration in human milk was also positively correlated with the frequency of consuming linseed oil, coconut oil, milk, and milk products. Milk fat consists mainly of SFA (59.4 g/100 g) with the dominating palmitic acid (~30.6% of total FAs) [63], which was predominant SFA in human milk in our study. However, no significant correlation between milk consumption and palmitic acid concentration in human milk was found ($r = 0.13$; $p = 0.29$).

We did not find a correlation between dietary intake of DHA based on the three-day dietary record and any omega-3 FAs (ALA, EPA, and DHA) concentrations in human milk. It confirmed the results of Nishimura et al. [55], who did not observe any association between dietary EPA and DHA intake during the postpartum period and their concentrations in human milk. However, they noted that the dietary EPA and DHA content during the third trimester of pregnancy was directly related to the content of these fatty acids in mature human milk. These results suggest that maternal body stocks of fatty acids have a greater influence on breast milk fatty acid composition than the estimated dietary intake during the postpartum period, validating previously reported results [2,64]. An additional possible explanation would be also the increased weight gain and body fat storage observed in women mainly during late pregnancy [65]. Fiddler et al. [66] reported that about 20% of the supplemented 0.2 g of DHA per day was secreted into milk and observed a pronounced increase in DHA content in human milk (% of total FAs), with a peak between 6–12 h after dietary DHA intake. Although 24-h human milk samples were collected in our study, these findings may explain the lack of correlation between the intake of omega-3 FAs measured with the three-day dietary record and its concentration in human milk. Further, while the synthesis of DHA from ALA may be sufficient for the healthy breastfeeding women, some non-physiological conditions, such as non-alcoholic fatty liver, which reduces the activity of δ-6-desaturase enzymes, could contrary affect the synthesis of DHA, decreasing the levels of omega-3 FAs in erythrocytes and then in human milk [67]. Moreover, the endogenous synthesis of PUFAs is mediated by genes controlling the elongation of very long chain fatty acids protein 2 and 5 (ELOVL2 and ELOVL5) [68]. It is also suggested that the genes encoding fatty acids desaturases 1 and 2 (FADS1 FADS2 gene cluster) were reported to be associated with omega-3 and omega-6 FA proportions in human plasma, tissues, and milk [69]. Moltó-Puigmartí et al. [69] found that DHA proportions were lower in women homozygous for the minor allele than in women who were homozygous for the major allele (DHA proportions in plasma phospholipids: $p < 0.01$; DHA proportions in milk: $p < 0.05$). Contrary to our and previous citied studies, a Greek observational study reported that human milk omega-3 PUFA content was positively correlated with maternal dietary intakes of total PUFAs ($r = 0.26$; $p < 0.05$), inversely correlated with maternal carbohydrates intake ($r = -0.29$; $p < 0.05$), and unrelated to maternal energy, total fat, SFA, MUFA, and protein intakes [37].

Bravi et al. [70] carried out a systemic review searching studies investigating the impact of maternal diet on their milk composition. Considering fatty acid profiles, most of the studies were based on supplements intervention or assessed maternal nutritional status with a food frequency questionnaire. As the authors concluded, the gathered results were scarce and diversified.

The time of incorporation of omega-3 FAs into plasma lipids is quite long, and similar to their half-life. The longest time of incorporation is in adipose tissue, which is an important source of plasma lipids. As the content of omega-3 FAs in human milk depends on its availability from the plasma [71], dietary intake assessment methods studying habitual intake are more appropriate to exhibit the discussed correlation. Thus, the survey based solely on a three-day dietary record, especially in this case, does not work. This is not only because of the described omega-3 FAs kinetics, but also because omega-3 FAs food sources (mainly fish) are not consumed every day, and whether they happen to occur in the three-day period is a matter of chance.

The strengths of this study are the use of milk collection protocol, which enabled minimizing possible errors in the measurement of human milk composition, and also an assessment of maternal diet with two techniques (food frequency questionnaire and three-day dietary record). Additionally, contrary to other authors [56], we assayed α-linolenic acid that can convert to DHA; therefore, a correlation between dietary intake of DHA and its concentration in human milk is more reliable. Our study has also limitations. First, there were the issues of convenience sampling and the modest number of participants, resulting from the complex nutrition questionnaire and milk collection protocol. Second, this study was conducted in a large city (the capital of Poland); all the participants had a university education and high material status. Third, genetic polymorphisms that influence the use of PUFAs from the diet have not been studied. These factors may limit the generalization of our findings to a broader population, because the availability and consumption patterns of food products (e.g., fish and seafood) is strongly related to the place of living and material status.

5. Conclusions

To summarize, this study shows that the women under study during breastfeeding had an adequate intake of foods that are natural sources of omega-3 FAs (fatty fish, seafood, vegetable oils), which resulted in a high concentration of DHA in their milk. However, it should be stressed that we did not observe a correlation between dietary intake of omega-3 FAs and their concentrations in human milk. On the other hand, we observed that the intake frequency of some food products affected omega-3 FAs concentrations in human milk. Considering these findings and the highest content of DHA in human milk being observed between 6–12 h after dietary DHA intake, a short-term assessment of omega-3 FAs intake (based on three-day dietary record) may be unreliable. A FFQ assessing nutritional habits in the last three months prior to the study seems to the more reliable tool, reflecting the habitual intake of omega-3 FAs.

Author Contributions: A.B.-J., A.C.-S., A.W., G.O. and D.S.-W. were responsible for the conception and design of this study. A.W. was involved in the funding acquisition. A.B.-J. was responsible for the data collection and statistical analysis. E.J.-M. and H.M. were responsible for fatty acids analysis of human milk (fat extraction and gas chromatography-mass spectrometry analysis). A.B.-J. was responsible for writing the manuscript. A.C.-S., A.W., G.O. and D.S.-W. were responsible for revising the manuscript critically for important intellectual content. The manuscript has been revised by all co-authors.

Funding: This research was funded by a Polish Society of Clinical Child Nutrition research grant obtained in 2015 for the project titled "Association between human milk composition and the nutritional status and body composition of lactating women".

Acknowledgments: All the authors have reviewed and approved the final manuscript. This study was supported by a Polish Society of Clinical Child Nutrition research grant obtained in 2015 for the project entitled "Association between human milk composition and the nutritional status and body composition of lactating women". We thank all the mothers from the Holy Hospital in Warsaw who participated in this study.

Conflicts of Interest: The authors declare no conflict of interest.

References

1. Innis, S.M. Impact of maternal diet on human milk composition and neurological development of infants. *Am. J. Clin. Nutr.* **2014**, *99*, 734–741. [CrossRef] [PubMed]
2. Koletzko, B.; Rodriguez-Palmero, M.; Demmelmair, H.; Fidler, N.; Jensen, R.; Sauerwald, T. Physiological aspects of human milk lipids. *Early Hum. Dev.* **2001**, *65*, 3–18. [CrossRef]
3. Birch, E.E.; Castaneda, Y.S.; Wheaton, D.H.; Birch, D.G.; Uauy, R.D.; Hoffman, D.R. Visual maturation of term infants fed long-chain polyunsaturated fatty acid-supplemented or control formula for 12 mo. *Am. J. Clin. Nutr.* **2005**, *81*, 871–879. [CrossRef] [PubMed]
4. Jones, P.J.H.; Rideout, T. Lipids, sterols, and their metabolites. In *Modern Nutrition in Health and Disease*, 11st ed.; Ross, A.C., Caballero, B., Cousins, R.J., Tucker, K.L., Ziegler, T.R., Eds.; Lippincott Williams & Wilkins: Baltimore, MD, USA, 2014.
5. Innis, S.M. Human milk: Maternal dietary lipids and infant development. *Proc. Nutr. Soc.* **2007**, *66*, 397–404. [CrossRef] [PubMed]
6. Weiser, M.J.; Butt, C.M.; Mohajeri, M.H. Docosahexaenoic acid and cognition throughout the lifespan. *Nutrients* **2016**, *8*, 99. [CrossRef] [PubMed]
7. Innis, S.M. Metabolic programming of long-term outcomes due to fatty acid nutrition in early life. *Matern. Child Nutr.* **2011**, *7*, 112–123. [CrossRef] [PubMed]
8. European Food Safety Authority (EFSA). Scientific Opinion on Nutrient Requirements and Dietary Intakes of Infants and Young Children in the European Union. 2013. Available online: https://efsa.onlinelibrary.wiley.com/doi/pdf/10.2903/j.efsa.2013.3408 (accessed on 1 October 2018).
9. Goyens, P.L.; Spilker, M.E.; Zock, P.L.; Katan, M.B.; Mensink, R.P. Conversion of alpha-linolenic acid in humans is influenced by the absolute amounts of alpha-linolenic acid and linoleic acid in the diet and not by their ratio. *Am. J. Clin. Nutr.* **2006**, *84*, 44–53. [CrossRef] [PubMed]
10. Pawlosky, R.J.; Hibbeln, J.R.; Novotny, J.A.; Salem, N. Physiological compartmental analysis of alpha-linolenic acid metabolism in adult humans. *J. Lipid Res.* **2001**, *42*, 1257–1265. [PubMed]
11. Arterburn, L.M.; Hall, E.B.; Oken, H. Distribution, interconversion, and dose response of n-3 fatty acids in humans. *Am. J. Clin. Nutr.* **2006**, *83*, 1467–1476. [CrossRef] [PubMed]
12. Greiner, R.C.; Winter, J.; Nathanielsz, P.W.; Brenna, J.T. Brain docosahexaenoate accretion in fetal baboons: Bioequivalence of dietary alpha-linolenic and docosahexaenoic acids. *Pediatr. Res.* **1997**, *42*, 826–834. [CrossRef] [PubMed]
13. Koletzko, B.; Lien, E.; Agostoni, C.; Bohles, H.; Campoy, C.; Cetin, I.; Decsi, T.; Dudenhausen, J.W.; Dupont, C.; Forsyth, S.; et al. The roles of long-chain polyunsaturated fatty acids in pregnancy, lactation and infancy: Review of current knowledge and consensus recommendations. *J. Perinat. Med.* **2008**, *36*, 5–14. [CrossRef]
14. Brenna, J.T.; Lapillonne, A. Background paper on fat and fatty acid requirements during pregnancy and lactation. *Ann. Nutr. Metab.* **2009**, *55*, 97–122. [CrossRef] [PubMed]
15. Szajewska, H.; Horvath, A.; Rybak, A.; Socha, P. Breastfeeding. A Position Paper by the Polish Society for Paediatric Gastroenterology, Hepatology and Nutrition. *Med. Standards/Ped.* **2016**, *13*, 9–24.
16. Oken, E.; Kleinman, K.P.; Olsen, S.F.; Rich-Edwards, J.W.; Gillman, M.W. Associations of seafood and elongated n-3 fatty acid intake with fetal growth and length of gestation: Results from a US pregnancy cohort. *Am. J. Epidemiol.* **2004**, *160*, 774–783. [CrossRef] [PubMed]
17. Friesen, R.W.; Innis, S.M. Dietary arachidonic acid to EPA and DHA balance is increased among Canadian pregnant women with low fish intake. *J. Nutr.* **2009**, *139*, 2344–2350. [CrossRef] [PubMed]
18. Denomme, J.; Stark, K.D.; Holub, B.J. Directly quantitated dietary (n-3) fatty acid intakes of pregnant Canadian women are lower than current dietary recommendations. *J. Nutr.* **2005**, *135*, 206–211. [CrossRef] [PubMed]
19. Sioen, I.; Devroe, J.; Inghels, D.; Terwecoren, R.; De Henauw, S. The influence of n-3 PUFA supplements and n-3 PUFA enriched foods on the n-3 LC PUFA intake of Flemish women. *Lipids* **2010**, *45*, 313–320. [CrossRef] [PubMed]
20. Coletto, D.; Morrison, J. Seafood Survey: Public Opinion on Aquaculture and a National Aquaculture Act. 2011. Available online: http://www.aquaculture.ca/files/CAIA-PUBLIC-REPORT-May-2011.pdf (accessed on 2 October 2018).

21. Oken, E.; Kleinman, K.P.; Berland, W.E.; Simon, S.R.; Rich-Edwards, J.W.; Gillman, M.W. Decline in fish consumption among pregnant women after a national mercury advisory. *Obstet. Gynecol.* **2003**, *102*, 346–351. [CrossRef] [PubMed]
22. Amaral, Y.N.; Marano, D.; Silva, L.M.; Guimarães, A.C.; Moreira, M.E. Are There Changes in the Fatty Acid Profile of Breast Milk with Supplementation of Omega-3 Sources? A Systematic Review. *Rev. Bras. Ginecol. Obstet.* **2017**, *39*, 128–141. [CrossRef] [PubMed]
23. Yang, Y.J.; Kim, M.K.; Hwang, S.H.; Ahn, Y.; Shim, J.E.; Kim, D.H. Relative validities of 3-day food records and the food frequency questionnaire. *Nutr. Res. Pract.* **2010**, *4*, 142–148. [CrossRef] [PubMed]
24. Liu, M.J.; Li, H.T.; Yu, L.X.; Xu, G.S.; Ge, H.; Wang, L.L.; Zhang, Y.L.; Zhou, Y.B.; Li, Y.; Bai, M.X.; et al. A correlation study of DHA dietary intake and plasma, erythrocyte and breast milk DHA concentrations in lactating women from coastland, Lakeland, and inland areas of China. *Nutrients* **2016**, *8*, 312. [CrossRef] [PubMed]
25. Serra-Majem, L.; Nissensohn, M.; Øverby, N.C.; Fekete, K. Dietary methods and biomarkers of omega 3 fatty acids: A systematic review. *Br. J. Nutr.* **2012**, *107*, 64–76. [CrossRef] [PubMed]
26. Euro Who. Available online: http://www.euro.who.int/en/health-topics/disease-prevention/nutrition/ahealthy-lifestyle/body-mass-index-bmi (accessed on 14 May 2019).
27. Bzikowska-Jura, A.; Czerwonogrodzka-Senczyna, A.; Olędzka, G.; Szostak-Węgierek, D.; Weker, H.; Wesołowska, A. Maternal Nutrition and Body Composition During Breastfeeding: Association with Human Milk Composition. *Nutrients* **2018**, *10*, 1379. [CrossRef] [PubMed]
28. Folch, J.; Lees, M.; Sloane, G.H.S. A simple method for the isolation and purification of total lipides from animal tissues. *J. Biol. Chem.* **1957**, *226*, 497–509. [PubMed]
29. Szponar, L.; Wolnicka, K.; Rychlik, E. *Album of Photographs of Food Products and Dishes*; National Food and Nutrition Institute: Warsaw, Poland, 2011.
30. Brenna, J.T.; Varamini, B.; Jensen, R.G.; Diersen-Schade, D.A.; Boettcher, J.A.; Arterburn, L.M. Docosahexaenoic and arachidonic acid concentrations in human breast milk worldwide. *Am. J. Clin. Nutr.* **2007**, *85*, 1457–1464. [CrossRef] [PubMed]
31. Nochera, C.L.; Goossen, L.H.; Brutus, A.R.; Cristales, M.; Eastman, B. Consumption of DHA + EPA by low-income women during pregnancy and lactation. *Nutr. Clin. Pract.* **2011**, *26*, 445–450. [CrossRef]
32. Hibbeln, J.R.; Davis, J.M.; Steer, C.; Emmett, P.; Rogers, I.; Williams, C.; Golding, J. Maternal seafood consumption in pregnancy and neurodevelopmental outcomes in childhood (ALSPAC study): An observational cohort study. *Lancet* **2007**, *369*, 578–585. [CrossRef]
33. Sontrop, J.; Avison, W.R.; Evers, S.E.; Speechley, K.N.; Campbell, M.K. Depressive symptoms during pregnancy in relation to fish consumption and intake of n-3 polyunsaturated fatty acids. *Paediatr. Perinat. Epidemiol.* **2008**, *22*, 389–399. [CrossRef]
34. Rodriguez-Bernal, C.L.; Ramon, R.; Quiles, J.; Murcia, M.; Navarrete-Munoz, E.M.; Vioque, J.; Ballester, F.; Rebagliato, M. Dietary intake in pregnant women in a Spanish Mediterranean area: As good as it is supposed to be? *Public Health Nutr.* **2013**, *16*, 1379–1389. [CrossRef]
35. Jia, X.; Pakseresht, M.; Wattar, N.; Wildgrube, J.; Sontag, S.; Andrews, M.; Begum Subhan, F.; McCargar, F.; Field, C. Women who take n-3 long-chain polyunsaturated fatty acid supplements during pregnancy and lactation meet the recommended intake. *Appl. Physiol. Nutr. Metab.* **2015**, *40*, 474–481. [CrossRef]
36. Aumeistere, L.; Ciprovica, I.; Zavadska, D.; Volkovs, V. Fish intake reflects DHA level in breast milk among lactating women in Latvia. *Int. Breastfeed. J.* **2018**, *13*. [CrossRef] [PubMed]
37. Antonakou, A.; Skenderi, K.P.; Chiou, A.; Anastasiou, C.A.; Bakoula, C.; Matalas, A.L. Breast milk fat concentration and fatty acid pattern during the first six months in exclusively breastfeeding Greek women. *Eur. J. Nutr.* **2013**, *52*, 963–973. [CrossRef] [PubMed]
38. Makrides, M.; Gibson, R.A.; McPhee, A.J.; Collins, C.T.; Davis, P.G.; Doyle, L.W.; Simmer, K.; Colditz, P.B.; Morris, S.; Smithers, L.G.; et al. Neurodevelopmental outcomes of preterm infants fed high-dose docosahexaenoic acid: A randomized controlled trial. *JAMA* **2009**, *301*, 175–182. [CrossRef] [PubMed]
39. Marangoni, F.; Agostoni, C.; Lammardo, A.M.; Bonvissuto, M.; Giovannini, M.; Galli, C.; Riva, E. Polyunsaturated fatty acids in maternal plasma and in breast milk. *Prostaglandins Leukot. Essent. Fat. Acids* **2002**, *66*, 535–540. [CrossRef]

40. Marangoni, F.; Agostoni, C.; Lammardo, A.M.; Giovannini, M.; Galli, C.; Riva, E. Polyunsaturated fatty acid concentrations in human hindmilk are stable throughout 12-months of lactation and provide a sustained intake to the infant during exclusive breastfeeding: An Italian study. *Br. J. Nutr.* **2000**, *84*, 103–109. [PubMed]
41. Rueda, R.; Ramirez, M.; Garcia-Salmeron, J.L.; Maldonado, J.; Gil, A. Gestational age and origin of human milk influence total lipid and fatty acid contents. *Ann. Nutr. Metab.* **1998**, *42*, 12–22. [CrossRef] [PubMed]
42. Sala-Vila, A.; Campoy, C.; Castellote, A.I.; Garrido, F.J.; Rivero, M.; Rodríguez-Palmero, M.; López-Sabater, M.C. Influence of dietary source of docosahexaenoic and arachidonic acids on their incorporation into membrane phospholipids of red blood cells in term infants. *Prostaglandins Leukot. Essent. Fat. Acids* **2006**, *74*, 143–148. [CrossRef] [PubMed]
43. Olafsdottir, A.S.; Thorsdottir, I.; Wagner, K.H.; Elmadfa, I. Polyunsaturated fatty acids in the diet and breast milk of lactating icelandic women with traditional fish and cod liver oil consumption. *Ann. Nutr. Metab.* **2006**, *50*, 270–276. [CrossRef]
44. Kovacs, A.F.S.; Marosvolgyi, T.; Burus, I.; Decsi, T. Fatty acids in early human milk after preterm and full-term delivery. *J. Pediatr. Gastroenterol. Nutr.* **2005**, *41*, 454–459. [CrossRef]
45. Jackson, K.H.; Harris, W.S. Should there be a target level of docosahexaenoic acid in breast milk? *Curr. Opin. Clin. Nutr. Metab. Care* **2016**, *19*, 92–96. [CrossRef]
46. Gibson, R.A.; Neumann, M.A.; Makrides, M. Effect of increasing breast milk docosahexaenoic acid on plasma and erythrocyte phospholipid fatty acids and neural indices of exclusively breast fed infants. *Eur. J. Clin. Nutr.* **1997**, *51*, 578–584. [CrossRef] [PubMed]
47. Presa-Owens, S.D.; Lopez-Subater, M.C.; Rivero-Urgell, M. Fatty acid composition of human milk in Spain. *J. Pediatr. Gastroenterol. Nutr.* **1996**, *22*, 180–185. [CrossRef] [PubMed]
48. Koletzko, B.; Thiel, I.; Abiodun, P.O. The fatty acid composition of human milk in Europe and Africa. *J. Pediatr.* **1992**, *120*, 62–70. [CrossRef]
49. Wu, T.C.; Lau, B.H.; Chen, P.H.; Wu, L.T.; Tang, R.B. Fatty acid composition of Taiwanese Human Milk. *J. Chin. Med Assoc.* **2010**, *73*, 581–588. [CrossRef]
50. ISSFAL. Recommendations for Intake of Polyunsaturated Fatty Acids in Healthy Adults. 2004. Available online: http://www.issfal.org/ (accessed on 31 May 2019).
51. Lassek, W.D.; Gaulin, S.J.C. Maternal milk DHA concentration predicts cognitive performance in a sample of 28 nations. *Matern. Child Nutr.* **2015**, *11*, 773–779. [CrossRef] [PubMed]
52. Bernard, J.Y.; Armand, M.; Peyre, H.; Garcia, C.; Forhan, A.; De Agostini, M.; Charles, M.A.; Heude, B. Breastfeeding, polyunsaturated fatty acid levels in colostrum and child intelligence quotient at age 5–6 years. *J. Pediatr.* **2017**, *183*, 43–50. [CrossRef] [PubMed]
53. Zielinska, M.; Hamulka, J.; Wesolowska, A. Carotenoid Content in Breastmilk in the 3rd and 6th Month of Lactation and Its Associations with Maternal Dietary Intake and Anthropometric Characteristics. *Nutrients* **2019**, *11*, 193. [CrossRef] [PubMed]
54. Lassek, W.D.; Gaulin, S.J. Linoleic and docosahexaenoic acids in human milk have opposite relationships with cognitive test performance in a sample of 28 countries. *Prostaglandins Leukot. Essent. Fat. Acids* **2014**, *91*, 195–201. [CrossRef] [PubMed]
55. Silva, M.H.L.; Silva, M.T.C.; Brandão, S.C.C.; Gomes, J.C.; Peternelli, L.A.; Franceschini, S.C.C. Fatty acid composition of mature breast milk in Brazilian women. *Food Chem.* **2005**, *93*, 297–303. [CrossRef]
56. Nishimura, R.Y.; de Castro, G.S.F.; Junior, A.A.J.; Sartorelli, D.S. Breast milk fatty acid composition of women living far from the coastal area in Brazil. *J. Pediatr.* **2013**, *89*, 263–268. [CrossRef]
57. Simopoulos, A.P. The importance of the ratio of omega-6/omega-3 essential fatty acids. *Biomed. Pharmacother.* **2002**, *56*, 365–379. [CrossRef]
58. Lauritzen, L.; Jorgensen, M.H.; Hansen, H.S.; Michaelsen, K.F. Fluctuations in human milk long-chain PUFA levels in relation to dietary fish intake. *Lipids* **2002**, *37*, 237–244. [CrossRef] [PubMed]
59. Juber, B.A.; Jackson, K.H.; Johnson, K.B.; Harris, W.S.; Baack, M.L. Breast milk DHA levels may increase after informing women: A community-based cohort study from South Dakota USA. *Int. Breastfeed. J.* **2016**, *12*, 7. [CrossRef] [PubMed]
60. Strobel, C.; Jahreis, G.; Kuhnt, K. Survey of n-3 and n-6 polyunsaturated fatty acids in fish and fish products. *Lipids Health Dis.* **2012**, *11*, 144. [CrossRef] [PubMed]

61. Quinn, E.A.; Kuzawa, C.W. A dose-response relationship between fish consumption and human milk DHA concentration among Filipino women in Cebu City. *Philippines Acta Paediatr.* **2012**, *101*, 439–445. [CrossRef] [PubMed]
62. United States Department of Agriculture (USDA). Available online: https://ndb.nal.usda.gov/ndb/search/list (accessed on 30 October 2018).
63. Kunachowicz, H.; Nadolna, I.; Przygoda, B.; Iwanow, K. *Food Composition Tables*; PZWL: Warsaw, Poland, 2005.
64. Torres, A.G.; Ney, J.G.; Meneses, F.; Trugo, N.M. Polyunsaturated fatty acids and conjugated linoleic acid isomers in breast milk are associated with plasma non-esterified and erythrocyte membrane fatty acid composition in lactating women. *Br. J. Nutr.* **2006**, *95*, 517–524. [CrossRef]
65. Haggarty, P. Fatty acid supply to the human fetus. *Annu. Rev. Nutr.* **2010**, *30*, 237–255. [CrossRef]
66. Fidler, N.; Sauerwald, T.; Pohl, A.; Demmelmair, H.; Koletzko, B. Docosahexaenoic acid transfer into human milk after dietary supplementation: A randomized clinical trial. *J. Lipid Res.* **2000**, *41*, 1376–1383.
67. Valenzuela, R.; Videla, L.A. The importance of the long-chain polyunsaturated fatty acid n-6/n-3 ratio in development of non-alcoholic fatty liver associated with obesity. *Food Funct.* **2011**, *2*, 644–648. [CrossRef]
68. Wu, Y.; Wang, Y.; Tian, H.; Lu, T.; Yu, M.; Xu, W.; Liu, G.; Xie, L. DHA intake interacts with ELOVL2 and ELOVL5 genetic variants to influence polyunsaturated fatty acids in human milk. *J. Lipid. Res.* **2019**, *60*, 1043–1049. [CrossRef]
69. Moltó-Puigmartí, C.; Plat, J.; Mensink, R.P.; Müller, A.; Jansen, E.; Zeegers, M.P.; Thijs, C. FADS1 FADS2 gene variants modify the association between fish intake and the docosahexaenoic acid proportions in human milk. *Am. J. Clin. Nutr.* **2010**, *91*, 1368–1376. [CrossRef] [PubMed]
70. Bravi, F.; Wiens, F.; Decarli, A.; Dal Pont, A.; Agostoni, C.; Ferraroni, M. Impact of maternal nutrition on breast-milk composition: A systematic review. *Am. J. Clin. Nutr.* **2016**, *104*, 646–662. [CrossRef] [PubMed]
71. Van, R.M.; Hunter, D. Biochemical Indicators of Dietary Intake. In *Nutritional Epidemiology*, 3rd ed.; Oxford University Press: Oxford, UK, 2013; pp. 150–212. ISBN -13: 978-0199754038.

© 2019 by the authors. Licensee MDPI, Basel, Switzerland. This article is an open access article distributed under the terms and conditions of the Creative Commons Attribution (CC BY) license (http://creativecommons.org/licenses/by/4.0/).

Article

Antenatal Influenza A-Specific IgA, IgM, and IgG Antibodies in Mother's Own Breast Milk and Donor Breast Milk, and Gastric Contents and Stools from Preterm Infants

Veronique Demers-Mathieu [1], Robert K. Huston [2], Andi M. Markell [2], Elizabeth A. McCulley [2], Rachel L. Martin [2] and David C. Dallas [1],*

[1] Nutrition Program, School of Biological and Population Health Sciences, College of Public Health and Human Sciences, Oregon State University, Corvallis, OR 97331, USA
[2] Department of Pediatrics, Randall Children's Hospital at Legacy Emanuel, Portland, OR 97227, USA
* Correspondence: Dave.Dallas@oregonstate.edu; Tel.: +1-541-737-1751

Received: 5 June 2019; Accepted: 9 July 2019; Published: 11 July 2019

Abstract: Antenatal milk anti-influenza antibodies may provide additional protection to newborns until they are able to produce their own antibodies. To evaluate the relative abundance of milk, we studied the antibodies specific to influenza A in feeds and gastric contents and stools from preterm infants fed mother's own breast milk (MBM) and donor breast milk (DBM). Feed (MBM or DBM) and gastric contents (MBM or DBM at 1 h post-ingestion) and stool samples (MBM/DBM at 24 h post-ingestion) were collected, respectively, from 20 preterm (26–36 weeks gestational age) mother-infant pairs at 8–9 days and 21–22 days of postnatal age. Samples were analyzed via ELISA for anti-H1N1 hemagglutinin (anti-H1N1 HA) and anti-H3N2 neuraminidase (anti-H3N2 NA) specificity across immunoglobulin A (IgA), immunoglobulin M (IgM), and immunoglobulin G (IgG) isotypes. The relative abundance of influenza A-specific IgA in feeds and gastric contents were higher in MBM than DBM at 8–9 days of postnatal age but did not differ at 21–22 days. Anti-influenza A-specific IgM was higher in MBM than in DBM at both postnatal times in feed and gastric samples. At both postnatal times, anti-influenza A-specific IgG was higher in MBM than DBM but did not differ in gastric contents. Gastric digestion reduced anti-H3N2 NA IgG from MBM at 21–22 days and from DBM at 8–9 days of lactation, whereas other anti-influenza A antibodies were not digested at either postnatal times. Supplementation of anti-influenza A-specific antibodies in DBM may help reduce the risk of influenza virus infection. However, the effective antibody dose required to induce a significant protective effect remains unknown.

Keywords: passive immunization; maternal immunoglobulins; lactation; prematurity; flu vaccine; human milk

1. Introduction

Infants are susceptible to influenza infections and cannot be vaccinated before six months of age [1,2]. Therefore, the Centers for Disease Control and Prevention recommends that all pregnant women receive the inactivated influenza virus vaccine during the second or third trimester [3]. Antenatal vaccination helps protect the infant against influenza infection [4–7]. Part of that protection stems from the vaccine-stimulated increased maternal blood influenza-specific immunoglobulin G (IgG), which allows increased transfer of influenza-specific IgG across the placenta and leads to increased persistence of these antibodies in the infant bloodstream post-birth [8]. In addition to the changes in blood, antenatal vaccination increases secretion of influenza-specific antibodies into breast milk [9]. Milk anti-influenza antibodies may provide additional protection to newborns until they are able to produce antibodies against the virus [9]. Regardless of whether a mother was recently vaccinated, their breast milk typically contains some influenza-specific Ig as they were likely exposed to the virus or vaccinated in the past [10]. These milk antibodies may be protective against influenza infection [9].

The site of action for milk antibodies is mainly thought to be within the gastrointestinal tract, whereas influenza A infects within the lung mucosal tissue. Thus, the relevance of human milk antibodies to influenza protection in unclear. However, milk influenza A-specific antibodies can be studied as a model for other milk antibodies to enteric pathogens for which the gut is their site of action. To neutralize enteric pathogens throughout the infant gastrointestinal tract, milk antibodies must survive the digestive protease actions. No oral supplementation studies have determined the percentage of remaining relative abundance of anti-influenza A IgA, IgM, and IgG during infant digestion.

In neonatal intensive care units (NICU), very preterm infants are often fed donor breast milk (DBM) to supplement an insufficient maternal supply of mother's own breast milk (MBM) [11,12]. Whether MBM and DBM differ in milk anti-influenza A antibody relative abundance is unknown. DBM typical processing includes pooling milks from different mothers, pasteurizing (Holder pasteurization, 62.5 °C for 30 min) and freezing and thawing at least twice, which could reduce the antibody abundance. Our previous study demonstrated that total IgA, secretory IgA (SIgA), total IgM, and IgG concentrations were higher in MBM than DBM and higher in gastric contents (1 h post-ingestion) when preterm infants were fed with MBM than when infants were fed DBM [13]. A previous study demonstrated that IgA survived intact after Holder pasteurization in non-centrifuged MBM (0% loss) whereas IgG concentration was reduced (34.3% loss) (IgM was not measured) [14]. In another study, IgA from centrifuged MBM was reduced 22% after Holder pasteurization whereas IgM concentration was completely destroyed [15].

The remaining relative abundance of milk antibodies specific to influenza A virus from MBM or DBM during gastric digestion is unknown and a comparison of mother's milk and donor milk have not been determined. The aim of this study was to determine the difference of antenatal influenza A-specific IgA, IgM, and IgG antibodies in MBM and DBM, and gastric contents and stools from preterm infants. Milk anti-influenza A neutralizing antibodies can be specific to subtypes of HA or NA [3]. In this study, we examined anti-H1N1 HA and anti-H3N2 NA antibodies.

2. Materials and Methods

2.1. Participants and Sample Collection

2.1.1. Participants and Enrollment

This study was approved by the Institutional Review Boards of The Legacy Health and Oregon State University (1402–2016, first approved on 05/03/17). Samples were collected from 20 premature-delivering mother-infant pairs ranging in gestational age (GA) at birth from 26 weeks to 36 weeks (Table S1) in the NICU. Parents of all eligible infants in the NICU were approached for participation and informed consent. Clinical data were collected for each mother-infant pair, including GA and postnatal age for both type of feeding (see Table S1). Eligibility criteria included having an indwelling naso/orogastric feeding tube, bolus feeding (<60 min infusion tolerated), feeding volumes of at least four milliliter, and mothers who could produce a volume of MBM adequate for one full-volume feed per day. Exclusion criteria included neonates with diagnoses that were incompatible with life, gastrointestinal system anomalies, major gastrointestinal surgery, severe genitourinary anomalies and significant metabolic, or endocrine diseases.

2.1.2. Feeding and Sampling

To compare the effects of DBM and MBM, two separate feedings of DBM and MBM without fortification were given to infants. Milk and gastric samples (2 mL) were collected on 8–9 and

21–22 days of postnatal age (Figure 1). At both sample time periods, each infant received two of the normal eight daily feedings as unfortified MBM or DBM on alternate days (randomized order). The order of feeding MBM and DBM was randomized to control for any potential effect of infant day of life on antibody digestion. The pool of DBM was acquired from two batches at Northwest Mother's Milk Bank. Three-liter batches were pasteurized and frozen in 50 mL doses so that only a small fraction was thawed for each infant feeding. The power analysis based in our previous study indicated that at least 15 infants were required to compare DBM and MBM-fed infants [16].

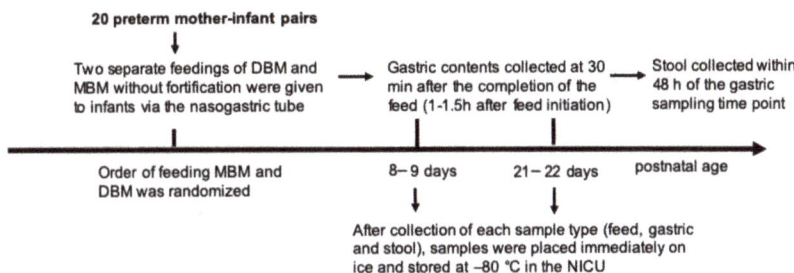

Figure 1. Schema of study design to determine the difference of antenatal influenza-specific antibodies in mother's breast milk (MBM) and donor breast milk (DBM) in gastric contents and stool samples from preterm infants.

The protocol of feeding is described in our previous study [13]. Milk (either MBM or DBM) was fed to the infant via the nasogastric tube with a feeding pump set to deliver the entire bolus over 30–60 min. A sample of the gastric fluid was collected 30 min after the completion of the feed (1–1.5 h after feed initiation). Stool (1 g) was collected within 48 h of the gastric sampling time point and was recovered from the diaper and scraped into a sterile jar. Stool sample collection was not specific to DBM/MBM and thus represents stools deriving from a mixture of DBM and MBM feeding. After collection of each sample type (feed, gastric and stool), samples were placed immediately on ice and stored at −80 °C in the NICU. Samples were transported on dry ice to the Dallas laboratory at Oregon State University for sample analysis.

2.2. Sample Preparation and ELISAs

Feed (MBM and DBM) and gastric samples were thawed at 4 °C, pH was determined and samples (1 mL) were centrifuged at 3500× g for 30 min at 4 °C. The infranate was collected, separated into aliquots (100 µL) and stored at −80 °C. Frozen stool samples (0.1 g) were diluted in 700 µL of phosphate-buffered saline pH 7.4 (Thermo Fisher Scientific, MA, USA) with 0.05% Tween-20 (Bio-Rad Laboratories, Irvine, CA, USA) (PBST) and 3% fraction V bovine serum albumin solution (Innovative Research, Novi, MI, USA). Diluted stool samples were mixed by vortex for 2 min and the vials were centrifuged at 3500× g for 20 min at 4 °C. The supernatant was collected, separated into aliquots (100 µL) and stored at −80 °C. Sample pH and protein concentration measurements were performed and results were published previously [13].

The spectrophotometric ELISAs were recorded with a microplate reader (Spectramax M2, Molecular Devices, Sunnyvale, CA, USA) with two replicates of blanks, standards and samples. Clear flat-bottom 96-well plates (Thermo Fisher Scientific, Waltham, MA, USA) were coated with 0.75 µg/mL (100 µL) of influenza A/Hong Kong/4801/2014 H3N2 neuraminidase (H3N2 NA) or influenza A/California/07/2009 H1N1 hemagglutinin (H1N1 HA) (Sino Biological US Inc., Wayne, PA, USA). These antigens were chosen based on the strains present in the vaccine (Flucelvax Quadrivalent, 2017–2018 formula, Seqirus, Holly Springs, NC, USA, Table S2) that mothers received for 2017–2018 when samples were collected. Plates were incubated overnight at 4 °C. After incubation, plates were washed three times with PBST. Blocking buffer (100 µL PBST with 3% fraction V bovine serum albumin solution) was added in all wells and incubated for 1 h at room temperature. Standards were prepared using human serum from a female adult that received the influenza vaccine (Flucelvax Quadrivalent, 2017–2018 formula, Seqirus, Holly Springs, NC, USA, Table S2) four months prior. Blood was collected into six BD vacutainer serum separation tubes (SSTTM) (Becton Dickinson, Franklin Lakes, NJ, USA), incubated at room temperature for 5 min, and centrifuged at 3900× g for 5 min. Serum was aliquoted into several vials and stored at −80 °C until used. The relative abundance of influenza A antibodies was derived by interpolation from the standard curve generated from human serum with the assigned quantity of anti-influenza A antibodies expressed in ELISA units/mL (EU/mL) (non-diluted serum (1x) = 50,000 arbitrary EU/mL). The standard curves were prepared using a dilution series of standard human serum in blocking buffer and the final concentration covered a range from 12.5× (4000 EU/mL) to 25,600× (1.95 EU/mL). Feed (MBM or DBM) and gastric samples were diluted 10× with blocking buffer for anti-influenza IgA, IgM, and IgG measurements whereas a 2× dilution was used for the prediluted stool samples. For each step (addition of 100 μL standards/samples and secondary antibodies at 1 μg/mL), washing and incubation for 1 h at room temperature were performed. The detection antibodies were goat anti-human IgA alpha-chain: horseradish peroxidase (HRP) for anti-influenza IgA, goat anti-human IgM mu-chain:HRP for anti-influenza IgM and goat anti-human IgG gamma-chain:HRP for anti-influenza IgG (Bio-Rad Laboratories). The substrate 3, 3′, 5, 5′-tetramethylbenzidine (Thermo Fisher Scientific) (1×, 100 μL) was added for 5 min at room temperature followed by addition of 2 N sulfuric acid (100 μL) to stop the reaction. Optical density was measured at 450 nm.

2.3. Pasteurization Effect on Antibodies Specific to Influenza A

To examine the effects of pasteurization directly, a breast milk sample was collected from one mother who was vaccinated with the 2017–2018 influenza vaccine at 25 weeks of pregnancy and gave birth at term. At 12 days of lactation, the mother pumped and collected 150 mL of milk into a sterile plastic bag. Six 1 mL aliquots were centrifuged at 4000× g for 30 min, 4 °C and infranates were collected (skim). Three of these aliquots were used for the control raw human milk (RHM) and 3 others were pasteurized at 62.5 °C for 30 min, referred to as the pasteurized skimmed human milk (PSHM). Three other 1-mL aliquots were pasteurized whole (referred to as pasteurized whole human milk, PWHM) and centrifuged at 4000× g for 30 min at 4 °C prior to ELISA.

2.4. Statistical Analyses

Wilcoxon matched-pairs signed-rank test in GraphPad Prism software (version 8) was used to compare relative abundance of each anti-influenza antibody isotype (EU/mL) in milk and in gastric contents between MBM and DBM (type of feeding) within the same mother-infant pairs at 8–9 days and 21–22 days of postnatal age. Student's t-tests were used to evaluate the effect of whether or not infants received antibiotics in gastric and stool samples from both feeding types (MBM and DBM) (Table S3). One-way ANOVA followed by Dunnett's multiple comparisons test was performed to compare RHM to PSHM and PWHM. Differences were designated significant at $p < 0.05$. Geometric mean relative abundances (GMRA) of these antibodies in samples at 8–9 days and 21–22 days of postnatal age were also calculated.

3. Results

3.1. Infant Demographics

Demographic details for the preterm-delivering mother-infant pairs are presented in Table S1.

3.2. Anti-Influenza Antibody Relative Abundance in Feeds

The relative abundance of H1N1 HA- (Figure 2) and H3N2 NA-specific IgA (Figure 2) were 3.6- and 2-fold higher, respectively, in MBM than DBM feeds given at 8–9 days of postnatal age but did not differ at 21–22 days. H1N1 HA- (10- and 3-fold) and H3N2 NA-specific IgM (13- and 8-fold) were higher in MBM than in DBM feeds given at 8–9 days and 21–22 days of postnatal age, respectively. H1N1 HA-specific IgG was 2-fold higher in MBM than DBM feeds given at 8–9 days and 21–22 days of postnatal age. H3N2 NA-specific IgG was 30% higher in MBM than DBM feeds given at 8–9 days but did not differ at 21–22 days. The GMRA of each anti-influenza isotype are summarized in Table S4.

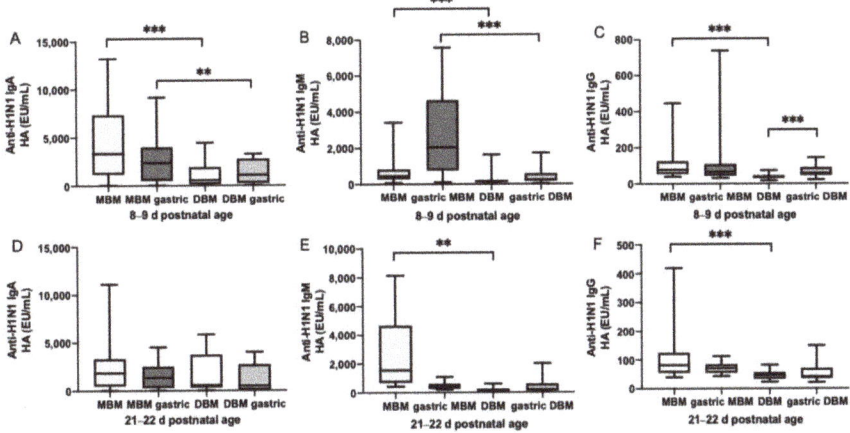

Figure 2. Milk antibodies specific to anti-H1N1 hemagglutinin (anti-H1N1 HA) in mother's own breast milk (MBM) and donor breast milk (DBM) feeds, and gastric samples from preterm infants fed MBM and DBM (26–36 wk of GA). Relative abundance of (**A**) anti-H1N1 HA IgA, (**B**) anti-H1N1 HA IgM, and (**C**) anti-H1N1 HA IgG at 8–9 days of postnatal age. Relative abundance (EU/mL) of (**D**) anti-H1N1 HA IgA, (**E**) anti-H1N1 HA IgM, and (**F**) anti-H1N1 HA IgG at 21–22 days of postnatal age. Values are min, max, and median, $n = 20$ for 8–9 days and $n = 16$ for 21–22 days of postnatal age. Asterisks show statistically significant differences between variables (*** $p < 0.001$; ** $p < 0.01$) using the Wilcoxon matched-pairs signed-rank test.

3.3. Anti-Influenza Antibody Relative Abundance in Gastric Contents

At 8–9 days of postnatal age, relative abundance of H1N1 HA- (Figure 2) and H3N2 NA-specific IgA (Figure 2) were 4- and 7-fold higher in gastric contents from infants fed MBM than those fed DBM. At 21–22 days, H3N2 NA-specific IgA was 4-fold higher in gastric contents after feeding MBM than after feeding DBM but did not differ for H1N1 HA-specific IgA. H1N1 HA- and H3N2 NA-specific IgM were 6- and 11-fold higher in gastric contents from infants fed MBM than those fed DBM at 8–9 days. At 21–22 days, H3N2 NA-specific IgM was 5-fold higher in MBM than DBM but did not differ for H1N1 HA-specific IgM. H1N1 HA- and H3N2 NA-specific IgG in gastric contents did not differ after feeding infants MBM and those fed DBM. The GMRA of each anti-influenza isotype are summarized in Table S5.

3.4. Gastric Digestion of Anti-Influenza Antibody from MBM and DBM

Relative abundance of H1N1 HA- (Figure 1) and H3N2 NA-specific IgA and IgM (Figure 3) did not significantly decrease from feed to gastric contents in infants fed MBM or DBM at 8–9 days or 21–22 days of postnatal age. H1N1 HA-specific IgG from MBM did not change from feed to gastric contents at 8–9 days or 21–22 days, whereas H1N1 HA-specific IgG from DBM slightly increased in gastric contents at 8–9 days but did not differ at 21–22 days of postnatal age. H3N2 NA-specific IgG from DBM decreased in the gastric contents at 8–9 days but did not differ at 21–22 days. H3N2 NA-specific IgG from MBM decreased in gastric contents at 21–22 days but not at 8–9 days. The GMRA abundances of each anti-influenza isotype are summarized in Table S6.

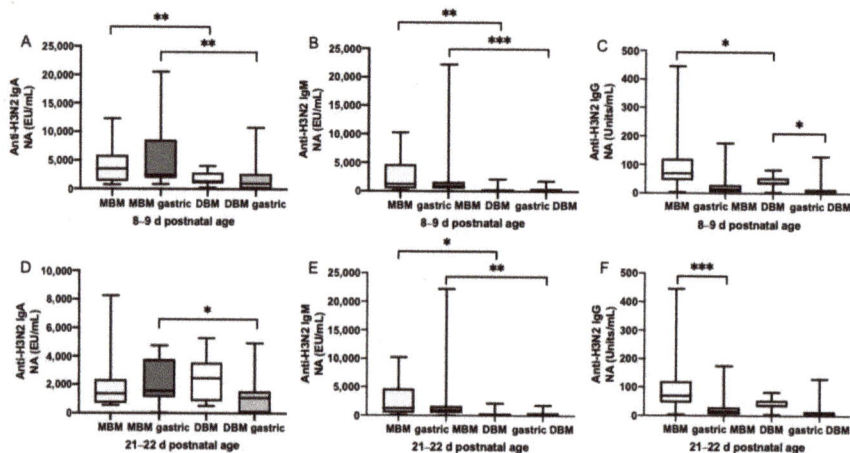

Figure 3. Milk antibodies specific to anti-H3N2 neuraminidase (anti-H3N2 NA) in mother's own breast milk (MBM) and donor breast milk (DBM) feeds, and gastric samples from preterm infants fed MBM and DBM (26–36 wk of GA). Relative abundance of (**A**) anti-H3N2 NA IgA, (**B**) anti-H3N2 NA IgM, and (**C**) anti-H3N2 NA IgG at 8–9 d postnatal age. Relative abundance (EU/mL) of (**D**) anti-H3N2 NA IgA, (**E**) anti-H3N2 NA IgM, and (**F**) anti-H3N2 NA IgG at 21–22 days of postnatal age. Values are min, max, and median, $n = 20$ for 8–9 days and $n = 16$ for 21–22 days of postnatal age. Asterisks show statistically significant differences between variables (*** $p < 0.001$; ** $p < 0.01$; * $p < 0.05$) using the Wilcoxon matched-pairs signed-rank test.

3.5. Postnatal Age Effects on Anti-Influenza A Antibodies in Preterm Infant Stools Derived from Mixed MBM/DBM

The relative abundance of H1N1 HA- and H3N2 NA-specific IgA, IgM and IgG in infant stools did not differ between 8–9 days and 21–22 days of postnatal age (Table S7).

3.6. Pasteurization Effects on Human Milk Anti-Influenza A Antibodies

The anti-influenza antibodies in RHM to PWHM and PSHM were compared to determine the effect of Holder pasteurization on them. Whole and skim breast milk were pasteurized because previous literature reported different results with and without delipidation before pasteurization [14,15]. The relative abundance of anti-H1N1 HA IgA, anti-H3N2 NA IgA, and anti-H1N1 HA IgG did not differ between RHM and PWHM or PSHM (Figure 4). Anti-H1N1 HA IgM, anti-H3N2 NA IgM, and anti-H3N2 NA IgG in RHM were 2.5-, 23-, and 2-fold higher than in PSHM. Anti-H3N2 NA IgM in RHM was 3-fold higher than in PWHM. Anti-H1N1 HA IgM, anti-H3N2 NA IgM, and anti-H3N2 NA IgG in PWHM were 2.5-, 7.3-, and 1.5-fold higher than PSHM but did not differ for other antibodies.

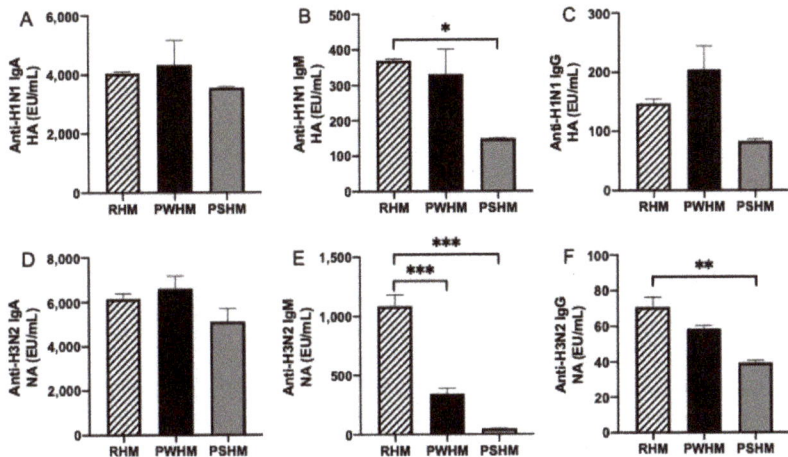

Figure 4. Comparison of immunoglobulin relative abundance between raw human milk (RHM) and pasteurized whole human milk (PWHM) or pasteurized skimmed human milk (PSHM) ($n = 3$). Relative abundance of anti-H1N1 HA (**A**) IgA, (**B**) IgM, (**C**) IgG, and of anti-H3N2 NA. (**D**) IgA, (**E**) IgM, and (**F**) IgG in mother's milk. Milk samples were from one mother who delivered one term infant at 38 wk of GA and collected milk at 12 days of postnatal age. Asterisks show statistically significant differences between variables (*** $p < 0.001$; ** $p < 0.01$; * $p < 0.05$) using one-way ANOVA followed by Dunnett's multiple comparisons test.

4. Discussion

This study aimed to determine whether the relative abundance of maternal milk anti-influenza A antibodies differed between MBM and DBM and across their digestion in gastric contents and in stool of preterm infants. Influenza A viruses vary based on their hemagglutinin (HA) and neuraminidase (NA) subtypes; each influenza A virus has a specific combination of an HA and an NA subtype [3]. Milk anti-influenza A neutralizing antibodies can be specific to subtypes of HA or NA. In this study, we examined anti-H1N1 HA and anti-H3N2 NA antibodies. Anti-influenza IgA level was previously evaluated in human milk via ELISA [9]. The anti-H1N1 HA IgA relative abundance measured herein was higher than those measured previously in term-delivering influenza-vaccinated mothers by Schlaudecker et al. [9]. However, abundance values of milk antibodies cannot be compared with those reported in other studies because different human serums were used to prepare the standard curve for each anti-influenza antibody isotype. No commercial standard exists for anti-influenza IgA, IgM, and IgG; therefore, determining absolute concentrations was not possible in this study. This research is the first to determine the anti-influenza A IgM and IgG relative abundance in breast milk.

Anti-influenza A H1N1 HA and H3N2 NA IgA in MBM were higher than in DBM during the first week of postnatal age for both feeds and gastric contents from preterm infants but did not differ during the third week for either feed or gastric contents. Neither anti-influenza A H1N1 HA nor H3N2 NA IgA was reduced by pasteurization of whole or skim breast milk. Therefore, we speculate that high relative abundance of anti-influenza IgA in MBM is related to early GA and early postnatal age of infants when mothers delivered prematurely. DBM is donated from mothers delivering mostly term infants and milk is usually collected later than three months of lactation, which could have different milk antibody levels compared with those from preterm-delivering mothers at early lactation times. Our recent clinical study demonstrated that antibody concentrations (total IgA and SIgA) were higher in MBM than DBM in feeds and gastric samples (1 h post-feeding) from preterm infants [13]. A previous study found that IgA concentration in preterm milk was higher than in term milk from 3 to 15 days of lactation time [17]. Total IgA concentration in colostrum from preterm-delivering mothers was higher than colostrum (1–3 days of lactation time) from term-delivering mothers in several studies [17–19]. However, our previous study showed that total IgA did not differ between preterm milk and term milk from 6 days to 28 days of lactation time [16]. No study has previously evaluated the difference in relative abundance of milk anti-influenza IgA antibodies between preterm- and term-delivering mothers or between early and late lactation time. Our results suggest that the provision of passive immune protection with milk anti-influenza A IgA was highest when infants are fed MBM during their first week of postnatal age.

Anti-influenza A H1N1 HA and H3N2 NA IgM in MBM were higher than in DBM during the first and third week of postnatal age in both feeds and in infant gastric samples. Among isotypes, anti-influenza IgM relative abundance differed most between DBM and MBM. Anti-H1N1 HA IgM was reduced when skim milk was pasteurized but not when whole breast milk was pasteurized. On the other hand, anti-H3N2 NA IgM was decreased by pasteurizing whole and skim milk. We hypothesize that the lower relative abundance of anti-influenza A IgM in DBM compared with MBM was due to the pasteurization of pooled donor milk after fat loss during freezing and thawing. The reduction of anti-influenza IgM was not likely due to the difference of GA or lactation time between MBM (preterm-delivering mothers) and DBM (term-delivering mothers). Anti-influenza IgM did not change across lactation time (1 weeks to 3 weeks) in milk from preterm-delivering mothers. Our recent clinical study demonstrated that total IgM concentration was higher in MBM than DBM in feeds and gastric samples (1 h post-feeding) from preterm infants [13]. Previous studies showed that IgM concentration in preterm-delivering mothers was lower than [16], higher than [17] or similar to [20,21] term-delivering mothers, suggesting a high variability of IgM concentration between mothers. A strong variation of anti-H3N2 NA IgM in MBM from preterm-delivering mothers (0 EU/mL to 4301 EU/mL at 1 week; 0–900 EU/mL at 3 weeks) was observed, which could be due to the time of vaccination during pregnancy and the mother's immune response to the vaccine.

Anti-H1N1 HA IgG was higher in MBM than in DBM fed during the first and third week of postnatal age. Pasteurization did not affect anti-H1N1 HA IgG in whole or skim breast milk. The higher relative abundance of anti-H1N1 HA IgG in MBM could be due to the difference of GA and lactation time between preterm-delivering mothers and term-delivering mothers (DBM). Anti-influenza A H1N1 HA and H3N2 NA IgG did not differ in gastric contents of infants fed MBM and DBM either at 1 week or 3 weeks of postnatal age. Interestingly, anti-H3N2 NA IgG was higher in MBM than in DBM during the first week but did not differ at third week of MBM lactation. Anti-H3N2 NA IgG relative abundance in skimmed milk but not whole milk was reduced by pasteurization. We speculate that anti-H3N2 NA IgG was low in DBM due to the time of lactation. Our recent clinical study demonstrated that total IgG concentration was higher in MBM than DBM in feed and gastric samples (1 h post-feeding) from preterm infants [13].

Gastric digestion reduced anti-H3N2 NA IgG from MBM at 21–22 days and from DBM at 8–9 days of postnatal age, but the other anti-influenza A antibodies were not digested in the gastric contents of preterm infants. A slight increase of anti-H3N2 NA IgG from DBM in gastric contents at 8–9 days of postnatal age was observed, which could be due to contamination of the gastric contents with residual from previous feeds based on MBM.

All influenza A antibodies most likely resisted preterm infant overall digestion to some extent as all were detected in stools 24 h post-feeding. Antibodies detected in stool samples could derive from the MBM and/or DBM feeding or potentially be generated by the infant. Anti-H1N1 and anti-H3N2 antibodies in stool samples did not differ between the first and third week of postnatal age.

Five infants among twenty received antibiotics (Table S1). Provision of antibiotics could change the microbiome and alter the survival of maternal influenza antibodies to the stool. However, no differences were detected in their survival in the gastric contents and stools between infants with and without antibiotics (Table S3).

Schlaudecker et al. [9] demonstrated that anti-influenza IgA level in breast milk correlated inversely with frequency of respiratory illness with fever in term infants. The authors hypothesized that orally ingested milk antibodies may contribute to immune protection within airways against respiratory infections [9]. However, this high protection effect could also be due to the high anti-influenza antibody concentration (IgG) in mother's blood that was transmitted to the infant across the placenta.

The function of milk anti-influenza antibodies may be to neutralize viruses in breast milk or in the gastrointestinal or respiratory tracts. Small amounts of breast milk can reach the infant respiratory tract as a result of regurgitation and inhalation of breast milk during and after feeding [22]. Therefore, milk influenza-specific antibodies may reach the respiratory tract through regurgitation and inhalation of breast milk and protect the mucosa of the respiratory tract against influenza A viruses, as previously hypothesized for milk anti-respiratory syncytial virus-specific Ig [23]. Nutritional intervention studies are needed to confirm whether the milk antibodies specific to influenza A can contribute to protection against respiratory pathogens.

Milk antibodies may also interact with immune cells within the infant intestinal lumen to protect against pathogens. Maternal milk-derived SIgA2 and IgA2 (but not IgA1, IgG or IgM) can adhere to the apical surfaces of mice intestinal M cells and be transported across the epithelial cell layer [24]. IgA1 has a heavily O-glycosylated hinge region, whereas the smaller IgA2 hinge region is not glycosylated. The investigators speculated that the extended hinge of IgA1 may interfere with binding to the M cell receptor [24]. A hypothetical IgA receptor on M cells could mediate the delivery of SIgA2 from the intestinal lumen to the mucosa-associated lymphoid tissues. SIgA2-antigen complexes absorbed across M cells could be sampled by subepithelial B lymphocytes and dendritic cells [25]. Indeed, SIgA was detected on the apical surface of M cells in the infant ileum, suggesting that they may be able to endocytose SIgA. Therefore, milk anti-influenza antibodies complexed with influenza antigens could hypothetically be absorbed by M cells, sampled by leukocytes and lead to altered mucosal immune response in the respiratory tract.

A limitation in our study was the absence of information on vaccination (vaccinated or not and dates) from recruited mothers in order to demonstrate a positive correlation between vaccinated mothers and anti-influenza antibody level in breast milk. However, this association was already demonstrated in previous a study for anti-influenza IgA in breast milk [9].

Another study limitation is that too few subjects were included to detect many differences between MBM and DBM feeds and gastric contents when grouped by GA (26–27 weeks, 30–31 weeks, and 35–36 weeks).

5. Conclusions

The present study suggests that passive immunization with milk anti-influenza A IgA may have been highest when infants were fed MBM rather than DBM during the first week of postnatal age. Anti-influenza A IgM was higher in MBM than in DBM during the first and third week of postnatal age in both type of samples (feeds and gastric samples). Anti-influenza IgM differed most between DBM and MBM, likely due to its sensitivity to pasteurization. Anti-influenza IgG relative abundance was higher in MBM than DBM feeds but did not differ in gastric contents at first and third week of postnatal age. Passive immune protection with maternal anti-influenza antibodies from MBM may provide infant protection. Milk antibodies could neutralize viruses within the mammary gland, in the infant mouth and after entering their respiratory system. Supplementing DBM with additional influenza A-specific antibodies could benefit infants.

Supplementary Materials: The following are available online at http://www.mdpi.com/2072-6643/11/7/1567/s1, Table S1. Demographics of preterm-delivering mother-infant pairs sampled for MBM and DBM feeds, gastric contents (1-h postprandial time) and stools (24-h post-feeding); Table S2. Description of influenza vaccine (Flucelvax Quadrivalent 2017-2018 Formula) given to mothers during pregnancy; Table S3. Statistical results (p-values) for Student's t-tests comparing antibody concentration between infants that received antibiotics ($n = 5$ at 8–9 days and $n = 5$ at 21–22 days of postnatal age) and infants that did not receive antibiotics ($n = 15$ at 8–9 days and $n = 11$ at 21–22 days) in gastric and stool samples. G, gastric contents; S, stool; Table S4. GMRA of anti-influenza A IgA, IgM and IgG (H1N1 HA and H3N2 NA) in MBM and DBM at 8–9 days and for 21–22 days of postnatal age. MBM samples were from 20 preterm-delivering mothers (26–36 wk of GA). P-values were calculated using the Wilcoxon matched-pairs signed-rank test; Table S5. GMRA of anti-influenza IgA, IgM, and IgG (H1N1 HA and

H3N2 NA) in gastric contents fed MBM and DBM feeds. Gastric contents from infants fed MBM or DBM were from 20 mother-preterm pairs (26–36 wk of GA at 8–9 days and for 21–22 days of postnatal age. p-values were calculated using the Wilcoxon matched-pairs signed-rank test; Table S6. p-value for the difference between MBM and gastric MBM or between DBM and gastric DBM on anti-influenza IgA, IgM, and IgG reactivities (H1N1 HA and H3N2 NA). Feed and gastric samples were from 20 mother-preterm pairs (26–36 wk of GA). p-values were calculated using the Wilcoxon matched-pairs signed-rank test; Table S7. GMRA for anti-influenza IgA, IgM and IgG antibodies (H1N1 HA and H3N2 NA) in stools from preterm infants. Stool samples were from 20 preterm infants at 26–36 wk of GA. p-values were calculated using the Wilcoxon matched-pairs signed-rank test between 8–9 days and 21–22 days of postnatal age.

Author Contributions: V.D.-M. conducted the ELISA analyses, analyzed data and conducted the statistical analysis. R.K.H., A.M.M., E.A.M., and R.L.M. provided milk and gastric samples. V.D.-M, D.C.D., R.K.H., A.M.M., E.A.M., and R.L.M. designed the clinical study. V.D.-M. designed the study and drafted the manuscript. V.D.-M. and D.C.D. have primary responsibility for the final content.

Funding: This research was funded by the K99/R00 Pathway to Independence Career Award, Eunice Kennedy Shriver Institute of Child Health & Development of the National Institutes of Health (R00HD079561) (D.C.D) and The Gerber Foundation (2017-1586).

Acknowledgments: We thank the Northwest Mother's Milk Bank for providing DBM for this study.

Conflicts of Interest: The authors declare no conflict of interest.

References

1. Glezen, W.P.; Taber, L.H.; Frank, A.L.; Gruber, W.C.; Piedra, P.A. Influenza virus infections in infants. *Pediatr. Infect. Dis. J.* **1997**, *16*, 1065–1068. [CrossRef] [PubMed]
2. Louie, J.K.; Schechter, R.; Honarmand, S.; Guevara, H.F.; Shoemaker, T.R.; Madrigal, N.Y.; Woodfill, C.J.; Backer, H.D.; Glaser, C.A. Severe pediatric influenza in California, 2003–2005: Implications for immunization recommendations. *Pediatrics* **2006**, *2005*, 1373. [CrossRef] [PubMed]
3. Cox, R.J.; Brokstad, K.A.; Ogra, P.L. Influenza virus: Immunity and vaccination strategies. Comparison of the immune response to inactivated and live, attenuated influenza vaccines. *Scand. J. Immunol.* **2004**, *59*, 1–15. [CrossRef] [PubMed]
4. Salam, R.A.; Das, J.K.; Soeandy, C.D.; Lassi, Z.S.; Bhutta, Z.A. Impact of Haemophilus influenzae type B (Hib) and viral influenza vaccinations in pregnancy for improving maternal, neonatal and infant health outcomes. *Cochrane Datab. Syst. Rev.* **2015**, *6*, 1–41. [CrossRef] [PubMed]
5. Omer, S.B.; Clark, D.R.; Aqil, A.R.; Tapia, M.D.; Nunes, M.C.; Kozuki, N.; Steinhoff, M.C.; Madhi, S.A.; Wairagkar, N. Maternal influenza immunization and prevention of severe clinical pneumonia in young infants. *Pediatr. Infect. Dis. J.* **2018**, *37*, 436–440. [CrossRef] [PubMed]
6. Zaman, K.; Roy, E.; Arifeen, S.E.; Rahman, M.; Raqib, R.; Wilson, E.; Omer, S.B.; Shahid, N.S.; Breiman, R.F.; Steinhoff, M.C. Effectiveness of maternal influenza immunization in mothers and infants. *N. Engl. J. Med.* **2008**, *359*, 1555–1564. [CrossRef] [PubMed]
7. Eick, A.A.; Uyeki, T.M.; Klimov, A.; Hall, H.; Reid, R.; Santosham, M.; O'Brien, K.L. Maternal influenza vaccination and effect on influenza virus infection in young infants. *Arch. Pediatr. Adolesc. Med.* **2011**, *165*, 104–111. [CrossRef]
8. Niewiesk, S. Maternal Antibodies: Clinical Significance, Mechanism of interference with immune responses, and possible vaccination strategies. *Front. Immunol.* **2014**, *5*, 446. [CrossRef]
9. Schlaudecker, E.P.; Steinhoff, M.C.; Omer, S.B.; McNeal, M.M.; Roy, E.; Arifeen, S.E.; Dodd, C.N.; Raqib, R.; Breiman, R.F.; Zaman, K. IgA and neutralizing antibodies to influenza A virus in human milk: A randomized trial of antenatal influenza immunization. *PLoS ONE* **2013**, *8*, e70867. [CrossRef]
10. Finn, A. *Clinical Trials of Influenza Vaccines: Special Challenges*; Humana Press: New York, NY, USA, 2019; pp. 567–573.
11. Sagrera, X.; Ginovart, G.; Raspall, F.; Rabella, N.; Sala, P.; Sierra, M.; Demestre, X.; Vila, C. Outbreaks of influenza A virus infection in neonatal intensive care units. *Pediatr. Infect. Dis. J.* **2002**, *21*, 196–200. [CrossRef]
12. Carroll, K.; Herrmann, K.R. The cost of using donor human milk in the NICU to achieve exclusively human milk feeding through 32 weeks postmenstrual age. *Breastfeed. Med.* **2013**, *8*, 286–290. [CrossRef] [PubMed]
13. Demers-Mathieu, V.; Huston, R.K.; Markell, A.M.; McCulley, E.A.; Martin, R.L.; Spooner, M.; Dallas, D.C. Differences in maternal immunoglobulins within mother's own breast milk and donor breast milk and across digestion in preterm infants. *Nutrients* **2019**, *11*, 920. [CrossRef] [PubMed]

14. Evans, T.J.; Ryley, H.C.; Neale, L.M.; Dodge, J.A.; Lewarne, V.M. Effect of storage and heat on antimicrobial proteins in human milk. *Arch. Dis. Child.* **1978**, *53*, 239–241. [CrossRef] [PubMed]
15. Ford, J.E.; Law, B.A.; Marshall, V.M.; Reiter, B. Influence of the heat treatment of human milk on some of its protective constituents. *J. Pediatr.* **1977**, *90*, 29–35. [CrossRef]
16. Demers-Mathieu, V.; Underwood, M.A.; Beverly, R.L.; Nielsen, S.D.; Dallas, D.C. Comparison of human milk immunoglobulin survival during gastric digestion between preterm and term infants. *Nutrients* **2018**, *10*, 631. [CrossRef] [PubMed]
17. Chandra, R.K. Immunoglobulin and protein levels in breast milk produced by mothers of preterm infants. *Nutr. Res.* **1982**, *2*, 27–30. [CrossRef]
18. Ballabio, C.; Bertino, E.; Coscia, A.; Fabris, C.; Fuggetta, D.; Molfino, S.; Testa, T.; Sgarrella, M.; Sabatino, G.; Restani, P. Immunoglobulin—A profile in breast milk from mothers delivering full term and preterm infants. *Int. J. Immunopathol. Pharmacol.* **2007**, *20*, 119–128. [CrossRef] [PubMed]
19. Montagne, P.; Cuillière, M.L.; Molé, C.; Béné, M.C.; Faure, G. Immunological and nutritional composition of human milk in relation to prematurity and mothers' parity during the first 2 weeks of lactation. *J. Pediatr. Gastroenterol. Nutr.* **1999**, *29*, 75–80. [CrossRef]
20. Gross, S.J.; Buckley, R.H.; Wakil, S.S.; McAllister, D.C.; David, R.J.; Faix, R.G. Elevated IgA concentration in milk produced by mothers delivered of preterm infants. *J. Pediatr.* **1981**, *99*, 389–393. [CrossRef]
21. Koenig, Á.; Diniz, E.M.D.A.; Barbosa, S.F.C.; Vaz, F.A.C. Immunologic factors in human milk: The effects of gestational age and pasteurization. *J. Hum. Lact.* **2005**, *21*, 439–443. [CrossRef]
22. Decarlo, J.O.; Tramer, A.; Startzman, H.N. Iodized oil aspiration in the newborn. *Arch. Pediatr. Adolesc. Med.* **1952**, *84*, 442–445. [CrossRef]
23. Laegreid, A.; OTNÆSS, A.B.K.; Ørstavik, I.; Carlsen, K.H. Neutralizing activity in human milk fractions against respiratory syncytial virus. *Acta Paediatr.* **1986**, *75*, 696–701. [CrossRef]
24. Mantis, N.J.; Cheung, M.C.; Chintalacharuvu, K.R.; Rey, J.; Corthésy, B.; Neutra, M.R. Selective adherence of IgA to murine Peyer's patch M cells: Evidence for a novel IgA receptor. *J. Immunol.* **2002**, *169*, 1844–1851. [CrossRef] [PubMed]
25. Iwasaki, A.; Kelsall, B.L. Unique functions of CD11b+, CD8α+, and double-negative Peyer's patch dendritic cells. *J. Immunol.* **2001**, *166*, 4884–4890. [CrossRef] [PubMed]

© 2019 by the authors. Licensee MDPI, Basel, Switzerland. This article is an open access article distributed under the terms and conditions of the Creative Commons Attribution (CC BY) license (http://creativecommons.org/licenses/by/4.0/).

Article

Longitudinal Analysis of Macronutrient Composition in Preterm and Term Human Milk: A Prospective Cohort Study

Céline J. Fischer Fumeaux [1], Clara L. Garcia-Rodenas [2], Carlos A. De Castro [3], Marie-Claude Courtet-Compondu [4], Sagar K. Thakkar [5], Lydie Beauport [1], Jean-François Tolsa [1] and Michael Affolter [4,*]

1. Clinic of Neonatology, Department Woman Mother Child, University Hospital of Lausanne, 1011 Lausanne, Switzerland
2. Nestlé Institute of Health Sciences, Nestlé Research, 1000 Lausanne, Switzerland
3. Clinical Development Unit, Nestlé Research Asia, Singapore 138567, Singapore
4. Nestlé Institute of Food Safety & Analytical Science, Nestlé Research, 1000 Lausanne, Switzerland
5. Nestlé Research Asia, Singapore 138567, Singapore
* Correspondence: Michael.Affolter@rdls.nestle.com; Tel.: +41-21-785-8966

Received: 2 May 2019; Accepted: 1 July 2019; Published: 4 July 2019

Abstract: Background: Mother's own milk is the optimal source of nutrients and provides numerous health advantages for mothers and infants. As they have supplementary nutritional needs, very preterm infants may require fortification of human milk (HM). Addressing HM composition and variations is essential to optimize HM fortification strategies for these vulnerable infants. Aims: To analyze and compare macronutrient composition in HM of mothers lactating very preterm (PT) (28 0/7 to 32 6/7 weeks of gestational age, GA) and term (T) infants (37 0/7 to 41 6/7 weeks of GA) over time, both at similar postnatal and postmenstrual ages, and to investigate other potential factors of variations. Methods: Milk samples from 27 mothers of the PT infants and 34 mothers of the T infants were collected longitudinally at 12 points in time during four months for the PT HM and eight points in time during two months for the T HM. Macronutrient composition (proteins, fat, and lactose) and energy were measured using a mid-infrared milk analyzer, corrected by bicinchoninic acid (BCA) assay for total protein content. Results: Analysis of 500 HM samples revealed large inter- and intra-subject variations in both groups. Proteins decreased from birth to four months in the PT and the T HM without significant differences at any postnatal time point, while it was lower around term equivalent age in PT HM. Lactose content remained stable and comparable over time. The PT HM contained significantly more fat and tended to be more caloric in the first two weeks of lactation, while the T HM revealed higher fat and higher energy content later during lactation (three to eight weeks). In both groups, male gender was associated with more fat and energy content. The gender association was stronger in the PT group, and it remained significant after adjustments. Conclusion: Longitudinal measurements of macronutrients compositions of the PT and the T HM showed only small differences at similar postnatal stages in our population. However, numerous differences exist at similar postmenstrual ages. Male gender seems to be associated with a higher content in fat, especially in the PT HM. This study provides original information on macronutrient composition and variations of HM, which is important to consider for the optimization of nutrition and growth of PT infants.

Keywords: human milk; preterm; term; neonate; infant; macronutrients; protein; fat; lactose; nutrition

1. Introduction

Human milk (HM) is a highly complex, dynamic and species-specific system that incorporates numerous nutritional and bioactive elements. Despite progress in analytical technologies enabling the growth of data over the past decade, HM composition remains only partially elucidated [1–4]. Recent studies revealed countless inter- and intra-individual factors influencing HM composition, such as maternal, circadian, pregnancy, delivery, infantile, chronological and environmental variables [5–9]. Among these factors, temporal changes of HM composition before and after term equivalent time are of particular interest considering the critical challenges of preterm (PT) infants' nutrition and the critical importance of HM for this vulnerable population [10,11].

As for healthy term (T) babies, breastfeeding provides numerous health advantages for PT infants and their mothers, and is thus strongly recommended [12–16]. In addition, for very PT at risk neonates, mother's own milk appears particularly protective against several severe complications, such as necrotizing enterocolitis, sepsis, bronchopulmonary dysplasia, retinopathy of prematurity, and reduces duration of hospital stay or incidence of rehospitalizations [17–20]. Furthermore, it may improve brain growth and the development of these fragile infants [21–24].

Despite these crucial protective effects, HM may not fully meet some specific additional nutritional requirements of very PT infants to avoid growth flattening, nutritional deficits, and related complications [25–27]. Therefore, human milk fortification is often recommended in very PT neonates, however controversies exist and optimal, feasible fortification strategies remain to be identified [11,16]. Nevertheless, even with the development of HM fortification practices, growth failure remains frequent in neonatal intensive care units (NICUs), affecting up to half of very low birthweight neonates [28].

In this context, we need to have better knowledge and understanding of HM composition and its variations, and especially temporal changes, in order to enhance nutritional enteral strategies, mainly human milk fortification and/or formula complements according to the situation [11]. Currently, studies on HM composition after very PT delivery remain scarce and are mostly based on cross-sectional studies. Only a few studies have longitudinally measured the trajectory of HM composition and its time variations in both T and PT infants, and the data remain controversial [10].

In this prospective cohort study, we aimed: (i) to quantify macronutrient composition in very PT and T HM over time; (ii) to compare macronutrient composition between PT and T milk, at similar lactation stages (postnatal ages) and gestational stages (postmenstrual ages); and (iii) to investigate other factors potentially associated with macronutrient variations in HM.

2. Materials and Methods

2.1. Study Design, Subjects, and Setting

This study was a part of a prospective, monocentric, cohort study aiming to analyze various components of HM over time and to compare their contents in the PT HM versus the T HM. We previously published the study design and the description of the population in detail [29].

The study was conducted between October 2013 and July 2014 at the University Hospital of Lausanne, Switzerland. Eligible mothers were those older than 18 years of age, intending to breastfeed their offspring, and who delivered either (i) very prematurely (PT group), from 28 0/7 to 32 6/7 weeks of gestational age (GA) or (ii) at term (T group), between 37 0/7 to 41 6/7 weeks of GA.

Exclusion criteria included any counter-indication to breastfeeding, maternal diabetes (type I or II) before pregnancy, alcohol or drug consumption, and insufficient French language skills.

The mothers who were included in the study were followed until postnatal week 16 for the PT group, and week 8 for the T group, or until lactation discontinuation (whichever came first).

2.2. Data Collection

The main clinical and sociodemographic maternal and neonatal characteristics were prospectively collected. All data were recorded in electronic case report forms in a dedicated database.

2.3. Milk Sampling and Processing

Sequential, iterative samples of milk were collected according to the following schedule, illustrated in Figure 1:

(i) Preterm group: 1–10 mL of milk was collected at the end of the 1st week, then weekly for the first 8 weeks, then every 2 weeks for an additional 8 weeks; this corresponded thus to a maximum of 12 samples during the 16 weeks.

(ii) Term group: 10 mL of milk were collected weekly for 8 weeks, starting at the end of week 1.

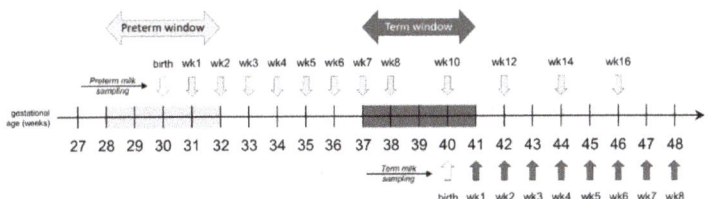

Figure 1. Milk sampling schedule (reprinted from Garcia-Rodenas et al. 2018, with permission from Elsevier).

Standardized milk sampling relied on a single sampling of a single breast in the morning (first morning expression). Milk samples were expressed between 6–12 a.m., using an electric breast pump (Symphony®, Medela, Baar, Switzerland). After the breast was entirely emptied, the milk was homogenized and an aliquot of maximum 10 mL of milk (1–10 mL for the first two time points in the PT group) was collected. Each sample aliquot was immediately transferred to dedicated freezing tubes (15 mL polypropylene tubes, Falcon™, Fisher Scientific, Reinach, Switzerland), and stored at −18 °C in the home freezer for a maximum of 1 week until frozen transfer to the hospital. At the hospital, samples were temporarily kept at −80 °C and then transferred to the Nestlé Research Centre (NRC, Lausanne, Switzerland). The frozen milk samples were thawed for splitting into 15 aliquots and stored at −80 °C until analysis of macronutrient and other HM components (reported in independent publications).

2.4. Measurement of HM Macronutrient Composition

The macronutrient content (total protein, fat, and carbohydrate, i.e., lactose) and energy density were measured using a human milk analyzer (MIRIS AB, Uppsala, Sweden) based on mid-infrared transmission spectroscopy (www.mirissolutions.com). One mL of each milk sample was thawed and warmed to 40 °C in a water bath. Prior to the measurement, each sample was homogenized using the MIRIS sonicator to avoid protein aggregation and lipid phase separation. A daily calibration check (MIRIS check solution) was performed according to the manufacturer's recommendations. Macronutrient quantities were measured in a single run and the energy content was calculated automatically by the MIRIS instrument (protein 4.4 kcal/g, carbohydrate 4 kcal/g, fat 9 kcal/g) [30].

We used a MIRIS measure validation process that had been previously published [31]. As in other literature reports, this validation showed that the MIRIS was not accurate enough for the measurement of total protein in human milk as compared with the Kjeldahl reference method (AOAC International, method number 991.22), while no differences were observed for fat and carbohydrate content as compared with the corresponding reference methods. To address this issue, we used colorimetric bicinchoninic acid (BCA, ThermoFisher Scientific) assay, which produced accurate human milk protein values as compared with those obtained with the reference Kjeldahl method. In order to include this more accurate protein information in the energy density value of the human milk samples, energy density was recalculated for each milk sample using the same formula as the MIRIS instrument: Energy (kcal/L) = protein (g/L) × 4.4 + fat (g/L) × 9.25 + carbohydrate (g/L) × 4.

2.5. Statistics

The paucity of quantitative data on the macronutrient content in the PT HM precluded a proper power calculation in this exploratory study. The study size was initially set at $n = 20$ subjects per group (preterm and term infant mothers) according to the estimated recruitment feasibility at the study center within a one-year period.

The temporal changes of macronutrient contents were compared in the PT and T HM at equivalent infant (1) postpartum ages and (2) postmenstrual ages. For both comparisons, mixed linear models were used to estimate the differences between preterm and term infants. The models used age (either postpartum or postmenstrual), term/preterm status, and interaction between age and term/preterm status. Within subject variability was accounted for by declaring the subject ID as a random effect. Contrast estimates of the model were calculated by comparing the PT and T HM groups at each time point. No imputation method was applied for missing data (both in between visits and loss to follow up) as the method used does not require a complete dataset. A conventional two-sided 5% error rate was used without adjusting for multiplicity as this exploratory trial is for hypothesis generation purposes.

Similar methods were used to analyze the effects of gender and delivery mode and their interaction with age (both postpartum and postmenstrual and also both in the PT and the T populations together and separately). Statistical analyses were done using SAS 9.3 (SAS Institute, Cary, North Carolina, USA) and R 3.2.1 (R Foundation for Statistical Computing, Vienna, Austria; https://www.R-project.org).

2.6. Ethics

The study was approved by the local Ethical Board (Commission cantonal d'éthique de la recherche sur l'être humain du Canton de Vaud) (Protocol 69/13, clinical study 11.39.NRC; April 9, 2013). Maternal written consent was obtained. The study was registered at ClinicalTrials.gov with the identifier NCT02052245.

3. Results

3.1. Study Population

The detailed study flow chart and study population description has been published elsewhere [29]. They are presented in the supplementary material (see Figure S1 and Table S1).

Sixty-one mothers were included in the study: 27 mothers of 33 PT neonates, and 34 mothers of 34 T neonates. In the PT group, 25/27 mothers (93%) completed the study (one neonatal death, one withdrawal). In the T group, 28/34 mothers (82.4%) completed the study (one maternal illness, two withdrawals, three early breastfeeding disruption). In all, we collected 500 HM samples (280 PT and 220 T samples).

Mothers of the two groups were comparable in age (mean ± SD: 32.4 ± 5.6 years in the PT group, versus 31.2 ± 4.2 years in the T group, $p = 0.3173$) and body mass index before pregnancy and at delivery. The mean ± SD gestational age and birthweight of the PT and T neonates differed as expected (30.8 ± 1.4 weeks versus 39.5 ± 1.0 weeks, $p < 0.0001$, and 1421 ± 373 g versus 3278 ± 354 g, $p < 0.0001$), and more PT neonates were born by caesarean section (63% versus 23%, $p = 0.0019$). The sex ratio was similar in the two groups (54% versus 53% of males, $p = 0.8952$).

3.2. HM Macronutrient Composition

Overall, we noticed a substantial intra- and inter-individual variability in data for all macronutrients, indicating a large heterogeneity between macronutrient content for milk samples. Maximal/minimal ratios reached up to 16 for fat, 5.4 for protein, 4.1 for energy, and 3.4 for lactose content.

Detailed numerical values of each macronutrient according to postnatal age (A) and postmenstrual age (B) are reported in the corresponding supplemental Tables.

3.2.1. Total Protein

The values of total protein content in the PT and T HM according to postnatal age are depicted in Figure 2A. The total protein content gradually decreases from birth to four months, from (mean ± SD) 2.2 ± 0.3 g/100 mL to 1.5 ± 0.5 g/100 mL in preterm and from 2.5 ± 0.8 g/100 mL to 1.7 ± 0.3 g/100 mL in T HM, without significant differences between the PT and T groups at any point in time.

Figure 2B shows the total protein content in the PT and T HM according to postmenstrual age. There were significant differences ($p < 0.005$) at gestational ages of 38–40 and 42 weeks, with T HM containing higher amounts of protein. These differences peaked at 39 weeks with T values up to 1.7 times higher than PT values. Later postmenstrual ages (>43 weeks) did not show any significant differences in protein content.

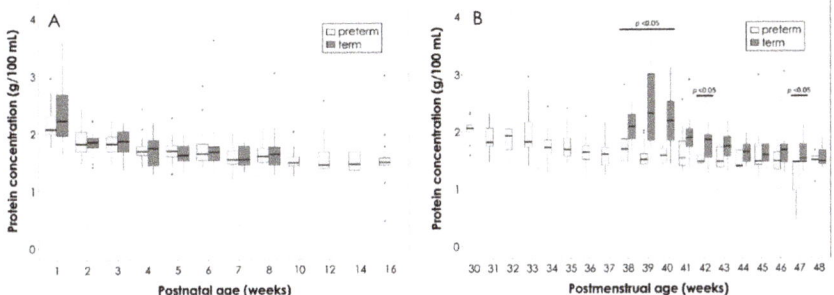

Figure 2. Total protein content in the preterm (PT) and the term (T) human milk (HM) according to postnatal age (**A**) and postmenstrual age (**B**). Box plots represent medians with 25th and 75th percentile, minimal-maximal range, and outliers (values >4.5 are excluded from the graph). Statistically significant *p*-values are indicated in the graph.

3.2.2. Total Fat

The mean values for total fat content in the PT and T HM slightly increase in early lactation and then remain constant over time (Figure 3A). As compared to the T HM, the PT HM contained significantly more fat (mean ± SD, 2.8 ± 1.1 g/100 mL in PT HM versus 2.1 ± 1.0 g/100 mL in T HM; $p = 0.04$) in the first week of lactation, but less in the later stages of lactation (week three to eight).

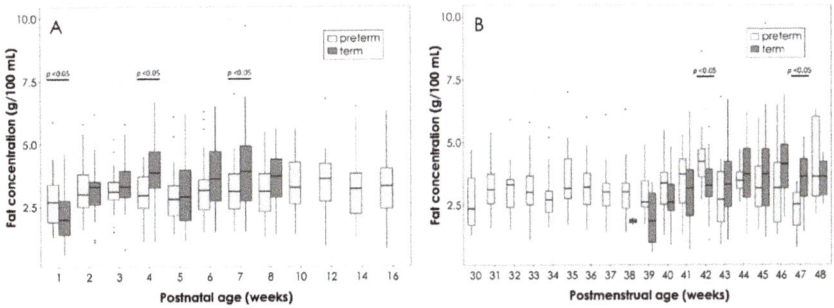

Figure 3. Total fat content in PT and T HM according to postnatal age (**A**) and postmenstrual age (**B**). Box plots represent medians with 25th and 75th percentile, minimal-maximal range, and outliers. Statistically significant *p*-values are indicated in the graph.

When comparing the fat content according to postmenstrual age (Figure 3B), two significant differences were observed for gestational age of 42 and 47 weeks, respectively, with higher fat content at week 42 and lower fat content at week 47 in the PT HM as compared to the T HM.

3.2.3. Total Lactose

The mean values for lactose content remained stable over postnatal time in both groups with a consistent, but not significant, higher content in the PT versus the T HM (mean ± SD over all lactation stages: 5.9 ± 0.2 g/100 mL versus 5.8 ± 0.1 g/100 mL) (Figure 4A). A significant difference was observed at week 7 with the PT HM containing higher amounts of lactose (6.1 ± 0.5 g/100 mL versus 5.6 ± 0.9 g/100 mL, $p = 0.0233$).

When comparing lactose contents according to gestational age (Figure 4B), significant variations were found for week 45 and 47, where lactose content was lower in the PT HM than in the T HM.

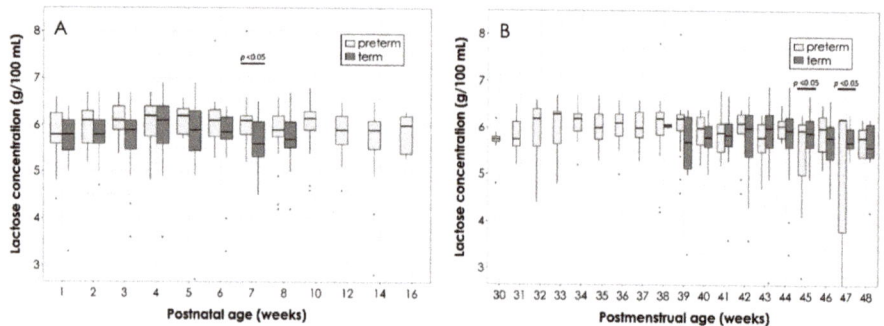

Figure 4. Total lactose content in the PT and T HM according to postpartum age (**A**) and postmenstrual age (**B**). Box plots represent medians with 25th and 75th percentile, minimal-maximal range, and outliers. Statistically significant *p*-values are indicated in the graph.

3.2.4. Energy Density

The mean ± SD values for energy density tended to be higher in the PT HM (58.7 ± 10.2 kcal/100 mL versus 53.1 ± 8.8 kcal/100 mL; $p = 0.08$) in the first week of lactation, whereas in weeks three to eight postpartum, energy density in the T HM was higher (Figure 5A).

When comparing the energy density according to postmenstrual age (Figure 5B), there were punctual variations between the PT and T HM, but trajectory over lactation remained relatively constant.

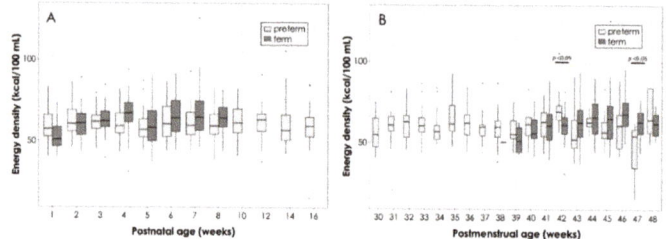

Figure 5. Calculated energy density in the PT and T HM according to postpartum age (**A**) and postmenstrual age (**B**). Box plots represent medians with 25th and 75th percentile, minimal-maximal range, and outliers. Statistically significant *p*-values are indicated in the graph.

3.2.5. Other Factors Influencing HM Macronutrient Composition

The composition of macronutrients was also compared over time in the population and in each T and PT subgroup separately, according to infant gender or mode of delivery (see Figures in supplementary material).

Milk of mothers with male infants was consistently higher in fat and energy as compared to milk of mothers with female infants, reaching statistical significance at postnatal weeks five and seven (estimated differences of 0.89 g/100 mL and 1.07 g/100 mL of fat, respectively). The difference at five weeks was more pronounced in the PT group (estimated difference of 1.04 g/100 mL) and the difference at seven weeks was more pronounced for the T group (1.58 g/100 mL). There was also a trend of increased content of total protein in milk for male as compared to female infants in both the T and the PT groups, but the differences were not significant. We did not observe a gender difference for lactose content. Regarding the mode of delivery, it was not associated with consistent nor significant changes in HM macronutrients contents in our population.

A mixed linear model, taking into account the variables T or PT, postpartum age, interaction between T/PT and postpartum age, gender, and mode of delivery, was assessed in order to have an indication of which variables affect macronutrient concentration. It was observed that in general, the postpartum age has a significant effect on the macronutrient concentration of milk. Carbohydrate content was significantly higher for premature infants globally (all time points combined) while differences in proteins, energy, and fat depended on the postpartum age. Energy and fat concentration were significantly affected by the gender of the infant.

4. Discussion

This study provides substantial longitudinal data on the macronutrient compositions of 500 HM samples from 61 mothers delivering either very prematurely ($n = 27$) or at term ($n = 34$), during a time period of up to 16 and eight weeks for a PT and a T group, respectively. The original milk sampling design enabled the comparison of the PT and T HM composition not only at similar postnatal ages, corresponding to similar maternal lactation stages, but also at similar postmenstrual ages, corresponding to a similar infant developmental stage.

In this cohort, HM iterative macronutrients analysis showed notably: (i) large inter- and intra-subject variations, most marked for fat content; (ii) little differences between the PT and T HM at similar postnatal ages/lactation stages, while there were more significant differences between the PT and T HM at similar postmenstrual ages/developmental stages; (iii) an overall decrease in total protein content over time, whereas lactose, fat, and energy density remained stable; and (iv) some gender differences in HM composition, as HM dedicated to male infants tended to be richer in fat and energy.

First, we have emphasis on the magnitude of the variations observed in our cohort. Differences between minimal and maximal measurements reached up to more than 10 times for fat. These variations are more important, on the whole, than those reported in other studies [10,11,32]. Extreme values were verified and confirmed. An effect of milk preservation or homogenization is unlikely, as the procedure was standardized [33]. An impact of the method of collection (first morning one single breast, rather than 24 h two breasts collections) is possible [5]. Nevertheless, these results confirm and underline how much a unique "standard" HM composition basis of calculation cannot fit all individual situations, and that more accurate alternatives for estimations of nutritional intake are required, notably for fortification issues [34].

Among factors of variations, temporal changes in HM macronutrient composition have been previously reported [10,35–37]. The study plan allowed us to investigate the HM composition changes according to both postnatal and gestational ages, related to adaptive and developmental stages, respectively. Interestingly, in our study, macronutrient composition changes were overall more pronounced according to postnatal time than to gestational age, which is in line with other recent observations [38]. Consequently, there were fewer differences between the PT and the T HM at similar postnatal ages, than at similar gestational ages. Its final implication, however, remains to be further explored. This finding should also be considered regarding the definition of nutritional needs of the PT infants and establishment of fortification targets.

Concerning the protein composition, the observed trajectory in our cohort follows a progressive decrease, which is consistent with previous observations [10,35,39]. Our protein values align well with published results for the first four weeks of lactation, and then seem to remain slightly higher than reported values (week 6 to 12) [10,37]. Unlike most existing data, we did not find a higher protein content in the PT HM than in the T HM, and there were no significant differences in the PT and T HM protein contents at any point. By contrast, the longitudinal trajectory of the protein content according to postmenstrual age revealed some significant differences between PT and T HM, which culminate between weeks 38 and 42. This implies that a preterm breastfed infant arriving at term equivalent age receives lower amounts of protein as compared with a breastfed term infant.

Concerning the fat composition, the changes of the PT and T HM contents follow a very similar trajectory as described previously, with an initial slight increase in fat content from week one to week two, followed by a constant level over the remaining lactation period. Initially at week one, the PT HM contained a higher fat content, and then, until the end of the sampling period, the T HM contained slightly higher fat levels, which also effectively matches published data [10].

Concerning the lactose content, we observed stable concentrations over time with a trend of higher content in the PT versus T HM. Our results are in line with most of the previous reported data, although some studies described an increase in lactose concentration during the first month of lactation [10,35,39]. As lactose constitutes the major energy source among total carbohydrates, it is an acceptable proxy for digestible carbohydrates [35].

Finally, the energy density was not directly measured in our study, but it was calculated and corrected for adjusted protein content for each milk sample, using the formula published by Polberg and Lönnerdal [30]. Overall, values of our cohort tended to be generally slightly lower than those reported in the literature [10,35,39]. The single morning HM sampling procedure could have contributed to this observation, as fat content has been reported to be higher in the evening [5].

Besides these temporal factors of variations, we also investigated the influence of other factors, such as gender and mode of delivery. We noticed that HM for male infants was more concentrated in fat and energy as compared with female infants. This was observed both in the T and PT group, but it was more marked in the PT group. A multivariate analysis confirmed that the infant's male gender was significantly and independently associated with higher concentrations of fat and energy in HM. So far, this phenomenon of "sex bias" milk synthesis, possibly already partly conditioned in utero, has mainly been studied and discussed in other mammals, such as cows, horses, monkeys, and hamsters [40,41]. In humans, by contrast, only a small number of studies have reported a possible gender specificity of the HM composition. In 2010, Powe et al. observed an increase in energy value of 25% for male infants of well-fed American women [42]. In a study conducted in Singapore, Thakkar et al. found similar differences in term newborns (energy + 24% and fat + 39% for boys), with a lipid profile that also varied by gender [43]. More recently, in a term delivering population of mothers living in Seoul, Hahn also observed an increased energy density in HM for male infants, related to a higher carbohydrates content. By contrast, Quinn et al. did not observe any gender difference in the milk of 103 Filipino women [7,44]. Interaction with environmental and/or nutritional conditions is possible, as suggested by Fujita et al. [45]. According to an evolutionary perspective, these authors hypothesize that economically disadvantaged mothers produce richer milk for girls, while well-nourished women produce richer milk for boys [45]. According to BMI indicators, our population was rather well nourished [29]. The reasons for these observed differences remain unclear. One can hypothesize that, due to their different growth trajectories and hormonal environment, male and female infants may have different energy requirements. The implications of such differences, if confirmed, could be important. On the other hand, the effect of delivery mode reported by Dizdar et al. was not verified in our population [8]. However, due to the relatively limited number of subjects and the important variability of data, these results should be interpreted cautiously.

This work has some limitations to consider: (i) as mentioned above, the method of sampling (first milk of the morning rather than pooled over 24 h) could have increased the measured variations, especially for lipids; (ii) variations in milk volumes, that may have also contributed to the important variability observed, were not recorded and thus could not be taken into account; (iii) the method of measuring macronutrients (MIRIS ©) is a method whose reliability is open to criticism [46]. We attempted to enhance its accuracy through validation methods and protein correction strategy [31]; (iv) finally, the small number of patients in each group limited the power of the study, and allowed only a limited number of associated factors to be analyzed. However, the influence of other factors, such as intrauterine growth restriction, twinning, antenatal steroids, smoking, and maternal comorbidities, for example, also deserves to be further investigated.

Despite this, our results emphasize that, although pragmatic, a unique definition of a "standard" nutritional composition of HM, as traditionally applied, can be insufficient and inaccurate for the nutritionally vulnerable infants, such as preterm infants [47]. These infants may require more individualized nutritional approaches, such as "targeted" fortification based on regular measurements of HM composition, or "adjusted" fortification according to serum urea values [11,30,48]. However, so far, targeted approaches remain challenging and expensive in daily care practices, without enthusiastic durable results on growth or development. Thus, future research should aim to develop optimized, efficient and feasible fortification strategies. Recent progresses in bioengineering should help in these issues [49]. Importantly, studies should also investigate other maternal and infant factors interacting with HM composition, including offspring gender.

Meanwhile, it would be good to consider changing the way nutritional intakes are calculated for clinical or research purposes. Currently, when direct measurements of HM composition are not available, a unique HM composition is generally assumed. However, it would be preferable to use a model that adjusts the assumed composition according to the T/PT status and/or postnatal age as recently proposed by an expert group [34].

5. Conclusions

Macronutrient composition and energy density in human milk change over time, with important inter- and intra- individual variations. The original design of this longitudinal study allowed us to compare term and preterm mother milk composition both at similar postnatal stages, corresponding to similar adaptive stages after birth, and at similar gestational ages, corresponding to similar developmental stages after conception. This original approach revealed that differences between the PT and T HM were more preponderant at equivalent gestational ages, especially around term, than at equivalent postpartum lactation stages. Moreover, our results suggest a possible gender adaptation of HM, with a more caloric milk for male infants. This is also a rather novel and compelling issue that deserves additional research.

Accordingly, there is definitely not one, but a multitude of human milk compositions and breastfeeding appears to be a powerful preventive and personalized medicine. Whereas HM use is strongly encouraged in preterm as well as in term infants, there is a need to further explore HM composition and variations to assess whether more individualized nutritional approaches may help in optimizing growth and development of more vulnerable infants.

Supplementary Materials: The following are available online at http://www.mdpi.com/2072-6643/11/7/1525/s1, Figure S1: Study flow chart, Figure S2: Macronutrients by gender, Figure S3: Macronutrients by delivery, Table S1: Maternal and infant characteristics, Table S2: Macronutrient data: postnatal, postmenstrual, gender, delivery.

Author Contributions: C.J.F.F. and M.A. contributed to the study design and to the development of the overall research plan, conducted the research, contributed to the interpretation of the results, and drafted the manuscript; C.L.G.-R. and S.K.T. contributed to the interpretation of the results, reviewed the manuscript, and approved the final version; C.A.D.C. performed the statistical analysis, participated in manuscript writing, and approved the final version; M.-C.C.-C. conducted the research, reviewed the manuscript, and approved the final version; L.B.

and J.-F.T. contributed to the study design and to the development of the overall research plan, conducted the research, reviewed the manuscript, and approved the final version.

Funding: This study was funded by Nestlé Research, Lausanne (Nestec Ltd.). Nestlé Research sponsored the study, participated to the protocol building, and provided the infrastructure and the material and human resources to analyze the HM samples and macronutrient data. Employees of Nestlé Research further participated in the interpretation of the results and the manuscript writing.

Acknowledgments: Sincere thanks to all mothers who enthusiastically participated in this study, to Nassima Grari, the study lactation nurse who admirably cared about all mothers during the full study duration, and to Céline Romagny and Emilie Darcillon who managed the organizational and operational part of the clinical study and the data management in perfection. We are grateful to Sean Austin for English language corrections and proof-reading of the manuscript.

Conflicts of Interest: C.L.G.-R., C.A.D.C., M.-C.C.-C., S.K.T., and M.A. are all employees of Nestec Ltd.; C.J.F.F., L.B. and J.-F.T. declare no conflict of interest.

References

1. Andreas, N.J.; Kampmann, B.; Mehring Le-Doare, K. Human breast milk: A review on its composition and bioactivity. *Early Hum. Dev.* **2015**, *91*, 629–635. [CrossRef] [PubMed]
2. Jakaitis, B.M.; Denning, P.W. Human breast milk and the gastrointestinal innate immune system. *Clin. Perinatol.* **2014**, *41*, 423–435. [CrossRef] [PubMed]
3. Ballard, O.; Morrow, A.L. Human milk composition: Nutrients and bioactive factors. *Pediatr. Clin. N. Am.* **2013**, *60*, 49–74. [CrossRef] [PubMed]
4. Picciano, M.F. Nutrient composition of human milk. *Pediatr. Clin. N. Am.* **2001**, *48*, 53–67. [CrossRef]
5. Moran-Lev, H.; Mimouni, F.B.; Ovental, A.; Mangel, L.; Mandel, D.; Lubetzky, R. Circadian Macronutrients Variations over the First 7 Weeks of Human Milk Feeding of Preterm Infants. *Breastfeed. Med.* **2015**, *10*, 366–370. [CrossRef] [PubMed]
6. Çetinkaya, A.K.; Dizdar, E.A.; Yarcı, E.; Sari, F.N.; Oguz, S.S.; Uras, N.; Canpolat, F.E. Does Circadian Variation of Mothers Affect Macronutrients of Breast Milk? *Am. J. Perinatol.* **2017**, *34*, 693–696. [PubMed]
7. Quinn, E.A. No evidence for sex biases in milk macronutrients, energy, or breastfeeding frequency in a sample of Filipino mothers. *Am. J. Phys. Anthropol.* **2013**, *152*, 209–216. [CrossRef] [PubMed]
8. Dizdar, E.A.; Sari, F.N.; Degirmencioglu, H.; Canpolat, F.E.; Oguz, S.S.; Uras, N.; Dilmen, U. Effect of mode of delivery on macronutrient content of breast milk. *J. Matern. Fetal Neonatal Med.* **2014**, *27*, 1099–1102. [CrossRef]
9. Affolter, M.; Garcia-Rodenas, C.L.; Vinyes-Pares, G.; Jenni, R.; Roggero, I.; Avanti-Nigro, O.; de Castro, C.A.; Zhao, A.; Zhang, Y.; Wang, P.; et al. Temporal Changes of Protein Composition in Breast Milk of Chinese Urban Mothers and Impact of Caesarean Section Delivery. *Nutrients* **2016**, *17*, 8. [CrossRef]
10. Gidrewicz, D.A.; Fenton, T.R. A systematic review and meta-analysis of the nutrient content of preterm and term breast milk. *BMC Pediatr.* **2014**, *14*, 216. [CrossRef]
11. Rochow, N.; Landau-Crangle, E.; Fusch, C. Challenges in breast milk fortification for preterm infants. *Curr. Opin. Clin. Nutr. Metab. Care* **2015**, *18*, 276–284. [CrossRef] [PubMed]
12. Gibertoni, D.; Corvaglia, L.; Vandini, S.; Rucci, P.; Savini, S.; Alessandroni, R.; Sansavini, A.; Fantini, M.P.; Faldella, G. Positive effect of human milk feeding during NICU hospitalization on 24 month neurodevelopment of very low birth weight infants: An Italian cohort study. *PLoS ONE* **2015**, *10*, e0116552. [CrossRef] [PubMed]
13. Section on Breastfeeding. Breastfeeding and the use of human milk. *Pediatrics* **2012**, *129*, e827–e841. [CrossRef] [PubMed]
14. WHO. Evidence on the Long-Term Effects of Breastfeeding. Systematic Reviews and Meta-Analyses. 2007. Available online: http://www.who.int/maternal_child_adolescent/documents/9241595230/en/ (accessed on 20 February 2019).
15. WHO; Edmond, K.; Bahl, R. Optimal Feeding of Low-Birth-Weight Infants. Technical Review. ISBN 9241595094. Available online: http://www.who.int/maternal_child_adolescent/documents/9241595094/en/ (accessed on 20 February 2019).

16. Moro, G.E.; Arslanoglu, S.; Bertino, E.; Corvaglia, L.; Montirosso, R.; Picaud, J.C.; Polberger, S.; Schanler, R.J.; Steel, C.; van Goudoever, J.; et al. XII. Human Milk in Feeding Premature Infants: Consensus Statement. *J. Pediatr. Gastroenterol. Nutr.* **2015**, *61*, S16–S19. [CrossRef] [PubMed]
17. Miller, J.; Tonkin, E.; Damarell, R.A.; McPhee, A.J.; Suganuma, M.; Suganuma, H.; Middleton, P.F.; Makrides, M.; Collins, C.T. A Systematic Review and Meta-Analysis of Human Milk Feeding and Morbidity in Very Low Birth Weight Infants. *Nutrients* **2018**, *31*, 10. [CrossRef] [PubMed]
18. Boquien, C.Y. Human Milk: An Ideal Food for Nutrition of Preterm Newborn. *Front. Pediatr.* **2018**, *6*, 295. [CrossRef] [PubMed]
19. Underwood, M.A. Human milk for the premature infant. *Pediatr. Clin. N. Am.* **2013**, *60*, 189–207. [CrossRef]
20. Menon, G.; Williams, T.C. Human milk for preterm infants: Why, what, when and how? *Arch. Dis. Child. Fetal Neonatal Ed.* **2013**, *98*, F559–F562. [CrossRef]
21. Vohr, B.R.; Poindexter, B.B.; Dusick, A.M.; McKinley, L.T.; Higgins, R.D.; Langer, J.C.; Poole, W.K.; National Institute of Child Health and Human Development National Research Network. Persistent beneficial effects of breast milk ingested in the neonatal intensive care unit on outcomes of extremely low birth weight infants at 30 months of age. *Pediatrics* **2007**, *120*, e953–e959. [CrossRef]
22. Belfort, M.B.; Anderson, P.J.; Nowak, V.A.; Lee, K.J.; Molesworth, C.; Thompson, D.K.; Doyle, L.Y.; Inder, T.F. Breast Milk Feeding, Brain Development, and Neurocognitive Outcomes: A 7-Year Longitudinal Study in Infants Born at Less Than 30 Weeks' Gestation. *J. Pediatr.* **2016**, *177*, 133–139. [CrossRef]
23. Belfort, M.B.; Ehrenkranz, R.A. Neurodevelopmental outcomes and nutritional strategies in very low birth weight infants. *Semin. Fetal Neonatal Med.* **2017**, *22*, 42–48. [CrossRef] [PubMed]
24. Blesa, M.; Sullivan, G.; Anblagan, D.; Telford, E.J.; Quigley, A.J.; Sparrow, S.A.; Serag, A.; Semple, S.I.; Bastin, M.E.; Boardman, J.P. Early breast milk exposure modifies brain connectivity in preterm infants. *NeuroImage* **2018**, *18*, 431–439. [CrossRef] [PubMed]
25. Henderson, G.; Anthony, M.Y.; McGuire, W. Formula milk versus maternal breast milk for feeding preterm or low birth weight infants. *Cochrane Database Syst. Rev.* **2007**, *4*. [CrossRef] [PubMed]
26. Kuschel, C.A.; Harding, J.E. Multicomponent fortified human milk for promoting growth in preterm infants. *Cochrane Database Syst. Rev.* **2004**, *1*. [CrossRef]
27. Tudehope, D.I. Human milk and the nutritional needs of preterm infants. *J. Pediatr.* **2013**, *162*, S17–S25. [CrossRef] [PubMed]
28. Horbar, J.D.; Ehrenkranz, R.A.; Badger, G.J.; Edwards, E.M.; Morrow, K.A.; Soll, R.F.; Buzas, J.S.; Bertino, E.; Gagliardi, L.; Bellù, R. Weight Growth Velocity and Postnatal Growth Failure in Infants 501 to 1500 Grams: 2000–2013. *Pediatrics* **2015**, *136*, e84–e92. [CrossRef] [PubMed]
29. Garcia-Rodenas, C.L.; De Castro, C.A.; Jenni, R.; Thakkar, S.K.; Beauport, L.; Tolsa, J.F.; Fischer-Fumeaux, C.J.; Affolter, M. Temporal changes of major protein concentrations in preterm and term human milk. A prospective cohort study. *Clin. Nutr. Edinb. Scotl.* **2018**, *38*, 1844–1852. [CrossRef]
30. Polberger, S.; Lönnerdal, B. Simple and rapid macronutrient analysis of human milk for individualized fortification: Basis for improved nutritional management of very-low-birth-weight infants? *J. Pediatr. Gastroenterol. Nutr.* **1993**, *17*, 283–290. [CrossRef]
31. Giuffrida, F.; Austin, S.; Cuany, D.; Sanchez-Bridge, B.; Longet, K.; Bertschy, E.; Sauser, J.; Thakkar, S.K.; Lee, L.Y.; Affolter, M. Comparison of macronutrient content in human milk measured by mid-infrared human milk analyzer and reference methods. *J. Perinatol.* **2018**, *14*, 497–503. [CrossRef]
32. Cooper, A.R.; Barnett, D.; Gentles, E.; Cairns, L.; Simpson, J.H. Macronutrient content of donor human breast milk. *Arch. Dis. Child. Fetal Neonatal. Ed.* **2013**, *98*, F539–F541. [CrossRef]
33. García-Lara, N.R.; Escuder-Vieco, D.; García-Algar, O.; De la Cruz, J.; Lora, D.; Pallás-Alonso, C. Effect of freezing time on macronutrients and energy content of breastmilk. *Breastfeed. Med.* **2012**, *7*, 295–301. [CrossRef] [PubMed]
34. Cormack, B.E.; Embleton, N.D.; van Goudoever, J.B.; Hay, W.W.; Bloomfield, F.H. Comparing apples with apples: It is time for standardized reporting of neonatal nutrition and growth studies. *Pediatr. Res.* **2016**, *79*, 810–820. [CrossRef] [PubMed]
35. Boyce, C.; Watson, M.; Lazidis, G.; Reeve, S.; Dods, K.; Simmer, K.; McLeod, G. Preterm human milk composition: A systematic literature review. *Br. J. Nutr.* **2016**, *116*, 1033–1045. [CrossRef] [PubMed]
36. Hsu, Y.C.; Chen, C.H.; Lin, M.C.; Tsai, C.R.; Liang, J.T.; Wang, T.M. Changes in preterm breast milk nutrient content in the first month. *Pediatr. Neonatol.* **2014**, *55*, 449–454. [CrossRef] [PubMed]

37. Bauer, J.; Gerss, J. Longitudinal analysis of macronutrients and minerals in human milk produced by mothers of preterm infants. *Clin. Nutr. Edinb. Scotl.* **2011**, *30*, 215–220. [CrossRef] [PubMed]
38. Maly, J.; Burianova, I.; Vitkova, V.; Ticha, E.; Navratilova, M.; Cermakova, E.; Premature Milk Study Group. Preterm human milk macronutrient concentration is independent of gestational age at birth. *Arch. Dis. Child. Fetal Neonatal Ed.* **2019**, *104*, F50–F56. [CrossRef] [PubMed]
39. Tsang, R.; Uauy, R.; Koletzko, B.; Zlotkin, S. Nutrition of the preterm infant. *Early Hum. Dev.* **2005**, *88*, S5–S7.
40. Hinde, K. Richer milk for sons but more milk for daughters: Sex-biased investment during lactation varies with maternal life history in rhesus macaques. *Am. J. Hum. Biol.* **2009**, *21*, 512–519. [CrossRef] [PubMed]
41. Hinde, K.; Carpenter, A.J.; Clay, J.S.; Bradford, B.J. Holsteins favor heifers, not bulls: Biased milk production programmed during pregnancy as a function of fetal sex. *PLoS ONE* **2014**, *9*, e86169. [CrossRef] [PubMed]
42. Powe, C.E.; Knott, C.D.; Conklin-Brittain, N. Infant sex predicts breast milk energy content. *Am. J. Hum. Biol.* **2010**, *22*, 50–54. [CrossRef] [PubMed]
43. Thakkar, S.K.; Giuffrida, F.; Cristina, C.H.; De Castro, C.A.; Mukherjee, R.; Tran, L.A.; Steenhout, P.; Lee, L.Y.; Destaillats, F. Dynamics of human milk nutrient composition of women from Singapore with a special focus on lipids. *Am. J. Hum. Biol.* **2013**, *25*, 770–779. [CrossRef] [PubMed]
44. Hahn, W.H.; Song, J.H.; Song, S.; Kang, N.M. Do gender and birth height of infant affect calorie of human milk? An association study between human milk macronutrient and various birth factors. *J. Matern. Fetal Neonatal Med.* **2017**, *30*, 1608–1612. [CrossRef] [PubMed]
45. Fujita, M.; Roth, E.; Lo, Y.J.; Hurst, C.; Vollner, J.; Kendell, A. In poor families, mothers' milk is richer for daughters than sons: A test of Trivers-Willard hypothesis in agropastoral settlements in Northern Kenya. *Am. J. Phys. Anthropol.* **2012**, *149*, 52–59. [CrossRef] [PubMed]
46. Fusch, G.; Rochow, N.; Choi, A.; Fusch, S.; Poeschl, S.; Ubah, A.O.; Lee, S.Y.; Raja, P.; Fusch, C. Rapid measurement of macronutrients in breast milk: How reliable are infrared milk analyzers? *Clin. Nutr. Edinb. Scotl.* **2015**, *34*, 465–476. [CrossRef] [PubMed]
47. American Academy of Pediatrics. Committee on Nutrition. Nutritional Needs of Preterm Infants. In *Pediatric Nutrition Handbook*; Kleinman, R.E., Ed.; American Academy of Pediatrics: Elk Grove Village, IL, USA, 2004.
48. Arslanoglu, S. IV. Individualized Fortification of Human Milk: Adjustable Fortification. *J. Pediatr. Gastroenterol. Nutr.* **2015**, *61*, S4–S5. [CrossRef] [PubMed]
49. Van Goudoever, J. VI. Bioengineering Human Milk: Why? *J. Pediatr. Gastroenterol. Nutr.* **2015**, *61*, S7–S9. [CrossRef] [PubMed]

© 2019 by the authors. Licensee MDPI, Basel, Switzerland. This article is an open access article distributed under the terms and conditions of the Creative Commons Attribution (CC BY) license (http://creativecommons.org/licenses/by/4.0/).

Article

Human Milk Oligosaccharides in the Milk of Mothers Delivering Term versus Preterm Infants

Sean Austin [1,*], Carlos A. De Castro [2], Norbert Sprenger [1], Aristea Binia [1], Michael Affolter [1], Clara L. Garcia-Rodenas [1], Lydie Beauport [3], Jean-François Tolsa [3] and Céline J. Fischer Fumeaux [3]

[1] Nestlé Research, Vers-Chez-Les-Blanc, 1000 Lausanne, Switzerland; norbert.sprenger@rdls.nestle.com (N.S.); Aristea.Binia@rdls.nestle.com (A.B.); michael.affolter@rdls.nestle.com (M.A.); Clara.Garcia@rdls.nestle.com (C.L.G.-R.)
[2] Nestle Research Singapore, 29 Quality Road, 618802 Singapore, Singapore; CarlosAntonio.DeCastro@rdsg.nestle.com
[3] Centre Hospitalier Universitaire Vaudois, 1011 Lausanne, Switzerland; Lydie.Beauport@chuv.ch (L.B.); Jean-Francois.Tolsa@chuv.ch (J.-F.T.); Celine-Julie.Fischer@chuv.ch (C.J.F.F.)
* Correspondence: sean.austin@rdls.nestle.com; Tel.: +41-21-785-8050

Received: 9 May 2019; Accepted: 4 June 2019; Published: 5 June 2019

Abstract: Human milk oligosaccharides (HMOs) are a major component of human milk, and play an important role in protecting the infant from infections. Preterm infants are particularly vulnerable, but have improved outcomes if fed with human milk. This study aimed to determine if the HMO composition of preterm milk differed from that of term milk at equivalent stage of lactation and equivalent postmenstrual age. In all, 22 HMOs were analyzed in 500 samples of milk from 25 mothers breastfeeding very preterm infants (< 32 weeks of gestational age, < 1500 g of birthweight) and 28 mothers breastfeeding term infants. The concentrations of most HMOs were comparable at equivalent postpartum age. However, HMOs containing α-1,2-linked fucose were reduced in concentration in preterm milk during the first month of lactation. The concentrations of a number of sialylated oligosaccharides were also different in preterm milk, in particular 3′-sialyllactose concentrations were elevated. At equivalent postmenstrual age, the concentrations of a number of HMOs were significantly different in preterm compared to term milk. The largest differences manifest around 40 weeks of postmenstrual age, when the milk of term infants contains the highest concentrations of HMOs. The observed differences warrant further investigation in view of their potential clinical impact.

Keywords: 2′-fucosyllactose (2′FL); 3′-sialyllactose (3′SL); disialyllacto-N-tetraose (DSLNT); human milk oligosaccharides (HMO); milk group; secretor; Lewis; lactation; preterm

1. Introduction

Human milk is the optimal source of nutrition for infants, and it is widely recommended that infants are exclusively or predominantly breastfed for the first 6 months of life [1,2]. In preterm neonates, human milk feeding is known to have several important specific protective actions and it is strongly encouraged too [1,3,4]. Preterm infants consuming human milk have notably improved immunity [5], are less likely to develop necrotizing enterocolitis (NEC) [6], have improved neurodevelopmental outcomes [1,7] and a better long term health outcome [8,9].

However, as nutritional needs of preterm infants are higher compared to those of term infants, human milk composition may not match the nutritional needs of very preterm infants (< 32 weeks of gestational age, < 1500g of birthweight) during the first weeks of life. In this context, human milk fortification in energy, proteins and minerals is commonly recommended in routine nutritional neonatal care of very preterm infants [4]. To optimize current practices, additional knowledge on human milk composition and opportunities in nutrient supplementation remain to be further explored.

While the composition of preterm milk has been investigated [10], and it is reported to be slightly different from that of term milk [11,12], there are relatively few studies focusing on the human milk oligosaccharide (HMO) composition of preterm milk.

HMOs are the 4th most abundant component of human milk after water, lipids and lactose, and may be present at concentrations up to 25 g/L in colostrum and between 10 to 15 g/L in mature milk [13–15]. This family of over 160 compounds [16] is postulated to play an important role in protecting infants from infection, by acting as decoy receptors or through modulation of the gut microbiota. They may also modulate the immune system through direct interactions [17–20]. Furthermore, HMOs may act as a dietary source of sialic acid [21,22], potentially important for learning and memory [23,24]. Recent evidence from animal studies suggests non-sialylated HMOs may also be important for learning and memory [25]. Today, randomized placebo controlled trials with HMO supplementation are scarce and only done with individual HMOs. Trials that have been carried out indicate that 2'FL is associated with infant immunity [26] and 2'FL together with lacto-N-neotetraose (LNnT) relate to protection from illnesses of the lower respiratory tract and a reduction in antibiotic use [27].

Of specific relevance to preterm infants, HMOs have been linked to the prevention of gut dysfunction [28] and development of NEC [28,29]. Disialyllacto-N-tetraose (DSLNT) and 2'FL have been shown to protect against the development of NEC in rat models [30,31]. It was also observed that DSLNT was present in lower concentrations in the milk fed to preterm infants who developed NEC compared to those who did not [32], leading to the proposal that DSLNT could be used as a marker to predict the likelihood of an infant developing NEC [32].

HMOs are built from lactose, the lactose can be elongated with residues of galactose and N-acetylglucosamine to produce at least 13 different core oligosaccharides [16,33,34]. The core oligosaccharides can be further decorated with sialic acid residues by the action of sialylltransferases, and fucose residues by the action of fucosyltransferases. Due to genetic polymorphisms, two fucosyltransferases, fucosyltransferase-2 (FUT2) and fucosyltransferase-3 (FUT3) are not active in 100% of the population. Fucosyltransferase-2 (FUT2) is responsible for the attachment of fucose to core oligosaccharides through an α-1,2-linkage creating HMOs such as 2'-fucosyllactose (2'FL) or lacto-N-fucopentaose-I (LNFP-I). When FUT2 is inactive such HMOs are absent from the milk. Fucosyltransferase-3 (FUT3), attaches fucose to the core oligosaccharides through an α-1,4 linkage creating oligosaccharides such as LNFP-II, and along with other fucosyltransferases HMOs containing α-1,3 linkages. When FUT3 is inactive HMOs containing α-1,4 linkages are absent from the milk. This results in there being four major milk groups with differing HMO compositions depending on the activities of the FUT2 and FUT3 enzymes:

- Milk group 1, in which both enzymes are active
- Milk group 2, in which FUT3 is active but FUT2 is not
- Milk group 3, in which FUT2 is active but FUT3 is not
- Milk group 4, in which both enzymes are inactive.

In general, milk group 1 seems to be the most common milk group, and milk group 4 is rare [35], however the distribution of milk groups in different populations varies depending on genetic background [36]. Based on the HMO composition of different milk samples, there is also evidence that further subgroups may exist [37,38].

There have been a few previous investigations looking at the HMO composition of preterm milk [17,32,36,39–43], in most cases it is reported that the HMO composition of preterm milk is comparable to that of term milk. However, Coppa [39] reported that preterm milk contained a higher total HMO concentration compared to term milk and De Leoz [36] reported that some HMO features were more variable in preterm milk than in term milk. Kunz [43] pointed out that the Lewis and secretor status should be taken in to account when making studies on HMOs, since these are important factors determining the HMO composition, but milk group has been considered in only a few studies [40–42].

Here we report the composition and trajectories of several different HMOs analyzed in milk collected from mothers of preterm infants and term infants. The HMO composition was compared at equivalent stage of lactation (i.e., equivalent infant postnatal age) and at equivalent postmenstrual age (i.e., equivalent infant developmental stage). In addition the milk group for each donor could be identified from the HMO composition of the milk, thus comparisons could be performed on samples with matched milk groups.

2. Materials and Methods

2.1. Trial Design

This research is part of a prospective cohort study to compare the nutritional composition of human milk from mothers giving birth at term or preterm. The detailed study design has already been published [44].

The study was conducted at the Neonatal Intensive Care Unit (NICU) and at the maternity ward of the University Hospital in Lausanne (CHUV), Switzerland, between October 2013 and July 2014. In this cohort, human milk was longitudinally collected from mothers of preterm infants (gestational age 28 0/7 weeks to 32 6/7 weeks) and mothers of term infants (gestational age 37 0/7 weeks to 41 6/7 weeks). A dedicated research nurse, qualified as a lactation consultant, followed and supported the subjects during the study period. Neonatal demographic and delivery data were collected from the medical charts upon enrollment.

2.2. Milk Collection & Storage

Preterm human milk samples were collected once per week at 7 ± 1 day intervals during the first 8 weeks after delivery. An additional 4 samples were collected until 16 weeks after delivery with sample collection at 14 ± 1 day intervals. Term human milk samples were collected once per week at 7 ± 1 day intervals during the first 8 weeks after delivery (Figure 1). With such a sampling design, it was possible to compare preterm milk with term milk at the same stage of lactation (in Figure 1 the preterm sample at wk1 with the term sample at wk1, etc) or at the equivalent postmenstrual age of the infant (in Figure 1 samples at 42, 44 and 46 weeks would be compared ie. preterm wk12 with term wk2, preterm wk14 with term wk4, etc.)

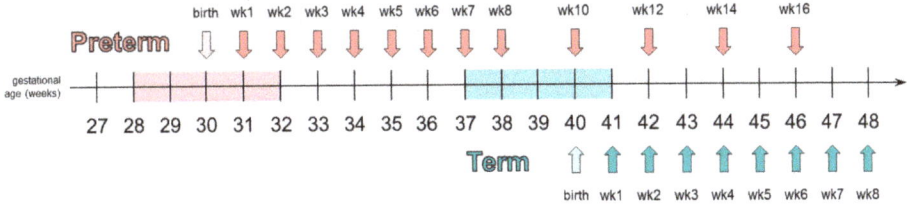

Figure 1. Example of milk sampling scheme from term and preterm mothers delivering at postmenstrual weeks 40 and 30 respectively. Reprinted from Garcia-Rodenas, et al. Clinical Nutrition, 2018 [44] with kind permission from Elsevier.

Samples were collected at the first milk expression in the morning between 06:00 and 12:00 using an electric, double breast pump (Symphony®, Medela, 6340 Baar Switzerland). Full milk expression from a single breast was collected, homogenized, and a 10 mL sample was taken for analysis (except for the first 2 visits, when the volume of sample for analysis was 1–3 mL). Milk for analysis was transferred to a 15 mL polypropylene tube and stored at −18 °C for up to 1 week until hospital transfer and then at −80 °C. Samples were thawed once, homogenized, and split in to 15 different aliquots for different analyses. Those aliquots were then stored at −80 °C until analysis or shipment. Aliquots for HMO analysis were shipped to Neotron S.p.A, Modena, Italy on dry ice.

2.3. Ethical & Legal Considerations

The study was conducted according to the guidelines in the Declaration of Helsinki. The study was approved by the local ethics committee (Commission cantonale d'éthique de la recherche sur l'être humain du Canton de Vaud, Switzerland; Protocol 69/13, clinical study 11.39.NRC). Written informed consent was obtained from all subjects participating in the study. The study was registered on ClinicalTrials.gov with the identifier NCT02052245.

2.4. Analytical Method

HMOs were analyzed by liquid chromatography with fluorescence detection after labelling with 2-aminobenzamide using the protocol described by Austin & Benet [45]. Ten HMOs were quantified against genuine standards with known purity; 2'-fucosyllactose (2'FL), 3-fucosyllactose (3FL), A-tetrasaccharide (A-Tetra), Lacto-N-tetraose (LNT), Lacto-N-neotetraose (LNnT), 3'-sialyllactose (3'SL), 6'-sialyllactose (6'SL), Lacto-N-fucopentaose-I (LNFP-I), LNFP-V and Lacto-N-neofucopentaose (LNnFP). All other HMOs were quantified against maltotriose with known purity, assuming equimolar response factors (graphical representations of the HMOs studied can be found in supplementary Figure S1). A pooled human milk sample (Lee Biosolutions, St Louis, USA) was analyzed with every batch of analysis and at least every 25 injections as a quality control (QC) sample to ensure the method was performing consistently between analytical batches.

2.5. Assignment to FUT2 and FUT3-Dependant Milk Group

The milk samples were assigned to one of 4 milk groups depending on the levels of 2'FL (a marker for FUT2 activity) and LNFP-II (a marker for FUT3 activity) in the sample at visit 2 (2 weeks postpartum). Samples in milk group 1 have high levels (> 25 mg/L) of 2'FL and (>35 mg/L) LNFP-II, samples in milk group 2 have high levels (> 35 mg/L) of LNFP-II and low levels (< 25 mg/L) of 2'FL, samples in milk group 3 have high levels (> 25 mg/L) of 2'FL and low levels (< 35mg/L) of LNFP-II, samples in milk group 4 have low levels (< 25mg/L) of 2'FL and (< 35 mg/L) LNFP-II.

2.6. Data Analysis

All statistical analyses were done with the statistical software R 3.2.3. Prior to statistical analysis all results below the method limit of quantification (LoQ) were set to a value of $0.5 \times$ LoQ. A mixed linear model was used in comparing the two groups (preterm and term) in which the group, infant age (postpartum or postmenstrual) and mode of delivery were considered as fixed effects. The "within subject" variability is taken into account by declaring the subjects as random effects. The main point of comparison is preterm vs term infants. An adjustment for mode of delivery was made because it is a confounding effect with term status given that there are higher proportion of preterm infants delivered by C-section. A logarithmic transformation was applied to some of the HMO concentrations when modelling as the distribution was skewed and according to Box-Cox and QQ plots a log-transformation seemed adequate. Contrast estimates of the model were used to assess significant differences between preterm and term infants at specified lactation and postmenstrual weeks.

3. Results

3.1. Subject Characteristics

This study included 27 mothers with a total of 33 preterm infants, and 34 mothers with a total of 34 term infants. Two of the 27 mothers with preterm infants and six of the 34 mothers with term infants dropped out of the study before completion. No serious adverse events were reported during the study. A total of 500 milk samples were collected and analyzed for HMO content; 280 preterm samples and 220 term samples. The study flow chart and detailed demographic and anthropometric data have already been reported [44]. In summary, baseline maternal characteristics were not different between the two groups, and all mothers in both groups were healthy. Delivery by cesarean section was more common in the preterm group, as were multiple deliveries (twins). Unsurprisingly, term and preterm infants differed in gestational age, birth weight, height and head circumference. There was no difference in gender distribution.

3.2. Milk Groups

The milk samples were assigned to one of 4 milk groups. Overall, 75% of samples were assigned to milk group 1, 19% to milk group 2, 4% to milk group 3 and 2% (1 individual) to milk group 4 (Figure 2). Milk group distribution was similar in the preterm and term populations.

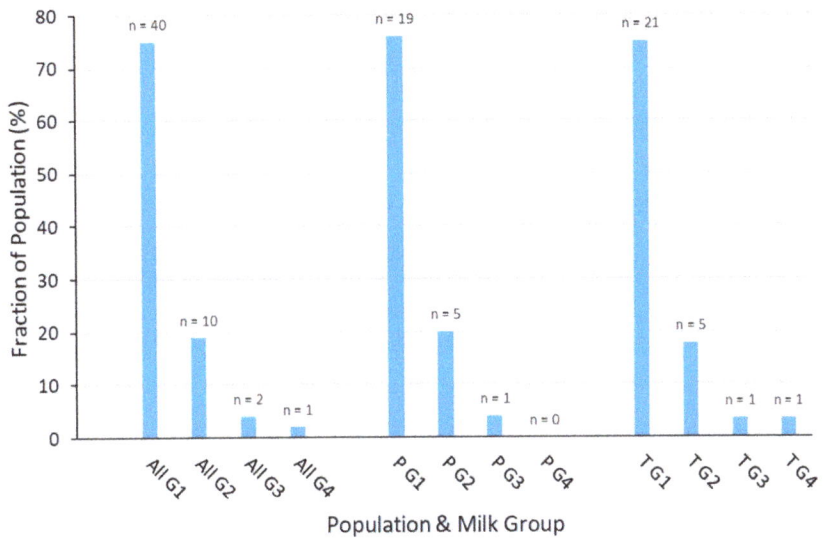

Figure 2. Distribution of milk groups in the different populations: All = all participants, P = participants with preterm infants, T = participants with term infants, G1 = milk group 1, G2 = milk group 2, G3 = milk group 3, G4 = milk group 4.

3.3. Changes in HMO Concentration During Lactation

For both term and preterm milk, the concentration of many of the HMOs decreased at later stages of lactation (Figure 3, and Supplementary Table S1), of those LNT, DSLNT, LNDFH-I, MFLNH-III and DFLNHa reached their maximum concentration after the first week postpartum (i.e., at weeks 2 or 3 postpartum). The only HMO to increase in concentration at later stages of lactation was 3FL.

3.4. HMO Concentration in Term versus Preterm Milk at Equivalent Postpartum Age (Lactation Stages)

Considering all samples, when the children were the same age postpartum, the concentration of each HMO in the mother's milk was generally comparable between term and preterm groups (Figure 3 and Supplementary Table S1). The main exceptions to this were 3'SL, LSTb and DSLNT all of which were at significantly higher concentrations in the preterm milk at postpartum weeks 2 to 8. The concentrations of LSTc (at weeks 1 to 4) and 6'SL (at weeks 2 to 4) were higher in term milk during the first month, but were not significantly different at later time points.

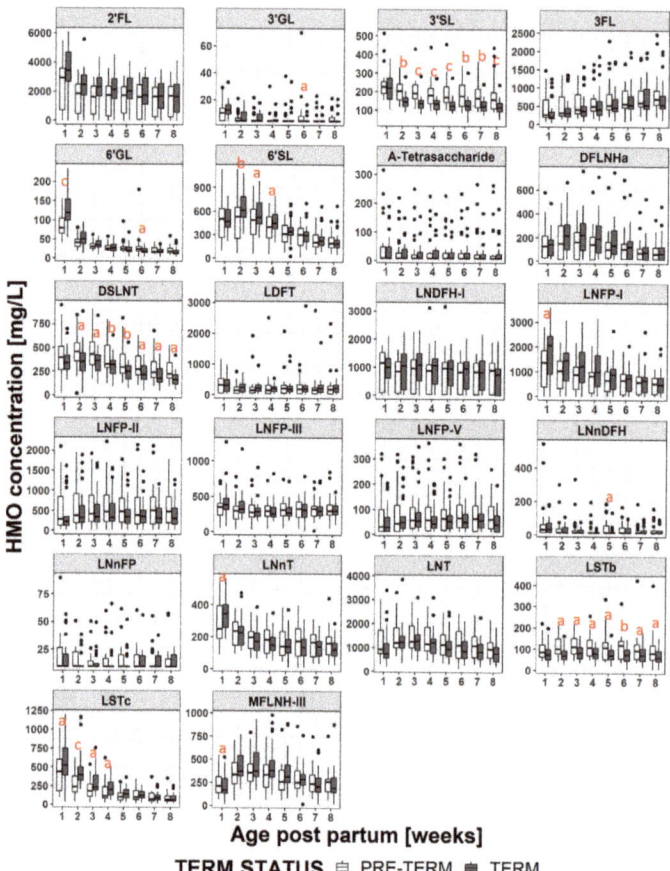

Figure 3. Mean concentration of each HMO at each visit for term (grey) and preterm (white) infants, letters indicate if difference between term and preterm is significant; a: $p < 0.05$, b: $p < 0.005$, c: $p < 0.0005$.

Comparing term and preterm milk at the same age postpartum, but restricting the comparison to milks within the same milk group the picture changes slightly (Figure 4, Figure 5, Supplementary Figure S2, and Supplementary Table S2). Within samples from milk group 1, LNFP-I concentrations are significantly higher in term milk at weeks 1-4, 2′FL concentrations are significantly higher in term milk samples at weeks 1-3 and DFLNHa concentrations are significantly higher in term milk samples at weeks 3-4 (Figure 4). These three HMO do not occur in group 2 or group 4 milks. The concentration of 3′SL remains significantly higher in group 1 preterm milk at weeks 2-8, but for DSLNT the difference between term and preterm milk is only significant at weeks 5 and 8 and for LSTb only at weeks 5, 6 and 8 (Figure 5). For 6′SL, the differences observed between term and preterm is lost when considering only group 1 milk and for LSTc the difference between term and preterm group 1 milk is only significant at weeks 2 to 3 (Figure 5). For the group 2 milks (Supplementary Figure S2) there are only 5 subjects in the term and 5 subjects in the preterm group so statistical significance was not tested at each visit. However from Figure 5 it can be seen that LNFP-V, 3′SL, 6′SL and DSLNT concentrations appear numerically much higher in preterm group 2 milk than term group 2 milk, while 6′SL and LSTc, appear numerically higher in term group 2 milk, especially in the first month. For group 3 milk there was only 1 subject in the term group and 1 subject in the preterm group so a comparison was not made, and for group 4 milk we had only one subject in the whole study.

3.5. HMO Concentration in Term versus Preterm Milk at Equivalent Postmenstrual Age (Developmental Status)

When the milks are compared at infants' equivalent postmenstrual age (Figure 6 and Supplementary Table S3), the concentrations of several HMOs (2′FL, 3′GL, 3′SL, 3FL, 6′GL, 6′SL, DFLNHa, DSLNT, LNFP-I, LNnDFH, LNnT, LNT, LSTc, MFLNH-III) are different at 2 or more visits in the preterm milk compared to term milk. The differences tend to manifest at weeks 38-41 where the concentration of several HMOs in term milk are at their maximum, and the concentration of 3FL is at the minimum. Thus, at equivalent developmental status, preterm infants consume higher concentrations of 3FL in their milk than term infants. However the preterm infants consume lower concentrations of most other HMOs. Making the same exercise only on milk group 1 samples (Figure 7 and Supplementary Table S4) doesn't change this a lot although the magnitude and the duration of some differences (especially for 2′FL and LNFP-I) does change. Similar plots for milk group 2 are reported in Supplementary Figure S3, but no statistical analyses have been performed due to the low number of subjects.

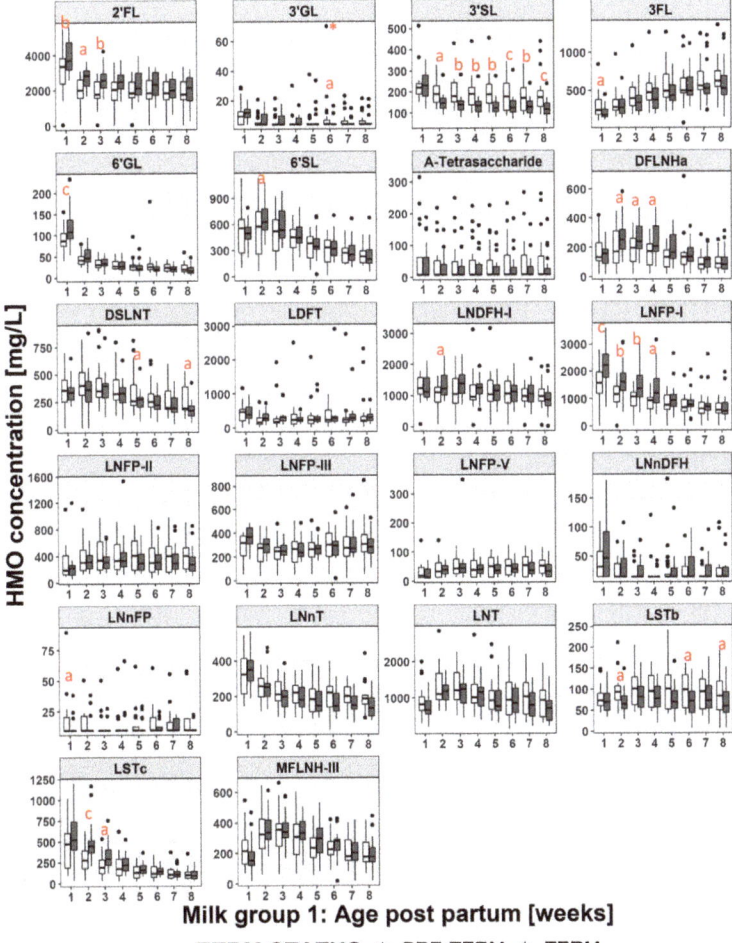

Figure 4. Comparison of HMO concentrations in group 1 term (grey bars) and preterm (white bars) milk. a: $p < 0.05$, b: $p < 0.005$ c: $p < 0.0005$.

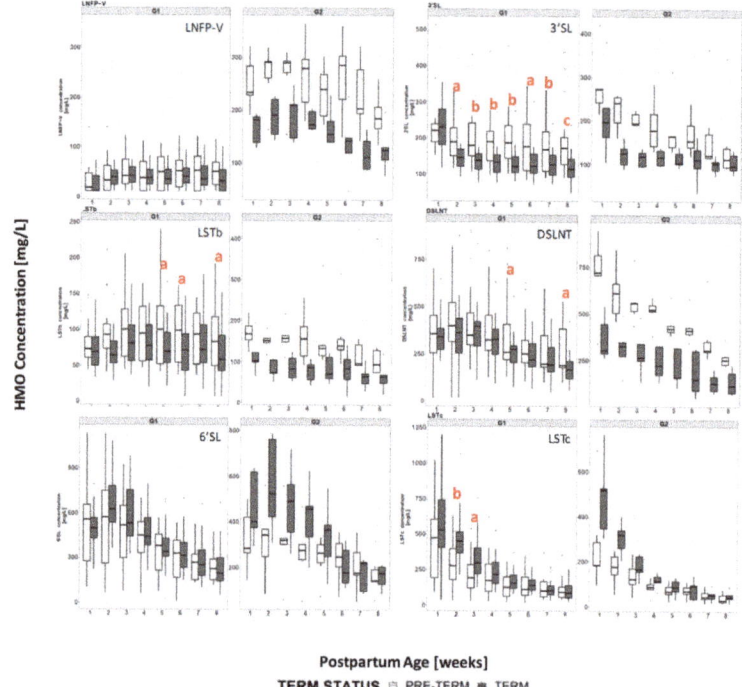

Figure 5. Comparison of selected HMO concentrations in group 1 (G1) and group 2 (G2) term (grey bars) and preterm (white bars) milk. a: $p < 0.05$, b: $p < 0.005$ c: $p < 0.0005$. Significance not tested in G2 milk due to low number of subjects (5 for each arm).

4. Discussion

In this study we have compared the content of 22 HMOs of 500 breast milk samples from mothers of preterm infants and term infants over time. The comparison has been performed both by equivalent stage of lactation (postpartum age) or equivalent postmenstrual age (developmental stage).

4.1. HMO in Term vs Preterm Milk at Equivalent Postpartum Age (Lactation Stages)

The HMO concentrations and trajectories over lactation were generally comparable in the milk of mothers giving birth preterm and term, but several important differences were observed, in particular for the sialylated oligosaccharides. This contrasts some of the previous studies [40,43,46] in which no significant differences were observed in HMO concentrations between term and preterm milk. Coppa et al. [39] reported that the sum of the measured HMOs in preterm milk was significantly higher than that of term milk at day 4 postpartum, but not at days 10 or 30 postpartum. Gabrielli et al. [41] also reported that concentrations of HMOs were higher in preterm milk, but did not go in to details. In this study, we observed the HMOs carrying sialic acid residues, 3'SL, LSTb and DSLNT, were present at significantly higher concentrations in preterm milk between weeks 2-8, while 6'SL (at weeks 2-4) and LSTc (at weeks 1-4) were higher in term milk. Wang et al. [47] reported that the total sialic acid content of preterm milk was significantly higher than that of term milk up to 45 days postpartum, but not beyond 92 days postpartum. In this study we did not collect milk from term infants beyond 56 days postpartum. Although our data appear to be in line with the observations of Wang et al. [47], they contradict the data of Kunz et al. [43] and Spevacek et al. [46] who did not report differences in concentrations of sialylated HMO between term and preterm milk. Variability in HMO concentration between individuals is high, and both of these studies were slightly smaller than ours, which is itself rather small. As pointed out by Kunz et al. [43] it is difficult to compare quantitative HMO results between studies, due to the lack of standardized methods of analysis, milk collection timing and milk collection methodology. Further, although some studies have considered the impact of milk group and stage of lactation on HMO concentrations, they have not matched both stage of lactation and milk group prior to comparing term vs preterm milk, which could mask (or enhance) any compositional differences.

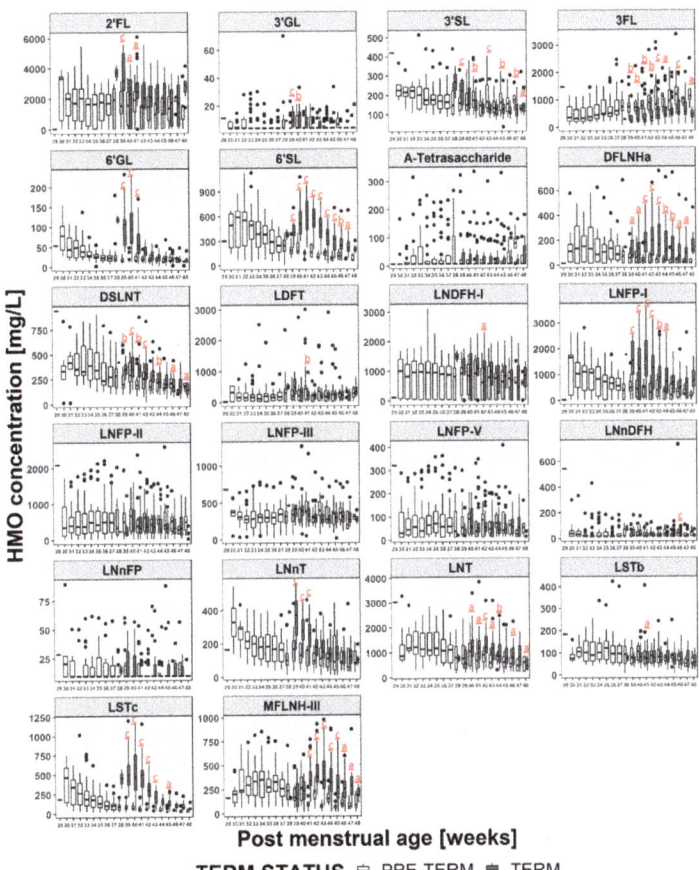

Figure 6. Comparison of the HMO concentration in the milk of mothers giving birth to term (grey bars) or preterm (white bars) infants at equivalent postmenstrual age. a: $p < 0.05$, b: $p < 0.005$ c: $p < 0.0005$.

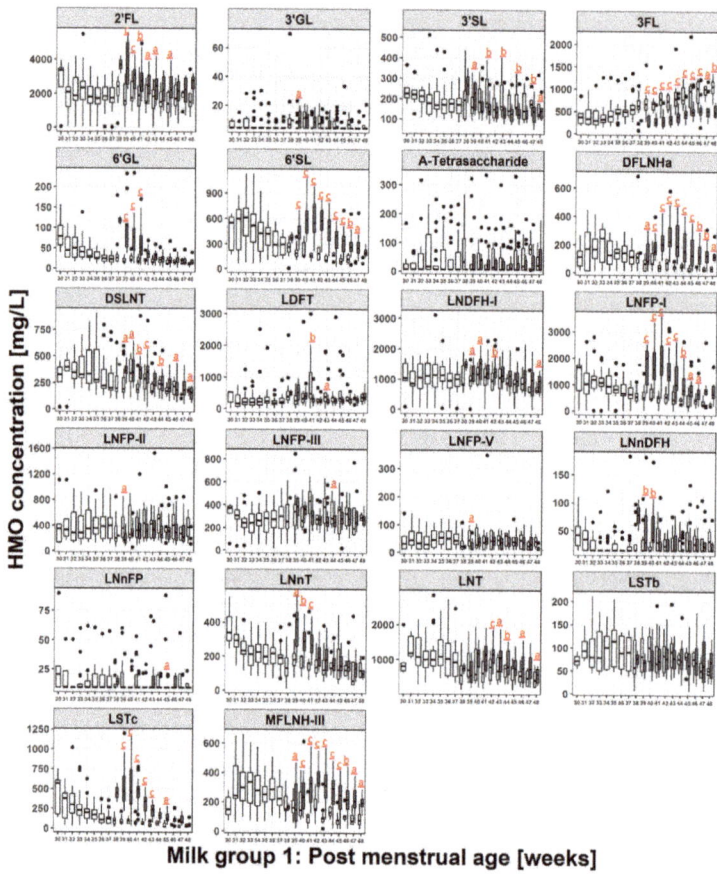

Figure 7. Comparison of the HMO concentration in group 1 milk of mothers giving birth to term (grey bars) or preterm (white bars) infants at equivalent gestational age. a: $p < 0.05$, b: $p < 0.005$ c: $p < 0.0005$.

As to why we observed higher concentrations of several sialylated HMOs in preterm milk and lower concentrations of other sialylated HMOs, we can only speculate. From a structural viewpoint, 6'SL and LSTc, both at higher concentrations in term milk, both contain the structural motif α-Neu5Ac(2→6)β-D-Gal- which is absent from 3'SL, LSTb and DSLNT. 3'SL, LSTb and DSLNT all contain the structural motifs α-Neu5Ac(2→6)β-D-GlcNAc- and/or α-Neu5Ac(2→3)β-D-Gal-. Therefore we may envisage that the sialyltransferase responsible for attaching sialic acid to galactose through an α-1,6-linkage has reduced activity after preterm delivery while those responsible for attaching sialic acid through an α-1,3 linkage to galactose or an α-1,6-linkage to N-acetylglucosamine have increased activity. During the final trimester of gestation, the ganglioside density of the cerebral cortex increases significantly [48], and sialic acids are an important component of the gangliosides. One may also postulate that at this stage of gestation the mother has an increased production of sialic acid to help with the baby's brain development. If the baby is born prematurely, and lactation begins early, the upregulation of the system to produce sialic acids may result in an increased production of sialylated HMOs. Dietary sources of sialic acid have been demonstrated to be processed by the body and the sialic acids incorporated in to glycoproteins [49]. Data from Wang et al. [24] have also demonstrated that supplementing pigs' diets with a dietary source of sialic acid improves learning and memory.

The higher concentration of sialylated HMOs in preterm milk may be a factor that contributes to the important benefits experienced by preterm infants consuming their own mothers' milk.

DSLNT has been observed to protect neonatal rats from development of NEC [30]. An observational study in preterm human infants also observed that infants receiving milk with higher levels of DSLNT were less likely to develop NEC [32] leading to the proposition that DSLNT concentrations could be used as a possible marker for risk of developing NEC. This work demonstrates that the mothers own milk of preterm infants is probably a good source of DSLNT. Our data also suggest children receiving group 2 milk may receive higher concentrations of DSLNT than those receiving group 1 milk, but a larger study would be needed to test if the difference is significant, and to determine if infants receiving group 2 milk may be better protected from NEC.

4.2. HMO at Equivalent Postmenstrual Age (Developmental Status)

As shown above, when comparisons were made at equivalent postpartum age, there were few differences in concentrations of most HMOs between term milk and preterm milk and the trajectories of concentration during lactation were comparable. This can indicate that birth sets off a program that defines the HMO trajectory with most HMOs showing a decrease in concentration with stage of lactation with the exception of 3FL, which increases.

Since the HMO concentration varies with stage of lactation, when the milk is compared at infants' equivalent postmenstrual age, the concentrations of most of the HMOs in preterm milk are lower than in the term milk, with the exception of 3FL (since it increases during lactation) and 3'SL (since it is generally in higher concentrations in preterm milk). The largest differences manifest around 40 weeks of postmenstrual age, which is when the milk for term infants contains the highest concentrations of HMOs. Interestingly, as the postmenstrual age increases, the differences reduce, and for some HMOs the concentrations are equivalent by 44-46 weeks postmenstrual age, while for some others a significant difference was maintained for the duration of the study. The biological relevance of these differences is unclear, and would be worthy of future investigation. Making the hypothesis that infants need to be exposed to a certain amount of HMOs at specific stages of their development, we can presume that preterm infants may be losing out by being exposed to lower amounts of HMOs than term infants as similar stages of development. In such a case supplementing preterm milk with HMOs is likely to be beneficial. Alternatively, perhaps the HMOs help the infant adjust to life outside the womb, and the concentrations in the mothers' milk are appropriate. However a very preterm infant may not receive full enteral feeding until several days or even weeks after delivery, meaning that the actual amount of HMOs they receive will be below what they would be receiving if they could be immediately fully breast fed. In such cases HMO supplementation may also be beneficial. Clinical evidence shows that premature infants greatly benefit from being fed own mother's milk [3,50], or, alternatively, being fed donor human milk [50]. For very preterm infants the milk needs to be fortified with energy, proteins and minerals [3]. Now that HMOs start to be commercially available, HMO fortification is also possible, but further work is needed to confirm the potential benefits and the appropriate timing and dosages for intervention.

4.3. Impact of Milk Group

In general, when comparing HMO compositions, it is important to consider the milk group. In this study the distribution of milk groups is comparable between mothers giving birth term or preterm, with the majority of mothers producing group 1 milk (milk containing oligosaccharides indicating both FUT2 and FUT3 were active). Milk group 2 was next most common with 5 mothers in each group. The equal distributions suggest that there is no link between the genetic factors controlling the milk groups, and the likelihood of giving birth prematurely, although this study is too small to properly address this question.

In this study, when combining data from all milk groups at the same postpartum stages it appears that there was relatively little difference in the concentrations of the neutral HMOs between term or preterm milk, similar to what has been observed previously [40,43,46]. However when only the group 1 milks were compared, three neutral HMOs, 2'FL, LNFP-I, and DFLNHa, were at significantly lower concentrations in preterm milk during the first month of lactation. 2'FL was 700-800 mg/L (20–27%) lower, LNFP-I was 420–590 mg/L (27–32%) lower and DFLNHa was 60-70 mg/L (26–30%) lower. These differences were hidden by the increased variability introduced by the inclusion of the data from the other milk groups, especially milk group 2 with inactive FUT2. All of these structures contain α-1,2-linked fucose residues and are present in significant concentrations only in milk of groups 1 and 3. The observation that these structures are at lower concentrations in group 1 preterm milk implies that the fucosyltransferase-2 enzyme (FUT2) is not fully active during the first month of lactation when an infant is born very preterm as suggested previously [36]. Higher levels of fucosylation have been shown to increase protection against infection [18] but the clinical relevance of the magnitude of the differences remains to be assessed.

Interestingly in group 1 milk, 3'SL is the only sialylated HMO that remains higher in preterm milk across weeks 2-8 (as observed when studying all milk groups together). The differences in concentrations of LSTb and DSLNT between term and preterm milks are largely lost, the differences being significant only at 2 or 3 non-consecutive time points. Likewise for 6'SL and LSTc, which were observed as being significantly higher in term milk when all milk groups were combined, the significance is largely lost when considering only milk group 1 data. Most of the differences in concentrations of sialylated structures appears to be driven by their concentrations in the group 2 milks (Figure 5). The reasons for this are not clear, but one may postulate that FUT2 outcompetes the other enzymes for access to the core structures, and so when FUT2 is active (even if slightly less active than normal for preterm milks), the difference in activity of the sialyltransferases is masked. However, when FUT2 is not active, the difference in sialyltransferase activities becomes more apparent.

Looking at differences in HMO concentrations at equivalent postmenstrual age, and considering only milk group 1 (Figure 7), does not much change the observations made considering all milk groups (Figure 6) with most of the differences being driven by the stage of lactation. However, it does emphasize the differences in the HMOs containing α-1,2-linked fucose. For example looking at all milk groups, 2'FL concentrations appear significantly different only between weeks 39-41, but when considering only milk group 1 then the concentration differences are significant between weeks 39-43 and at week 45. Similar phenomena can be observed for LNFP-I and LNDFH-I. This highlights the importance of considering milk group when making comparisons between different milk samples.

4.4. Strengths and Limitations of the Study

The main limitations of this study are the limited sample size and the monocentric design of the study. One may also argue that collecting milk at only one time of the day, the first milk expression in the morning, may mean the data are not representative of the milk composition during the whole day. Conversely one may see this as a strength of the study, since the main aim was to compare term versus preterm milk and sampling at different times of the day could have introduced additional variation not specifically related to the comparison of term versus preterm milk. Certainly a strength of the study was the very tight collection windows at each stage of lactation, each window being only ±1 day. It is well established that the HMO concentration changes at different stages of lactation [51–56], especially during the first few months of lactation. Keeping the milk collection window tight reduces variability due to differences in sampling day. The study protocol places quite a burden on the mothers who need to collect the full milk expression from a single breast at every sampling point. Despite this, we had very few drop outs from the study, the support provided to the mothers by the lactation nurse was surely a key factor in the successful retention of the volunteers. Importantly, the analytical method used for determination of the HMOs has been extensively validated [45]. To ensure the method performance was maintained day to day, a reference sample of pooled human milk was analysed with every batch of analyses, and at least every 25 samples. Such rigor assured that the variability introduced by the analytical method was also kept to a minimum. Milk group assignments have been done based on the concentrations of 2'FL and LNFP-II in the milk. Appropriate cut-off concentrations for making milk group assignments have not yet been established, so we have based our cut-offs based on the performance of the analytical method. Unfortunately, due to the small sample size, it was not possible to make statistical comparisons between term and preterm milk of milk groups 2,3 or 4. In this study we have only determined the concentrations of 22 out of over 160 HMOs. Although those 22 include the most abundant HMOs, concentrations of a large number of HMOs and how they are impacted by preterm birth remains unknown.

5. Conclusions

The HMO composition and concentration trajectory of term and preterm milk are largely comparable at equivalent infant postpartum age, suggesting that birth triggers a program that defines the HMO trajectory during lactation. Consequently, significant differences exist between preterm and term infants milk when comparing HMO concentrations at equivalent postmenstrual ages. Nevertheless some differences were observed when the comparison was made at the same postpartum age, in particular for the sialylated oligosaccharides and oligosaccharides containing α-1,2-linked fucose. The data suggest that when an infant is born preterm, FUT2 is not fully active during the first month of lactation, leading to reduced concentrations of HMOs such as 2'FL and LNFP-I. In addition the expression of enzymes responsible for sialyllation of HMOs is also perturbed leading to differences in concentrations of the sialylated HMOs. Interestingly, these observations are not fully consistent with previous studies in preterm infants. This study was small, thus the observations need corroboration in larger cohorts. The possible physiological significance of the observations remain to be determined.

Supplementary Materials: The following are available online at http://www.mdpi.com/2072-6643/11/6/1282/s1, Figure S1: Structures of the Determined HMO Depicted Using Symbol Nomenclature, Figure S2: Mean Concentration of each HMO in Group 2 Milk for Term and Preterm Infants at Equivalent Stage of Lactation, Figure S3: Mean Concentration of each HMO in Group 2 Milk for Term and Preterm Infants at Equivalent Postmenstrual Age, Table S1: Concentration of Human Milk Oligosaccharides in Term or Preterm Milk At Different Weeks Postpartum, Table S2: Concentration of Human Milk Oligosaccharides in Term or Preterm Milk At Different Weeks Postpartum Separated By Milk Group, Table S3: Concentration of Human Milk Oligosaccharides in Term or Preterm Milk At Specified Postmenstrual Age, Table S4: Concentration of Human Milk Oligosaccharides in Term or Preterm Milk At Specified Postmenstrual Age Separated By Milk Group.

Author Contributions: Conceptualization, C.L.G.-R., M.A., S.A., L.B., J.-F.T. and C.J.F.F.; Formal Analysis, C.A.D.C.; Investigation, L.B., J.-F.T., C.J.F.F., S.A., N.S. and A.B.; Resources, L.B., J.-F.T. and C.J.F.F.; Writing-Original Draft Preparation, S.A. Writing-Review & Editing, S.A., C.A.D.C., N.S., A.B., M.A., C.L.G.-R., L.B., J.-F.T., and C.J.F.F.; Visualization, C.A.D.C. and S.A.; Supervision, C.L.G.-R., M.A., and C.J.F.F.; Project Administration, M.A.; Funding Acquisition, C.L.G.-R. and M.A.

Funding: This research was funded by Nestlé Research, Lausanne (Nestec Ltd.).

Acknowledgments: The authors would like to thank all the mothers and their infants who kindly donated their milk for this study, Nassima Grari, the study lactation nurse who cared for all mothers for the duration of the study, and the staff at Neotron S.p.A., in particular Marco Meschiari for performing the analysis of all the milk samples in their laboratory. Thanks also to Céline Romagny and Emilie Darcillon who managed the organizational and operational part of the clinical study and the data management.

Conflicts of Interest: S.A., C.A.D.C., N.S., A.B., M.A., and C.L.G.R. are all employees of Nestec Ltd. L.B., J.F.T. and C.J.F.F. declare no conflict of interest.

References

1. Eidelman, A.I.; Schanler, R.J. Breastfeeding and the use of human milk. *Pediatrics* **2012**, *129*, e827–e841. [CrossRef]
2. Fewtrell, M.; Bronsky, J.; Campoy, C.; Domellof, M.; Embleton, N.; Fidler Mis, N.; Hojsak, I.; Hulst, J.M.; Indrio, F.; Lapillonne, A.; et al. Complementary feeding: A position paper by the european society for paediatric gastroenterology, hepatology, and nutrition (espghan) committee on nutrition. *J. Pediatr. Gastroenterol. Nutr.* **2017**, *64*, 119–132. [CrossRef] [PubMed]
3. Boquien, C.Y. Human milk: An ideal food for nutrition of preterm newborn. *Front. Pediatr.* **2018**, *6*, 295. [CrossRef]

4. Agostoni, C.; Buonocore, G.; Carnielli, V.P.; De Curtis, M.; Darmaun, D.; Decsi, T.; Domellof, M.; Embleton, N.D.; Fusch, C.; Genzel-Boroviczeny, O.; et al. Enteral nutrient supply for preterm infants: Commentary from the european society of paediatric gastroenterology, hepatology and nutrition committee on nutrition. *J. Pediatr. Gastroenterol. Nutr.* **2010**, *50*, 85–91. [CrossRef] [PubMed]
5. Lewis, E.D.; Richard, C.; Larsen, B.M.; Field, C.J. The importance of human milk for immunity in preterm infants. *Clin. Perinatol.* **2017**, *44*, 23–47. [CrossRef]
6. Corpeleijn, W.E.; Kouwenhoven, S.M.; Paap, M.C.; van Vliet, I.; Scheerder, I.; Muizer, Y.; Helder, O.K.; van Goudoever, J.B.; Vermeulen, M.J. Intake of own mother's milk during the first days of life is associated with decreased morbidity and mortality in very low birth weight infants during the first 60 days of life. *Neonatology* **2012**, *102*, 276–281. [CrossRef]
7. Lechner, B.E.; Vohr, B.R. Neurodevelopmental outcomes of preterm infants fed human milk: A systematic review. *Clin. Perinatol.* **2017**, *44*, 69–83. [CrossRef]
8. Lucas, A. Long-term programming effects of early nutrition—Implications for the preterm infant. *J. Perinatol.* **2005**, *25* (Suppl. 2), S2–S6. [CrossRef]
9. Lewandowski, A.J.; Lamata, P.; Francis, J.M.; Piechnik, S.K.; Ferreira, V.M.; Boardman, H.; Neubauer, S.; Singhal, A.; Leeson, P.; Lucas, A. Breast milk consumption in preterm neonates and cardiac shape in adulthood. *Pediatrics* **2016**, *138*. [CrossRef]
10. Boyce, C.; Watson, M.; Lazidis, G.; Reeve, S.; Dods, K.; Simmer, K.; McLeod, G. Preterm human milk composition: A systematic literature review. *Br. J. Nutr.* **2016**, *116*, 1033–1045. [CrossRef]
11. Ballard, O.; Morrow, A.L. Human milk composition: Nutrients and bioactive factors. *Pediatr. Clin. N. Am.* **2013**, *60*, 49–74. [CrossRef]
12. Gidrewicz, D.A.; Fenton, T.R. A systematic review and meta-analysis of the nutrient content of preterm and term breast milk. *BMC Pediatr.* **2014**, *14*, 216. [CrossRef]
13. Urashima, T.; Asakuma, S.; Leo, F.; Fukuda, K.; Messer, M.; Oftedal, O.T. The predominance of type I oligosaccharides is a feature specific to human breast milk. *Adv. Nutr.* **2012**, *3*, 473S–482S. [CrossRef]
14. Kunz, C.; Rodriguez-Palmero, M.; Koletzko, B.; Jensen, R. Nutritional and biochemical properties of human milk, part i: General aspects, proteins, and carbohydrates. *Clin. Perinatol.* **1999**, *26*, 307–333. [CrossRef]
15. Coppa, G.V.; Gabrielli, O.; Pierani, P.; Catassi, C.; Carlucci, A.; Giorgi, P.L. Changes in carbohydrate composition in human milk over 4 months of lactation. *Pediatrics* **1993**, *91*, 637–641.
16. Urashima, T.; Hirabayashi, J.; Sato, S.; Kobata, A. Human milk oligosaccharides as essential tools for basic and application studies on galectins. *Trends Glycosci. Glycotechnol.* **2018**, *30*, SE51–SE65. [CrossRef]
17. Kunz, C.; Rudloff, S. Biological functions of oligosaccharides in human milk. *Acta Paediatr.* **1993**, *82*, 903–912. [CrossRef]
18. Newburg, D.S.; Ruiz-Palacios, G.M.; Morrow, A.L. Human milk glycans protect infants against enteric pathogens. *Annu. Rev. Nutr.* **2005**, *25*, 37–58. [CrossRef]
19. Morozov, V.; Hansman, G.; Hanisch, F.G.; Schroten, H.; Kunz, C. Human milk oligosaccharides as promising antivirals. *Mol. Nutr. Food Res.* **2018**, *62*, 1700679. [CrossRef]
20. Bode, L. The functional biology of human milk oligosaccharides. *Early Hum. Dev.* **2015**, *91*, 619–622. [CrossRef]
21. Jacobi, S.K.; Yatsunenko, T.; Li, D.; Dasgupta, S.; Yu, R.K.; Berg, B.M.; Chichlowski, M.; Odle, J. Dietary isomers of sialyllactose increase ganglioside sialic acid concentrations in the corpus callosum and cerebellum and modulate the colonic microbiota of formula-fed piglets. *J. Nutr.* **2016**, *146*, 200–208. [CrossRef]
22. Sprenger, N.; Duncan, P.I. Sialic acid utlization. *Adv. Nutr.* **2012**, *3*, 392S–397S. [CrossRef]
23. Oliveros, E.; Vazquez, E.; Barranco, A.; Ramirez, M.; Gruart, A.; Delgado-Garcia, J.M.; Buck, R.; Rueda, R.; Martin, M.J. Sialic acid and sialylated oligosaccharide supplementation during lactation improves learning and memory in rats. *Nutrients* **2018**, *10*, 1519. [CrossRef]
24. Wang, B.; Yu, B.; Karim, M.; Hu, H.; Sun, Y.; McGreevy, P.; Petocz, P.; Held, S.; Brand-Miller, J. Dietary sialic acid supplementation improves learning and memory in piglets. *Am. J. Clin. Nutr.* **2007**, *85*, 561–569. [CrossRef]
25. Oliveros, E.; Ramirez, M.; Vazquez, E.; Barranco, A.; Gruart, A.; Delgado-Garcia, J.M.; Buck, R.; Rueda, R.; Martin, M.J. Oral supplementation of 2′-fucosyllactose during lactation improves memory and learning in rats. *J. Nutr. Biochem.* **2016**, *31*, 20–27. [CrossRef]

26. Goehring, K.C.; Marriage, B.J.; Oliver, J.S.; Wilder, J.A.; Barrett, E.G.; Buck, R.H. Similar to those who are breastfed, infants fed a formula containing 2′-fucosyllactose have lower inflammatory cytokines in a randomized controlled trial. *J. Nutr.* **2016**, *146*, 2559–2566. [CrossRef]
27. Puccio, G.; Alliet, P.; Cajozzo, C.; Janssens, E.; Corsello, G.; Sprenger, N.; Wernimont, S.; Egli, D.; Gosoniu, L.; Steenhout, P. Effects of infant formula with human milk oligosaccharides on growth and morbidity: A randomized multicenter trial. *J. Pediatr. Gastroenterol. Nutr.* **2017**, *64*, 624–631. [CrossRef]
28. Bering, S.B. Human milk oligosaccharides to prevent gut dysfunction and necrotizing enterocolitis in preterm neonates. *Nutrients* **2018**, *10*, 1461. [CrossRef]
29. Moukarzel, S.; Bode, L. Human milk oligosaccharides and the preterm infant a journey in sickness and in health. *Clin. Perinatol.* **2017**, *44*, 193–207. [CrossRef]
30. Jantscher-Krenn, E.; Zherebtsov, M.; Nissan, C.; Goth, K.; Guner, Y.S.; Naidu, N.; Choudhury, B.; Grishin, A.V.; Ford, H.R.; Bode, L. The human milk oligosaccharide disialyllacto-N-tetraose prevents necrotising enterocolitis in neonatal rats. *Gut* **2012**, *61*, 1417–1425. [CrossRef]
31. Autran, C.A.; Schoterman, M.H.; Jantscher-Krenn, E.; Kamerling, J.P.; Bode, L. Sialylated galacto-oligosaccharides and 2′-fucosyllactose reduce necrotising enterocolitis in neonatal rats. *Br. J. Nutr.* **2016**, *116*, 294–299. [CrossRef]
32. Autran, C.A.; Kellman, B.P.; Kim, J.H.; Asztalos, E.; Blood, A.B.; Spence, E.C.; Patel, A.L.; Hou, J.; Lewis, N.E.; Bode, L. Human milk oligosaccharide composition predicts risk of necrotising enterocolitis in preterm infants. *Gut* **2017**, *67*, 1064–1070. [CrossRef]
33. Blank, D.; Dotz, V.; Geyer, R.; Kunz, C. Human milk oligosaccharides and Lewis blood group: Individual high-throughput sample profiling to enhance conclusions from functional studies. *Adv. Nutr.* **2012**, *3*, 440S–449S. [CrossRef]
34. Kobata, A. Structures and application of oligosaccharides in human milk. *Proc. Jpn. Acad. Ser. B Phys. Biol. Sci.* **2010**, *86*, 731–747. [CrossRef]
35. Thurl, S.; Henker, J.; Siegel, M.; Tovar, K.; Sawatzki, G. Detection of four human milk groups with respect to Lewis blood group dependent oligosaccharides. *Glycoconj. J.* **1997**, *14*, 795–799. [CrossRef]
36. de Leoz, M.L.; Gaerlan, S.C.; Strum, J.S.; Dimapasoc, L.M.; Mirmiran, M.; Tancredi, D.J.; Smilowitz, J.T.; Kalanetra, K.M.; Mills, D.A.; German, J.B.; et al. Lacto-N-tetraose, fucosylation, and secretor status are highly variable in human milk oligosaccharides from women delivering preterm. *J. Proteome Res.* **2012**, *11*, 4662–4672. [CrossRef]
37. van Leeuwen, S.S.; Stoutjesdijk, E.; ten Kate, G.A.; Schaafsma, A.; Dijck-Brouwer, J.; Muskiet, F.A.J.; Dijkhuizen, L. Regional variations in human milk oligosaccharides in vietnam suggest FucTx activity besides FucT2 and FucT3. *Sci. Rep.* **2018**, *8*, 16790. [CrossRef]
38. Elwakiel, M.; Hageman, J.A.; Wang, W.; Szeto, I.M.; van Goudoever, J.B.; Hettinga, K.A.; Schols, H.A. Human milk oligosaccharides in colostrum and mature milk of Chinese mothers: Lewis positive secretor subgroups. *J. Agric. Food Chem.* **2018**, *66*, 7036–7043. [CrossRef]
39. Coppa, G.V.; Pierani, P.; Zampini, L.; Gabrielli, O.; Carlucci, A.; Catassi, C.; Giorgi, P.L. Lactose, oligosaccharide and monosaccharide content of milk from mothers delivering preterm newborns over the first month of lactation. *Minerva Pediatr.* **1997**, *49*, 471–475.
40. Nakhla, T.; Daotian, F.; Zopf, D.; Brodsky, N.L.; Hurt, H. Neutral oligosaccharide content of preterm human milk. *Br. J. Nutr.* **1999**, *82*, 361–367. [CrossRef]
41. Gabrielli, O.; Zampini, L.; Galeazzi, T.; Padella, L.; Santoro, L.; Peila, C.; Giuliani, F.; Bertino, E.; Fabris, C.; Coppa, G.V. Preterm milk oligosaccharides during the first month of lactation. *Pediatrics* **2011**, *128*, e1520–e1531. [CrossRef]
42. Van Niekerk, E.; Autran, C.A.; Nel, D.G.; Kirsten, G.F.; Blaauw, R.; Bode, L. Human milk oligosaccharides differ between HIV-infected and HIV-uninfected mothers and are related to necrotizing enterocolitis incidence in their preterm very-low-birth-weight infants. *J. Nutr.* **2014**, *144*, 1227–1233. [CrossRef]
43. Kunz, C.; Meyer, C.; Collado, M.C.; Geiger, L.; Garcia-Mantrana, I.; Bertua-Rios, B.; Martinez-Costa, C.; Borsch, C.; Rudloff, S. Influence of gestational age, secretor and Lewis blood group status on the oligosaccharide content of human milk. *J. Pediatr. Gastroenterol. Nutr.* **2017**, *64*, 789–798. [CrossRef]
44. Garcia-Rodenas, C.L.; De Castro, C.A.; Jenni, R.; Thakkar, S.K.; Beauport, L.; Tolsa, J.F.; Fischer-Fumeaux, C.J.; Affolter, M. Temporal changes of major protein concentrations in preterm and term human milk. A prospective cohort study. *Clin. Nutr.* **2018**, *38*, 1844–1852. [CrossRef]

45. Austin, S.; Benet, T. Quantitative determination of non-lactose milk oligosaccharides. *Anal. Chim. Acta* **2018**, *1010*, 86–96. [CrossRef]
46. Spevacek, A.R.; Smilowitz, J.T.; Chin, E.L.; Underwood, M.A.; German, J.B.; Slupsky, C.M. Infant maturity at birth reveals minor differences in the maternal milk metabolome in the first month of lactation. *J. Nutr.* **2015**, *145*, 1698–1708. [CrossRef]
47. Wang, B.; Brand-Miller, J.; McVeagh, P.; Petocz, P. Concentration and distribution of sialic acid in human milk and infant formulas. *Am. J. Clin. Nutr.* **2001**, *74*, 510–515. [CrossRef]
48. Schnaar, R.L.; Gerardy-Schahn, R.; Hildebrandt, H. Sialic acids in the brain: Gangliosides and polysialic acid in nervous system development, stability, disease, and regeneration. *Physiol. Rev.* **2014**, *94*, 461–518. [CrossRef]
49. Tangvoranuntakul, P.; Gagneux, P.; Diaz, S.; Bardor, M.; Varki, N.; Varki, A.; Muchmore, E. Human uptake and incorporation of an immunogenic nonhuman dietary sialic acid. *Proc. Natl. Acad. Sci. USA* **2003**, *100*, 12045–12050. [CrossRef]
50. Bhatia, J. Human milk and the premature infant. *Ann. Nutr. Metab.* **2013**, *62* (Suppl. 3), 8–14. [CrossRef]
51. Chaturvedi, P.; Warren, C.D.; Altaye, M.; Morrow, A.L.; Ruiz-Palacios, G.; Pickering, L.K.; Newburg, D.S. Fucosylated human milk oligosaccharides vary between individuals and over the course of lactation. *Glycobiology* **2001**, *11*, 365–372. [CrossRef]
52. Kunz, C.; Rudloff, S.; Baier, W.; Klein, N.; Strobel, S. Oligosaccharides in human milk: Structural, functional, and metabolic aspects. *Annu. Rev. Nutr.* **2000**, *20*, 699–722. [CrossRef]
53. Xu, G.; Davis, J.C.; Goonatilleke, E.; Smilowitz, J.T.; German, J.B.; Lebrilla, C.B. Absolute quantitation of human milk oligosaccharides reveals phenotypic variations during lactation. *J. Nutr.* **2017**, *147*, 117–124. [CrossRef]
54. Austin, S.; De Castro, C.A.; Benet, T.; Hou, Y.; Sun, H.; Thakkar, S.K.; Vinyes-Pares, G.; Zhang, Y.; Wang, P. Temporal change of the content of 10 oligosaccharides in the milk of Chinese urban mothers. *Nutrients* **2016**, *8*, 346. [CrossRef]
55. Erney, R.M.; Malone, W.T.; Skelding, M.B.; Marcon, A.A.; Kleman-Leyer, K.M.; O'Ryan, M.L.; Ruiz-Palacios, G.; Hilty, M.D.; Pickering, L.K.; Prieto, P.A. Variability of human milk neutral oligosaccharides in a diverse population. *J. Pediatr. Gastroenterol. Nutr.* **2000**, *30*, 181–192. [CrossRef]
56. Sprenger, N.; Lee, L.Y.; De Castro, C.A.; Steenhout, P.; Thakkar, S.K. Longitudinal change of selected human milk oligosaccharides and association to infants' growth, an observatory, single center, longitudinal cohort study. *PLoS ONE* **2017**, *12*, e0171814. [CrossRef]

© 2019 by the authors. Licensee MDPI, Basel, Switzerland. This article is an open access article distributed under the terms and conditions of the Creative Commons Attribution (CC BY) license (http://creativecommons.org/licenses/by/4.0/).

Communication

Milk Therapy: Unexpected Uses for Human Breast Milk

Malgorzata Witkowska-Zimny *, Ewa Kamińska-El-Hassan and Edyta Wróbel

Department of Biophysics and Human Physiology, Medical University of Warsaw, Chalubinskiego 5, 02-004 Warsaw, Poland; malirob@poczta.onet.pl (E.K.-E.-H.); mwzgenetyka@onet.pl (E.W.)
* Correspondence: mwitkowska@wum.edu.pl; Tel.: +48-22-628-63-34; Fax: +48-22-628-78-46

Received: 27 March 2019; Accepted: 25 April 2019; Published: 26 April 2019

Abstract: Background: Human breast milk provides a child with complete nutrition but is also a popular therapeutic remedy that has been used in traditional, natural pharmacopeia, and ethnomedicine for many years. The aim of this current review is to summarize studies of non-nutritional uses of mothers' milk. Methods: Two databases (PubMed and Google Scholar) were searched with a combination of twelve search terms. We selected articles that were published between 1 January 2010, and 1 January 2019. The language of publication was limited to English. Results: Fifteen studies were included in the systematic review. Ten of these were randomized controlled trials, one was a quasi-experimental study, two were in vitro studies, and four employed an animal research model. Conclusions: Many human milk components have shown promise in preclinical studies and are undergoing active clinical evaluation. The protective and treatment role of fresh breast milk is particularly important in areas where mothers and infants do not have ready access to medicine.

Keywords: human milk benefits; colostrum; milk therapy; bioactive factors

1. Introduction

Human breast milk (HBM) is perhaps the most important functional food known. It is a dynamic food with both nutritional and health benefits for neonates and infants. Human milk has powerful immunological properties, protecting infants from respiratory diseases, middle ear infections, and gastro-intestinal diseases. It is now appreciated that human breast milk has health impacts that are lifelong, with breastfeeding showing protective effects against diabetes mellitus, obesity, hyperlipidemia, hypertension, cardiovascular diseases, autoimmunity, and asthma [1]. However, human milk is also a popular therapeutic remedy that has been applied as a part of traditional, natural pharmacopeia, and ethnomedicine for many years. Public health nurses have reported on the effects of fresh colostrum and human milk as a treatment for conjunctivitis, chapped nipples, rhinitis, infections of the skin and soft tissues. The discovery of growth factors, cytokines, and a heterogeneous population of cells—including stem cells, probiotic bacteria, and the HAMLET complex (human alpha-lactalbumin made lethal to tumor cells)—in human milk has led to researchers' increased interest in human breast milk as a natural medicine. In recent years, human milk has been the focus of many types of evidence-based research. There have been a number of reports on the topical application of human milk as an effective treatment for diaper rash, atopic eczema, diaper dermatitis, and umbilical cord separation [2–4]. The protective and treatment role of human milk is particularly important in areas where mothers and infants do not have ready access to medicine, such as in developing countries. In these situations, milk therapy is often a determining factor of infant recovery and survival. For this reason, more clinical trials and research into mothers' milk come from low and middle-income countries in Africa and Asia. Many human milk components have shown promise in preclinical studies and are undergoing active clinical evaluation. A few milk-derived therapeutic preparations are available to clinicians. The study of human milk has resulted in abundant opportunities for translational medicine. However, complementary and alternative medicine (CAM) therapies often fare unfavorably under the scrutiny of evidence-based practice (EBP), due to the lack of or shortage of research and the inherent differences in healing ideology. Yet excess medicalization and pharmaceuticalization can lead to the extension of the medical gaze to human conditions. Physically maladaptive outliers have been treated as diseases and pulled into the realm of medicine. Many "non-disease" states can creep up into medicine, and with time become medicalized due to the redefining of many conditions that were long considered social or psychological phenomena as disease states. Processes regarded as natural are now looked at medical problems or diseases. Unnecessary medicalization leads to great social and financial cost, as well as increased anxiety and the risk for complication from further workups for incidental or clinically unimportant findings. The growth of research and reflections on medicalization has led to the proposal of other parallel concepts, such as biomedicalization. These tools could be useful in the analysis of human enhancement and can be defined as contributing to a "bionic society". Medicalization risks neglecting the role of social determinants, natural therapies, and ethnomedicine in shaping human health. On the other hand, where individuals do not have ready access to medicine—particularly in developing countries—knowledge, skills, and practices based on the theories and experiences indigenous to different cultures, whether explicable or not, are used in the maintenance of health as well as in the prevention, improvement, and treatment of illness.

There is no doubt that incorporating traditional and modern evidence-based medicine (EBM) as integral parts of the formal health-care system is important and likely to be achieved in many countries. The aim of the current review is to summarize studies of non-nutritional usage of mothers' milk. Due to its low cost, wide availability, and lack of undesirable effects, mothers' milk has the potential to play a role in human health and in evidence-based therapy.

2. Methods

The literature review was performed by conducting an electronic search of PubMed and Google Scholar. No filter or limitation was used during the search. Reference lists from selected studies were manually scanned to identify any other relevant studies. The electronic search used the following keywords and medical subject headings (MeSH) terms: Human milk; breast milk, mother milk, colostrum, atopic eczema, diaper dermatitis, nipple pain, breast inflammation, umbilical cord separation, neonatal conjunctivitis, HAMLET, topical treatment. The two authors independently searched databases and reviewed articles. Bibliographic references to retrieved reviews and studies were searched for additional articles. We included articles that were published between 1 January 2010, and 1 January 2019. The language of publication was limited to English.

From total of 1503 initially screened articles, only 15 fulfilled all the inclusion criteria. The following criteria needed to be fulfilled for a study to be included in this review: (1) Topical application of human milk or human milk active factors versus control; (2) in the case of human studies, participants had to be newborns; (3) in the case of animal models, HBM in vitro assessment had to be included. Articles that did not provide sufficient information from the title and abstract were included for further evaluation, and reading was done in full. Records were first selected after which 1469 were excluded based on excluded criteria: Did not report the data of interest, the language of publication, no access to the full text, conference proceeding, reviews. There was no restriction of study designs included. In order to limit bias in the inclusion–exclusion process, the selection was made on the basis of the consensus of two authors. The Cohen's kappa index was calculated to assess the agreement between the two reviewers and any discrepancy was resolved by consensus or by a third reviewer. The reviewers were not blind to author, institutions, or manuscript journals. Data extraction and analysis were performed by the same two reviewers. Figure 1 presents a flow diagram of the review process while Table 1 summarizes the studies that were included.

Table 1. Articles included in this review, with study details.

References	Intervention/Study Type	Sample Size	The Dropout Rate	Study Design	Outcomes	Main Findings
Berents et al. (2015) [5]	Atopic eczema/Randomized clinical trial	18 participants	33%	HBM or emollient was applied on the spot, three times a day for four weeks. The severity and area of the eczema spots was calculated weekly by SCORAD.	Both control and intervention areas of the eczema spots were increased during the intervention. At inclusion mean SCORAD (SCORing Atopic Dermatitis) was 35 and at the end of study 34.	No effect with topical application of HBM was found.
Kasrae et al. (2015) [6]	Atopic eczema/Randomized clinical trial	116 participants	10%	HBM or hydrocortisone 1% was applied twice a day for 21 days on the affected area. Efficiency of the treatment was defined by SCORAD index.	The frequency of healed infants was 81.5% and 76% in HBM and 1% hydrocortisone groups on day 21, respectively ($p < 0.24$).	HBM was as effective as 1% hydrocortisone.
Farahani et al. (2013) [7]	Diaper dermatitis/Randomized clinical trial	152 participants	4.6%	HBM or hydrocortisone 1% was applied for 7 days on the affected area. The efficiency of the treatment was evaluated at 3 and 7 days by a 6-point scale.	The severity score was not different between the topical HBM and hydrocortisone 1% groups at 3 and 7 days ($p < 0.95$).	HBM was as effective as 1% hydrocortisone.
Gozen et al. (2014) [8]	Diaper dermatitis/Randomized clinical trial	70 participants	10%	HBM and barrier cream containing 40% zinc oxide and cod liver oil was applied on diaper dermatitis change for 5 days and the postlesion score was establish by a 4-point scale.	The condition of dermatitis was improved in 60% of infant from HBM group and 93.6% treated with barrier cream. The postlesion score of barrier cream group was lower than HBM group ($p = 0.002$).	Barrier cream was more effective than HBM.
Seifi et al. (2017) [9]	Diaper dermatitis/Randomized clinical trial	30 participants	0	Infants suffering from diaper dermatitis assigned to HBM group and the control group were followed up for 5 days and the efficiency of the treatment was evaluated by a 5-point scale rash severity.	In the control group 26.1% infants showed improvement, in HBM group—80%. HBM has decreased the incidence of anal dermatitis rash ($p = 0.009$).	A positive effect with topical application of HBM was found.
Abou-Dakn et al. (2011) [10]	Painful and damaged nipples/No full randomized clinical trial	84 participants	14%	The efficacy of HBM and lanolin on pain and damage nipples was assessed on the 10-range Visual Analog Scale (VAS) and the Nipple Trauma Score (NTS) over 14 days after delivery.	Lanolin was more effective than HBM, including faster healing of nipple trauma and reducing nipple pain ($p = 0.043$).	No positive effect with topical application of HBM was found.
Golshan and Hossein (2013) [11]	Umbilical cord care/Randomized clinical trial	316 participants	5%	The neonates were divided into three groups: Topical ethanol or HBM application twice a day, the control group kept the stump dry. Umbilical separation time and local infection frequency were considered.	Umbilical separation time in human milk group was significantly shorter (6.5 days) than in ethanol (8.94 days) ($p < 0.0001$) and drying groups (7.54 days) ($p < 0.003$).	A positive effect with topical application of HBM was found.
Aghamohammadi et al. (2012) [12]	Umbilical cord care/Randomized clinical trial	152 participants	14.5%	The umbilical separation time was compared in the group of topical HBM application (three times a day) and dry cord care for 10 days.	Median time of cord separation in human milk application group (150.95 ± 28.68 h) was significantly shorter than dry cord care group (180.93 ± 37.42 h) ($p < 0.001$).	A positive effect with topical application of HBM was found.
Abbaszadeh et al. (2016) [13]	Umbilical cord care/Randomized clinical trial	174 participants	6.9%	The infant from HBM group received topical application of milk and group 2 chlorhexidine solution 4% to the umbilical stump 2 times a day. Follow-up and visit at home were done.	The mean cordseparation time in the human milk group (7.14 ± 2.15 days) was shorter than the chlorhexidine group (13.28 ± 6.79 days) ($p < 0.001$).	A positive effect with topical application of HBM was found.

Table 1. Cont.

References	Intervention/Study Type	Sample Size	The Dropout Rate	Study Design	Outcomes	Main Findings
Ghaemi et al. (2014) [14]	Neonatal conjunctivitis/Randomized clinical trial	300 preterm neonates	10.6%	The intervention group with culture-negative eye swab received two drops of HBM in each eye or erythromycin ointment (0.5%), control group—no treatment. All neonates were followed for the occurrence of clinical conjunctivitis for 28 days.	The beneficial preventive effects of colostrum against neonatal conjunctivitis ($p = 0.036$).	A positive effect with application of HBM was found.
Asena et al. (2017) [15]	Corneal epithelial wound/Randomized trial on mice model	24 female experimental corneal epithelial defect mice model	0	A central corneal epithelial defect was created in mice and HBM, autologous serum, artificial tears four times a day was applied for 3 days. Histopathological and electron microscopy examination was performed.	Topical human breast milk drops caused faster and better healing of central corneal epithelial defect than serum drops, artificial tears and in the control group ($p < 0.001$).	A positive effect with application of HBM was found.
Beynham et al. (2013) [16]	Antimicrobial effect on pediatric conjunctivitis/in vitro study	milk from 23 women/9 bacterial species tested	not applicable	The inhibitory effects of HBM against three common ocular pathogens were assessed. Zones of inhibition by milk samples, sterile saline, and trimethoprim ophthalmic solution were measured	Growth of N gonorrhoeae, M catarrhalis, M viridans was significantly inhibited ($p \leq 0.01$) by human milk samples.	A positive effect with application of HBM was found.
Diego et al. (2016) [17]	Dry eye syndrome/Animal in vivo study	91 BALB/c mice	0	The animals with dry eye syndrome were treated with HBM, nopal, nopal extract derivatives, or cyclosporine four times daily for 7 days. Punctate staining and preservation of corneal epithelial thickness were used as indices of therapeutic efficacy.	Reduction in corneal epithelial thickness was largely prevented by administration of HBM (33.2 ± 2.5 μm).	HBM decreased epithelial damage.
Mossberg et al. (2010) [18]	Bladder cancer treatment/animal model and in vitro studies	6 C57BL mice bladder cancer model	0	Bladder tumors cells and bladder mice cancel models were instilled by HAMLET. Effects of HAMLET on tumor size and apoptosis were analyzed.	HAMLET caused a dose dependent decrease in MB49 cell viability in vitro. Five intravesical HAMLET instillations significantly decreased tumor size and delayed development in vivo compared to controls.	HAMLET from HBM delays bladder cancer development.
Puthia et al. (2014) [19]	Colon cancer prevention and treatment/animal model and in vitro studies	Apc$^{Min/+}$ mice colorectal tumors model	0	HAMLET was given in therapeutic and prophylactic regimens. Tumor burden and animal survival were compared, and biochemistry and molecular methods were used to determine effects on colon cancer cells.	Peroral HAMLET administration reduced tumor progression and mortality in Apc$^{Min/+}$ mice.	HAMLET from HBM delays colon cancer development.

HBM: Human breast milk; HAMLET: human alpha-lactalbumin made lethal to tumor cells.

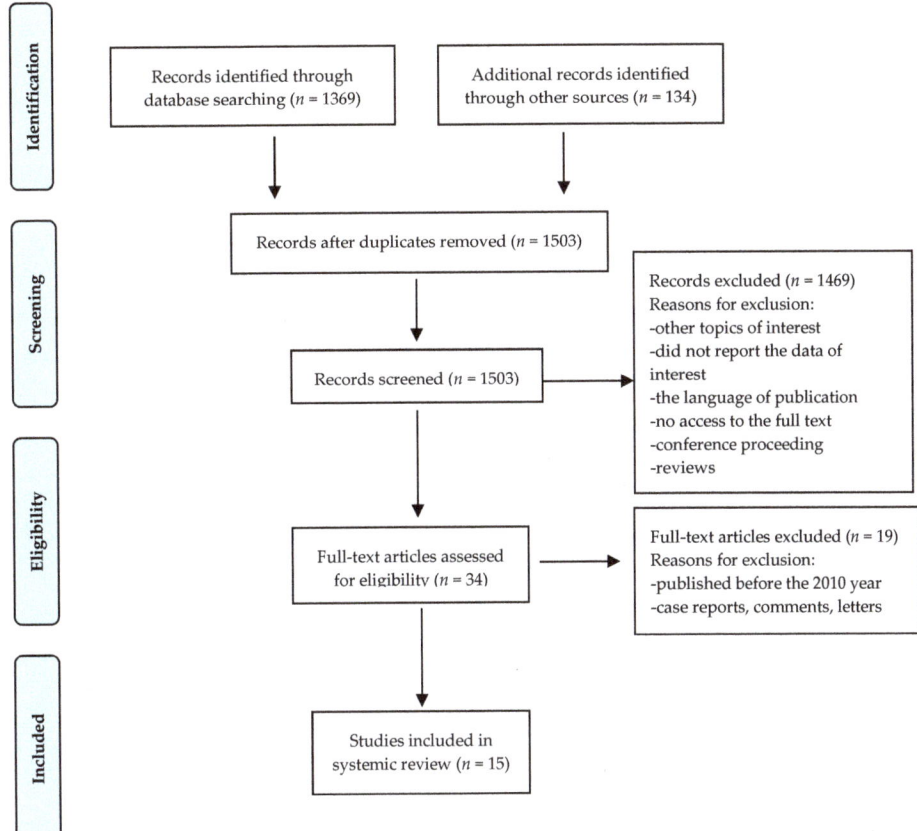

Figure 1. Flow diagram.

We present a narrative summary of studies, rather than a meta-analysis, because of the heterogeneity in measurement tools, populations, interventions, and design (whether qualitative or observational). The reporting in this review follows the Preferred Reporting Items for Systematic Reviews and MetaAnalysis (PRISMA) guidelines, where warranted.

3. Results

Fifteen studies were of sufficient quality to be included in the systematic review. All of these had been published in peer-reviewed journals. The agreement between the reviewers was substantial: κ = 0.625 (p = 0.02). Seven studies were conducted in Iran, two in the USA, two in each of Turkey and Sweden, and one in each of Norway and Germany (Table 1).

Ten of these fifteen studies were randomized controlled trials (two on experimental mice models), and one was a quasi-experimental study. We also considered two in vitro studies and four with animal research models. We decided to include these because they have practical implication for clinical trials and, in our opinion, are examples of translational science.

Below we briefly describe the studies by medical problem.

3.1. Skin Problems: Atopic Eczema and Diaper Dermatitis

Recently, a few studies have been published on the topical anti-inflammatory effects of human breast milk in the treatment of skin problems, such as atopic eczema and diaper dermatitis [6,20,21].

Our systematic review included five randomized clinical trials devoted to infants' skin problems, but the results were not consistent.

Berents and colleagues, in a small pilot study, did not find any effect on eczema spots treated with topical application of fresh human milk. However, this clinical trial has some limitations. First, it had a very small study population of six children; second, two of them were treated with their mother's milk produced for a younger sibling. The mean age of the children was 18.5 months (ranging from 4 to 32) [5].

Kasrae et al. randomized 104 Iranian infants with atopic dermatitis for 21 days of treatment with 1% hydrocortisone versus human milk. The frequency of healed infants was 81.5% and 76% in the human milk and 1% hydrocortisone groups on day 21, respectively. The findings suggest that human milk can improve atopic eczema with similar results and is as easy to apply as 1% hydrocortisone ointment ($p < 0.001$), but without the side-effects and cost [6].

The effects of topical application of human milk and 1% hydrocortisone were also compared in the treatment of diaper dermatitis [7]. The randomized group consisted of 141 infants (aged 0–24 months). Parents received general advice about diaper rash care and were instructed to apply the medication for seven days. The mothers assigned to use milk were asked to rub milk gently on the affected area at the end of each breastfeeding session. Hydrocortisone 1% in an ointment base was applied sparingly to clinically affected areas twice a day. The children were reassessed on days 3 and 7 of the study. The presence of diaper rash was noted daily using a six-point scale, and in both groups, was not significantly different after the topical application of either tested medication. HBM was as effective and safe as hydrocortisone 1% ointment alone ($p < 0.001$).

In the randomized controlled trials by Seifi et al., 30 Iranian infants (between 0 and 12 months of age) suffering from diaper dermatitis were divided into two matched groups: One applying their mother's milk three times a day to the affected area, and a control without any application for five days. The findings revealed positive effects of human milk on the healing of diaper dermatitis and a significant difference between both groups. Out of 15 infants with mild or moderate erythema, 80% improved during the five-day study, whereas in the control group 26.1% infants showed improvement ($p = 0.009$) [9].

Gozen et al. tested the effectiveness of human milk and barrier cream (40% zinc oxide with cod liver oil formulation) on the healing of diaper dermatitis. The population of the trial included 63 term and preterm newborns with developed diaper dermatitis in neonatal intensive care units, divided into two groups. There were no statistically significant differences between the groups in terms of the mean number of clinical improvement days, but the postlesion score in the barrier cream group was lower than in human milk group ($p = 0.002$) [8]. As the researchers stated, neonatal intensive care units typically host infants with disorders and who are on antibiotics. Hence, negative findings can be difficult to discuss and compare with other study data.

3.2. Nipple Problems

A common breastfeeding difficulty for mothers is painful nipples. One traditional non-pharmacological intervention to reduce nipple pain in breastfeeding women is topical treatment with expressed breast milk. Abou-Dakn et al. carried out clinical trials to evaluate the efficacy of lanolin versus breast milk on painful and damaged nipples during lactation [10]. They evaluated 84 lactating mothers from Berlin who developed nipple pain while breastfeeding within 72h after delivery. The first group was instructed after each feed to express a few drops of breast milk and massage them into the nipples and areola, allowing to air-dry. The second group patted the nipples dry after each feeding session and applied a pea-sized amount of lanolin to the nipple and areola, keeping this area covered. During a number of visits over two weeks, the nipple trauma score was used to evaluate healing rates and the visual analog scale (VAS) was employed to judge the pain intensity. Significantly lower pain levels were detected in the lanolin group, and these decreased with the continuation of treatment. Lanolin was more effective than HBM, including faster healing of nipple trauma ($p = 0.043$). According to many studies, the women who applied expressed breast milk had significantly lower perceptions of nipple pain following four to five days of treatment than the women who applied lanolin. However, this beneficial effect was not maintained after six to seven days of treatment. At no assessment were there any group differences in nipple pain perceptions between the women who applied expressed breast milk and the women who had applied lanolin, warm compresses, or nothing [22].

3.3. Eye Problems

Treatment of ocular surface disease with human milk is documented in ancient Egyptian, Roman, Greek, and Byzantine texts and was traditionally used by mothers to treat infectious conjunctivitis. Ghaemi et al. have shown the beneficial preventive effects of colostrum against neonatal conjunctivitis on 89 breastfeeding neonates. The 269 preterm neonates were randomly divided into three groups: The first ($n = 89$) received two drops of colostrum in each eye, the second ($n = 82$) was treated with topical erythromycin ointment (0.5%), and the control group (97) received no treatment. The frequency of conjunctivitis was higher in the control group, followed by the group receiving topical application of colostrum and antibiotic ($p = 0.003$) [14]. However, colostrum does not have potential hazards or side effects and is easily accessible without extra cost. Diego et al. observed that human milk was able to preserve corneal epithelial thickness in the dry-eye mouse model. Epithelial damage, reflected in the punctate scores, decreased over four days of treatment with milk. This was the first study to demonstrate that human milk can preserve corneal epithelial thickness in a dry-eye model, and that preservation of corneal epithelial thickness was comparable to topical cyclosporine [17]. According to the study of Asena et al., topical human breast milk drops caused faster and better healing of central corneal epithelial defects in a mouse model than in the case of treatment with serum drops or artificial tears or in the control group ($p < 0.001$) [15]. They concluded that the rich contents of human breast milk may be an alternative to epithelial healers and artificial tears.

3.4. Umbilical Cord Care

Breast milk is widely reported to be used for umbilical cord care in developing countries, as evidenced by numerous publications. Since 1998, the World Health Organization (WHO) has advocated the use of dry umbilical cord care in high-resource settings but has also recommended research into the use of breast milk and colostrum in umbilical cord care [23]. There are several reports of the effectiveness of applying mother's milk in assisting umbilical cord separation [12,24]. However, in the PubMed database between 2010 and 2018, only three studies were randomized controlled trials. All were published in peer-reviewed journals and were conducted in Iran (see Table 1).

Aghamohammadi et al. randomized 130 singletons, all mature and healthy newborns born in hospital, and compared the effect of topical application of human milk and dry cord care on umbilical cord separation time. Newborns were breastfed. Mothers were asked not to cover the cord with diapers and not to bath the child until the cord had separated. All mothers received instructions, a form for recording symptoms of infection and cord bleeding, and a form for observation on the progress of care. The human milk group dropped milk on the remaining part of the cord and the cut edge, letting it dry, three times a day for two days after separation of the cord. Two days after cord separation, a physician checked the cord. The median time of the cord separation was 150.95 ± 28.68 hours in the human milk topical application group and 180.93 ± 37.42 hours in the dry cord care group ($p = 0.001$). The median number of bleeding days after cord separation was 1.2 ± 2.33 and 3.1 ± 3.77, respectively [12].

In the study of Golshan et al., 300 healthy neonates, delivered normally or by Cesarean section, were divided into three random groups, in which ethanol, their mother's milk, or dry cord care was applied. In the milk group, mothers washed the umbilical cord stump with their milk two times a day. Umbilical separation time in neonates of the human milk group was 6.5 ± 1.93 days, whereas in the ethanol and dry care group this was 8.94 ± 2.39 and 7.54 ± 2.37 days, respectively. The frequency of omphalitis was not significantly different between the three groups. Umbilical separation time in the human milk group was significantly shorter than in the ethanol ($p < 0.0001$) and drying groups ($p < 0.003$) [11].

In a clinical trial by Abbaszadeh et al., 162 healthy, hospital-borne neonates were assigned to two groups, where cord care was performed using human milk or chlorhexidine. Human milk was applied to the umbilical cord every twelve hours for days after separation. The shortest cord separation time was 3 days in the topical human milk group (7.14 ± 2.15), while the longest was 53 days in chlorhexidine group (13.28 ± 6.79) ($p < 0.001$) [13].

All three studies recommend topical application of mother's milk for umbilical cord stump care, which leads to shorter cord separation time and can be used as an easy, cheap, natural, and noninvasive means of cord care.

3.5. The Antitumoricidal Effect of HAMLET

One example of translational medicine is the topical application of α-lactalbumin-oleic acid, a natural product from breast milk. The complex called human alpha-lactalbumin made lethal to tumor cells (HAMLET) was discovered by the Svanborg group when they were studying antiadhesive molecules in human milk [25].

HAMLET is formed during low-pH precipitation of the casein fraction, which allows for partial unfolding of the α-lactalbumin structure and binding with the fatty acid. HAMLET triggers rapid carcinoma cell detachment in vitro and in cancer patients after topical administration of the lyophilized complex. To form HAMLET, α-lactalbumin is obtained from human milk by chromatography. The partially unfold protein is subsequently bound to oleic acid on an ion-exchange matrix, and the complex is eluted with salt, purified, lyophilized, and frozen in aliquots for instillation [26].

The therapeutic efficacy of HAMLET has been demonstrated through in vitro research in animal models of glioblastoma, bladder cancer, and intestinal cancer; and in clinical studies targeting bladder cancers and skin papillomas [27,28]. Local HAMLET infusion was shown to delay the development of tumors and prolong survival in animal models of human cancer. In the study of Mossberg et al., groups of C57BL/6 mice with MB49-implanted murine bladder cancer cells were given instillations of HAMLET or phosphate buffered saline PBS for eight days. The HAMLET treated mice lacked detectable tumors more often than the controls (33% vs. 0%, $p < 0.02$) and the tumors were significantly reduced (mean score 1.9 vs. 2.5, respectively; $p < 0.02$) [18]. Puthia et al. tested whether HAMLET could be used for colon cancer therapy. Peroral HAMLET administration caused a significant reduction in the number of small intestinal tumors and in tumor size ($p < 0.0001$ for total tumor count), reduced polyps by about 58% ($p = 0.0001$), and improved survival ($p = 0.0103$) over the control group mice.

Through gene set enrichment analysis, the researchers concluded that the prophylactic and therapeutic effects of HAMLET are accompanied by well-defined series of stable changes in gene expression, affecting Wnt signaling and β-catenin, glycolysis, oxidative phosphorylation, and lipid metabolism in tumor tissue [19].

The same complex showed strong bactericidal activity against specific pathogens of the oral cavity and respiratory track, with the highest activity against the gram-positive organism *Streptococcus pneumoniae* by cell shrinkage, DNA condensation, and DNA degradation [29].

4. Discussion

The transfer of traditional medical knowledge is an ongoing process. It is important both for the preservation of traditional natural medicine, but also in the search for novel agents in treatment. Home remedies are generally believed to be natural ways to cure minor illnesses or conditions. They are usually cultural practices, traditions, customs, or folk remedies that have been passed down from generation to generation or from person to person. However, it should be kept in mind that there is not necessarily any medical proof that any of these treatments work, or whether they can cause more harm than good. Human milk is considered to be the gold standard in infant nutrition, providing optimal nutrients for normal growth and development. Apart from its nutritional benefits, human milk contains multiple bioactive and immunomodulatory components. The latter of these include growth and immunological factors, as well as micro-RNAs, cellular components such as leukocytes, epithelial cells, progenitor cells, and stem cells [30]. Furthermore, breast milk is also a continuous source of commensal and beneficial bacteria, including lactic acid bacteria and bifidobacteria [31,32]. The discovery of stem cells, the HAMLET complex, and probiotic bacteria in breast milk has resulted in increased interest in human breast milk as a natural medicine. The studies described here suggest safe and cost-effective non-nutritional uses of mothers' breast milk, though further evaluation of effectively is needed. Breast milk has natural antibacterial properties, so it can be used to treat a range of skin problems, including cuts and scrapes.

Common skin problems may appear during lactation and breastfeeding, particularly affecting the nipple, areola, and breast. Some medications used in the treatment of skin conditions are unsuitable during lactation. It has been shown that expressing a few drops of milk and rubbing them gently into the sore nipples, then allowing it to dry naturally, takes advantage of the healing properties of human milk. Many studies have indicated that bioactive components of human milk and microbiota have promise as adjuvants for wound healing [33,34]. From lesions of the corneal epithelium to lacerations of the skin, milk-treated groups healed faster than controls.

Breast milk is used in many cultures for skin irritations. Breast milk involves no risk of allergy, contains antibodies, epidermal growth factor (EGF), and erythropoietin, which may promote the growth and repair of skin cells. Human milk is a source of commensal bacteria that can play an anti-infectious, immunomodulatory role. Their possible function in the acceleration of conditions for skin biofilm formation can open new perspectives for the prevention and treatment of skin and wound healing diseases. Interestingly, the analysis of Simpson et al. showed that miRNAs are possible mediators of the observed preventative effects of atopic dermatitis [35].

The concentration, regulation, and individual variation between bioactive element, immune factors, various progenitor and mature cell types, and stage of lactation are not well established. Complexity and variability in human milk composition, and infants' responses to many human milk constituents may also explain some of the conflicting results of studies evaluating the effects of non-nutritional uses of human milk.

The studies considered here vary in methodology and in definition of outcomes, which leads to considerable heterogeneity. Human milk composition varies both within and between individuals, and this may partially explain the conflicting data.

However, the use of breast milk in the treatment of inflamed or injured eyes is not applicable in all cases. It should rather serve as complementary therapy, and not the only mode of treatment. At our present level of knowledge, non-nutritional uses of breast milk are certainly better suited to prevention than to a medicated process. However, in areas where mothers and infants do not have ready access to medicine, such as in developing countries, the application of breast milk is often a determining factor in infant recovery and survival.

4.1. Implications for Future Research

Fresh whole human milk and its components have potential as a novel therapeutic tool in the treatment of many diseases. In our opinion, the future implications of non-nutritional application are associated with particular components of breast milk, rather than with the whole milk.

In Hakansson's study of the antimicrobial activity of human milk, a complex of α-lactalbumin and oleic acid induced apoptosis in tumor cells without affecting healthy differentiated cells. HAMLET is a tumoricidal protein–lipid complex from human milk with broad effects against cancer cells of different origins. The mechanism of its action is unusual, as the complex interacts with a number of molecular targets and cellular components. Importantly, HAMLET does not have any toxic effects on healthy tissues in the treated patients and animals. HAMLET has been shown to be safe and effective in humans in two proof-of-concept human clinical trials: Convincing therapeutic efficacy was demonstrated in a topical skin papilloma study and in patients with bladder cancer (ClinicalTrials.gov

Identifier: NCT03560479) [33]. Publications on HAMLET are related to the establishment of a Swedish company, HAMLET Pharma, and the increased number of patents regarding this molecule. Based on these discoveries, HAMLET Pharma is developing natural tumor-killing drugs based on molecules with tumor selectivity. A HAMLET patent portfolio has been established with a number of patents issued worldwide. Intellectual property rights include patents covering the manufacturing and use of HAMLET and substances derived from HAMLET "second generation drug candidates".

Many studies highlight progenitor and breast milk stem cells. The presence of stem cells in human milk poses numerous questions and implications for breastfeeding, newborn, and maternal health, but also opens a new perspective of future potential applications of these cells in personal and regenerative medicine. The goals of future research should be to assess the function, potency, and therapeutic value of breast milk cells—including cell therapy for future applications—and should determine the direct or indirect effects of breast milk cell components on promoting immunological tolerance and newborn development, and also on providing effective and complementary treatment of diaper rash, atopic eczema, diaper dermatitis, umbilical cord separation, and eye problems. Without any doubt, future research on these topics will need to involve evidence-based medicine and clinical trials.

4.2. Limitations

Due to the limited and heterogeneous body of evidence that included animal studies, human intervention studies, and observational human studies, the risk of bias assessment for individual studies was not performed. The overall body of evidence was narratively discussed.

This article is limited by its emphasis on papers published in English in journal databases, so many useful local ethnomedical studies may have been missed. Evidence-based medicine focuses more on new approaches than on developing traditional folk and ethnobiological data, so many promising intervention studies are not published in papers indexed in PubMed with high impact factor. It is unclear whether a lack of institutional support and funding for clinical trials of natural products might be critical in the low number of studies.

4.3. Conclusions

The findings of this review provide information about possible non-nutritional uses of breast milk in postnatal care. Breast milk is a natural agent and is biologically suitable for the body, having no side effects; it is always available and can be used in all social and economic groups of society. The health implications of milk components—such as macronutrients, biologically active factors, and somatic cells—remain unknown or not well understood. The positive effects of HBM found by in vitro and animal studies must be substantiated by findings from clinical studies. The most reliable clinical studies for assessing the benefits of HBM are randomized, double-blinded, multicenter controlled trials but to date, they are very scarce.

Non-nutritional uses of human breast milk can be considered examples of personalized medicine. Further research including developed countries is recommended to find or confirm the results and to evaluate the effects of traditional therapies. This knowledge may also have the effect of convincing mothers to continue to breastfeed with their own milk, as a substance that possesses extraordinary properties, not only for nutrition.

Author Contributions: M.W.-Z. designed the model and analyzed and processed the data, drafted the manuscript; E.K.-E.-H. analyzed and processed the data, performed the analysis; E.W. designed the figures. All authors provided critical feedback and helped shape the research, analysis, and manuscript.

Funding: The APC was funded by Medical University of Warsaw.

Conflicts of Interest: Authors declare no conflict of interest.

References

1. Kramer, M.S. "Breast is best": The evidence. *Early Hum. Dev.* **2010**, *86*, 729–732. [CrossRef] [PubMed]

2. Allam, N.; Megrin, W.; Talat, A. The effect of topical application of mother milk on separation of umbilical cord for newborn babies. *Am. J. Nurs. Sci.* **2015**, *4*, 288–296. [CrossRef]
3. Mahrous, E.S.; Darwish, M.M.; Dabash, S.A.; Marie, I.; Abdelwahab, S.F. Topical application of human milk reduces umbilical cord separation time and bacterial colonization compared to ethanol in newborns. *Transl. Biomed.* **2012**, *3*, 4.
4. Arroyo, R.; Martín, V.; Maldonado, A.; Jiménez, E.; Fernández, L.; Rodríguez, J.M. Treatment of infectious mastitis during lactation: Antibiotics versus oral administration of Lactobacilli isolated from breast milk. *Clin. Infect. Dis.* **2010**, *50*, 1551–1558. [CrossRef]
5. Berents, T.L.; Rønnevig, J.; Søyland, E.; Gaustad, P.; Nylander, G.; Løland, B.F. Topical treatment with fresh human milk versus emollient on atopic eczema spots in young children: A small, randomized, split body, controlled, blinded pilot study. *BMC Dermatol.* **2015**, *15*, 7. [CrossRef] [PubMed]
6. Kasrae, H.; AmiriFarahani, L.; Yousefi, P. Efficacy of topical application of human breast milk on atopic eczema healing among infants: A randomized clinical trial. *Int. J. Dermatol.* **2015**, *54*, 966–971. [CrossRef]
7. Farahani, L.A.; Ghobadzadeh, M.; Yousefi, P. Comparison of the effect of human milk and topical hydrocortisone 1% on diaper dermatitis. *Pediatr. Dermatol.* **2013**, *30*, 725–729. [CrossRef]
8. Gozen, D.; Caglar, S.; Bayraktar, S.; Atici, F. Diaper dermatitis care of newborns human breast milk or barrier cream. *J. Clin. Nurs.* **2014**, *23*, 515–523. [CrossRef]
9. Seifi, B.; Jalali, S.; Heidari, M. Assessment Effect of Breast Milk on Diaper Dermatitis. *Dermatol. Rep.* **2017**, *9*, 7044. [CrossRef]
10. Abou-Dakn, M.; Fluhr, J.W.; Gensch, M.; Wöckel, A. Positive effect of HPA lanolin versus expressed breastmilk on painful and damaged nipples during lactation. *Skin Pharmacol. Physiol.* **2011**, *24*, 27–35. [CrossRef]
11. Golshan, M.; Hossein, N. Impact of ethanol, dry care and human milk on the time for umbilical cord separation. *J. Pak. Med. Assoc.* **2013**, *63*, 1117–1119. [PubMed]
12. Aghamohammadi, A.; Zafari, M.; Moslemi, L. Comparing the effect of topical application of human milk and dry cord care on umbilical cord separation time in healthy newborn infants. *Iran J. Pediatr.* **2012**, *22*, 158–162. [PubMed]
13. Abbaszadeh, F.; Hajizadeh, Z.; Jahangiri, M. Comparing the Impact of Topical Application of Human Milk and Chlorhexidine on Cord Separation Time in Newborns. *Pak. J. Med. Sci.* **2016**, *32*, 239–243. [CrossRef] [PubMed]
14. Ghaemi, S.; Navaei, P.; Rahimirad, S.; Behjati, M.; Kelishadi, R. Evaluation of preventive effects of colostrum against neonatal conjunctivitis: A randomized clinical trial. *J. Educ. Health Promot.* **2014**, *3*, 63. [CrossRef]
15. Asena, L.; Suveren, E.H.; Karabay, G.; Dursun-Altinors, D. Human Breast Milk Drops Promote Corneal Epithelial Wound Healing. *Curr. Eye Res.* **2017**, *42*, 506–512. [CrossRef] [PubMed]
16. Baynham, J.T.; Moorman, M.A.; Donnellan, C.; Cevallos, V.; Keenan, J.D. Antibacterial effect of human milk for common causes of paediatric conjunctivitis. *Br. J. Ophthalmol.* **2013**, *97*, 377–379. [CrossRef] [PubMed]
17. Diego, J.L.; Bidikov, L.; Pedler, M.G.; Kennedy, J.B.; Quiroz-Mercado, H.; Gregory, D.G.; Petrash, J.M.; McCourt, E.A. Effect of human milk as a treatment for dry eye syndrome in a mouse model. *Mol. Vis.* **2016**, *22*, 1095–1102. [PubMed]
18. Mossberg, A.K.; Hou, Y.; Svensson, M.; Holmqvist, B.; Svanborg, C. HAMLET treatment delays bladder cancer development. *J. Urol.* **2010**, *183*, 1590–1597. [CrossRef] [PubMed]
19. Puthia, M.; Storm, P.; Nadeem, A.; Hsiung, S.; Svanborg, C. Prevention and treatment of colon cancer by peroral administration of HAMLET (human α-lactalbumin made lethal to tumour cells). *Gut* **2014**, *63*, 131–142. [CrossRef] [PubMed]
20. Chaumeil, C.; Liotet, S.; Kogbe, O. Treatment of severe eye dryness and problematic eye lesions with enriched bovine colostrum lactoserum. *Adv. Exp. Med. Biol.* **1994**, *350*, 595–599.
21. Mohammadzadeh, A.; Farhat, A.; Esmaeily, H. The effect of breast milk and lanolin on sore nipples. *Saudi Med. J.* **2005**, *26*, 1231–1234. [PubMed]
22. Pugh, L.C.; Buchko, B.L.; Bishop, B.A.; Cochran, J.F.; Smith, L.R.; Lerew, D.J. A comparison of topical agents to relieve nipple pain and enhance breastfeeding. *Birth* **1996**, *23*, 88–93. [CrossRef] [PubMed]
23. World Health Organization. *Care of the Umbilical Cord: A Review of the Evidence*; WHO: Geneva, Switzerland, 1998; p. 19.
24. Karumbi, J.; Mulaku, M.; Aluvaala, J.; English, M.; Opiyo, N. Topical umbilical cord care for prevention of infection and neonatal mortality. *Pediatr. Infect. Dis. J.* **2013**, *32*, 78–83. [CrossRef]

25. Håkansson, A.; Zhivotovsky, B.; Orrenius, S.; Sabharwal, H.; Svanborg, C. Apoptosisinduced by a human milkprotein. *Proc. Natl. Acad. Sci. USA.* **1995**, *92*, 8064–8068. [CrossRef] [PubMed]
26. Trulsson, M.; Yu, H.; Gisselsson, L.; Chao, Y.; Urbano, A.; Aits, S.; Mossberg, A.K.; Svanborg, C. HAMLET Binding to α-Actinin Facilitates Tumor Cell Detachment. *PLoS ONE* **2011**, *6*, e17179. [CrossRef] [PubMed]
27. Ho, J.C.S.; Nadeem, A.; Svanborg, C. HAMLET—A protein-lipid complex with broad tumoricidal activity. *Biochem. Biophys. Res. Commun.* **2017**, *482*, 454–458. [CrossRef]
28. Rath, A.E.; Duff, A.P.; Håkansson, A.P.; Vacher, C.S.; Li, G.J.; Knott, R.B.; Church, W.B. Structure and potential cellular targets of HAMLET-like anti-cancer compounds made from milk components. *J. Pharm. Pharm. Sci.* **2015**, *18*, 773–824. [CrossRef]
29. Hakansson, A.P.; Roche-Hakansson, H.; Mossberg, A.K.; Svanborg, C. Apoptosis-like death in bacteria induced by HAMLET, a human milk lipid-protein complex. *PLoS ONE* **2011**, *6*, e17717. [CrossRef]
30. Hassiotou, F.; Geddes, D.T. Immune cell-mediated protection of the mammary gland and the infant during breastfeeding. *Adv. Nutr.* **2015**, *6*, 267–275. [CrossRef]
31. McGuire, M.K.; McGuire, M.A. Human milk: Mother nature's prototypical probiotic food. *Adv. Nutr.* **2015**, *6*, 112–123. [CrossRef]
32. Witkowska-Zimny, M.; Kaminska-El-Hassan, E. Cells of human breast milk. *Cell Mol. Biol. Lett.* **2017**, *22*, 11. [CrossRef] [PubMed]
33. Gustafsson, L.; Leijonhufvud, I.; Aronsson, A.; Mossberg, A.K.; Svanborg, C. Treatment of skin papillomas with topical alpha-lactalbumin-oleic acid. *N. Engl. J. Med.* **2004**, *350*, 2663–2672. [CrossRef] [PubMed]
34. Heikkila, M.P.; Saris, P.E. Inhibition of Staphylococcus aureus by the commensal bacteria of human milk. *J. Appl. Microbiol.* **2003**, *95*, 471–478. [CrossRef]
35. Simpson, M.R.; Brede, G.; Johansen, J.; Johnsen, R.; Storrø, O.; Sætrom, P.; Øien, T. Human Breast MilkmiRNA, Maternal Probiotic Supplementation and Atopic Dermatitis in Offspring. *PLoS ONE* **2015**, *10*, e0143496. [CrossRef] [PubMed]

© 2019 by the authors. Licensee MDPI, Basel, Switzerland. This article is an open access article distributed under the terms and conditions of the Creative Commons Attribution (CC BY) license (http://creativecommons.org/licenses/by/4.0/).

Article

Factors Associated with Increased Alpha-Tocopherol Content in Milk in Response to Maternal Supplementation with 800 IU of Vitamin E

Amanda de Sousa Rebouças [1], Ana Gabriella Costa Lemos da Silva [2], Amanda Freitas de Oliveira [2], Lorena Thalia Pereira da Silva [2], Vanessa de Freitas Felgueiras [2], Marina Sampaio Cruz [3], Vivian Nogueira Silbiger [3], Karla Danielly da Silva Ribeiro [2,*] and Roberto Dimenstein [1]

1. Department of Biochemistry, Federal University of Rio Grande do Norte, 59078-970 Natal-RN, Brazil; amandasousar2@hotmail.com (A.d.S.R.); rdimenstein@gmail.com (R.D.)
2. Department of Nutrition, Federal University of Rio Grande do Norte, 59078-970 Natal-RN, Brazil; gabriella_lemos_06@yahoo.com.br (A.G.C.L.d.S.); amanddad_freitas@outlook.com (A.F.d.O.); lorenathaliaps@gmail.com (L.T.P.d.S.); vanessadffelgueiras@gmail.com (V.d.F.F.)
3. Department of Pharmacy, Federal University of Rio Grande do Norte, 59012-570 Natal-RN, Brazil; marinasmcruz@gmail.com (M.S.C.); viviansilbiger@hotmail.com (V.N.S.)
* Correspondence: karladaniellysr@yahoo.com.br; Tel.: +55-084-99127-7204

Received: 26 February 2019; Accepted: 9 April 2019; Published: 22 April 2019

Abstract: Background: Vitamin E supplementation might represent an efficient strategy to increase the vitamin E content in milk. The present study aimed to evaluate the impact of supplementation with 800 IU RRR-alpha-tocopherol on the alpha-tocopherol content of milk and the factors associated with the increase in vitamin E. Methods: Randomized clinical trial with 79 lactating women from Brazil, who were assigned to the control group, or to the supplemented group (800 IU of RRR-alpha-tocopherol). Milk and serum were collected between 30 and 90 days after delivery (collection 1), and on the next day (collection 2). Alpha-tocopherol was analyzed using high-performance liquid chromatography. Results: In the supplemented group, the alpha-tocopherol content in serum and milk increased after supplementation ($p < 0.001$). In the multivariate analysis, only alpha-tocopherol in milk (collection 1) was associated with the level of this vitamin in milk after supplementation ($\beta = 0.927$, $p < 0.001$), and binary logistic regression showed that the dietary intake was the only determinant for the greater effect of supplementation in milk. Conclusion: The pre-existing vitamin level in milk and diet are determinants for the efficacy of supplementation in milk, suggesting that in populations with vitamin E deficiency, high-dose supplementation can be used to restore its level in milk.

Keywords: clinical trial; lactation; infants; breastfeeding; lactating women

1. Introduction

Breast milk contains all the essential nutrients and factors for the growth and development of the infant's gastrointestinal, cerebral and immune system [1,2]. Thus, exclusive breastfeeding is recommended during the first six months of life [3]. Among the vitamins present in milk, vitamin E, is an antioxidant responsible for protecting the lipoproteins and polyunsaturated fatty acids present in the cellular membranes against peroxidation [4]. Vitamin E deficiency in children and newborns, including preterm infants (birth <37 gestational weeks), can lead to intracranial hemorrhage, chronic pulmonary diseases, hemolytic anemia, retinopathy and childhood cognitive deficits [4]. The prevalence of vitamin E deficiency (SVD) in newborns can be up to 77% [5–7] and in Brazil, a study found low vitamin levels (<500 µg/dL) in 90% of newborns [8]. The transfer of vitamin E to breast milk depends on circulating lipoproteins, and this mechanism can be influenced by maternal factors, both intrinsic and extrinsic [2,9,10]. In colostrum milk, the actions of pregnancy hormones, such as estrogen, contributes to the increase in circulating lipoproteins, ensuring a greater transfer of vitamin E into milk [2,11]. However, in mature milk the vitamin E content decreases because of changes in fat globules, and other characteristics, such as maternal age, gestational age of delivery, and the fatty acid profile might that influence the vitamin E content in milk [8,9,12,13].

Studies analyzing this micronutrient in mature milk observed that even in lactating women with vitamin E deficiency, its concentration in milk was maintained, which suggested a possible mobilization of alpha-tocopherol from the adipose tissue, which is considered the largest extrahepatic vitamin E reserve [2,8,14,15].

One strategy to increase the concentrations of vitamin E in milk is maternal supplementation [9,10]. Garcia et al. [16] found that at 24 h after supplementation, alpha-tocopherol levels in the colostrum increased. Other studies [17,18] found that vitamin E supplementation in its naturally occurring form (RRR-alpha-tocopherol) is more efficient to increase its content in milk compared with supplementation with the synthetic form or with a blend of natural and synthetic forms. In the natural form, the lateral chain has the RRR conformation, whereas the synthetic form can present isomers with 2R- (RRR-, RSR-, RSS- and RRS-) and 2S- (SRR-, SRS-, SSR-, SSS-) conformations. This structural difference results in increased bioavailability of the RRR form because of its higher affinity for the liver alpha-tocopherol transfer protein (alpha-TPP) [18,19].

Single-dose supplementation with 400 IU RRR-alpha-tocopherol in the immediate postpartum period caused an increase in the vitamin in the transitional milk (between 7 and 15 days after delivery), but not in the mature milk [20,21]. The authors suggested that a higher dose of vitamin E could influence the duration of the response. This identified the need to investigate the effect of higher doses, because the studies only used 400 IU of alpha-tocopherol, and suggested that this supplementation should be provided in the mature milk phase, which comprises a period of greater stability in milk nutritional composition.

Interestingly, different responses to supplementation have been noted, where the same treatment caused a greater increase of the vitamin in the milk in some studies [20–22] and a smaller effect in others [17], however, these studies lacked an analysis of the factors that influenced this response. These observations should be considered, because maternal milk with a low alpha-tocopherol content has been found, which could expose infants to vitamin E deficiency (VED) [20,21,23–25]. By contrast, studies of vitamin E supplementation in a single dose and in greater quantity could reveal the previously unknown mechanism of vitamin transfer to the mammary gland.

Thus, given that maternal supplementation with vitamin E is an effective measure to increase this vitamin content in milk [21,22], the mother-child binomial should be protected from the adverse effects of VED, and that there are differences in the response to this supplementation, but no understanding of which characteristics contribute to this response. The objective of the present study was to evaluate the impact of supplementation with 800 IU RRR-alpha-tocopherol on the alpha-tocopherol level in mature milk and the factors associated with the increase, with the aim of improving our understanding of the mechanism the transfer of vitamin E in the lactation period.

2. Materials and Methods

2.1. Participants and Intervention

The study was a randomized, parallel-group trial. Participants were recruited at the Pediatric Ambulatory Care of the Onofre Lopes University Hospital (HUOL), Natal-RN, Brazil, and data collection took place between October 2017 and July 2018.

The present study was approved by the Ethics Committee of the Federal University of Rio Grande do Norte (UFRN), under the protocol number 2.327.614, CAAE 76779217.1.0000.5537, and was also registered in the Brazilian Registry of Clinical Trials—ReBec, under the code RBR-38nfg2, available at http://www.ensaiosclinicos.gov.br/rg/RBR-38nfg2/.

The sample calculation was performed using GPower software, Version 3.1.9 [26] considering two independent groups tested using one way analysis of variance (ANOVA) for repeated measures among factors, with alpha parameters equal to 5%, expected power at 80%, and the effect measure value equal to 0.25 [27]. The analysis showed that each group should have at least 33 individuals, totaling 66 participants.

The eligibility criteria included women between 30 and 90 days after delivery; who were breastfeeding their children, either exclusively or partially; who were residents of Natal, RN and its metropolitan regions; who were not diagnosed with a diseases (hypertension, diabetes, neoplasms, heart disease, diseases of the gastrointestinal and hepatic tract, syphilis or were HIV-positive); who were non-smokers; no multiple births and whose infants were not malformed. Exclusion criteria were women who did not have sufficient milk or blood for analysis of vitamin levels, users of illicit drugs, and those who made daily use of vitamin supplements containing vitamin E during lactation.

The eligible participants were informed of the study's objectives and those who agreed to participate signed the consent form. At recruitment, they were allocated in one of the study groups, depending on the day of the week: Monday and Thursday for the supplemented group and Tuesday and Wednesday for the control group, where only the supplemented group ingested two capsules containing 400 IU of RRR-alpha-tocopherol consecutively, totaling 800 IU (588 mg of alpha-tocopherol). The capsule contained 98% RRR-alpha-tocopherol acetate, as assessed according to the method of Lira (2017) [21]. The study complied with the Consolidated Standards of Reporting Trials—CONSORT (Figure 1).

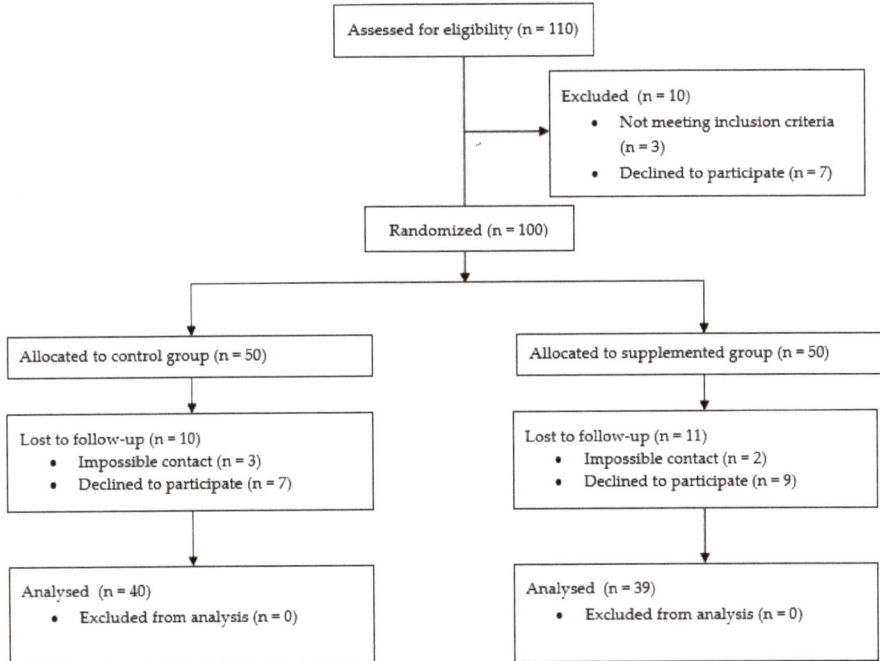

Figure 1. Consolidated Standards of Reporting Trials flow diagram (CONSORT).

2.2. Data Collection

A semi-structured questionnaire was used to collect data on socioeconomic aspects, such as family income, schooling and maternal age, as well as information on gestational age and type of delivery. Maternal height and current weight were also assessed and used to calculate the body mass index (BMI).

Milk and serum were collected from the participants at two time points. Collection 1 was performed at the hospital and collection 2 was performed at the participant's home the day after collection 1. In the supplemented group, supplementation with 800 IU of RRR-alpha-tocopherol was performed immediately after collection 1 of milk and serum.

A 2 mL sample of breast milk was collected by manual expression from a single breast that had not breastfed recently, and 5 mL of blood were collected by venipuncture. All the biological samples were collected after a 4 to 6 h fast, stored in polypropylene tubes packed in aluminum foil, and transported in refrigerated units. The breast milk was stored at −20 °C until the time of analysis. Before storage, the blood samples were centrifuged for 10 min (at 4000 rpm), to separate the serum for analysis of vitamin E and lipoproteins.

The dietary intake of vitamin E was evaluated by means of the 24 h dietary recall (24HR) applied at the two collection time points. Participants were asked about all foods, supplements and beverages consumed the day before the interview.

2.3. Determination of Alpha-Tocopherol and Lipid Profile in Biological Samples

The extraction of alpha-tocopherol from milk and serum was performed according to the method adapted by Lira et al. (2013) [28]. Ethanol (95%) was used to precipitate proteins (Vetec, Rio de Janeiro, Brazil), and hexane PA (Vetec, Rio de Janeiro, Brazil) was used as an extraction reagent. After evaporation in nitrogen, serum and milk residues were dissolved, respectively, in 250 µL of absolute ethanol (Vetec, Rio de Janeiro, Brazil) and 250 mL of dichloromethane (Vetec, Rio de Janeiro, Brazil): methanol (Sigma-Aldrich, St. Louis, Missouri, EUA) (2:1; v/v). The aliquots were then analyzed using high-performance liquid chromatography (HPLC).

HPLC consisted of an LC-20AT (Shimadzu, Kyoto, Japan) pump coupled to a CBM 20A communicator and an SPD-10A UV-VIS detector (Shimadzu, Kyoto, Japan). AC18 reversed phase column (LiChroCART 250-4, Merck, Darmstadt, Germany) was used for chromatographic separation. The mobile phase was 100% methanol in an isocratic system, with a flow rate of 1 mL/min and a wavelength of 292 nm was used to detect alpha-tocopherol. The identification and quantification of the vitamin in the samples were established by comparing the area of the peak obtained in the chromatogram with the area of the alpha-tocopherol standard (Sigma-Aldrich, São Paulo, Brazil). The concentration of the standard was confirmed by the specific extinction coefficient for alpha-tocopherol (e1%, alpha-tocopherol, 1 cm = 75.8 to 292 nm) in absolute ethanol (Vetec, Rio de Janeiro, Brazil) [29]. Women with serum alpha-tocopherol values less than 12 µmol/L were considered as deficient in vitamin E [30].

The transport of vitamin E from serum to breast milk involves lipoproteins; therefore, serum cholesterol and high-density lipoprotein (HDL) levels were analyzed with using a commercial kit (Labtest) and enzymatic colorimetric methods, by using an automatic biochemistry analyzer (Labmax plenno). Low-density lipoprotein (LDL) was quantified using the equation proposed by Martin et al. (2013) [31].

2.4. Dietary Intake of Vitamin E

Vitamin E intake was obtained using two 24 h dietary recall (24HR), applied using face-to-face interviews at the two data collection times in both groups. During this interview, the participants were asked about the food (and its preparation), supplements and beverages consumed in the last 24 h before the interview, in which the home measures described were converted to grams or milliliters [32,33] and the amount of vitamin E consumed was analyzed using the software Virtual Nutri Plus [34], from the database constructed by Rodrigues (2016) [8]. The dietary intake of vitamin E was corrected for total energy intake. The resulting values were obtained using SPSS, version 21.0 for Windows (SPSS Inc., Chicago, IL, USA) employing the residual method.

2.5. Statistical Analysis

Statistical analysis was performed using the statistical software IBM SPSS version 21.0 for Windows (SPSS Inc., Chicago, IL, USA). The Kolmogorov–Smirnov normality test was applied. Numerical data were expressed as the mean (standard deviation, SD), and categorical results were reported as absolute and relative frequencies. Student's *t*-test for dependent samples was used to verify intragroup differences, and the *t*-test for independent samples was used to analyze the differences between the groups. To evaluate the relationship between serum, breast milk and dietary vitamin E intake, the Pearson correlation coefficient was calculated. Linear multiple regression analysis was used to verify the ratio between alpha-tocopherol in milk after vitamin supplementation and in serum, the lipid profile, vitamin E intake and other maternal factors. The factors associated with the effect of supplementation on milk were also investigated. For this, the lactating women in the supplemented group were divided into quartiles according to the percentage increase in the milk alpha-tocopherol content between collection 1 and collection 2, being classified into a smaller effect (quartile 1) and greater effect (quartiles 2–4). The quartile categorization was used to identify the participants who presented lower effect and greater effects, because all participants should present higher alpha-tocopherol in milk values after supplementation. In addition to providing an analysis of the possible determinants for the milk supplementation response. The association of maternal variables with the effect of supplementation was evaluated according to binary multiple regression. All differences were considered significant when $p \leq 0.05$.

3. Results

3.1. General Characteristics of the Population

The socioeconomic characteristics of the lactating women are presented in Table 1. The mean age of the participants was 27 years, and the majority had completed high school. About 40% of the women were overweight according to their BMI values, and exclusive breastfeeding was predominant (>84%) in both groups. The dietary intake of vitamin E was equivalent to 8.7 mg/day in the control, which was below the recommended intake (16 mg/day) [30] and there was no difference between the groups in terms of dietary intake of vitamin E ($p = 0.901$).

3.2. Effect of Vitamin E Supplementation on Serum and Breast Milk

At collection 1, the maternal serum alpha-tocopherol concentrations were similar between the control and supplemented groups, at 26.37 (4.6) µmol/L and 26.38 (5.4) µmol/L, respectively ($p = 0.996$). In the control group, there was no difference in the alpha-tocopherol concentrations between collection 1 and collection 2 ($p > 0.05$). Neither group contained cases of VED (<12 µmol/L). In addition, the lipid profiles were similar between the collections and between the groups ($p > 0.05$) (Table 2).

After supplementation with 800 IU RRR-alpha-tocopherol, a 183% increase in serum alpha-tocopherol was observed in the supplemented group (collection 2), reaching 48.27 µmol/L ($p < 0.001$) (Table 2).

For the alpha-tocopherol content in mature milk, the control group presented 6.91 (1.81) µmol/L and the supplemented group presented 6.98 (2.18) µmol/L ($p = 0.883$). One day after supplementation (collection 2), milk from the supplemented group presented higher levels of alpha-tocopherol (15 µmol/L) compared with that in the control group (6.94 µmol/L) ($p < 0.001$), an increase equivalent to 124% in the post-supplementation milk.

3.3. Factors Associated with Alpha-Tocopherol in Breast Milk after Supplementation

In the supplemented group, after Pearson correlation analysis, milk from collection 1, dietary intake of vitamin E, and alpha-tocopherol in serum from collection 2 were identified as positively related to alpha-tocopherol levels in milk from collection 2 (Figure 2). These variables were included in the multiple linear regression analysis to evaluate the factors associated with the alpha-tocopherol concentration in breast milk after supplementation. Only alpha-tocopherol in the milk before supplementation was a determinant for the increase in the vitamin content in the milk after administration of 800 IU alpha-tocopherol ($\beta = 0.927$, $p < 0.001$, 95% CI 1.925–2.396). Thus, the higher the vitamin concentration in milk, the greater the transfer of the vitamin to the mammary gland.

When dividing the participants of the supplemented group according to the effect of supplementation (quartile 1 and quartiles 2–4), where quartile 1 is equivalent to 83% of the vitamin E increase percentage in milk after supplementation, we observed that the dietary intake of vitamin E was a determinant that caused a greater response to supplementation ($p = 0.020$, 95% CI 0.209–0.877), which suggested that the higher the intake, the greater the effect of supplementation (Table 3). The characteristics of the participants divided by the effect of supplementation are described in Table 4, which showed that the consumption of calories, alpha-tocopherol and total fat was higher in the group showing a higher effect of supplementation ($p = 0.001$, $p = 0.013$, $p = 0.033$, respectively).

Table 1. Characterization of the 79 lactating women randomized into the control and supplemented groups of the study. Natal, Rio Grande do Norte, Brazil, 2017–2018.

Characteristics	Control Group $n = 40$	Supplemented Group $n = 39$	p-Value
Maternal age (years), mean (SD)	27 (6.8)	27 (6.8)	0.833 *
Postpartum age (days), mean (SD)	57 (25.8)	56 (23.7)	0.833 *
Education level n, (%)			
Incomplete primary education	4 (10.0)	5 (12.8)	
Complete primary education	3 (7.5)	2 (5.1)	
Incomplete secondary education	14 (35.0)	6 (15.4)	0.149
Complete secondary education	16 (40.0)	22 (56.4)	
Complete higher education	3 (7.5)	4 (10.3)	
Family income level n, (%) [a]			
<1 Minimum wage	16 (40.0)	23 (59.0)	0.092
>1 Minimum wage	24 (60.0)	16 (41.0)	
Type of delivery n, (%)			
Vaginal	15 (37.5)	13 (33.3)	0.699
Caesarian	25 (62.5)	26 (66.7)	
Parity status n, (%)			
Primiparous	21 (52.5)	17 (43.6)	0.405
Multtiparou	19 (47.5)	22 (56.4)	
BMI classification (kg/m^2), (%) [b]			
Low weight	1 (2.5)	0 (0)	
Normal	15 (37.5)	18 (46.2)	0.735
Overweight	16 (40.0)	13 (33.3)	
Obese	8 (20.0)	8 (20.5)	
Type of maternal breastfeeding n, (%)			
Exclusive maternal breastfeeding	35 (87.5)	33 (84.6)	0.711
Maternal breast milk and other milks	5 (12.5)	6 (15.4)	
Calorie intake (Kcal/day), mean (SD)	3248.4 (711.2)	3270.7 (868.4)	0.970 *
Intake of alpha-tocopherol (mg/day), mean (SD)	8.7 (3.4)	8.8 (3.5)	0.901 *
Intake of total fat (g/dia), mean (SD)	69.2 (23.6)	69.9 (25.8)	0.905 *

n: number; BMI: Body Mass Index; SD: Standard deviation; Chi-square test. * Student's t-test for independent samples used for the variables maternal age, postpartum age, calorie consumption, alpha-tocopherol and total fat; [a] Brazilian minimum wage per month = US$ 291.5; [b] WHO classification, 2000. p-value = level of significance ($p < 0.05$ = statistically significant).

Table 2. Maternal biochemical indicators of the control and supplemented groups in collections 1 and 2, performed in the study. Natal, Rio Grande do Norte, Brazil, 2017–2018.

Biochemical Indicators Evaluated	Control Group				Supplemented Group				Differences between Control Group and Supplemented Group p-Value **	
	Collection 1	Collection 2	Change	p-Value *	Collection 1	Collection 2	Change	p-Value *	Collection 1 CG × SG	Collection 2 CG × SG
Serum alpha-tocopherol (µmol/L)	26.37 (4.6)	26.34 (4.92)	0.03 (2.5)	0.876	26.38 (5.4)	48.27 (10.5)	21.89 (7.4)	0.001	0.996	<0.001
Alpha-tocopherol in breast milk (µmol/L)	6.91 (1.8)	6.94 (2.0)	0.03 (1.2)	0.935	6.98 (2.2)	15.00 (5.1)	8.02 (4.2)	<0.001	0.883	<0.001
Serum cholesterol (mg/dL)	177 (41.0)	178 (42.0)	1.70 (33.5)	0.750	179 (44.0)	173 (36.0)	6.28 (16.0)	0.190	0.834	0.498
Serum triglycerides (mg/dL)	143 (99.0)	130 (86.0)	13.38 (57.8)	0.151	129 (66.0)	109 (51.0)	19.23 (32.9)	0.08	0.439	0.195
HDL (mg/dL)	40 (14.0)	41 (15.0)	0.25 (12.9)	0.903	42 (11.0)	42 (10.0)	0.31 (8.0)	0.811	0.574	0.724
LDL (mg/dL)	111 (35.0)	114 (40.0)	3.28 (34.4)	0.535	113 (43.0)	110 (35.0)	3.74 (14.7)	0.129	0.762	0.603

HDL: High-density lipoprotein; LDL: Low-density lipoprotein; ( ) Standard deviation; * Student's t-test for dependent samples; ** Student's t-test for independent samples. CG: Control group; SG: Supplemented group.

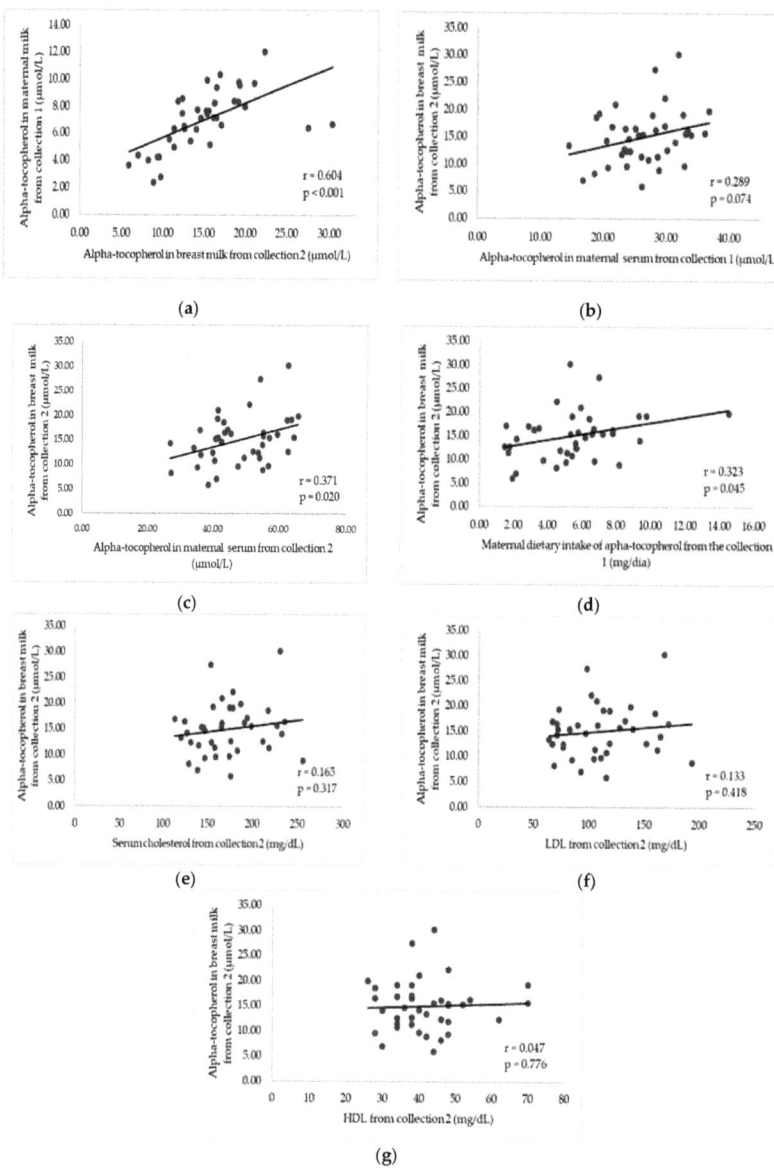

Figure 2. Correlations between alpha-tocopherol in breast milk obtained in collection 2 and maternal variables of the supplemented group. Natal, Rio Grande do Norte, Brazil, 2017–2018. (**a**) Correlation between alpha-tocopherol in breast milk from collection 1 and 2; (**b**) correlation between alpha-tocopherol in breast milk from collection 2 and serum from collection 1; (**c**) correlation between alpha-tocopherol in breast milk in serum from collection 2; (**d**) correlation between alpha-tocopherol in breast milk of collection 2 and the vitamin E intake of collection 1; (**e**) correlation between alpha-tocopherol in breast milk from collection 2 and serum cholesterol from collection 2; (**f**) correlation between alpha-tocopherol in breast milk from collection 2 and low-density lipoprotein (LDL) from collection 2; (**g**) correlation between alpha-tocopherol in breast milk from collection 2 and high-density lipoprotein (HDL) from collection 2. r = Pearson's correlation coefficient; p value = level of significance ($p < 0.05$ = statistically significant).

Table 3. Binary logistic regression model for variables associated with greater effect of 800 IU alpha-tocopherol supplementation in breast milk in the supplemented group.

Variables	Greater Effect of Supplementation (Quartiles 2–4) *	
	95% CI	p-Value
Alpha-tocopherol in milk collection 1 (μmol/L)	0.998–1.024	0.104
Alpha-tocopherol in serum collection 1 (μmol/L)	0.991–1.005	0.565
Alpha-tocopherol in serum collection 2 (μmol/L)	0.995–1.002	0.387
Dietary intake of vitamin E collection 1 (mg/day)	0.209–0.877	0.020 **
Serum cholesterol collection 1 (mg/dL)	0.937–1.163	0.431
LDL collection 1 (mg/dL)	0.846–1.051	0.289

* Above 83% of the vitamin E increase percentage in milk after supplementation. LDL: Low-density lipoprotein. p-value = level of significance. ** Significant difference.

Table 4. Characterization of the 39 lactating women randomized into the supplemented group, divided by the effect of milk supplementation (lower effect: quartile 1, equivalent to 83% of the vitamin increase percentage in the milk, and greater effect: quartiles 2–4). Natal, Rio Grande do Norte, Brazil, 2017–2018.

Characteristics	Quartile 1 n = 9	Quartiles 2–4 n = 30	p-Value
Maternal age (years), mean (SD)	27 (6.5)	27 (6.9)	0.922 *
Postpartum age (days), mean (SD)	58 (30.0)	55 (22.1)	0.749 *
Education level n, (%)			
Incomplete primary education	2 (22.2)	3 (10.0)	
Complete primary education	1 (11.1)	1 (3.3)	
Incomplete secondary education	1 (11.1)	5 (16.7)	0.343
Complete secondary education	3 (33.3)	19 (63.3)	
Complete higher education	2 (22.2)	2 (6.7)	
Family income level n, (%) [a]			
<1 Minimum wage	4 (44.4)	19 (63.3)	0.312
>1 Minimum wage	5 (55.6)	11 (36.7)	
Type of delivery n, (%)			
Vaginal	2 (22.2)	11 (36.7)	0.420
Caesarian	7 (77.8)	19 (63.3)	
Parity status n, (%)			
Primiparous	6 (66.7)	11 (36.7)	0.111
Multtiparous	3 (33.3)	19 (63.3)	
BMI classification (kg/m^2), (%) [b]			
Low weight	0 (0)	0 (0)	
Normal	3 (33.3)	14 (46.7)	0.754
Overweight	4 (44.5)	10 (33.3)	
Obese	2 (22.2)	6 (20.0)	
Type of maternal breastfeeding n, (%)			
Exclusive maternal breastfeeding	8 (88.9)	25 (83.3)	0.685
Maternal breast milk and other milks	1 (11.1)	5 (16.7)	
Calory intake (Kcal/day), mean (SD)	2624.6 (453.7)	3464.5 (873.4)	0.001 *
Intake of alpha-tocopherol (mg/day), mean (SD)	6.8 (2.1)	9.3 (3.6)	0.013 *
Intake of total fat (g/dia), mean (SD)	58.5 (12.8)	73.3 (27.8)	0.033 *

n: number. BMI: Body Mass Index; SD: Standard deviation; Chi-square test. * t-test for independent samples used for the variables maternal age, postpartum age, calorie consumption, alpha-tocopherol and total fat; [a] Brazilian minimum wage per month = US$ 291.5; [b] WHO classification, 2000. p-value = level of significance ($p < 0.05$ = statistically significant).

4. Discussion

Mature milk is the most stable stage of lactation, in which the content of alpha-tocopherol is not influenced by pregnancy-related factors, as occurs in the colostrum [2,7]. It should be emphasized that the mature milk presents a higher concentration of lipids and in contrast, there is a lower secretion of alpha-tocopherol, suggesting that there are distinct mechanisms involved in the transfer of this vitamin into breast milk [2,35,36].

In the present study, the lactating women had adequate vitamin E status, in accordance with other studies considering the same stage of lactation [8,21]. However, a low dietary intake of vitamin E was noted (Table 1), which could trigger the mobilization of alpha-tocopherol from maternal reserves, such as the adipose tissue, into breast milk [15,19]. This low consumption of vitamin E was also reported in other populations in Brazil, Greece and Poland [21,25,36], which suggests a frequent inadequacy in vitamin E consumption during lactation.

Even in situations of inadequate consumption, the vitamin E concentration in milk was not influenced by diet and circulating maternal levels [8,20,21,37–40]. However, this present study was the first to identify a positive association between alpha-tocopherol levels in milk and ingested vitamin E (diet + supplementation) and with serum alpha-tocopherol (Figure 2d,c). This suggested that in high-consumption situations, the ingested and circulating maternal levels are the main factors responsible for the vitamin level in milk. It is likely that in situations of low vitamin E consumption (as found in the studies cited), milk vitamin E might originate from other sources, such as the body's reserve [4], explaining the absence of a relationship between those variables.

To prevent of VED in infants, it is necessary for breast milk to contain adequate levels of vitamin E, so that children can obtain the benefits of the micronutrient, through the creation of vitamin reserves in the body and its antioxidant action [4,19]. Some studies that evaluated this vitamin in mature milk observed values below the nutritional requirements of infants [20,23,36,37], which suggested that maternal vitamin E supplementation could be an important strategy to increase milk vitamin contents [17,20,21].

In this study, supplementation with 800 IU of alpha-tocopherol caused a 183% increase in serum alpha-tocopherol, and a 124% increase in breast milk alpha-tocopherol (Table 2). Other clinical trials using a lower dose (400 IU alpha-tocopherol) found an increase of 60% to 80% in the vitamin content in milk after supplementation [18,20–22], showing a reduced effect compared with that shown in the present study. These findings suggested that the response to supplementation might be influenced by both the dosage used and the determinant factors. However, trials have not evaluated the factors associated with the different responses to large vitamin E doses [17,20], being important to understand how this vitamin is transferred into the mammary gland.

When analyzing the factors associated with a better response in milk after supplementation, it is important to highlight that only the content of this vitamin in the basal milk (before supplementation) and the dietary intake were demonstrated to increase the levels of this micronutrient in milk (Table 3), which suggested that the higher the consumption of vitamin E and its levels in milk, the greater the transfer of alpha-tocopherol from the supplement to the mammary gland.

Assessment of the profile of lactating women in the supplemented group, showed that the participants with the highest supplementation effect (quartiles 2–4) had a higher intake of calories, alpha-tocopherol and total fat (Table 4). These findings suggested that the amount of fat available in the diet may improve the bioavailability of the vitamin in the body, such as its absorption and distribution to tissues, and in this case, to the mammary gland, as reported in [41–43].

Such evidence also demonstrated that in mature milk, the Michaelis–Menten kinetic theory could not be applied, as it proposes that the transfer of vitamin to milk occurs through active transport, characterized by a saturation of the lipoprotein receptors in the breast tissue in situations of large contents of vitamin E, which would prevent the continuous transfer of the vitamin into the breast milk after supplementation [38,44]. Notably, in a study of dairy cows, Weiss and Wyatt (2003) [45] suggested that the ability of lipoproteins to carry alpha-tocopherol could determine the uptake of this vitamin by the breast tissue, and that the limiting factor for this mechanism would be the maximum content of vitamin E in the lipoproteins.

To further investigate this relationship between lipoproteins and vitamin E transport, we determined the circulating lipoproteins and the serum cholesterol and triglycerides profiles; however, no relation between them and the response to supplementation was found (Table 2). In fact, the mechanism of transport of vitamin E to the mammary gland is poorly understood [2,46,47]. Circulating lipoproteins are responsible for this transfer, with LDL being the main carrier [44,48]; however, transport may occur in the presence or absence of its receptors in the mammary gland [2,44]. Other receptors are found in breast tissue, such as Scavenger Receptor B-1 (SR-B1), which has binding sites for both LDL and HDL, and CD36, which has high affinity binding sites for HDL, LDL, and very low-density lipoprotein (VLDL) [2,48]. It has also been suggested the participation of lipoprotein lipase (LPL), which may show increased activity during lactation, contributes to the greater circulation of alpha-tocopherol and its uptake [49].

These findings provide important information to understand the mechanisms by which vitamin E is transferred into the mammary gland, demonstrating that, in situations of supplementation with 800 IU of vitamin E and its effect in mature milk, the better the vitamin E status (considering the milk and dietary intake), the more effective uptake into the mammary gland will be, regardless of receptor saturation. Further investigation into how this transport occurs is required, by means of in vitro and in vivo studies and using labeled isotopes of alpha-tocopherol, for example, to investigate its biotransformation. Notably, in populations with dietary inadequacy and low contents of vitamin E in milk, supplementation with higher doses of the vitamin, such as 800 IU alpha-tocopherol, might be required to obtain a more effective intervention. The analysis of a single dose during the day allowed us to investigate possible factors that could interfere with the response to supplementation; however, it is necessary to analyze how long the effect of this supplementation could be sustained, its contribution to maternal and infant nutritional status, and the use of smaller daily doses.

Therefore, supplementation associated with an adequate intake of vitamin E is an effective strategy to increase vitamin E levels in breast milk and prevent cases of vitamin E deficiency in infants, especially premature infants [2,4,11].

5. Conclusions

Vitamin E supplementation increased vitamin levels in milk and in maternal serum, and a positive relationship was found between alpha-tocopherol levels in milk, serum and dietary intake of vitamin E. Factors associated with the increase in alpha-tocopherol contents in milk after maternal supplementation with 800 IU of vitamin E were the basal levels of alpha-tocopherol in milk and the dietary intake of vitamin E.

Author Contributions: A.d.S.R. contributed to article writing, review and editing. A.G.C.L.d.S., A.F.d.O., L.T.P.d.S., V.d.F.F. contributed to data collection and analysis of the samples. M.S.C., V.N.S. contributed to the lipid profile analysis. K.D.d.S.R. contributed to selection of the journal and participated in the orientation, design, analysis and review of the study. R.D. contributed to the orientation and design of the study.

Funding: This study was financed in part by the Coordenação de Aperfeiçoamento de Pessoal de Nível Superior—Brasil (CAPES)—Finance Code 001.

Acknowledgments: We thank the study participants and the nursing team of the Pediatric Ambulatory Care of the Onofre Lopes University Hospital (HUOL), Natal-RN, Brazil.

Conflicts of Interest: There was no conflict of interest reported by the authors.

References

1. Victora, C.G.; Bahl, R.; Barros, A.J.D.; França, G.V.A.; Horton, S.; Krasevec, J.; Murch, S.; Sankar, M.J.; Walker, N.; Rollins, N.C. Breastfeeding in the 21st century: Epidemiology, mechanisms, and lifelong effect. *Lancet* **2016**, *387*, 475–490. [CrossRef]
2. Debier, C. Vitamin E during pre- and postnatal periods. In *Vitamins & Hormones*; Elsevier: Amsterdam, The Netherlands, 2007; Volume 76, pp. 357–373. ISBN 9780123735928.
3. Brasil. *Saúde da Criança: Aleitamento Materno e Alimentação Complementar*, 2rd ed.; Ministério da Saúde: Brasília, Brazil, 2015; ISBN 9788533422902.
4. Traber, M.G. Vitamin E. In *Present Knowledge in Nutrition*, 10th ed.; Erdman, J.W., Jr., Macdonald, I.A., Zeisel, S.H., Eds.; ILSI Press: Washington, DC, USA, 2012; pp. 214–229. ISBN 978-0-470-95917-6.
5. Schulpis, K.H.; Michalakakou, K.; Gavrili, S.; Karikas, G.A.; Lazaropoulou, C.; Vlachos, G.; Bakoula, C.; Papassotiriou, I. Maternal-neonatal retinol and alpha-tocopherol serum concentrations in Greeks and Albanians. *Acta Paediatr.* **2004**, *93*, 1075–1080. [CrossRef] [PubMed]
6. Fares, S.; Feki, M.; Khouaja-Mokrani, C.; Sethom, M.M.; Jebnoun, S.; Kaabachi, N. Nutritional practice effectiveness to achieve adequate plasma vitamin A, E and D during the early postnatal life in Tunisian very low birth weight infants. *J. Matern.-Fetal Neonatal Med.* **2015**, *28*, 1324–1328. [CrossRef] [PubMed]
7. Kositamongkol, S.; Suthutvoravut, U.; Chongviriyaphan, N.; Feungpean, B.; Nuntnarumit, P. Vitamin A and E status in very low birth weight infants. *J. Perinatol.* **2011**, *31*, 471–476. [CrossRef]
8. Rodrigues, K.D.S.R. Estado Nutricional em Vitamina E de Mães e Crianças Pré-Termo e Termo do Nascimento aos 3 Meses Pós-Parto. Ph.D. Thesis, Departamento de Bioquímica, Universidade Federal do Rio Grande do Norte, Natal, Brazil, June 2016; 148p.
9. Lima, M.S.R.; Dimenstein, R.; Ribeiro, K.D.S. Vitamin E concentration in human milk and associated factors: A literature review. *J. Pediatria* **2014**, *90*, 440–448. [CrossRef]
10. Hampel, D.; Shahab-Ferdows, S.; Islam, M.M.; Peerson, J.M.; Allen, L.H. Vitamin Concentrations in Human Milk Vary with Time within Feed, Circadian Rhythm, and Single-Dose Supplementation. *J. Nutr.* **2017**, *147*, 603–611. [CrossRef] [PubMed]
11. Debier, C.; Pottier, J.; Goffe, C.; Larondelle, Y. Present knowledge and unexpected behaviours of vitamins A and E in colostrum and milk. *Livest. Prod. Sci.* **2005**, *98*, 135–147. [CrossRef]
12. Tijerina-Sáenz, A.; Innis, S.; Kitts, D. Antioxidant capacity of human milk and its association with vitamins A and E and fatty acid composition. *Acta Paediatr.* **2009**, *98*, 1793–1798. [CrossRef] [PubMed]
13. Stuetz, W.; Carrara, V.; Mc Gready, R.; Lee, S.; Sriprawat, K.; Po, B.; Hanboonkunupakarn, B.; Grune, T.; Biesalski, H.; Nosten, F. Impact of Food Rations and Supplements on Micronutrient Status by Trimester of Pregnancy: Cross-Sectional Studies in the Maela Refugee Camp in Thailand. *Nutrients* **2016**, *8*, 66. [CrossRef]
14. Szlagatys-Sidorkiewicz, A.; Zagierski, M.; Jankowska, A.; uczak, G.; Macur, K.; Bączek, T.; Korzon, M.; Krzykowski, G.; Martysiak-Żurowska, D.; Kamińska, B. Longitudinal study of vitamins A, E and lipid oxidative damage in human milk throughout lactation. *Early Hum. Dev.* **2012**, *88*, 421–424. [CrossRef]
15. Olafsdottir, A.S.; Wagner, K.-H.; Thorsdottir, I.; Elmadfa, I. Fat-Soluble Vitamins in the Maternal Diet, Influence of Cod Liver Oil Supplementation and Impact of the Maternal Diet on Human Milk Composition. *Ann. Nutr. Metab.* **2001**, *45*, 265–272. [CrossRef] [PubMed]
16. Garcia, L.R.S.; Ribeiro, K.D.d.S.; de Araújo, K.F.; Azevedo, G.M.M.; Pires, J.F.; Batista, S.D.; Dimenstein, R. Níveis de alfa-tocoferol no soro e leite materno de puérperas atendidas em maternidade pública de Natal, Rio Grande do Norte. *Revista Brasileira de Saúde Materno Infantil* **2009**, *9*, 423–428. [CrossRef]
17. Clemente, H.A.; Ramalho, H.M.M.; Lima, M.S.R.; Grilo, E.C.; Dimenstein, R. Maternal Supplementation with Natural or Synthetic Vitamin E and Its Levels in Human Colostrum. *J. Pediatr. Gastroenterol. Nutr.* **2015**, *60*, 533–537. [CrossRef] [PubMed]
18. Gaur, S.; Kuchan, M.J.; Lai, C.-S.; Jensen, S.K.; Sherry, C.L. Supplementation with *RRR*- or *all-rac* -α-Tocopherol Differentially Affects the α-Tocopherol Stereoisomer Profile in the Milk and Plasma of Lactating Women. *J. Nutr.* **2017**, *147*, 1301–1307. [CrossRef] [PubMed]
19. Traber, M.G. Vitamin E Inadequacy in Humans: Causes and Consequences. *Adv. Nutr.* **2014**, *5*, 503–514. [CrossRef]

20. Medeiros, J.F.P.; da Silva Ribeiro, K.D.; Lima, M.S.R.; das Neves, R.A.M.; Lima, A.C.P.; Dantas, R.C.S.; da Silva, A.B.; Dimenstein, R. α-Tocopherol in breast milk of women with preterm delivery after a single postpartum oral dose of vitamin E. *Br. J. Nutr.* **2016**, *115*, 1424–1430. [CrossRef] [PubMed]
21. Lira, L.Q. Efeito de Dois Protocolos de Suplementação Materna com Alfa-Tocoferol Sobre o Soro e o Leite de Lactantes até 60 Dias Pós-Parto. Ph.D. Thesis, Departamento de Bioquímica, Universidade Federal do Rio Grande do Norte, Natal, Brazil, December 2017; 127p.
22. de Melo, L.R.M.; Clemente, H.A.; Bezerra, D.F.; Dantas, R.C.S.; Ramalho, H.M.M.; Dimenstein, R. Effect of maternal supplementation with vitamin E on the concentration of α-tocopherol in colostrum. *J. Pediatria* **2017**, *93*, 40–46. [CrossRef]
23. Cortês da Silva, A.L.; da Silva Ribeiro, K.D.; Miranda de Melo, L.R.; Fernandes Bezerra, D.; Carvalho de Queiroz, J.L.; Santa Rosa Lima, M.; Franco Pires, J.; Soares Bezerra, D.; Osório, M.M.; Dimenstein, R. Vitamina e no leite humano e sua relação com o requerimento nutricional do recém-nascido a termo. *Revista Paulista de Pediatria* **2017**, *35*, 158–164. [CrossRef]
24. Ma, D.; Ning, Y.; Gao, H.; Li, W.; Wang, J.; Zheng, Y.; Zhang, Y.; Wang, P. Nutritional Status of Breast-Fed and Non-Exclusively Breast-Fed Infants from Birth to Age 5 Months in 8 Chinese Cities. *Asia Pac. J. Clin. Nutr.* **2014**, *23*, 282–292. [CrossRef]
25. Antonakou, A.; Chiou, A.; Andrikopoulos, N.K.; Bakoula, C.; Matalas, A.-L. Breast milk tocopherol content during the first six months in exclusively breastfeeding Greek women. *Eur. J. Nutr.* **2011**, *50*, 195–202. [CrossRef]
26. GPower Software. Available online: http://www.gpower.hhu.de (accessed on 19 June 2017).
27. Faul, F.; Erdfelder, E.; Lang, A.-G.; Buchner, A. G*Power 3: A flexible statistical power analysis program for the social, behavioral, and biomedical sciences. *Behav. Res. Methods* **2007**, *39*, 175–191. [CrossRef] [PubMed]
28. de Lira, L.Q.; Lima, M.S.R.; de Medeiros, J.M.S.; da Silva, I.F.; Dimenstein, R. Correlation of vitamin A nutritional status on alpha-tocopherol in the colostrum of lactating women: Relationship of serum retinol and alpha-tocopherol in colostrum. *Matern. Child Nutr.* **2013**, *9*, 31–40. [CrossRef]
29. Nierenberg, D.W.; Nann, S.L. A method for determining concentrations of retinol, tocopherol, and five carotenoids in human plasma and tissue samples. *Am. J. Clin. Nutr.* **1992**, *56*, 417–426. [CrossRef]
30. *Dietary Reference Intakes for Vitamin C, Vitamin E, Selenium, and Carotenoids*; National Academies Press: Washington, DC, USA, 2000; ISBN 9780309069359.
31. Martin, S.S.; Blaha, M.J.; Elshazly, M.B. Comparison of a novel method vs. the Friedewald equation for estimating low-density lipoprotein cholesterol levels from the standard lipid profile. *JAMA* **2013**, *310*, 2061–2068. [CrossRef]
32. Araújo, M.O.D.; Guerra, T.M. *Alimentos per Capita*, 3rd ed.; Editora Universitária—UFRN: João Pessoa, Brazil, 2007; p. 323. ISBN 9788572733.
33. Tomita, L.Y.; Cardoso, M.A. *Relação de Medidas Caseiras, Composição Química e Receitas de Alimentos Nipo-Brasileiros*; Metha: São Paulo, Brazil, 2002; p. 85. ISBN 9788588888012.
34. Virtual Nutri Plus. Available online: http://www.virtualnutriplus.com.br/ (accessed on 10 January 2018).
35. Schweigert, F.J.; Bathe, K.; Chen, F.; Boscher, U.; Dudenhausen, J.W. Effect of the stage of lactation in humans on carotenoid levels in milk, blood plasma and plasma lipoprotein fractions. *Eur. J. Nutr.* **2004**, *43*, 39–44. [CrossRef]
36. Didenco, S.; Gillingham, M.B.; Go, M.D.; Leonard, S.W.; Traber, M.G.; McEvoy, C.T. Increased vitamin E intake is associated with higher α-tocopherol concentration in the maternal circulation but higher α-carboxyethyl hydroxychroman concentration in the fetal circulation. *Am. J. Clin. Nutr.* **2011**, *93*, 368–373. [CrossRef] [PubMed]
37. Xue, Y.; Campos-Gimenez, E.; Redeuil, K.M.; Leveques, A.; Actis-Goretta, L.; Vinyes-Pares, G.; Zhang, Y.; Wang, P.; Thakkar, S.K. Concentrations of Carotenoids and Tocopherols in Breast Milk from Urban Chinese Mothers and Their Associations with Maternal Characteristics: A Cross-Sectional Study. *Nutrients* **2017**, *9*, 1229. [CrossRef] [PubMed]
38. Dimenstein, R.; Medeiros, A.C.P.; Cunha, L.R.F.; Araújo, K.F.; Dantas, J.C.O.; Macedo, T.M.S.; Stamford, T.L.M. Vitamin E in human serum and colostrum under fasting and postprandial conditions. *J. Pediatria* **2010**, *86*, 345–348. [CrossRef] [PubMed]
39. Jiang, J.; Xiao, H.; Wu, K.; Yu, Z.; Ren, Y.; Zhao, Y.; Li, K.; Li, J.; Li, D. Retinol and α-tocopherol in human milk and their relationship with dietary intake during lactation. *Food Funct.* **2016**, *7*, 1985–1991. [CrossRef]

40. Martysiak-Żurowska, D.; Szlagatys-Sidorkiewicz, A.; Zagierski, M. Concentrations of alpha- and gamma-tocopherols in human breast milk during the first months of lactation and in infant formulas: Tocopherols in human milk and infant formulas. *Matern. Child Nutr.* **2013**, *9*, 473–482. [CrossRef] [PubMed]
41. Leonard, S.W.; Good, C.K.; Gugger, E.; Traber, M.G. Vitamin E bioavailability from fortified breakfast cereal is greater than that from encapsulated supplements. *Am. J. Clin. Nutr.* **2004**, *79*, 86–92. [CrossRef] [PubMed]
42. Bruno, R.S.; Leonard, S.W.; Park, S.I.; Zhao, Y.; Traber, M.G. Human vitamin E requirements assessed with the use of apples fortified with deuteriumlabeled alpha-tocopheryl acetate. *Am. J. Clin. Nutr.* **2006**, *83*, 299–304. [CrossRef] [PubMed]
43. Traber, M.G.; Leonard, S.W.; Bobe, G.; Fu, X.; Saltzman, E.; Grusak, M.A.; Estande, S.L. α-Tocopherol Disappearance Rates from Plasma Depend on Lipid Concentrations: Studies Using Deuterium-Labeled Collard Greens in Younger and Older Adults. *Am. J. Clin. Nutr.* **2015**, *101*, 752–759. [CrossRef] [PubMed]
44. Jensen, S.K.; Johannsen, A.K.B.; Hermansen, J.E. Quantitative secretion and maximal secretion capacity of retinol, β-carotene and α-tocopherol into cows' milk. *J. Dairy Res.* **1999**, *66*, 511–522. [CrossRef] [PubMed]
45. Weiss, W.P.; Wyatt, D.J. Effect of Dietary Fat and Vitamin E on α-Tocopherol in Milk from Dairy Cows. *J. Dairy Sci.* **2003**, *86*, 3582–3591. [CrossRef]
46. Lauridsen, C.; Engel, H.; Jensen, S.K.; Craig, A.M.; Traber, M.G. Lactating Sows and Suckling Piglets Preferentially Incorporate RRR- over All-rac-alpha-Tocopherol into Milk Plasma and Tissues. *J. Nutr.* **2002**, *132*, 1258–1264. [CrossRef]
47. Wang, Y.; Tong, J.; Li, S.; Zhang, R.; Chen, L.; Wang, Y.; Zheng, M.; Wang, M.; Liu, G.; Dai, Y.; et al. Over-Expression of Human Lipoprotein Lipase in Mouse Mammary Glands Leads to Reduction of Milk Triglyceride and Delayed Growth of Suckling Pups. *PLoS ONE* **2011**, *6*, 208–295. [CrossRef]
48. Monks, J.; Huey, P.U.; Hanson, L.; Eckel, R.H.; Neville, M.C.; Gavigan, S.A. lipoprotein-containing particle is transferred from the serum across the mammary epithelium into the milk of lactating mice. *J. Lipid Res.* **2001**, *42*, 686–696.
49. Mardones, P.; Rigotti, A. Cellular mechanisms of vitamin e uptake: Relevance in α-tocopherol metabolism and potential implications for disease. *J. Nutr. Biochem.* **2004**, *15*, 252–260. [CrossRef]

© 2019 by the authors. Licensee MDPI, Basel, Switzerland. This article is an open access article distributed under the terms and conditions of the Creative Commons Attribution (CC BY) license (http://creativecommons.org/licenses/by/4.0/).

Article

Online Video Instruction on Hand Expression of Colostrum in Pregnancy is an Effective Educational Tool

Therese A. O'Sullivan [1,*], Joy Cooke [2], Chris McCafferty [3] and Roslyn Giglia [4]

[1] School of Medical and Health Sciences, Edith Cowan University, 270 Joondalup Drive, Joondalup, WA 6027, Australia
[2] Maternity Unit, Glengarry Private Hospital, Duncraig, WA 6023, Australia; joy.cooke@westnet.com.au
[3] School of Nursing and Midwifery, Edith Cowan University, Joondalup, WA 6027, Australia; c.mccafferty@ecu.edu.au
[4] Telethon Kids Institute, University of Western Australia, Subiaco, WA 6008, Australia; roslyn.giglia@telethonkids.org.au
* Correspondence: t.osullivan@ecu.edu.au; Tel.: +61-8-6304-3529

Received: 22 March 2019; Accepted: 14 April 2019; Published: 19 April 2019

Abstract: The use of antenatal colostrum expression in the weeks prior to birth may help improve long-term breastfeeding, but few large-scale studies exist. Typically, antenatal colostrum expression instruction relies on face-to-face education, making large interventions costly. We aimed to determine whether an expert online instructional video can improve knowledge and confidence around antenatal colostrum expressing. Pregnant women were asked to complete a questionnaire pre- and post-watching the instructional video online. Ninety five pregnant women completed both pre- and post-questionnaires. Total antenatal colostrum expression knowledge scores improved after watching the video, from a mean of 3.05 ± 1.70 correct out of a maximum of 7, to 6.32 ± 0.76 ($p < 0.001$). Self-reported confidence around hand expressing in pregnancy also improved from an average ranking of not confident (2.56 ± 1.17, out of a possible 5) to confident (4.32 ± 0.80, $p < 0.001$). Almost all women (98%) reported that they would recommend the video to a friend or family member if antenatal colostrum expression was suggested by their healthcare provider. Findings suggest that the use of an online expert video is an acceptable and effective way to educate pregnant women in antenatal colostrum expression.

Keywords: antenatal; expressing; pregnancy; video instruction; colostrum; breastfeeding; online

1. Introduction

During pregnancy, breasts begin to produce the first milk, colostrum. Historically, antenatal colostrum expression (ACE) was performed as a means of preparing the breasts for breastfeeding after birth. More recently, ACE is performed to collect a supply of colostrum. Colostrum can be collected using a syringe, and safely stored in a freezer. The colostrum can then be defrosted and given to the baby. The stored colostrum can be used to treat hypoglycaemia in the infant after birth, or if breastfeeding issues occur. The collection and storage of colostrum can ensure that an infant receives its own mother's milk as the first feed after birth, instead of reliance on formula during this time. In women intending to exclusively breastfeed, formula supplementation while in hospital has been associated with a two- to three-fold risk of early breastfeeding cessation [1]. Some have proposed that ACE can also promote the onset of Lactogenesis II, the "coming in" of breastmilk, however there is no evidence to support this effect, and the efficacy of the ACE practice is yet to be established in this regard [2].

In addition, there has been some concern that antenatal nipple stimulation while teaching hand expression will act as a precursor for the release of oxytocin, and may induce premature labour [3]. In a Cochrane review of the literature [4], breast stimulation was shown to reduce the number of women in labour after 72 h (i.e., had given birth), along with reduced postpartum haemorrhage rates. However, the authors went on to recommend not promoting this practice in high-risk women and noted that more information on maternal satisfaction was required. More recently, evidence suggests a role for nipple stimulation at term as a non-pharmacological method of labour induction [5,6]. In 2017, results of the Diabetes and Antenatal Milk Expressing (DAME) trial of 635 pregnant women with low-risk diabetes demonstrated that antenatal expressing did not result in a lower gestational age or increased admissions to the Neonatal Intensive Care Unit [7]. The authors concluded that there is no harm in advising women with diabetes in pregnancy at low risk of complications to express colostrum for the last few weeks of gestation. Other results from the DAME trial show a non-significant trend toward improvements in breastfeeding outcomes for women in the antenatal expressing intervention group, although the study was not powered for breastfeeding outcomes, and indicators for collecting infant feeding methods differed from standardised national indicators [8].

The question remains as to the effectiveness of ACE in promoting positive breastfeeding outcomes, and the safety of its practice as part of regular antenatal education at 37 weeks gestation in healthy, low-risk pregnant women. In our own medical record audit of a secondary, general hospital in Greater Perth, Western Australia, we compared the inclusion of ACE in antenatal education in relation to birthing outcomes in a hospital which had previously not included ACE in this education, using a retrospective control group. Results showed that inclusion of ACE in antenatal care was not associated with an increased risk of premature birth or admission to special care nursery [9].

Very few randomised controlled trials have been published in the area of ACE, with none in healthy women, and most using a small sample size (<100). Due to a lack of research investigating the safety and efficacy of antenatal colostrum expression in the general population, ACE is therefore not uniformly encouraged across Australian maternity health services. One major barrier to large scale trials in Australia is the cost of high-quality ACE instruction. A one-on-one session with a midwife (or lactation consultant) provides optimal ACE instruction, however it is not routinely taught as part of antenatal education. In our experience, detailed instruction and assistance with hand expressing is most often provided postnatally, commonly when women are experiencing breastfeeding difficulties (e.g., to relieve engorgement), are sleep deprived, and are potentially anxious about their ability to feed their babies.

Online videos can be an effective and cheap tool for instruction, in areas where the internet is easily accessible. The expert expressing instructional videos that are currently available demonstrate breastmilk expressing in women who have already given birth. Given that expressing in postnatal women is conceptually different, this type of video is not well suited to an antenatal setting. Colostrum has a different look and consistency to mature milk, and the presence of the baby often stimulates milk "let down". Feedback from West Australian midwives and lactation consultants, along with our local consumer reference group, indicated that a video of a lactation consultant-led demonstration of ACE using pregnant women would be very valuable. We have therefore developed an online instructional video on ACE specifically for pregnant women. To our knowledge, no such expert-informed video exists nationally or internationally that provides specific enough detail to suit requirements for use in a large-scale trial or in our health setting. The 20 min video featured a real-life situation, with two pregnant women in a session with a lactation consultant. The lactation consultant explained how to use ACE and how to collect and store the colostrum, and guided both women through the process in a practical demonstration (Appendix A, Figures A1–A3). The women had differing prior breastfeeding experience (one was primiparous, the other had previously breastfed multiple children), and differing breast sizes and nipples. The women also asked questions as they went through the process.

The aim of this study was to investigate whether the instruction on antenatal expression of colostrum via this instructional online video is an effective way to increase knowledge around ACE and confidence in using it, in pregnant women. If so, the use of an instructional video for ACE will provide an economical and feasible education option for future research.

2. Materials and Methods

2.1. Study Setting

This was a pre- and post-test study, using online questionnaires and video. Data collection for the study was completed online between August 2017 and April 2018.

2.2. Study Sample

Pregnant women were recruited through social media via local university and research institutions, and infant and mother organizations, along with personal contacts. The only inclusion criteria were being currently pregnant (any stage) and having access to the internet. Potential participants were provided with an online information letter and were requested to provide online consent prior to the start of the questionnaires. Ethics approval was granted from the Edith Cowan University Human Research Ethics Committee (17501).

2.3. Measurement

Two linked online questionnaires were developed using Qualtrics survey software (Provo, UT, USA). The first was a pre-video questionnaire which contained the ACE video at the end, followed by a second questionnaire which participants completed after viewing the video (post-video questionnaire). The pre-video questionnaire asked about age, gestation of current pregnancy, previous children and previous breastfeeding experience. The post-video questionnaire asked for feedback on the video, including aspects that were liked and disliked (free text response), the length of the video (too long, too short or appropriate), use of real pregnant women in the video (free text), and satisfaction on the depth of information covered, clarity of voices, image quality and overall satisfaction (assessed on a continuous 1–5 scale) from very unsatisfied to very satisfied. Respondents were also asked whether they would recommend the video to pregnant family and friends if ACE was suggested by their healthcare provider, and whether they had any suggestions for improvement of the video. Knowledge and confidence were assessed in both the pre- and post-questionnaires. Knowledge was assessed over seven areas and also as a total knowledge score (one point for each correct answer). To assess confidence about hand expressing in pregnancy, respondents were asked to move a sliding scale which ranged from 1–5, representing very unconfident to very confident.

2.4. Data Analysis

Descriptive statistics were used to report on subject characteristics, with means and standard deviations used for normally distributed results, and medians and interquartile ranges used to report results that were not normally distributed. The related samples McNemar statistical test was used to compare the number of correct answers in knowledge questions in the pre-video and post-video questionnaires, and the paired samples *t*-test was used to compare total knowledge scores and confidence around using ACE. Significance was set at $p < 0.05$.

For free text responses, information was grouped into themes and quotes that were identified as appropriate.

3. Results

A total of 171 pregnant women completed at least one aspect of the survey. Of these, 112 completed the full pre-questionnaire, with 95 going on to also complete the post-questionnaire after watching the video. Subject characteristics for the women who completed both questionnaires are shown in Table 1.

Knowledge of ACE improved significantly for all areas after watching the video (Table 2), as did the total knowledge score (for all $p < 0.001$, Figure 1). Confidence in hand expressing during pregnancy also significantly increased from a mean score of 2.56 ± 1.17 to 4.32 ± 0.80 ($n = 93$, $p < 0.001$) (Figure 2). IOR: Inter quartile range.

Table 1. Characteristics of the pregnant women participating in the video evaluation ($n = 95$).

Characteristic/Response	Mean (SD) or Median (IQR)/n (%)
Age (years)	30.8 (4.7)
Gestation of current pregnancy (weeks)	31.0 (23.0–37.0)
Number of previous children:	
0	103 (60.2)
1	40 (23.4)
2	2 (12.3)
3+	7 (4.1)
Have previously breastfed an infant (yes)	56 (32.7%)

Table 2. Knowledge questions asked in pre- and post-questionnaires, with responses ($n = 95$). The related samples McNemar statistical test was used to compare the number of correct answers pre- and post-questionnaire for each question, and the paired t-test was used to compare the total knowledge scores.

Knowledge Questions and Responses * Indicates Correct Response to the Question	Pre n (%)	Post n (%)	p
If you wanted to do hand expressing in pregnancy, approximately when should you start? (with healthcare provider permission)			
After 28 weeks gestation	5 (5.3)	1 (1.1)	
After 32 weeks gestation	2 (2.1)	0	<0.001
After 35 weeks gestation	16 (16.8)	0	
After 37 weeks gestation *	40 (42.1)	94 (98.9)	
Not sure	32 (33.7)	0	
Click on the picture below that shows the correct finger placement for hand expressing (set of three photos displayed)			
Photo showing narrow finger placement	3 (3.2)	0	
Photo showing correct placement *	48 (50.5)	78 (82.1)	<0.001
Photo showing wide finger placement	16 (16.8)	17 (17.9)	
Not sure	28 (29.5)	0	
If you were doing hand expressing in pregnancy, how often should you express?			
Once per day	16 (16.8)	1 (1.1)	
Two or three times per day *	25 (26.3)	90 (94.7)	<0.001
Four times a day	2 (2.1)	2 (2.1)	
Not sure	52 (54.7)	2 (2.1)	
Can you reuse the same syringe for collecting colostrum over the day?			
Yes, can be left at room temperature	0	1 (1.1)	
Yes, as long as syringe is refrigerated *	30 (31.6)	91 (95.8)	<0.001
No, syringes are for a single expressing use only	30 (31.6)	3 (3.2)	
Not sure	35 (36.8)	0	
How much colostrum would you expect to collect in the first session of hand expressing?			
A lot (>10 mL)	0	0	
A moderate amount (between 1–10 mLs)	0	2 (2.1)	
A little (about 1 mL)	25 (26.3)	23 (24.23)	<0.001
Generally nothing at all	8 (8.4)	0	
Everybody will be different *	42 (44.2)	70 (73.7)	
Not sure	20 (21.1)	0	
If you can't gather any colostrum during expressing, does this mean that you will have trouble breastfeeding?			
Yes	1 (1.1)	0	
Probably	3 (3.2)	0	<0.001
Probably not	13 (13.7)	9 (9.5)	
No *	53 (55.8)	85 (89.5)	
Not sure	25 (26.3)	1 (1.1)	
Should hand expressing be painful?			
Yes sometimes	7 (7.4)	1 (1.1)	
No *	52 (54.7)	92 (96.8)	<0.001
Not sure	36 (37.9)	2 (2.1)	
TOTAL KNOWLEDGE SCORE (mean ± SD correct/7)	3.05 ± 1.70	6.32 ± 0.76	<0.001

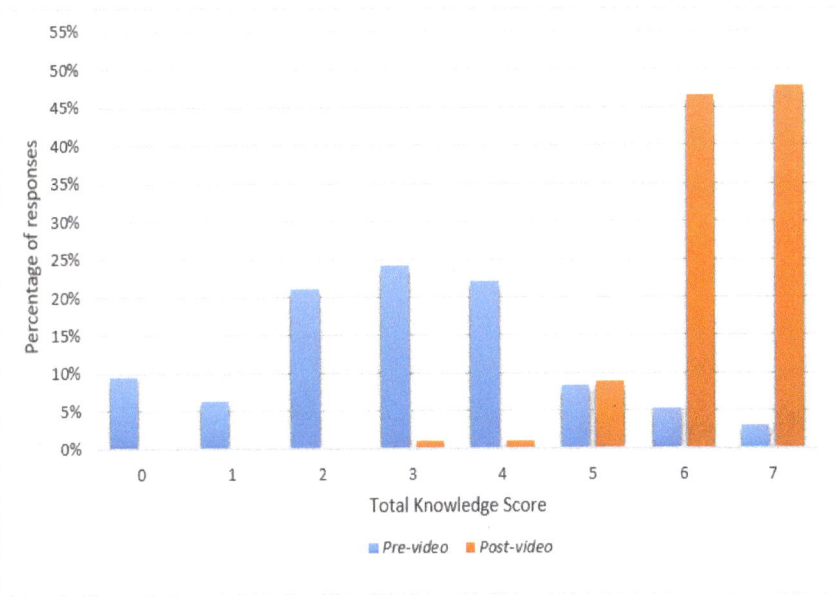

Figure 1. Chart of antenatal expressing of colostrum total knowledge scores scale pre- and post-video (minimum possible score = 0; maximum possible score = 7, n = 95).

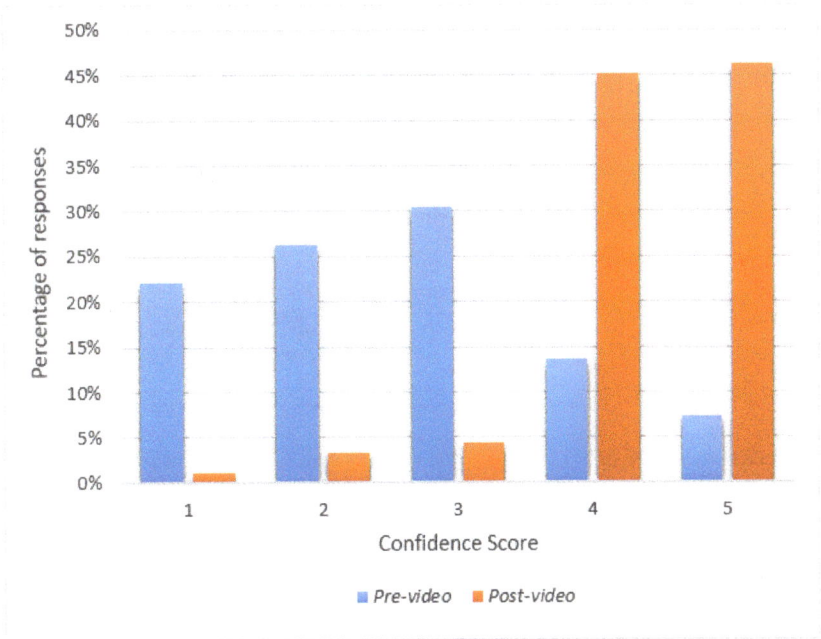

Figure 2. Chart of confidence scale pre- and post-video: How confident do you feel at the moment about hand expressing in pregnancy? (Minimum possible score = 1, representing very unconfident; maximum possible score = 5, representing very confident, n = 93).

When asked what they liked most about the video, 90 out of 95 respondents chose to write a full text response, which was subsequently grouped into one of five categories. The most common response was that the video was informative/interesting/detailed (32.6%). The inclusion of a practical demonstration (22.1%) in a real situation, and the use of real pregnant women (20.0%) were the next most frequently cited responses. Respondents also commented that the video was easy to understand and used clear explanations (15.8%). Others liked the video being "unrushed", "comfortable" or "relaxed" (3.2%). One respondent answered that they liked "everything" (1.1%). Specific quotes relating to the responses are shown in Table 3.

Table 3. Most common full text responses provided by pregnant women when giving their evaluation of the antenatal colostrum expressing video.

Response	Illustrative Quote
Aspects most liked about the video (n = 90/95 responded with full text response)	
The video was informative and interesting	"Very informative, covering all aspects." (#44, 26 yo, 2 previous children, has breastfed previously) "I found it very interesting and helpful. I will definitely be asking my midwife if I can do it when I turn 37 weeks." (#32, 36 yo, 1 previous child, has breastfed previously)
A practical demonstration using real women	"Real women showing how to actually do it rather than cartoons or crochet breasts. Also that one woman had flat nipples like mine and was able to express which makes me hopeful." (#23, 30 yo, 1 previous child, no previous breastfeeding) "It was good to see the spectrum of colostrum colour and volume. Good to see techniques. Good to see a primip and multip." (#52, 39 yo, no previous children). "Close ups of colostrum and technique as well as how to collect on your own." (#60, 35 yo, no previous children).
Easy to understand	"Easy to understand. And the first video I've watched where they've shown how to collect one handed using the syringe too." (#22, 32 yo, 1 previous child, has breastfed previously) "Clear and explained well. Went over important bits twice" (#21, 28 yo, 1 previous child, no previous breastfeeding)
Feelings about the use of real pregnant women in the video (n = 92/95 responded with full text response)	
More comforted about the process of antenatal expressing	"I felt like I could relate to them more." (#32, 36 yo, 1 previous child, has breastfed previously) "It was much more reassuring seeing real pregnant mums" (#13, 33 yo, 2 previous children, has breastfed before) "Comforting to know it was real people learning the skill" (#73, 34 yo, no previous children)
A realistic approach	"Excellent and helpful because both were different i.e.) different sized breasts, one had breastfed before the other was a first time mum" (#1, 29 yo, no previous children) "It was good to see colostrum actually being expressed rather than someone imitating the process be it through use of an actor or prop." (#76, 35 yo, no previous children).

refers to subject number. Yo = years old.

When asked what they didn't like about the video, the most common response was that it was too long (8.4%). This was followed by requests for more information (7.4%), including requests for information on why colostrum would be used, a visual example of when colostrum is not able to be collected using breast massage, when to stop collecting, and a reference diagram. Two women opposed the term "girls" used as a colloquial term in the video by the lactation consultant. Other comments included that it was hard to read the accompanying text and watch at the same time, that there was a need for more eye contact, and that it was hard to hear in places (1.1% for all).

When specifically asked about the length of the video, 78.9% of women found it appropriate, while 17.9% found it too long and 1.1% found it too short.

A total of 92 out of 95 women responded to the free text answer question about how they felt about the use of real women in the video, and all responded positively. Comments reflected women feeling more comforted about the process, even for those who had breastfed before, and finding a realistic approach helpful from a learning perspective (see Table 3).

A total of 93 out of 95 women responded to questions on video satisfaction which were determined on a continuous sliding scale of 1 (very unsatisfied), to 5 (very satisfied). All three aspects and overall satisfaction scored a median of 5.00, with a small interquartile range (IQR): depth of information covered (IQR 4.10–5.00, with 53.8% of respondents scoring 5), clarity of voices (IQR 4.10–5.00, 53.8% scoring 5), quality of images (IQR 4.25–5.00, 59.1% scoring 5), overall satisfaction (IQR 4.20–5.00, 53.8% scoring 5). Free text comments in this section indicated some minor dissatisfaction with the volume of voices, being told breastfeeding is best, a preference for non-scrolling text, a request for more information around how to administer the collected colostrum to a baby, the video being too long and a request for more specific information on hand placement.

When asked for additional suggestions for improvement of the video, three responses were received. These were the inclusion of a diagram of exactly where to place the hands with distance from the nipple and direction of pressure improvement, better music, and more focus on the syringe during the demonstration.

After viewing the video, approximately a fifth of women still had questions about ACE. Most questions related to safety, current health service recommendations for ACE and evidence around using ACE:

"Do all hospitals recommend doing this?"
"Can I start at 36 weeks?"
"Is it recommended for all uncomplicated pregnancies?"
"Have any studies been done on women who have previously experienced low supply to see if antenatal hand expressing helps?"
"Will this bring on early labour?"
"Will it help with establishing a good milk supply?"

This issue of safety and evidence can be better clarified as more research is done in the area.

Other questions focused on the use of stored colostrum, and have since been addressed in a revised version of the video:

"Should I give my baby the expressed colostrum even if he is feeding well?"
"How long can you store it frozen?"
"How do you defrost it?"

There were also questions specifically regarding expressing with gestational diabetes mellitus (GDM):

"What is the best time of day to express if you have gestational diabetes—does it matter if it's done before or after a meal?"
"What are the benefits for GDM?"

In addition, one mother asked what to do if ACE was hurting, and another asked what to do if they try ACE and do not produce any colostrum.

After viewing the video, women were also asked whether they would be likely to recommend the video to pregnant family or friends, if hand expression of colostrum was suggested by their healthcare provider. Responses were gauged using a five-point Likert scale ranging from "definitely yes" to "definitely not". Almost all women responded "definitely yes" (59.1%) or "probably yes" (37.6%), with 2.2% responding "might or might not" and 1.1% responding "probably not". No women selected "definitely not".

4. Discussion

The online video was effective in significantly increasing both knowledge and confidence around antenatal expressing of colostrum. The average knowledge score doubled in the post-video questionnaire, from 3.05 ± 1.70 to 6.32 ± 0.76, from a possible total score of 7. After viewing the video, the majority of women were able to correctly answer each knowledge question. The knowledge areas showing the largest improvements were when to start ACE (from 42.1% correct to 98.9% correct), and how often to express per day (from 26.3% correct to 94.7% correct). This observed increase in knowledge validates the most popular comments regarding what was liked about the video, which focused on the video being informative, detailed and interesting. If a learner finds the subject content interesting, this facilitates the learning process and makes them more likely to retain the information [10].

Reported confidence in using ACE also showed a substantial increase, from a mean score of 2.56 to 4.32 out of a possible 5. It is important to see an increase in confidence alongside an increase in knowledge, as knowledge alone is not necessarily enough to result in a change to practice. Self-confidence in applying skills, also known as a sense of competence, is recognised in the literature as being important when implementing health-related behaviour changes [11]. Importantly, self-confidence, along with expectations, can influence a mother's judgement on her ability to breastfeed [12]. The use of real pregnant women in the video, as opposed to the use of breast models or actors, likely helped women to feel more confident—it was noted by the respondents as being the second most common liked aspect of the video. As in real life, the women in the video had differently shaped breasts and nipples, and this resonated with some respondents. Observational learning improves maternal breastfeeding confidence postpartum [12], and our results suggest that observational learning via video is also valuable during the prenatal period for improving confidence with hand expressing. Self-confidence encompasses both self-efficacy and self-esteem. Self-efficacy refers to an individual's belief in his or her capacity to perform behaviours necessary to accomplish a specific task or achieve a certain outcome. Bandura's theoretical sources of self-efficacy includes vicarious experience (alongside enactive mastery experience, verbal/social persuasion, and physiological and affective states), which refers to the learning of behaviour from observing others, such as watching videos of the behaviour [13,14]. This theory supports the concept that women who watch the ACE video and see women similar to themselves successfully expressing are more likely to succeed, due to an increased belief that they too possess the capabilities to master the skill of expressing in pregnancy.

The most common criticisms were that the video was too long and that it lacked some minor information, although these were an exception, with less than 20% of women expressing dislikes when prompted. When specifically asked about the length of the video, 17% reported that it was too long. The length of the video allowed for a complete demonstration and a relaxed atmosphere, which was noted as positive by some women, however it was possible to shorten it by speeding up some sections, particularly around the demonstration. Other aspects of video satisfaction indicated that it was well-received in terms of clarity of voices, quality of images and depth of information covered.

We have since updated the video to cut out unnecessary and repetitive sections, reducing it to 15 min. We have also added additional information in the form of text notes at relevant times throughout the video.

Overall, 97% of women in this video evaluation stated that they would be likely to recommend it to pregnant family or friends, if hand expression of colostrum was suggested by their healthcare provider. Family and friend recommendations can have an important impact on public health strategies. Eighty-three percent of online respondents across 60 different countries said that they completely or somewhat trusted the recommendations of friends and family, according to Nielsen's Global Trust in Advertising Report in 2015 [15]. The high level of recommendation demonstrated in our findings suggests that women valued the information they received from the video enough to share it with others.

An online video provides unique advantages for instruction over a traditional face-to-face format. Being online allows pregnant women to view the video at a time and location convenient for them. This is particularly valuable for mothers who already have children, along with those located in remote areas. Another advantage of a video is the ability to pause and replay certain sections according to the viewer's needs and understanding. Videos have become increasingly popular in education, and many universities now offer online instruction, incorporating videos and recorded lectures. Disadvantages also exist with video instruction, notably technical issues with glitches in internet bandwidth and connectivity [16]. In 2018, about 88% of the Australian population were active internet users, with an average of 5 h and 34 min spent on the internet daily [17]. A high-quality video with professional sound and lighting is key to better engage learners, as is audio/visual elements to the video that support the content above and beyond the use of text [18]. Results of our evaluation suggested that our ACE video is likely to be viewed as a valuable tool by pregnant women.

Risk of self-selection bias was an important potential limitation of this study, with women who were interested in ACE being more likely to participate. In addition, the women who participated in our study had a relatively narrow age range. These factors limited our ability to generalise our results to a wider population.

Our findings have important implications for the use of ACE video instruction for future research, and potential public health initiatives leading on from this. Many mothers introduce formula due to a perception of poor milk supply [19], and cite a previously unsuccessful breastfeeding experience, poor breastmilk or difficulties breastfeeding as reasons for ceasing breastfeeding early [20]. A mother's attitude towards breastfeeding has been shown to be a strong predictor of breastfeeding duration. This is shaped by personal factors including personal experience, exposure to positive role models and societal value [21]. Performing ACE and the storage of frozen colostrum in the weeks prior to birth has been demonstrated to build confidence with breastfeeding [22]. Instruction on ACE during pregnancy, alongside general instruction on breastfeeding routinely provided in antenatal classes, may help to build a positive attitude toward breastfeeding. Our video has been specifically designed for this purpose, and results from this study suggest that it is an acceptable and economical way to allow the efficacy and safety of ACE to be tested in future research. Future research may also investigate a potential role of this video in the education of healthcare practitioners working in maternal and infant health, who also have a need to increase knowledge, skills, and attitudes to assist breastfeeding families.

Author Contributions: R.G., J.C. and T.A.O. were involved in the study design and video scripting, C.M. produced and edited the video, T.A.O. analysed the data, T.A.O. and R.G. interpreted the findings, T.A.O. drafted the manuscript. All authors critically reviewed the manuscript and approved the final version submitted for publication.

Funding: This work was supported by Soroptimist International of Perth Inc. Soroptimist International is a global volunteer movement working to transform the lives of women and girls locally and internationally through education, empowerment and opportunities.

Acknowledgments: The authors would like to thank the pregnant mothers who participated in this evaluation. We would also like to acknowledge our valuable community reference group including Raenee Polmear, Pia Kaczor and Emma Connolly. We are particularly grateful to the two pregnant mothers who graciously agreed to feature in our video and made this project possible—thank-you!

Conflicts of Interest: The authors declare no conflict of interest. The funders had no role in the design of the study; in the collection, analyses, or interpretation of data; in the writing of the manuscript, or in the decision to publish the results.

Appendix A Video Stills

Figure A1. The lactation consultant discusses syringes and storage.

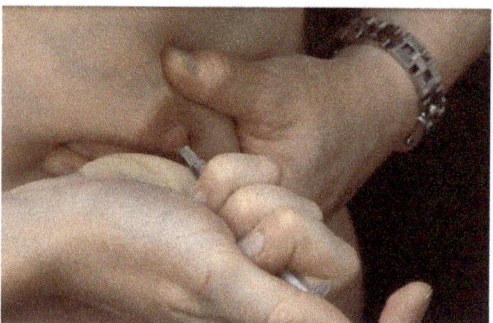

Figure A2. Close up of the colostrum collection during one of the demonstrations.

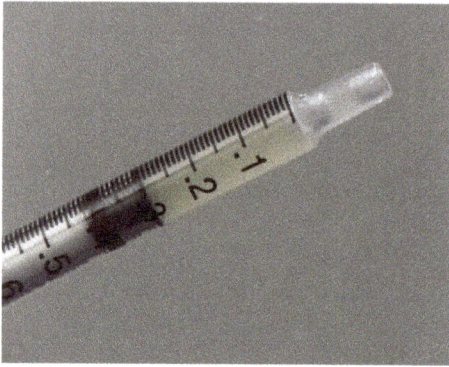

Figure A3. Close up of collected colostrum.

References

1. Chantry, C.J.; Dewey, K.G.; Peerson, J.M.; Wagner, E.A.; Nommsen-Rivers, L.A. In-hospital formula use increases early breastfeeding cessation among first-time mothers intending to exclusively breastfeed. *J. Pediatr.* **2014**, *164*, 1339–1345.e1335. [CrossRef] [PubMed]

2. Chapman, T.; Pincombe, J.; Harris, M. Antenatal breast expression: A critical review of the literature. *Midwifery* **2013**, *29*, 203–210. [CrossRef] [PubMed]
3. Cox, S. Expressing and storing colostrum antenatally for use in the newborn period. *Breastfeed. Rev.* **2006**, *14*, 11–16. [PubMed]
4. Kavanagh, J.; Kelly, A.; Thomas, J. Breast stimulation for cervical ripening and induction of labour. *Cochrane Database Syst. Rev.* **2005**, *20*, CD003392.
5. Singh, N.; Tripathi, R.; Mala, R.; Yedla, N. Breast stimulation in low-risk primigravidas at term: Does it aid in spontaneous onset of labour and vaginal delivery? A pilot study. *Biomed Res. Int.* **2014**, 695037. [CrossRef] [PubMed]
6. Demirel, G.; Guler, H. The effect of uterine and nipple stimulation on induction with oxytocin and the labor process. *Worldviews Evid.-Based Nurs.* **2015**, *12*, 273–280. [CrossRef] [PubMed]
7. Forster, D.A.; Moorhead, A.M.; Jacobs, S.E.; Davis, P.G.; Walker, S.P.; McEgan, K.M.; Opie, G.F.; Donath, S.M.; Gold, L.; McNamara, C.; et al. Advising women with diabetes in pregnancy to express breastmilk in late pregnancy (diabetes and antenatal milk expressing [DAME]): A multicentre, unblinded, randomised controlled trial. *Lancet* **2017**, *389*, 2204–2213. [CrossRef]
8. Binns, C.W.; Fraser, M.L.; Lee, A.H.; Scott, J. Defining exclusive breastfeeding in Australia. *J. Paediatr. Child Health* **2009**, *45*, 174–180. [CrossRef] [PubMed]
9. Connolly, E.L.; Reinkowsky, M.; Giglia, R.; Sexton, B.; Lyons-Wall, P.; Lo, J.; O'Sullivan, T.A. Education on antenatal colostrum expression and the baby friendly health initiative in an Australian hospital: An audit of birth and breastfeeding outcomes. *Breastfeed. Rev.* **2019**, 21–30.
10. Hidi, S. Interest and its contribution as a mental resource for learning. *Rev. Educ. Res.* **1990**, *60*, 549–571. [CrossRef]
11. Ryan, R.M.; Patrick, H.; Deci, E.L.; Williams, G.C. Facilitating health behaviour change and its maintenance: Interventions based on self-determination theory. *Eur. Health Psychol.* **2008**, *10*, 2–5.
12. Dennis, C.-L. Theoretical underpinnings of breastfeeding confidence: A self-efficacy framework. *J. Hum. Lact.* **1999**, *15*, 195–201. [CrossRef] [PubMed]
13. Bandura, A. Social learning through imitation. In *Nebraska Symposium of Motivation*; Jones, M.R., Ed.; University of Nebraska Press: Lincoln, OR, USA, 1962; pp. 211–269.
14. Bandura, A. *Social Foundations of Thought and Action: A Social Cognitive Theory*; Prentice Hall: Englewood Cliffs, NJ, USA, 1986.
15. The Nielsen Company. The Nielsen Global Survey of Trust in Advertising. Available online: http://www.nielsen.com/content/dam/nielsenglobal/apac/docs/reports/2015/nielsen-global-trust-in-advertising-report-september-2015.pdf (accessed on 21 May 2018).
16. Hartsell, T.; Yuen, S. Video streaming in online learning. *AACE J.* **2006**, *14*, 31–43.
17. Despinola, C. 2018 Digital Report—Australia. Available online: https://wearesocial.com/au/blog/2018/02/2018-digital-report-australia (accessed on 21 September 2018).
18. Hibbert, M. What Makes an Online Instructional Video Compelling? Available online: https://er.educause.edu/articles/2014/4/what-makes-an-online-instructional-video-compelling (accessed on 21 May 2018).
19. Inch, S.; Renfrew, M.J. Common breastfeeding problems. In *Effective Care in Pregnancy and Childbirth*, 2nd ed.; Enkin, M., Keirse, M., Renfrew, M.J., Eds.; Oxford University Press: Oxford, UK, 1995; Volume 2.
20. Australian Institute of Health and Welfare. *2010 Australian National Infant Feeding Survey: Indicator Results*; AIHW: Canberra, Australia, 2011.
21. Cox, K.; Giglia, R.; Binns, C.W. The influence of infant feeding attitudes on breastfeeding duration: Evidence from a cohort study in rural Western Australia. *Int. Breastfeed. J.* **2015**, *10*, 1–9. [CrossRef] [PubMed]
22. Brisbane, J.; Giglia, R. Experiences of expressing and storing colostrum antenatally: A qualitative study of mothers in regional Western Australia. *J. Child Health Care* **2015**, *19*, 206–215. [CrossRef] [PubMed]

© 2019 by the authors. Licensee MDPI, Basel, Switzerland. This article is an open access article distributed under the terms and conditions of the Creative Commons Attribution (CC BY) license (http://creativecommons.org/licenses/by/4.0/).

Article

Variability of Serum Proteins in Chinese and Dutch Human Milk during Lactation

Mohèb Elwakiel [1,2], Sjef Boeren [3], Jos A. Hageman [4], Ignatius M. Szeto [5], Henk A. Schols [2] and Kasper A. Hettinga [1,*]

1. Food Quality and Design Group, Wageningen University & Research, Bornse Weilanden 9, 6708 WG Wageningen, The Netherlands; moheb.elwakiel@wur.nl
2. Laboratory of Food Chemistry, Wageningen University & Research, Bornse Weilanden 9, 6708 WG Wageningen, The Netherlands; henk.schols@wur.nl
3. Laboratory of Biochemistry, Wageningen University & Research, Stippeneng 4, 6708 WE Wageningen, The Netherlands; sjef.boeren@wur.nl
4. Biometris-Applied Statistics, Wageningen University & Research, Droevendaalsesteeg 1, 6708 PB Wageningen, The Netherlands; jos.hageman@wur.nl
5. Inner Mongolia Yili Industrial Group Co., Ltd., Jinshan Road 8, Hohhot 010110, China; szeto@yili.com
* Correspondence: kasper.hettinga@wur.nl; Tel.: +31-3-1748-2401

Received: 30 January 2019; Accepted: 24 February 2019; Published: 27 February 2019

Abstract: To better understand the variability of the type and level of serum proteins in human milk, the milk serum proteome of Chinese mothers during lactation was investigated using proteomic techniques and compared to the milk serum proteome of Dutch mothers. This showed that total milk serum protein concentrations in Chinese human milk decreased over a 20-week lactation period, although with variation between mothers in the rate of decrease. Variation was also found in the composition of serum proteins in both colostrum and mature milk, although immune-active proteins, enzymes, and transport proteins were the most abundant for all mothers. These three protein groups account for many of the 15 most abundant proteins, with these 15 proteins covering more than 95% of the total protein concentrations, in both the Chinese and Dutch milk serum proteome. The Dutch and Chinese milk serum proteome were also compared based on 166 common milk serum proteins, which showed that 22% of the 166 serum proteins differed in level. These differences were observed mainly in colostrum and concern several highly abundant proteins. This study also showed that protease inhibitors, which are highly correlated to immune-active proteins, are present in variable amounts in human milk and could be relevant during digestion.

Keywords: mammary gland; immune-active proteins; proteases; protease inhibitors; digestive tract

1. Introduction

Human milk is the best source of nutrition for babies, enhances children's immune system and influences the microbiota [1–3]. Health benefits have been linked to the presence and concentration of human milk components like oligosaccharides and proteins [4,5]. There are two distinct groups of proteins in human milk; caseins and milk serum proteins [6]. Human milk in early lactation consists of approximately 30% caseins and 70% serum proteins, with a 50:50 ratio typically found after a six month lactation period [6].

Serum proteins in human milk have been categorized according to their main and highly diverse biological functions [7,8]. It was found that immune-related proteins, transport proteins, and enzymes were present in the largest quantities, and their concentrations generally decrease over lactation [7,8]. Immune-active proteins not only protect infants against pathogenic microorganisms, but also confer passive immunity to the neonate until its own immune system has been fully developed [9–11].

Serum proteins in human milk also include an array of blood coagulation proteins, membrane proteins, signaling proteins, and protease inhibitors [9–11]. Protease inhibitors play a key role in the blood coagulation cascade and complement pathway [12–14], and might protect proteins against degradation by proteases in the mammary gland and even in the infant's gastrointestinal tract [12–18].

There is a wide range of proteins (e.g., α_{S1}-, β-, and κ-casein, lactoferrin, immunoglobulins, serum albumin, and α-lactalbumin) in relatively high concentrations in human milk [19]. Most milk proteins are synthesized in the mammary gland, except for immunoglobulins and serum albumin [19]. Serum albumin can enter milk via the paracellular pathway and immunoglobulins are transported from blood through mammary epithelial cells by a receptor-mediated mechanism [19]. Caseins are transport proteins that form micelles, and these micelles are capable of binding—and thereby transporting—minerals. Caseins can easily be digested in the infant's gastrointestinal tract [15–18], being a valuable source of amino acids and minerals, which can easily be absorbed. Milk serum proteins such as lactoferrin, immunoglobulins, serum albumin, and α-lactalbumin cover 90% of the milk serum proteome in abundance [20]. The milk serum protein α-lactalbumin is required for the synthesis of lactose, supplies infants with large amounts of tryptophan, and facilitates the absorption of essential minerals [21]. Several other milk serum proteins, like lactoferrin and immunoglobulins, protect infants against pathogens and decrease the risk of having acute or chronic diseases [21,22]. Lactoferrin, a globular glycoprotein of the transferrin family, ends up in the infant's feces, and was shown to influence the microbiota composition of neonates [22]. Human milk is also a rich source of antibodies or immunoglobulins, which are able to recognize and bind to unique epitopes of pathogens, preventing their colonization [23–25]. Serum albumin is a protein mainly involved in the transportation of hormones, fatty acids, and other milk components [21].

Individual differences in milk serum proteins between mothers have been reported, where it was found that there was a large overlap in identified proteins in human milk among mothers, whereas there were also major quantitative changes, both between mothers and over time [7]. Given the various potential benefits of milk serum proteins, it would be of interest to obtain insights in the variability of serum proteins in human milk from mothers from other geographical and ethnic origin.

Therefore, the main objective of this study was to investigate the milk serum proteome of seven Chinese mothers and to investigate the variability in type and level of serum proteins in Chinese human milk over a 20-week lactation period using liquid chromatography and mass spectrometry (LC-MS/MS). Additionally, the type and level of serum proteins in Chinese human milk were compared to those in colostrum and mature milk from Dutch mothers.

2. Materials and Methods

2.1. Study Setup and Sample Collection

Chinese participants were recruited in the Hohhot region, China, between August 2014 and November 2015 by the Yili Innovation Center (Hohhot, China). Yili organized the collection of the human milk, including sampling using a human milk pump. For every time point, a volume of 10 mL was collected in a polypropylene bottles. Milk bottles were shaken gently, aliquoted directly into 2 mL Eppendorf tubes, and stored at -20 °C. Milk samples from seven healthy mothers who delivered term (38–42 weeks) infants were assessed in weeks 1, 2, 4, 8, 12, and 20 postpartum. Human milk collection was approved by the Chinese Ethics Committee of Registering Clinical Trials (ChiECRCT-20150017). Written informed consent was obtained from all mothers. Milk collection and analysis of the milk of four Dutch mothers over a 24-week lactation period was described preciously and was a collaboration with the Dutch Human Milk Bank (Amsterdam, The Netherlands) [7]. Healthy women who delivered singleton term infants (38–42 weeks) were eligible for that study. The data from these analyses were re-used and made compatible with the Chinese data within this research to facilitate direct comparison, as explained further in Section 2.4 (Data Analysis).

2.2. Milk Serum Preparation and Concentrations

Human milk samples (5 mL) were fractionated, as described previously [10]. Briefly, the milk fat was removed by centrifugation (10 min, 1500 g, 4 °C) and the obtained skim milk was transferred to ultracentrifuge tubes. After ultracentrifugation (90 min, 100,000 g, 4 °C), the top layer represented the remaining milk fat still present, the middle layer was milk serum (with some free soluble caseins), and the bottom layer consisted of micellar casein. The free soluble caseins are part of the milk serum proteome. A comparative study previously showed that ultracentrifugation is the most effective method to separate caseins from serum proteins [26], although it is not possible to rule out low amounts of serum proteins in the casein pellet [6]. Milk serum concentrations were measured in duplicate using the bicinchoninic acid (BCA) protein assay kit (Thermo Scientific Pierce, Massachusetts, U.S.), to ensure that the same amount of protein (10 µg) was used for further sample preparation. Bovine serum albumin was used as standard for making a BCA calibration curve.

2.3. Sample Preparation, Dimethyl Labeling, Protein Digestion, and Peptide Analysis

Milk serum samples were prepared for protein analysis using filter-aided sample preparation and dimethyl labeling, as described previously [27]. Milk serum (20 µL) was mixed with a buffer containing sodium dodecyl sulfate (SDS) for protein denaturation and dithiothreitol (DTT) to reduce the disulfide bridges in proteins, after which the samples were loaded on a Pall 3 K omega filter (10–20 kDa cutoff, OD003C34, Pall, Washington, U.S.) for protein digestion. The lysis buffer contained 0.1 M Tris/HCl pH 8.0 + 4% SDS + 0.1 M DTT to get a 1 µg/µL protein solution. Next, 180 µL of 0.05 M iodoacetamide/urea (0.1 M Tris/HCl pH 8 + 8 M urea) was used for protein alkylation. Samples were washed three times with 100 µL of 8 M urea, using centrifugation, followed by 110 µL of 50 mM ammonium bicarbonate (ABC). Then 0.5 µg trypsin in 100 µL ABC was added, followed by overnight incubation at room temperature while mildly shaking, and centrifuged to separate peptides from undigested material. The trypsin digested samples were then labeled, using distinct combinations of isotopic isomers of formaldehyde and cyanoborohydride, leading to a unique stable isotope composition of labeled peptide doublets with different masses [27]. After dimethyl labeling, the prepared samples were analyzed using LC-MS/MS, as described before [7]. For LC-MS/MS, a Prontosil 300-3-C18Hmagic C18AQ 200 Å analytical column was used, and the full scan FTMS spectra were measured in positive mode between m/z 380 and 1400 on a Thermo LTQ-Orbitrap XL. CID fragmented MS/MS scans of the four most abundant doubly- and triply-charged peaks in the FTMS scan were recorded in data-dependent mode in the linear trap (MS/MS threshold = 5.000).

2.4. Data Analysis

The MS/MS spectra obtained were processed by the software package Maxquant 1.3.0.5 with the Andromeda search engine, as described previously [28]. Protein identification and quantification was done according to the literature [7]. Maxquant created a decoy database consisting of reversed sequences to calculate the false discovery rate (FDR). The FDR was set to 0.01 at the peptide and protein levels. The minimum required peptide length was six amino acids, and proteins were identified based on a minimum of two distinct peptides. The intensity–based absolute quantification (iBAQ) values were selected, representing the total peak intensity as determined by Maxquant for each protein and their values were corrected for the number of measurable peptides [7]. The iBAQ values have been reported to have a good correlation with known absolute protein amounts over at least four orders of magnitude [29]. For data normalization, iBAQ values for each protein were transformed into BCA equivalent milk serum protein concentrations, by dividing the iBAQ values of each protein in a sample by the summed iBAQ values of all protein within a sample, there were then multiplied with the corresponding milk serum protein concentration based on the BCA assay. To facilitate direct comparison between Chinese and Dutch data within this research, BCA equivalent values at time points 12 and 20 weeks postpartum were compared to weeks 16 and 24, respectively. Biological functions were assigned to all the serum proteins using the online UniprotKB database, as done previously [7]. To assign a specific function to multifunctional proteins, DAVID Bioinformatics Resource 6.7 was used additionally for further protein biological function classification and clarification [30].

2.5. Statistical Analysis

Statistical analysis was performed based upon previously described methods [7], with modifications. For the BCA equivalent values of each protein in Chinese and Dutch human milk over lactation, a regression line was fitted using R (Lucent Technologies, New York, NY, U.S.A.), summarizing the profile over time for each protein into an intercept and slope. The calculated intercepts are the protein BCA equivalent values at week 1, while the calculated slopes indicate the decrease or increase in BCA equivalent values per week. To determine the significant different milk serum proteins over the course of lactation per country, a comparison was made based on the calculated slope. Only BCA equivalent values of the common serum proteins found in both Chinese and Dutch human milk were used for comparison. The common serum proteins in Chinese and Dutch human milk were then evaluated based on the calculated intercept and slope using a two-tailed t-test, with a significance level set at $\alpha = 0.05$. Next, these common milk serum proteins were compared in Chinese and Dutch human milk using a two-tailed t-test in Perseus [31], separately for each lactation week, with correction for multiple testing based on permutation-based FDR. The BCA equivalent values of serum proteins in Chinese and Dutch human milk were also summed per function and were then compared using a two-tailed t-test. To quantify the relation between biological function groups, Pearson correlation coefficients were calculated for summed BCA equivalent values and visualized in correlation matrix plots. Pearson correlation coefficients of >0.5 were considered good. All the serum proteins in Chinese and Dutch human milk were plotted in a graph in order to visualize the differences in serum proteins over the course of lactation.

3. Results

The objective of this study was to investigate the variability in the type and level of serum proteins in Chinese human milk over a 20-week lactation period. For this, the milk serum proteome of seven mothers over the course of lactation was investigated using LC-MS/MS.

3.1. Level and Type of Milk Serum Proteins in Chinese Human Milk

The total milk serum protein concentrations in Chinese human milk of the seven mothers over the course of lactation are presented in Figure 1. Concentrations ranging from 12 to 25 g/L decreased significantly ($\alpha < 0.05$) over a 20-week lactation period, although with large individual variations (Figure 1).

Serum proteins in human milk were grouped based on their main biological functions (Supplementary Supporting information, data file). Not only the total protein concentrations, but also the protein composition differed among mothers and over lactation as measured after protein digestion and subsequent LC-MS/MS analysis (Figure 2). The figure shows that immune-active proteins, transport proteins, and enzymes were the most abundant for all mothers (Figure 2). The percentage of total protein attributable to these main biological functions, however, varied widely among mothers (Figure 2). Although the BCA equivalent values were always higher in colostrum than in mature milk, the rate of decline for the three main groups varied among mothers (Figure 2).

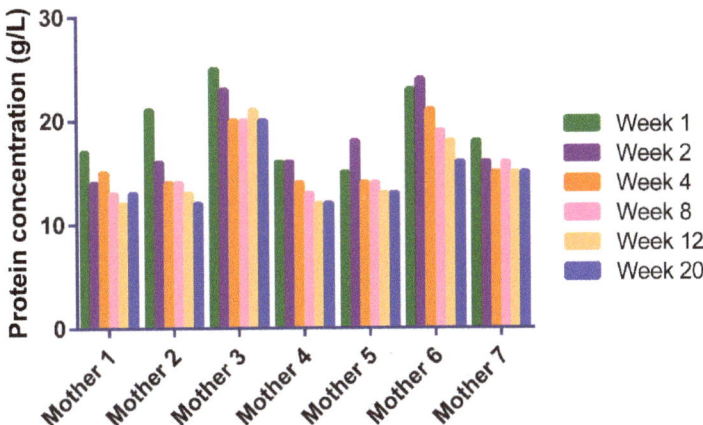

Figure 1. Total bicinchoninic acid (BCA) serum protein concentrations (g/L) in Chinese human milk per mother over a 20-week lactation period.

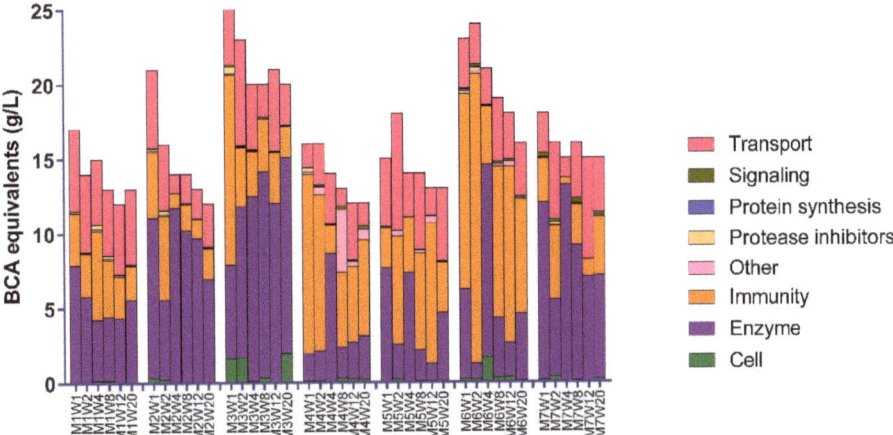

Figure 2. Serum protein composition in human milk of seven Chinese mothers over a 20-week lactation period, based on BCA equivalent values (g/L). The number after the M indicates the mother, and the numbers after the W (1 to 20) indicates the number of weeks postpartum.

To facilitate the comparison between Chinese and Dutch human milk, data were averaged among mothers, as shown in Figure 3. The average total BCA equivalent values in Chinese human milk for enzymes, immune-active proteins, and transport proteins ranged over 4.5–10.0 g/L, 2.9–7.8 g/L, and 2.9–5.0 g/L, respectively (Figure 3).

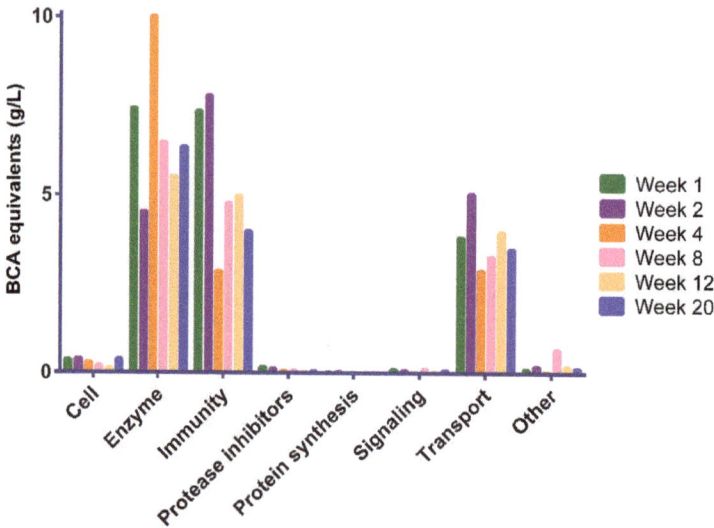

Figure 3. Averaged BCA equivalent values (g/L) of serum proteins for human milk from seven Chinese mothers categorized per biological function over a 20-week lactation period.

3.2. Comparison of the Chinese and Dutch Milk Serum Proteomes

The type and level of serum proteins in Chinese human milk were also compared to those in Dutch human milk. The raw data on Dutch human milk were reprocessed to be compatible with the Chinese data. The total BCA milk serum protein concentrations in Dutch human milk per mother and over the course of lactation are available as supplementary information (Figure S1). The total BCA equivalent values in Dutch human milk decreased over a 24-week lactation period from 21.6 to 13.6 g/L (Figure S2). Enzymes, immune-active proteins, and transport proteins were also the most abundant in Dutch human milk over the course of lactation (Figure S2). The BCA equivalent values for the groups enzymes, immune-active proteins, and transport proteins in Dutch human milk ranged over 4.5–9.0 g/L, 3.8–5.6 g/L, and 4.8–6.8 g/L, respectively. Although different patterns in Chinese and Dutch human milk can be observed, the difference was not significant between the same group of biological functions (data not shown), except for cell and signaling, where levels were higher in Chinese human milk.

The relations between the levels of different biological function groups of serum proteins within the Chinese and within the Dutch human milk populations were visualized in a correlation matrix plot (Figure 4).

3.3. Individual Milk Serum Proteins

Totals of 469 and 200 serum proteins were measured in Chinese and Dutch human milk, respectively. The milk serum proteomes of different Chinese and Dutch mothers were compared based on 166 common milk serum proteins. The overall 15 most abundant milk serum proteins can be found in Table 1.

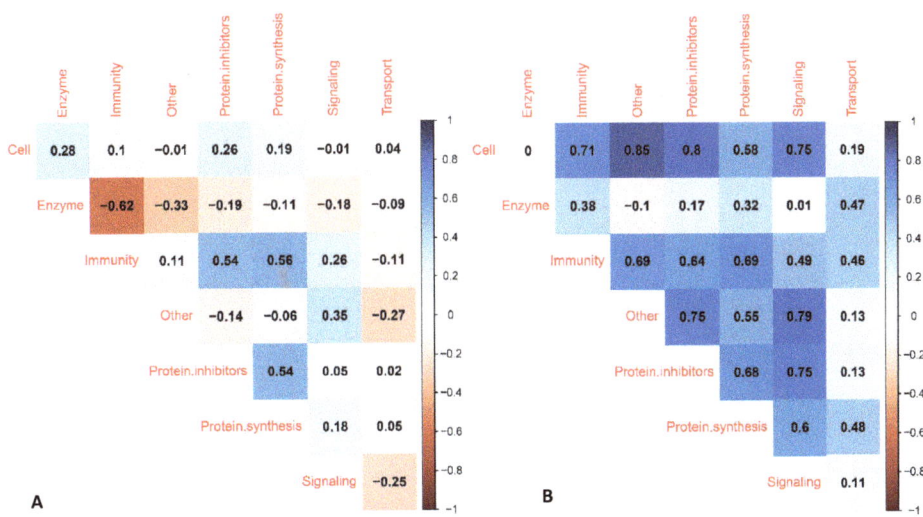

Figure 4. Calculated Pearson correlation coefficients between the different functional groups of serum proteins in Chinese and Dutch human milk, using the summed BCA equivalent values (g/L) over lactation. (**A**) Chinese human milk and (**B**) Dutch human milk.

Table 1. The 15 most abundant serum proteins categorized per function in both Chinese and Dutch human milk during lactation, with their corresponding BCA equivalent values (g/L) values at week 1.

Function	Protein Name	BCA Equivalent Values (g/L)	
		Chinese	Dutch
Enzyme	α-lactalbumin	6.98	8.73
	Bile salt-activated lipase	0.29	0.19
Immunity	Lactoferrin	3.74	2.10
	Ig $α_1$-chain c-region	0.91	0.71
	Ig $λ_2$-chain c-region	0.47	0.54
	Ig κ-chain c-region	0.39	0.90
	Polymeric immunoglobulin receptor	0.41	0.39
	Clusterin	0.23	0.17
	Osteopontin	0.17	0.19
	$β_2$-microglobulin	0.16	0.16
Protease inhibitors	$α_1$-antichymotrypsin	0.11	0.08
Transport	β-casein †	1.17	3.91
	$α_{S1}$-casein †	1.33	1.34
	Serum albumin	0.93	1.06
	κ-casein †	0.23	0.29
	Fatty acid-binding protein	0.07	0.13

† Micellar caseins were completely removed, while this was not the case for the free soluble part of the caseins.

In Dutch human milk, $α_1$-antichymotrypsin belongs to the top 15 serum proteins instead of the transport protein fatty acid-binding protein (Table 1). Within the group enzymes, the highly abundant α-lactalbumin and bile salt-activated lipase are mainly responsible for the changes in this group in human milk over the course of lactation (Table 1). Many immune-active proteins, like lactoferrin, osteopontin, different types of immunoglobulins, polymeric immunoglobulin receptor, and clusterin, belong to the most abundant serum proteins in human milk (Table 1). The changes within the group of transport proteins over the course of lactation can mainly be explained by the caseins (Table 1). The caseins in Table 1 probably refer to the free, non-micellar casein, as the micellar casein should have been removed during the sample preparation (Table 1). With the majority of the caseins in milk being part of the micellar fraction, the caseins in Table 1 therefore do not reflect the levels of total casein.

The differences in protein patterns between Chinese and Dutch human milk were examined by comparison of both the intercept (representing colostrum) and slope (representing the decline over lactation) of curves, fitted for the 166 common milk serum proteins. The *p*-values for these differences after using a two-tailed *t*-test are shown in Figure 5.

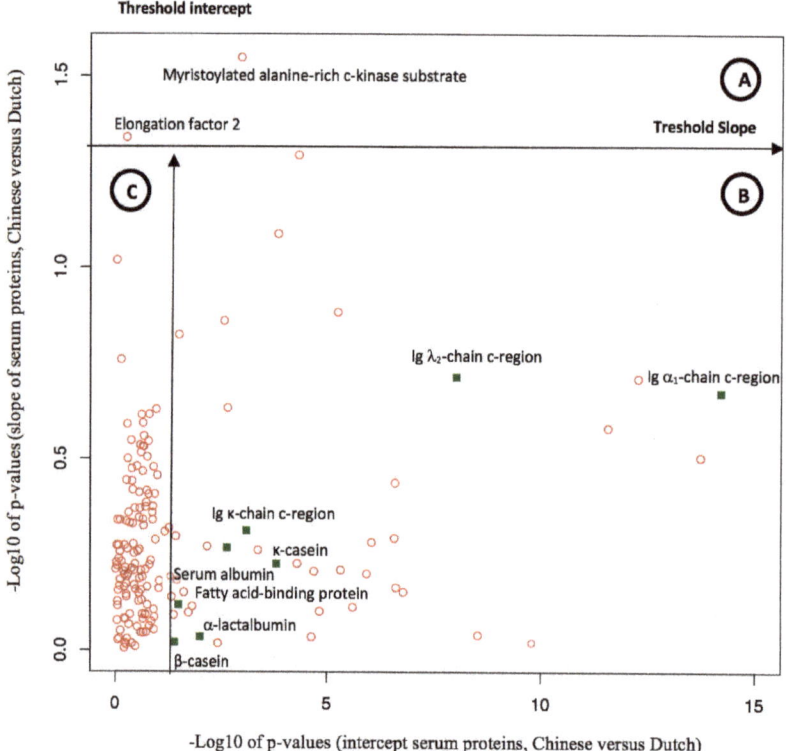

Figure 5. Comparison of the common serum proteins in Chinese and Dutch human milk during lactation. Green squares indicate the proteins displayed in Table 1. For each serum protein in Chinese and Dutch human milk over the course of lactation, a regression line was fitted, summarizing the profile for each protein into an intercept (representing week 1) and slope (representing rate of change over lactation). These profiles were used for comparison between Chinese and Dutch human milk, and the *p*-values for differences between them were plotted. (**A**) Significantly different proteins in Chinese and Dutch human milk over the course of lactation, based on difference in slope; (**B**) significantly different proteins in Chinese and Dutch human milk at week 1, based on intercept; and (**C**) no significant difference.

The levels of two serum proteins (elongation factor 2 and myristoylated alanine-rich c-kinase substrate) varied in the Chinese and Dutch human milk over the course of lactation, as shown by the significantly different slope (Figure 5, area A). Next to that, the levels of 35 serum proteins varied in intercept (Figure 5, area B), including several proteins from the top 15 (Table 1), as shown in green. The complete list of significantly different serum proteins in Chinese and Dutch human milk is shown in Table 2, grouped according to their biological function.

Table 2. Significantly different serum proteins in Chinese and Dutch human milk, with *p*-values for week 1 (intercept) and over the course of lactation (slope).

Function *	Protein Name	*p*-Values of Serum Proteins (Chinese Versus Dutch)	
		Intercept	Slope
Cell	Actin	0.002 *	0.540
	Calreticulin	0.000 *	0.620
	Follistatin-related protein 1	0.003 *	0.140
	MARCKS-like protein 1	0.004 *	0.959
	Protein deglycase DJ-1	0.000 *	0.051
	Peroxiredoxin 2	0.002 *	0.233
Enzyme	4-trimethylaminobutyraldehyde dehydrogenase	0.000 *	0.590
	α-lactalbumin	**0.010 ***	**0.922**
	Fructose-bisphosphate aldolase A	0.000 *	0.710
	Isocitrate dehydrogenase 1	0.000 *	0.310
	L-lactate dehydrogenase A	0.000 *	0.772
	Nucleoside diphosphate kinase A	0.000 *	0.082
	Protein disulfide-isomerase	0.000 *	0.685
	Transketolase	0.023 *	0.707
	Triosephosphate isomerase	0.000 *	0.912
	Tryptophan-tRNA ligase	0.000 *	0.131
	UTP-glucose 1-phosphate uridylyltransferase	0.000 *	0.630
Immunity	Complement C4B	0.000 *	0.200
	Ig α$_1$-chain c-region	**0.000 ***	**0.210**
	Ig γ$_3$ chain c-region	0.000 *	0.190
	Ig κ-chain c-region	**0.001 ***	**0.490**
	Ig λ$_2$-chain c-region	**0.045 ***	**0.640**
	Granulins	0.018 *	0.800
	Lysozyme C	0.000 *	0.937
	Monocyte differentiation antigen CD14	0.015 *	0.770
Protease inhibitors	Inter-α-trypsin inhibitor heavy chain H2	0.000 *	0.522
Protein synthesis	Elongation factor 2	0.547	0.050 *
Signaling	14-3-3 protein β/α	0.000 *	0.372
Transport	Apolipoprotein E	0.036 *	0.500
	β-casein †	**0.000 ***	**0.590**
	Fatty acid-binding protein	**0.000 ***	**0.790**
	Heat shock protein HSP 90-beta	0.040 *	0.810
	κ-casein †	**0.038 ***	**0.950**
	Selenium-binding protein 1	0.006 *	0.536
	Serum albumin	**0.031 ***	**0.760**
	Transcobalamin 1	0.000 *	0.509
Other	Myristoylated alanine-rich c-kinase substrate	0.001	0.028 *

Bold type indicates the proteins also displayed in Table 1. † Micellar caseins were completely removed, while this was not the case for the free soluble part of the caseins. * Corresponding *p*-values (two-tailed *t*-test, α < 0.05)

The levels of the 166 common milk serum proteins in the Chinese and Dutch populations that increased or decreased over the course of lactation, can be found as supporting information (Table S1). The levels of 17 (10%) and 21 (12%) of the 166 common milk serum proteins changed over the course of lactation in Chinese and Dutch human milk, respectively. In addition, the 166 common serum proteins were compared between Chinese and Dutch human milk for each week separately (Table S2). This showed that 16 of 17 proteins that significantly differed in week 1 were also significantly differing in one or more of the other weeks.

4. Discussion

4.1. The Level and Type of Serum Proteins in Chinese Human Milk

The total protein concentrations decrease significantly over a 20-week lactation period in each mother, although with individual variations (Figure 1). These milk serum protein concentrations match with those observed in earlier studies, ranging from 12 to 25 g/L [7,32–34], although other studies report lower values from 7 to 16 g/L over the course of lactation [3,24,35,36]. These differences may be explained by the BCA method [37,38], which generally overestimates the total protein in human milk by about 25–40% [37,38]. The serum protein levels in this study should thus be regarded as semi-quantitative, although this did not influence the comparisons reported here, as they are all based on the BCA method. Although the protein content seems high for milk serum, it should be taken into account that the samples with the highest protein content are actually those in early lactation. These samples are known to have higher protein and relatively lower casein contents [6], leading to higher milk serum protein contents. In addition, part of the casein remained in the sample after sample preparation and therefore also counted towards the BCA protein content.

As described previously [5], human milk becomes fully mature between 4 and 6 weeks postpartum, with the amounts of bioactive components decreasing relative to the nutrients. In early life, infants have an immature intestinal immune system, making them more vulnerable to infection by opportunistic pathogens [5]. The high levels of immune-related milk serum proteins in colostrum (Figure 3) may provide protection to the infant in this sensitive stage of development.

It was also observed that a large variability exists in the milk serum protein composition in colostrum among Chinese mothers (Figure 2). The results in this study comprising milk from seven mothers shows that immune-active proteins, enzymes, and transport proteins are highly abundant in Chinese human milk (Figure 3), which can also be observed from the individual data of mothers (Figure 2). Earlier studies had already shown that immune-active proteins, enzymes, and transport proteins were present in the largest quantities over the course of lactation [7,9,11].

4.2. The 15 Most Abundant Milk Serum Proteins

The large quantities of immune-active proteins are especially driven by the abundance of lactoferrin, immunoglobulins, polymeric immunoglobulin receptor, clusterin, osteopontin and β_2-microglobulin (Table 1), which may protect infants against pathogenic microorganisms, and confer passive immunity to the neonate until its own immune system has been developed [9–11]. As shown in Table 1, transport proteins, like free soluble caseins, serum albumin, and fatty acid binding protein were present in large quantities during lactation. Free soluble caseins could not be removed from the milk, unlike the micellar casein that can be pelleted by ultracentrifugation—a phenomenon that has also been reported by others [7,19,24]. Free soluble and micellar caseins belong to the most abundant proteins in human milk, and these proteins mainly supply infants with amino acids and minerals needed for their growth [23–25]. It can also be observed from Table 1 that enzymes are the largest group of proteins across lactation. The large quantities of enzymes in human milk can be explained by the presence of α-lactalbumin, which is known to be the most abundant milk serum protein (Table 1). This enzyme is required for the synthesis of lactose, the main macronutrient in milk [5,21]. It should be noticed that α-lactalbumin does not have enzymatic activity on its own. Besides α-lactalbumin, bile salt-activated lipase belongs to the 15 most important enzymes in Chinese and Dutch human milk during lactation (Table 1). Bile salt-activated lipase supports the digestion of fats in the immature infant digestive tract, and facilitates the absorption of cholesterol, vitamin A, and triacylglycerols [7]. The protease inhibitor α_1-antichymotrypsin is also among the 15 most abundant human milk serum proteins, and, like other protease inhibitors and proteases, might play a key role in the digestion of human milk [12–14]. Overall, the 15 most abundant proteins identified in this study were in levels dominating the entire milk composition, covering more than 95% of both the Chinese and Dutch milk serum proteomes.

4.3. Proteases and Protease Inhibitors

Proteases may play a key role in the digestion of human milk. Although trypsin was the most abundant protease in Chinese and Dutch human milk, many other proteases (e.g., cytosol aminopeptidase, elastase, kallikrein, plasmin, cathepsins) were found, albeit to a lesser extent (Supplementary Information, data file). As described by others, proteases might be present in human milk to hydrolyze proteins in the mammary gland to regulate casein micelle size [14,15]. Protein digestion in human milk by proteases target specific proteins (e.g., caseins, polymeric immunoglobulin receptor, osteopontin) that do not have an extensive tertiary structure and are thus more accessible to proteolytic cleavage [16,18]. These proteins were, in this study, part of the overall 15 most abundant proteins in Chinese and Dutch human milk during lactation (Table 1). In particular, the caseins are well digested [16–18], which indicates that proteases and bile salt-activated lipase in human milk aids overall in the digestion of two of its main macronutrients, fats and proteins [19].

Besides proteases, human milk also contains protease inhibitors. The ratio between protease inhibitors and proteases in colostrum is circa 10:1. The most abundant protease inhibitors were α_1-antichymotrypsin, α_1-antitrypsin, cystatin C, and phosphatidyletanolamine-binding protein (Supplementary Information, data file). As described by others, α_1-antichymotrypsin binds to chymotrypsin and other chymotrypsin-like serine proteases in human milk, while α_1-antitrypsin inhibits proteases, such as trypsin, elastase, plasmin, and thrombin, and irreversibly deactivates trypsin in vitro [12–15]. A correlation was found between protease inhibitors and immune-active proteins in Chinese and Dutch human milk (Figure 4). Previous literature focused specifically on the relation between serine protease inhibitors and immunoglobulins [7], which also in our data showed stronger correlations than for all protease inhibitors and all immune proteins (Figure S3). A correlation higher than 0.7 was also found in both Chinese and Dutch milk between proteases and protease inhibitors specifically (data not shown). A previous study presented an overview of the proteolytic system network in human milk [15], which consists of several proteases, protease inhibitors, and blood coagulation proteins, indicating that these protein groups share a common biochemical pathway; this may explain their correlations.

Where some of the major proteins are partially digested by milk proteases in human milk, most immune-active proteins are less sensitive to digestion by these proteases, due to their compact folded globular structure, that cannot be as easily digested [16]. For these immune-active proteins to have an immune-activating role in the small intestine, they must be protected against intestinal digestion, because they are sensitive to chymotrypsin and trypsin [17,18]. That might be the reason why protease inhibitors present in human milk seem to target intestinal enzymes, specifically blocking trypsin, chymotrypsin, and other proteases [17,18], especially through the relative abundant α_1-antichymotrypsin and α_1-antitrypsin. Overall, protease inhibitors may thus ensure that specific proteins stay intact in the infant's digestive tract. This may also explain previous findings that several immune-active proteins (e.g., lactoferrin, lysozyme, immunoglobulins) and protease inhibitors (e.g., α_1-antichymotrypsin, α_1-antitrypsin) can be found intact in the stool of breastfed infants [17,18]. The intact proteins in the infant's stool may also be related to the simultaneous decrease in the content of immune-active proteins and protease inhibitors over lactation. Protection is less necessary later in lactation due to the development of the infant's immune system and digestive tract over time, while digestion becomes important for the release of nutrients later in lactation.

4.4. Comparison of High- and Low- Abundance Serum Proteins in Chinese and Dutch Human Milk

It appears that the milk serum proteomes of Chinese and Dutch mothers are similar (Figure 3 and Figure S2). The main purpose of this study was to evaluate the common serum proteins in Chinese and Dutch human milk over the course of lactation. Totals of 469 and 200 serum proteins were found in Chinese and Dutch human milk, respectively. Although a lower number of serum proteins was identified in Dutch human milk, there was still an overlap of 166 serum proteins with Chinese human milk, which represents more than 95% of the milk serum proteome in term of concentrations. The reason for the higher number of serum proteins found in Chinese human milk might be due to the larger sample size (48 versus 24 human milk samples), which generally leads to more identified proteins [28].

In total, 22% (37 out of 166) of the common serum proteins in human milk differed between Chinese and Dutch mothers either at week 1 or over the course of lactation. The levels of 35 of the 166 (circa 21%) common serum proteins varied between Chinese and Dutch mothers in week 1 (Figure 5, area B). This, together with the results presented in Table 2 and Table S2, indicates that the differences between Chinese and Dutch human milk serum proteins were mainly in their level throughout lactation, and not in their changes over lactation, as the levels of only 2 of the 166 (circa 1%) common serum proteins identified in this study (myristoylated alanine-rich c-kinase substrate and elongation factor 2) differed over the course of lactation (Figure 5, area A, showing difference in slope). Overall, the main differences in the milk serum proteomes between Chinese and Dutch human milk were observed in the level of individual proteins, and not in rate of changes over lactation.

5. Conclusions

The milk serum proteome of Chinese and Dutch mothers were similar in term of relative the abundance of different functional groups as well as the most abundant proteins. Some quantitative differences were found, especially in absolute levels and not in rates of change over lactation. Human milk contains enzymes that can assist the digestion of milk proteins and lipids in the immature infant's digestive tract. Protease inhibitors, which are highly correlated to the immune-active proteins, are present in variable amounts in human milk; they could be relevant during digestion and might be involved in controlling protein breakdown in the infant's intestinal tract.

Supplementary Materials: The following are available online at http://www.mdpi.com/2072-6643/11/3/499/s1, Figure S1: Total BCA serum protein concentrations (g/L) in Dutch human milk per mother over a 24-week lactation period. Raw data from Dutch human milk were re-used [7]; Figure S2: BCA equivalent values (g/L) of serum proteins in human milk of 4 Dutch mothers categorized per biological function over a 24-week lactation period. Raw data from Dutch human milk were re-used [7]; Figure S3: Correlations between the functional groups consisting of protease inhibitors (including serine and non-serine protease inhibitors) and immune-active proteins (including immunoglobulins and non-immunoglobulins) in Chinese human milk, using BCA equivalent values (g/L) over a 20-week lactation period; Table S1: Significantly different serum proteins in Chinese and Dutch human milk over the course of lactation, based on the BCA equivalent values (g/L) over lactation (slope); Table S2: Serum proteins that were significantly different in at least one of the lactation weeks. Numbers are the p-value for the difference between the Chinese human milk serum proteins and Dutch human milk serum proteins. To facilitate direct comparison between Chinese and Dutch data within this research, the time points 12 and 20 weeks postpartum were compared to week 16 and 24, respectively; Supporting Information, data file: Serum proteins in human milk of Chinese mothers over a 20-week lactation period. The columns described in the next tab are the individual proteins, their functions and their iBAQ values averaged for all mothers at weeks 1, 2, 4, 8, 12, and 20 postpartum.

Author Contributions: M.E., I.M.S., H.A.S., and K.A.H. conceived and planned the experiments. M.E. carried out the experiments. S.B. contributed to sample preparation and analysis. M.E. and J.H. calculated the statistics and all authors contributed to the interpretation of the results. M.E. worked on all the data and took the lead in writing the manuscript. All authors, especially H.A.S. and K.A.H. provided critical feedback and helped shape the research, analysis, and manuscript.

Funding: This study was been financially supported by Inner Mongolia Yili Industrial Group Co., Ltd.

Conflicts of Interest: This study was been financially supported by Inner Mongolia Yili Industrial Group Co., Ltd.

References

1. Kramer, M.; Kakuma, R. The optimal duration of exclusive breastfeeding: A systematic review. *Adv. Exp. Med. Biol.* **2004**, *554*, 63–77. [PubMed]
2. Liao, Y.; Alvarado, R.; Phinney, B.; Lönnerdal, B. Proteomic characterization of human milk whey proteins during a twelve-month lactation period. *J. Proteome Res.* **2011**, *10*, 1746–1756. [CrossRef] [PubMed]

3. Martin, C.; Ling, P.; Blackburn, G. Review of infant feeding: Key features of breastmilk and infant formula. *Nutrients* **2016**, *8*, 279. [CrossRef] [PubMed]
4. Gartner, L.; Morton, J.; Lawrence, R.; Naylor, A.; O'Hare, D.; Schanler, R.; Eidelman, A. Breastfeeding and the use of human milk. *Pediatrics* **2005**, *115*, 496–506. [PubMed]
5. Elwakiel, M.; Hageman, J.A.; Wang, W.; Szeto, I.M.; van Goudoever, J.B.; Hettinga, K.A.; Schols, H.A. Human milk oligosaccharides in colostrum and mature milk of Chinese mothers: Lewis positive secretor subgroups. *J. Agric. Food Chem.* **2018**, *66*, 7036–7043. [CrossRef] [PubMed]
6. Kunz, C.; Lönnerdal, B. Re-evaluation of the whey protein/casein ratio of human milk. *Acta Paediatr.* **1992**, *81*, 107–112. [CrossRef] [PubMed]
7. Zhang, L.; de Waard, M.; Verheijen, H.; Boeren, S.; Hageman, J.; van Hooijdonk, T.; Vervoort, J.; van Goudoever, J.; Hettinga, K. Changes over lactation in breast milk serum proteins involved in the maturation of immune and digestive system of the infant. *J. Proteomics* **2016**, *147*, 40–47. [CrossRef] [PubMed]
8. Zhang, L.; van Dijk, A.; Hettinga, K. An interactomics overview of the human and bovine proteome over lactation. *Proteome Sci.* **2016**, *15*, 1–14. [CrossRef] [PubMed]
9. Hettinga, K.; Van Valenberg, H.; De Vries, S.; Boeren, S.; Van Hooijdonk, T.; Van Arendonk, J.; Vervoort, J. The host defense proteome of human and bovine milk. *PLoS ONE* **2011**, *6*, 1–8. [CrossRef] [PubMed]
10. Hettinga, K.; Reina, F.; Boeren, S.; Zhang, L.; Koppelman, G. Difference in the breast milk proteome between allergic and non-allergic mothers. *PLoS ONE* **2015**, *10*, 1–11. [CrossRef] [PubMed]
11. Beck, K.; Weber, D.; Phinney, B.; Smilowitz, J.; Hinde, K.; Lönnerdal, B.; Korf, I.; Lemay, D. Comparative proteomics of human and macaque milk reveals species-specific nutrition during postnatal development. *J. Proteome Res.* **2015**, *14*, 2143–2157. [CrossRef] [PubMed]
12. Chowanadisai, W.; Lönnerdal, B. α_1-Antitrypsin and antichymotrypsin in human milk: Origin, concentrations, and stability. *Am. J. Clin. Nutr.* **2002**, *76*, 828–833. [CrossRef] [PubMed]
13. Lindberg, T. Protease inhibitors in human milk. *Pediatr. Res.* **1982**, *16*, 479–483. [CrossRef] [PubMed]
14. Dallas, D.; Murray, N.; Gan, J. Proteolytic systems in milk: Perspectives on the evolutionary function within the mammary gland and the infant. *J. Mammary Gland Biol. Neoplasia* **2015**, *20*, 133–147. [CrossRef] [PubMed]
15. Kelly, A.; O'Flaherty, F.; Fox, P. Indigenous proteolytic enzymes in milk: A brief overview of the present state of knowledge. *Int. Dairy J.* **2006**, *16*, 563–572. [CrossRef]
16. Dingess, K.; de Waard, M.; Boeren, S.; Vervoort, J.; Lambers, T.; van Goudoever, J.; Hettinga, K. Human milk peptides differentiate between the preterm and term infant and across varying lactational stages. *Food Funct.* **2017**, *8*, 3769–3782. [CrossRef] [PubMed]
17. Dallas, D.; Guerrero, A.; Khaldi, N.; Borghese, R.; Bhandari, A.; Underwood, M. A peptidomic analysis of human milk digestion in the infant stomach reveals protein-specific degradation patterns. *J. Nutr.* **2014**, *144*, 815–820. [CrossRef] [PubMed]
18. Su, M.; Broadhurst, M.; Liu, C.; Gathercole, J.; Cheng, W.; Qi, X. Comparative analysis of human milk and infant formula derived peptides following in vitro digestion. *Food Chem.* **2017**, *221*, 1895–1903. [CrossRef] [PubMed]
19. Lönnerdal, B. Human milk proteins: Key components for the biological activity of human milk. *Adv. Exp. Med. Biol.* **2004**, *554*, 11–25.
20. Lönnerdal, B. Bioactive proteins in human milk: Mechanisms of action. *J. Pediatr.* **2010**, *156*, 26–30. [CrossRef] [PubMed]
21. Newburg, D. Bioactive components of human milk: Evolution, efficiency and protection. *Adv. Exp. Med. Biol.* **2001**, *501*, 3–10. [PubMed]
22. Spik, G.; Brunet, B.; Mazurier-Dehaine, C.; Fontaine, G.; Montreuil, J. Characterization and properties of the human and bovine lactotransferrins extracted from the faeces of newborn infants. *Acta Pædiatr. Scand.* **1982**, *71*, 979–985. [CrossRef] [PubMed]
23. Davidson, L.; Lönnerdal, B. Persistence of human milk proteins in the breastfed infant. *Acta Pædiatr.* **1987**, *76*, 733–740. [CrossRef]
24. Lönnerdal, B. Nutritional and physiologic significance of human milk proteins. *Am. J. Clin. Nutr.* **2003**, *77*, 1537–1543. [CrossRef] [PubMed]
25. Jakaitis, B.; Denning, P. Human breast milk and the gastrointestinal innate immune system. *Clin. Perinatol.* **2014**, *41*, 423–435. [CrossRef] [PubMed]

26. Jensen, H.; Poulsen, N.; Moller, H.; Stensballe, A.; Larsen, L. Comparative proteomic analysis of casein and whey as prepared by chymosin-induced separation, isoelectric precipitation or ultracentrifugation. *J. Dairy Res.* **2012**, *79*, 451–458. [CrossRef] [PubMed]
27. Lu, J.; Boeren, S.; de Vries, S.; van Valenberg, H.; Vervoort, J. Filter-aided sample preparation with dimethyl labeling to identify and quantify milk fat globule membrane proteins. *J. Proteomics* **2011**, *75*, 34–43. [CrossRef] [PubMed]
28. Cox, J.; Mann, M. MaxQuant enables high peptide identification rates, individualized ppb-range mass accuracies and proteome-wide protein quantification. *Nature* **2008**, *26*, 1367–1372.
29. Schwanhausser, B.; Busse, D.; Li, N.; Dittmar, G.; Schuchhardt, J.; Wolf, J.; Chen, W.; Selbach, M. Global quantification of mammalian gene expression control. *Nature* **2011**, 337–342. [CrossRef] [PubMed]
30. Huang, D.; Sherman, B.; Lempicki, R. Systematic and integrative analysis of large gene lists using DAVID Bioinformatics Resources. *Nat. Protoc.* **2009**, *4*, 44–57. [CrossRef] [PubMed]
31. Tyanova, S.; Temu, T.; Sinitcyn, P.; Carlson, A.; Hein, M.; Geiger, T.; Mann, M.; Cox, J. The Perseus computational platform for comprehensive analysis of proteomics data. *Nat. Methods* **2016**, *13*, 731–740. [CrossRef] [PubMed]
32. Breckwoldt, J.; Neulen, J.; Keck, C. Lactation. From cellular mechanisms to integration. In *Comprehensive Human Physiology*; Greger, R., Windhorst, U., Eds.; Springer: Berlin, Germany, 1996; Volume 1, pp. 2365–2373.
33. Monaco, M.; Donavan, S. Human milk: Nutritional properties. In *Nutrition in Pediatrics: Basic Science and Clinical Applications*; Walker, W., Watkins, J., Duggan, C., Eds.; BC Decker Inc.: Hamilton, ON, Canada, 2008; Volume 4, pp. 341–353.
34. Trend, S.; Strunk, T.; Hibbert, J.; Kok, C.; Zhang, G.; Doherty, D.; Richmond, P.; Burgner, D.; Simmer, K.; Davidson, D.; et al. Antimicrobial protein and peptide concentrations and activity in human breast milk consumed by preterm infants at risk of late-onset neonatal sepsis. *PLoS ONE* **2015**, *10*, 1–20. [CrossRef] [PubMed]
35. Liao, Y.; Weber, D.; Xu, W.; Durbin-Johnson, B.; Phinney, B.; Lönnerdal, B. Absolute quantification of human milk caseins and the whey/casein ratio during the first year of lactation. *J. Proteome Res.* **2017**, *16*, 4113–4121. [CrossRef] [PubMed]
36. Lönnerdal, B.; Erdmann, P.; Thakkar, S.; Sauser, J.; Destaillats, F. Longitudinal evolution of true protein, amino acids and bioactive proteins in breast milk: A developmental perspective. *J. Nutr. Biochem.* **2017**, *41*, 1–11. [CrossRef] [PubMed]
37. Wu, X.; Jackson, R.; Khan, S.; Ahuja, J.; Pehrsson, P. Human milk nutrient composition in the United States: Current knowledge, challenges, and research needs. *Curr. Dev. Nutr.* **2018**, *2*, 1–18. [CrossRef] [PubMed]
38. Perrin, M.; Fogleman, A.; Newburg, D.; Allen, J. A longitudinal study of human milk composition in the second year postpartum: Implications for human milk. *Matern. Child Nutr.* **2017**, *13*, 1–12. [CrossRef] [PubMed]

© 2019 by the authors. Licensee MDPI, Basel, Switzerland. This article is an open access article distributed under the terms and conditions of the Creative Commons Attribution (CC BY) license (http://creativecommons.org/licenses/by/4.0/).

Article

Do a Few Weeks Matter? Late Preterm Infants and Breastfeeding Issues

Beatrice Letizia Crippa [1,2,*], Lorenzo Colombo [1,2], Daniela Morniroli [1,2], Dario Consonni [3], Maria Enrica Bettinelli [4], Irene Spreafico [5], Giulia Vercesi [1,2], Patrizio Sannino [5], Paola Agnese Mauri [2,6], Lidia Zanotta [1,2], Annalisa Canziani [2,6], Paola Roggero [1,2], Laura Plevani [1], Donatella Bertoli [5], Stefania Zorzan [5], Maria Lorella Giannì [1,2] and Fabio Mosca [1,2]

1. Neonatal Intensive Care Unit, Fondazione IRCCS Cà Granda Ospedale Maggiore Policlinico, via Commenda 12, 20122 Milan, Italy; lorenzo.colombo@mangiagalli.it (L.C.); daniela.morniroli@gmail.com (D.M.); giulia.vercesi@policlinico.mi.it (G.V.); lidia.zanotta@policlinico.mi.it (L.Z.); paola.roggero@unimi.it (P.R.); laura.plevani@mangiagalli.it (L.P.); maria.gianni@unimi.it (M.L.G.); fabio.mosca@mangiagalli.it (F.M.)
2. Department of Clinical Sciences and Community Health, University of Milan, Via San Barnaba 8, 20122 Milan, Italy; paola.mauri@unimi.it (P.A.M.); annalisacanziani.ac@gmail.com (A.C.)
3. Epidemiology Unit, Fondazione IRCCS Ca' Granda Ospedale Maggiore Policlinico, Via San Barnaba 8, 20122 Milan, Italy; dario.consonni@unimi.it
4. Mother and Child Unit, Università degli Studi di Milano, ATS Città Metropolitana di Milano, 20122 Milan, Italy; MBettinelli@ats-milano.it
5. Direzione Professioni Sanitarie, Fondazione IRCCS Cà Granda Ospedale Maggiore Policlinico, via Commenda 12, 20122 Milan, Italy; irenespreafico@gmail.com (I.S.); patrizio.sannino@policlinico.mi.it (P.S.); donatella.bertoli@policlinico.mi.it (D.B.); stefania.zorzan@policlinico.mi.it (S.Z.)
6. Dipartimento Donna-Bambino-Neonato, Fondazione IRCCS Cà Granda Ospedale Maggiore Policlinico, via Commenda 12, 20122 Milan, Italy
* Correspondence: beatriceletizia.crippa@gmail.com; Tel.: +39-333-929-9116

Received: 12 December 2018; Accepted: 29 January 2019; Published: 1 February 2019

Abstract: The late preterm infant population is increasing globally. Many studies show that late preterm infants are at risk of experiencing challenges common to premature babies, with breastfeeding issues being one of the most common. In this study, we investigated factors and variables that could interfere with breastfeeding initiation and duration in this population. We conducted a prospective observational study, in which we administered questionnaires on breastfeeding variables and habits to mothers of late preterm infants who were delivered in the well-baby nursery of our hospital and followed up for three months after delivery. We enrolled 149 mothers and 189 neonates, including 40 pairs of twins. Our findings showed that late preterm infants had a low rate of breastfeeding initiation and early breastfeeding discontinuation at 15, 40 and 90 days of life. The mothers with higher educational levels and previous positive breastfeeding experience had a longer breastfeeding duration. The negative factors for breastfeeding were the following: Advanced maternal age, Italian ethnicity, the feeling of reduced milk supply and having twins. This study underlines the importance of considering these variables in the promotion and protection of breastfeeding in this vulnerable population, thus offering mothers tailored support.

Keywords: breastfeeding; late preterm; protective factors; promotion of breastfeeding; breastfeeding support

1. Introduction

Late preterms are defined as infants born from 34 0/7 to 36 6/7 weeks of gestational age (GA) and comprise approximately 75% of all preterm births [1]. Late preterm births are increasing overall due to the increased use of reproductive technologies and, therefore, the occurrence of multiple pregnancies, in combination with advances in obstetric surveillance and interventions, which could lead to the choice of preterm delivery for babies at risk of perinatal complications (intrauterine growth restriction, placental insufficiency or monochorionic multiple pregnancies) [2,3]. Compared with infants born at term, late preterm infants are at increased risk of neonatal morbidity (i.e., hypoglycemia, hypothermia, jaundice, delayed oral feeding, readmission to the hospital, transient tachypnoea, neuro-developmental delays and mortality) [2]. This population has immature organs and systems, including the brain; a compromised immunomodulatory response; and an increased susceptibility to inflammatory injury and oxidative stress [4,5]. For this reason, breast milk with its bioactive components and its combination of nutritional, anti-infective, anti-inflammatory, anti-oxidative, epigenetic, and gut-colonizing substances is particularly important for these infants [4–7].

Despite these benefits, late preterm infants have a decreased likelihood of breastfeeding initiation and shortened breastfeeding duration compared to term newborns [2,4,8]. To date, few studies have investigated the modifiable factors affecting breastfeeding and discussed the best approaches to promoting and supporting breastfeeding in this vulnerable population [2,4,8]. Kair et al. determined barriers and facilitators of breastfeeding continuation among a large cohort of late preterm infants and compared preterm infants admitted to the well-baby nursery to those admitted to the neonatal intensive care unit (NICU), concluding that NICU admission was not associated with early breastfeeding cessation [6]. Moreover, the experience of breastfeeding late preterm infants was investigated in a cohort of 44 mothers who were interviewed by phone up to 12 months after delivery [9]. Reduced milk supply, difficulties in latching, feelings of failure, and inadequate lactation support from health care providers after discharge have been identified as the most challenging difficulties encountered by the mothers [9].

The purpose of our research is to investigate the variables that could affect breastfeeding duration in a population of late preterm infants admitted to a well-baby nursery and followed up for three months after birth.

2. Materials and Methods

A prospective, observational, single-center study was carried out in the well-baby nursery of Fondazione IRCCS Ca' Granda Ospedale Maggiore Policlinico in Milan, Italy. The Ethics Committee of the "Fondazione Istituto di Ricovero e Cura a Carattere Scientifico Cà Granda Ospedale Maggiore Policlinico" endorsed the present study, and parents provided written informed consent.

We enrolled all consenting mothers with an adequate comprehension of the Italian language who had given birth to late preterm infants admitted to the well-baby nursery from October to December 2017. According to our internal protocol, late preterm infants with a GA from 34 0/7 to 36 6/7 weeks (estimated based on the last menstrual period) and birth weight $\geq$ 1800 g were admitted to the well-baby nursery, provided that their clinical conditions were stable. This means no need of intravenous infusions, noninvasive respiratory support, or any type of invasive assistance, whereas monitoring of vital parameters and/or the need of incubator were possible in the nursery in the first 24 h of life. We did not enroll a control group of term neonates. The exclusion criteria were the following: Hospitalization in the NICU, congenital anomalies, genetic syndromes, respiratory diseases, neurologic problems, metabolic disorders, or gastrointestinal problems. Before discharge, a structured questionnaire for each infant, including twins, was administered to mothers by one of five health care professionals. The questionnaire used was created by the public health organization of the city of Milan (ATS Città di Milano) for monitoring and comparing breastfeeding habits during the Baby-Friendly Hospital Initiative (BFHI) accreditation [10] of the city's hospitals. It included closed-ended questions and took fifteen minutes to administer.

The questionnaire assessed sociodemographic characteristics (maternal age, education, ethnicity), basic characteristics of newborns (GA, birth weight), previous experiences (participation in a prenatal class and previous experience with breastfeeding), type of delivery, peripartum experiences (skin-to-skin contact for at least two hours after birth and rooming-in, defined as the baby being kept in the mother's room for at least 23 h a day), factors affecting lactation (latching difficulties; use of a pacifier; the feeling of reduced milk supply, defined as a subjective maternal perception of having not enough milk to satisfy the baby; breastfeeding on demand or scheduled breastfeeding), and mode of feeding.

Our breastfeeding policy was based on the principles of the BFHI [10]. For this reason, the process of breastfeeding on demand was always explained to mothers and actively promoted by all health care professionals. In contrast, scheduled feeding, defined as feeding the baby at set intervals, was always the mother's choice. The mode of breastfeeding was reported according to the World Health Organization (WHO) definitions [11]: "Exclusive breastfeeding" indicates no food or drink other than breast milk; and "predominant breastfeeding" indicates that the infant's predominant source of nourishment was breast milk but that the infant also received other liquids (i.e., water and water-based drinks). "Mixed feeding" refers to the use of both breast milk and formula.

We used obstetric charts and infants' computerized medical charts (Neocare, i & t Informatica e Tecnologia Srl, Italy) to collect the basic characteristics of the newborns (i.e., birth weight and GA), mode of breastfeeding, and accurate reports of the skin-to-skin contact time and rooming-in time. All the other data were obtained from the questionnaires.

The mothers completed the first follow-up questionnaire during the visit after discharge, within the first seven days of their infants' lives. Then, structured phone interviews were conducted at 15, 40, and 90 days after birth. The data on the variables subject to changes over time (i.e., latching difficulties, use of a pacifier, mode of feeding) were also collected during each follow-up period. These interviews lasted approximately fifteen minutes. The health care providers in charge of collecting the data during the hospitalization were the same providers responsible for the administration of the questionnaires and phone interviews. With regard to twins, mothers completed the first follow-up questionnaire and the subsequent phone interviews separately for each infant.

To take into account intra-individual correlations over time and the probability of exclusive breastfeeding according to selected variables (including those related to sociodemographic features, clinical characteristics, delivery, peripartum, and past or current lactation determinants), odds ratios (ORs) and 95% confidence intervals (CIs) were calculated with univariate and multiple generalized estimation equation (GEE) logistic regression models. The statistical analyses were performed using Stata 15 (StataCorp LLC, College Station, TX, USA, 2017).

3. Results

The total eligible population included 149 mothers and 189 neonates, including 40 pairs of twins. Seventeen percent of these neonates (32 neonates) were born at 34 weeks of GA, 33.3% (63 neonates) were born at 35 weeks of GA, and 49.7% (94 neonates) were born at 36 weeks of GA. Neonatal birth weight ranged from 2050 g to 2990 g. Only 119 mothers completed the study, with a final dropout rate of 20%. The basic characteristics of the mothers are shown in Table 1.

The minimum maternal age was 20 years, and the maximum age was 49 years. Italians represented 82.5% of the study population: Among them, 56.8% had a degree and 51.2% attended a prenatal class. However, in the population of foreign mothers, there was a lower educational level (30.8% had a degree and 34.6% attended a prenatal class). In contrast to the Italian mothers, foreign mothers were more likely to breastfeed on demand (27% vs. 18%) and to have had positive breastfeeding experiences (57.7% vs. 45.5%). Most of the population did not experience skin-to-skin contact (95.5%) or rooming-in (88.6%).

Table 2 shows the proportion of different types of feeding over the first three months of life.

At discharge, only 16.8% of mothers exclusively breastfed their babies, and the rate increased to a maximum of 40.3% at 15 days of life. Mixed feeding peaked at discharge (78.8%) and decreased over time, while the rate of formula feeding progressively increased. Among the lactation factors also reported in Table 2, incorrect latching of the baby on the breast (even if only reported by the mother) was more common during the first days after birth, while the perception of reduced milk supply occurred later.

Table 1. Population features.

	Foreign Mothers (n = 26)		Italian Mothers (n = 123)		Total (n = 149)	
	n	%	n	%	n	%
Sociodemographic features						
Maternal age						
20–29 years	8	30.8	8	6.5	16	10.7
30–34 years	10	38.5	37	30.1	47	31.5
35–39 years	5	19.2	41	33.3	46	30.9
40–49 years	3	11.5	37	30.1	40	26.8
Education						
High school diploma	18	69.2	53	43.1	71	47.7
Degree	8	30.8	70	56.9	78	52.3
Previous experiences						
Prenatal classes						
Yes	9	34.6	63	51.2	72	48.3
No	17	65.4	60	48.8	77	51.7
Previous breastfeeding experience						
Positive	7	27	22	18	29	19.4
None	19	73	88	71.5	107	71.8
Negative	0	0	13	10.5	13	8.8
Delivery and peripartum experiences						
Mode of delivery						
Eutocic	8	30.8	35	28.4	43	28.9
Vacuum/forceps	0	0	6	4.9	6	4.0
Emergency cesarean section	8	30.8	36	29.3	44	29.5
Elective cesarean section	10	38.4	46	37.4	56	37.6
Twins	8	20	32	80	40	100
Skin-to-skin contact						
Yes	2	7.7	5	4.1	7	4.7
No	24	92.3	118	95.9	142	95.3
Rooming-in						
Yes	5	19.2	12	9.8	17	11.4
No	21	80.1	111	90.2	132	88.6
Lactation factors						
Type of breastfeeding						
Scheduled	11	42.3	67	54.5	78	52.3
On demand	15	57.7	56	45.5	71	47.7

Table 2. Follow-up findings.

Variable	Before Discharge * (n = 149)	First Visit after Discharge (n = 138)	15 Days of Life (n = 134)	40 Days of Life (n = 127)	90 Days of Life (n = 119)
	n (%)	n (%)	n (%)	n (%)	n (%)
Lactation factors					
Latching difficulty	44 (29.5)	29 (21)	28 (20.9)	13 (10.2)	8 (6.7)
Pacifier	42 (28.1)	45 (32.6)	51 (38)	69 (54.3)	72 (60.5)
Feeling of reduced milk supply	5 (3.3)	6 (4.4)	10 (7.4)	12 (9.4)	5 (4.2)
Type of feeding					
Exclusive breastfeeding	25 (16.8)	53 (38.4)	54 (40.3)	43 (33.8)	37 (31.1)
Predominant breastfeeding	0 (0.0)	0 (0.0)	0 (0.0)	2 (1.7)	6 (5.0)
Mixed feeding	113 (75.8)	71 (51.5)	64 (47.8)	46 (36.2)	25 (21.0)
Formula feeding	11 (7.4)	14 (10.1)	16 (11.9)	36 (28.3)	51 (42.9)
Lost to follow up	0	11 (7.3)	15 (10)	22 (14.8)	30 (20)

* "Before discharge" refers to the 24 h period before discharge.

For each investigated variable, no difference was found in the answers provided by the mothers with regard to twins.

According to univariate and multivariate regression (Table 3), the factors found to be protective for exclusive breastfeeding were a previous positive experience of breastfeeding and maternal education. Maternal age over 35 years, Italian ethnicity, the feeling of reduced milk supply, and twin pregnancies were risk factors for the early cessation of breastfeeding.

Table 3. Factors affecting breastfeeding: results of univariate and multivariate logistic regressions.

	OR Crude (95% CI)	OR Adjusted * (95% CI)
Sociodemographic features		
Maternal age		
Maternal age < 30 years	1.0 (reference)	1.0 (reference)
Maternal age 30–34 years	1.5 (0.6–3.9)	0.8 (0.2–2.9)
Maternal age 35–39 years	0.5 (0.2–1.4)	0.3 (0.1–0.9)
Maternal age > 40 years	0.3 (0.1–0.8)	0.2 (0.0–0.7)
Education		
High school diploma	1.0 (reference)	1.0 (reference)
Degree	1.2 (0.6–2.1)	2.2 (1.0–5.0)
Ethnicity		
Foreign mothers	1.0 (reference)	1.0 (reference)
Italian ethnicity	0.4 (0.2–0.7)	0.2 (0.0–0.7)
Previous experiences		
Prenatal classes		
No prenatal classes	1.0 (reference)	1.0 (reference)
Prenatal classes	1.5 (0.8–2.9)	1.5 (0.7–3.3)
Clinical history		
No breastfeeding experience	1.0 (reference)	1.0 (reference)
Previous negative experience with breastfeeding	0.7 (0.2–2.2)	0.9 (0.2–2.8)
Previous positive experience with breastfeeding	3.1 (1.4–7.0)	3.0 (1.0–8.5)
Delivery and peripartum experiences		
Mode of delivery		
Eutocic delivery	1.0 (reference)	1.0 (reference)
Vacuum assisted delivery	2.4 (0.5–11.5)	1.9 (0.4–10.9)
Emergency cesarean delivery	0.4 (0.2–0.9)	0.8 (0.3–2.5)
Elective cesarean delivery	0.3 (0.2–0.8)	0.5 (0.2–1.5)
Twins		
Singleton	1.0 (reference)	1.0 (reference)
Twins	0.2 (0.1–0.4)	0.2 (0.1–0.5)
Skin-to-skin contact		
No skin-to-skin contact	1.0 (reference)	1.0 (reference)
Skin to skin contact	3.2 (0.7–15.2)	9.9 (0.8–119.0)
Rooming-in		
No rooming-in	1.0 (reference)	1.0 (reference)
Rooming-in	1.1 (0.4–3.1)	0.7 (0.2–2.6)
Lactation Factors		
Type of breastfeeding		
Scheduled breastfeeding	1.0 (reference)	1.0 (reference)
On demand	1.8 (0.9–3.3)	1.4 (0.6–2.9)
Latching		
No difficulties	1.0 (reference)	1.0 (reference)
Attachment difficulties	0.8 (0.5–1.3)	0.8 (0.5–1.2)
Pacifier		
Use of pacifier	1.0 (reference)	1.0 (reference)
No use of pacifiers	1.3 (0.9–1.7)	1.2 (0.7–2.0)
Milk supply		
No reduced milk supply	1.0 (reference)	1.0 (reference)
Feeling of reduced milk supply	0.4 (0.2–0.7)	0.3 (0.1–0.5)

* Each factor was adjusted for the others. Odds ratios were from univariate and multiple generalized estimation equation logistic regression models. OR: odds ratio; CI: confidence interval.

4. Discussion

Late preterms are increasing in number overall, adding substantially to the impact on health care services, both in the acute, and primary health care settings [2,12]. This study contributed to expanding the current body of knowledge by increasing awareness of how to best target support for mothers of late preterm infants.

Our data confirm that compared to infants born at term in the same hospital setting and in a similar sociodemographic context, late preterm infants were less likely to be exclusively breastfed in the hospital and after discharge, and required more time to acquire breastfeeding skills [13]. Accordingly, the breastfeeding rate in the late preterm population progressively increased to 40.3% at 15 days of life, whereas Colombo et al. reported the peak rate of exclusive breastfeeding to be 75.3% in term newborns at discharge [13]. The present study identified an increased maternal age, Italian ethnicity, low maternal education, lack of previous positive breastfeeding experiences, having twins, and the feeling of reduced milk supply as risk factors for the early discontinuation of exclusive breastfeeding.

Little is known about the effect of maternal age on breastfeeding rates for late preterm infants, and the results are conflicting when different social contexts are considered [14]. Nevertheless, the most developed regions have seen an increase in maternal age in the last two decades [15], and our data indicate that mothers aged 35 years or older are less likely to breastfeed. Accordingly, increased maternal age has been associated with a negative effect on breastfeeding among mothers of term infants in the same hospital setting with a similar sociodemographic context [13].

Italian mothers were less likely to breastfeed than were foreign mothers, unlike what was observed in a population of mothers of term newborns [13]. It can be hypothesized that the limited number of foreign mothers and the high percentage of Italian mothers older than 35 years could have affected this result.

Regarding the mode of delivery, the literature has shown that cesarean is negatively associated with breastfeeding [2,16,17]. Studies established a negative association between cesarean delivery, the onset of lactation, milk transfer and milk production [2,16,17]. Accordingly, Ayton et al. reported that late preterm infants delivered by cesarean section were 80% less likely to initiate breastfeeding within one hour of birth and identified the combination of cesarean delivery with prematurity as a synergic risk factor that creates a complex breastfeeding scenario [2]. In the univariate analysis, our data confirmed the known negative association between cesarean section and breastfeeding. However, the significance was lost in multivariate analysis, probably because of the influence of other variables; increased maternal age and/or having had an iterative elective cesarean section could have implied a previous maternal breastfeeding experience.

Rooming-in and skin-to-skin contact at least two hours after birth are well-known factors associated with successful breastfeeding [18]. Skin-to-skin contact encourages breastfeeding behavior, stimulates innate reflexes in newborns [19], accelerates neurophysiological development [20], and is positively associated with breastfeeding duration in preterm infants [19]. However, in our study, we could not find any significant associations among skin-to-skin contact, rooming-in and breastfeeding, probably due to the very small percentage of mothers who experienced one or both of these practices. The limited implementation of skin-to-skin contact and the partial implementation of rooming-in in our population may be attributable to the fact that late preterm infants frequently need an incubator for the first hours or possibly days of life.

The use of pacifiers could be justifiable in certain situations, particularly in a NICU setting and when the neonate is separated from the mother. The pacifier could support breastfeeding in low-birth-weight and premature infants by promoting the maintenance and maturation of the sucking reflex; it also helps to achieve a neurobehavioral organization and could relieve pain and decrease stress [21]. Other studies have reported that, both in term and preterm infants, the minimization of the use of a pacifier positively affected exclusive breastfeeding [13,19]. However, in our late preterm population in a well-baby nursery setting, avoiding the use of pacifiers was not found to be associated with breastfeeding.

Twin pregnancies were strongly correlated with the early cessation of breastfeeding in our population. Our possible explanation could be the greater maternal fatigue in simultaneously managing and feeding more than one premature infant. Few studies examining breastfeeding among preterm infants include or differentiate among multiple pregnancies. As reported by Giannì et al., having twins was reported by the mother's of late preterm infants to be a barrier to breastfeeding [8]. In contrast, Demirci et al. observed a higher rate of breastfeeding initiation among mothers of late preterm multiples than among mothers of singleton infants [14].

Maternal education was positively associated with breastfeeding. This finding was not surprising, and it was likewise previously described as a predictor of a longer duration of breastfeeding in both term [13] and late preterm infants [6,22]. With regard to lactation factors, the feeling of reduced milk supply and latching difficulties negatively affected breastfeeding, while breastfeeding on demand was strongly associated with the continuation of breastfeeding, according to the literature [6,23]. The maternal perception of inadequate milk supply, breastfeeding difficulties, and concerns that breast milk alone did not satisfy the infants were reported as the top three reasons for breastfeeding discontinuation in the study of Kair et al., which surveyed a total of 2530 mothers of late preterm infants [6]. Moreover, the same barriers were reported in two qualitative [9,24] studies that illustrated a common phenomenon of maternal frustration associated with breastfeeding difficulties.

Breastfeeding a late preterm infant could be challenging due to the peculiar characteristics of this population, such as relative hypotonia and consequent ineffective sucking [2] and immaturity in coordinating swallowing and breathing during breastfeeding [25]. In this scenario, the availability of expert lactation support is crucial [8,26] and affects mothers' self-efficacy, which could be crucial in the establishment of breastfeeding, even in late preterm populations [27,28]. We also believe that mothers' self-efficacy may be correlated with personal experiences. Previous encouraging breastfeeding experiences were strongly positively associated with breastfeeding initiation and duration, whereas a previous negative breastfeeding experience affected breastfeeding initiation and duration much more than having no experience at all. These findings are consistent with those concerning term infants [13] and reinforce the need to closely explore mothers' backgrounds to identify any possible breastfeeding issues and specific groups of mothers that will need extra help.

The present study has some limitations. First, the data were collected from a single institution where internal procedures are applied and thus it is not possible to generalize our findings. Moreover, the results were not stratified according to gestational age. However, it addressed a relatively large number of mothers who received the same modalities of support, and all infants were late preterm infants without comorbidities and with a low-risk birthweight, making the population homogenous and composed of infants with other risk factors, except being born as a late preterm.

5. Conclusions

Many studies have demonstrated that, even when considered healthy, this population is at risk of having issues typical of premature babies, with breastfeeding problems being the most predictable. Considering the importance of mothers' milk and its beneficial effects for both term and preterm infants, the main significance of our study was to underline the need for all health care professionals to be aware of the support required for breastfeeding late preterm infants. The implementation of specific and timely breastfeeding support measures for mothers (especially if old, having twins, with low education, with no previous positive breastfeeding experiences, or with the perception of reduced milk supply), could facilitate to give their preterm infants the best start in life.

Author Contributions: Conceptualization: L.C., M.E.B., L.Z., D.B., M.L.G. and F.M.; data curation: B.L.C., D.M., D.C., I.S., G.V. and A.C.; formal analysis: B.L.C. and D.C.; methodology: L.C.; resources: I.S., G.V. and A.C.; supervision: M.E.B., P.S., P.A.M., L.Z., P.R., L.P., D.B., S.Z., M.L.G. and F.M.; validation: D.C., M.E.B., P.S., P.A.M., L.Z., P.R., L.P., D.B., S.Z. and F.M.; visualization: M.E.B., P.S., P.A.M., L.Z., P.R., L.P., D.B. and S.Z.; writing of the original draft: B.L.C. and D.M.; writing, the review and editing: B.L.C., L.C., D.M. and M.L.C.

Funding: This research received no external funding.

Conflicts of Interest: The authors declare no conflicts of interest.

References

1. Kugelman, A.; Colin, A.A. Late Preterm Infants: Near Term but Still in a Critical Developmental Time Period. *Pediatrics* **2013**, *132*, 741–751. [CrossRef] [PubMed]
2. Ayton, J.; Hansen, E.; Quinn, S.; Nelson, M. Factors associated with initiation and exclusive breastfeeding at hospital discharge: Late preterm compared to 37 week gestation mother and infant cohort. *Int. Breastfeed. J.* **2012**, *7*, 16. [CrossRef] [PubMed]
3. Engle, W.A.; Tomashek, K.M.; Wallman, C. and the Committee on Fetus and Newborn. "Late Preterm" Infants: A Population at Risk. *Pediatrics* **2007**, *120*, 1390–1401. [CrossRef] [PubMed]
4. Dosani, A.; Hemraj, J.; Premji, S.S.; Currie, G.; Reilly, S.M.; Lodha, A.K.; Young, M.; Hall, M. Breastfeeding the late preterm infant: Experiences of mothers and perceptions of public health nurses. *Int. Breastfeed. J.* **2016**, *12*, 23. [CrossRef] [PubMed]
5. Meier, P.; Patel, A.L.; Wright, K.; Engstrom, J.L. Management of Breastfeeding During and After the Maternity Hospitalization for Late Preterm Infants. *Clin. Perinatol.* **2013**, *40*, 689–705. [CrossRef] [PubMed]
6. Kair, L.R.; Colaizy, T.T. Breastfeeding Continuation among Late Preterm Infants: Barriers, Facilitators, and Any Association with NICU Admission? *Hosp. Pediatr.* **2016**, *6*, 261–268. [CrossRef] [PubMed]
7. Victora, C.G.; Bahl, R.; Barros, A.J.D.; França, G.V.A.; Horton, S.; Krasevec, J.; Murch, S.; Sankar, M.J.; Walker, N.; Rollins, N.C. Lancet Breastfeeding Series Group Breastfeeding in the 21st century: Epidemiology, mechanisms, and lifelong effect. *Lancet* **2016**, *387*, 475–490. [CrossRef]
8. Giannì, M.L.; Bezze, E.; Sannino, P.; Stori, E.; Plevani, L.; Roggero, P.; Agosti, M.; Mosca, F. Facilitators and barriers of breastfeeding late preterm infants according to mothers' experiences. *BMC Pediatr.* **2016**, *16*, 1–8. [CrossRef]
9. Kair, L.R.; Flaherman, V.J.; Newby, K.A.; Colaizy, T.T. The Experience of Breastfeeding the Late Preterm Infant: A Qualitative Study. *Breastfeed. Med.* **2015**, *10*, 102–106. [CrossRef]
10. WHO. Baby-Friendly Hospital Initiative. Revised, Updated and Expanded for Integrated Care. Geneva. 2009. Available online: http://www.who.int/nutrition/topics/bfhi/en/ (accessed on 11 November 2018).
11. WHO. *WHO Indicators for Assessing Infant and Young Child Feeding Practices*; WHO: Geneva, Switzerland, 2010; pp. 1–19.
12. Moster, D.; Lie, R.T.; Markestad, T. Long-term medical and social consequences of preterm birth. *N. Engl. J. Med.* **2008**, *359*, 262–273. [CrossRef]
13. Colombo, L.; Crippa, B.L.; Consonni, D.; Bettinelli, M.E.; Agosti, V.; Mangino, G.; Bezze, E.N.; Mauri, P.A.; Zanotta, L.; Roggero, P.; et al. Breastfeeding determinants in healthy term newborns. *Nutrients* **2018**, *10*, 48. [CrossRef]
14. Demirci, J.R.; Sereika, S.M.; Bogen, D. Prevalence and Predictors of Early Breastfeeding among Late Preterm Mother-Infant Dyads. *Breastfeed. Med.* **2013**, *8*, 277–285. [CrossRef] [PubMed]
15. Kitano, N.; Nomura, K.; Kido, M.; Murakami, K.; Ohkubo, T.; Ueno, M.; Sugimoto, M. Combined effects of maternal age and parity on successful initiation of exclusive breastfeeding. *Prev. Med. Rep.* **2016**, *3*, 121–126. [CrossRef] [PubMed]
16. Dewey, K.G.; Nommsen-rivers, L.A.; Heinig, M.J.; Cohen, R.J. Onset of Lactation and Excess Neonatal Weight Loss. *Pediatrics* **2015**, *112*, 607–619. [CrossRef]
17. Scott, J.A.; Binns, C.W.; Oddy, W.H. Predictors of delayed onset of lactation. *Matern. Child Nutr.* **2007**, *3*, 186–193. [CrossRef]
18. Åkerström, S.; Asplund, I.; Norman, M. Successful breastfeeding after discharge of preterm and sick newborn infants. *Acta Paediatr. Int. J. Paediatr.* **2007**, *96*, 1450–1454. [CrossRef] [PubMed]
19. Maastrup, R.; Hansen, B.M.; Kronborg, H.; Bojesen, S.N.; Hallum, K.; Frandsen, A.; Kyhnaeb, A.; Svarer, I.; Hallström, I. Breastfeeding progression in preterm infants is influenced by factors in infants, mothers and clinical practice: The results of a national cohort study with high breastfeeding initiation rates. *PLoS ONE* **2014**, *9*, e108208. [CrossRef]
20. Scher, M.S.; Ludington-Hoe, S.; Kaffashi, F.; Johnson, M.W.; Holditch-Davis, D.; Kenneth, A.L. Neurophysiologic assessment of brain maturation after an eight-week trial of skin to skin contact on preterm infants. *Clin. Neurophysiol.* **2009**, *120*, 1812–1818. [CrossRef]

21. Lubbe, W.; ten Ham-Baloyi, W. When is the use of pacifiers justifiable in the baby-friendly hospital initiative context? A clinician's guide. *BMC Pregnancy Childbirth* **2017**, *17*, 1–10. [CrossRef]
22. Brown, C.R.L.; Dodds, L.; Attenborough, R.; Bryanton, J.; Rose, A.E.; Flowerdew, G.; Langille, D.; Lauzon, L.; Semenic, S. Rates and determinants of exclusive breastfeeding in first 6 months among women in Nova Scotia: A population-based cohort study. *CMAJ Open* **2013**, *1*, E9–E17. [CrossRef]
23. Callen, J.; Pinelli, J. A review of the literature examining the benefits and challenges, incidence and duration, and barriers to breastfeeding in preterm infants. *Adv. Neonatal Care* **2005**, *5*, 89–92. [CrossRef]
24. Demirci, R.J.; Happ, M.B.; Bogen, D.L.; Albrecht, S.A.; Cohen, S.M. Weighing worth against uncertain work: The interplay of exhaustion, ambiguity, hope and disappointment in mothers breastfeeding late preterm infants. *Matern. Child Nutr.* **2015**, *11*, 59–72. [CrossRef] [PubMed]
25. Santos, I.S.; Matijasevich, A.; Silveira, M.F.; Sclowitz, I.K.T.; Barros, A.J.D.; Victora, C.G.; Barros, F.C. Associated factors and consequences of late preterm births: Results from the 2004 Pelotas birth cohort. *Paediatr. Perinat. Epidemiol.* **2008**, *22*, 350–358. [CrossRef] [PubMed]
26. Rayfield, S.; Oakley, L.; Quigley, M.A. Association between breastfeeding support and breastfeeding rates in the UK: A comparison of late preterm and term infants. *BMJ Open* **2015**, *5*. [CrossRef] [PubMed]
27. Brockway, M.; Benzies, K.; Hayden, K.A. Interventions to Improve Breastfeeding Self-Efficacy and Resultant Breastfeeding Rates: A Systematic Review and Meta-Analysis. *J. Hum. Lact.* **2017**, *33*, 486–499. [CrossRef] [PubMed]
28. Brockway, M.; Benzies, K.M.; Carr, E.; Aziz, K. Breastfeeding self-efficacy and breastmilk feeding for moderate and late preterm infants in the Family Integrated Care trial: A mixed methods protocol. *Int. Breastfeed. J.* **2018**. [CrossRef] [PubMed]

© 2019 by the authors. Licensee MDPI, Basel, Switzerland. This article is an open access article distributed under the terms and conditions of the Creative Commons Attribution (CC BY) license (http://creativecommons.org/licenses/by/4.0/).

Article

Quantification of Human Milk Phospholipids: The Effect of Gestational and Lactational Age on Phospholipid Composition

Ida Emilie Ingvordsen Lindahl [1], Virginia M. Artegoitia [1], Eimear Downey [2], James A. O'Mahony [2], Carol-Anne O'Shea [3], C. Anthony Ryan [3], Alan L. Kelly [2], Hanne C. Bertram [1] and Ulrik K. Sundekilde [1,*]

[1] Department of Food Science, Aarhus University, 5792 Årslev, Denmark; idaemilielindahl@food.au.dk (I.E.I.L.); virginia.artegoitia@food.au.dk (V.M.A.); hannec.bertram@food.au.dk (H.C.B.)
[2] School of Food and Nutritional Sciences, University College Cork, T12 YN60 Cork, Ireland; eimeardowney@gmail.com (E.D.); sa.omahony@ucc.ie (J.A.O.); a.kelly@ucc.ie (A.L.K.)
[3] Department of Paediatrics and Child Health, University College Cork, T12 YN60 Cork, Ireland; Ca.OShea@ucc.ie (C.-A.O.); tony.ryan@ucc.ie (C.A.R.)
* Correspondence: uksundekilde@food.au.dk; Tel.: +45-87158318

Received: 28 November 2018; Accepted: 17 January 2019; Published: 22 January 2019

Abstract: Human milk (HM) provides infants with macro- and micronutrients needed for growth and development. Milk phospholipids are important sources of bioactive components, such as long-chain polyunsaturated fatty acids (LC-PUFA) and choline, crucial for neural and visual development. Milk from mothers who have delivered prematurely (<37 weeks) might not meet the nutritional requirements for optimal development and growth. Using liquid chromatography tandem-mass spectrometry, 31 phospholipid (PL) species were quantified for colostrum (<5 days postpartum), transitional ($\geq$5 days and $\leq$2 weeks) and mature milk (>2 weeks and $\leq$15 weeks) samples from mothers who had delivered preterm (n = 57) and term infants (n = 22), respectively. Both gestational age and age postpartum affected the PL composition of HM. Significantly higher concentrations ($p < 0.05$) of phosphatidylcholine (PC), sphingomyelin (SM) and total PL were found in preterm milk throughout lactation, as well as significantly higher concentrations ($p < 0.002$) of several phosphatidylethanolamine (PE), PC and SM species. Multivariate analysis revealed that PLs containing LC-PUFA contributed highly to the differences in the PL composition of preterm and term colostrum. Differences related to gestation decreased as the milk matured. Thus, gestational age may impact the PL content of colostrum, however this effect of gestation might subside in mature milk.

Keywords: Human Milk; Preterm infant; Phospholipids; Lipidomics; Milk Fat Globule Membrane

1. Introduction

Human milk (HM) is deemed the ideal choice when it comes to nutrition of infants and young children (<6 months of age). HM provides fat, carbohydrate, protein and micronutrients needed for postnatal growth and development, contributes to passive immunity of the infant and promotes healthy microbial colonization [1,2]. During the first six months postpartum, lipids make up 45–55% of the total energy content of HM, equivalent to a mean fat content of approximately 3.7–9.1 g/100 kcal [3,4]. Fatty acids (FA) in the form of triacylglycerol (TAG) constitutes the vast majority of milk lipids (< 98%) [5] and are emulsified in structures called milk fat globules (MFG). A distinct trilayer phospholipid (PL) membrane makes up the milk fat globule membrane (MFGM) and contributes to approximately 60% of the PLs found in HM, whereas the remainder is found as MFGM aggregates and PL assemblies termed extracellular vesicles in the skim milk fraction [6,7].

Albeit constituting a mere 1% of the total lipid content of milk [6], dietary PLs play an important role in digestion, absorption and transport of TAG [8]. The glycerophospholipids (GPL) phosphatidylcholine (PC) and phosphatidylethanolamine (PE) and the sphingolipid (SL) sphingomyelin (SM) are the three major PL classes found in HM. Provision of important bioactive components, for example, choline and long-chain polyunsaturated fatty acids (LC-PUFA) such as arachidonic acid (AA), eicosapentaenoic acid (EPA) and docosahexaenoic acid (DHA) is crucial for neural and visual development and growth of the neonate [9–11].

Days postpartum (lactational stage), gestational age and maternal diet are factors known to influence milk composition [12]. Several studies have investigated the macronutrient [13] and micronutrient [14] composition of milk from mothers who delivered prematurely and at term and suggested that preterm milk might not meet the nutritional requirements of the preterm infant, potentially increasing the risk of impaired growth and development. Metabolomic studies on preterm and term milk have furthermore revealed significant differences in metabolites important for infant development, including phosphocholine, citrate and lactose [15], cholesterol, saturated FA and monounsaturated FA [16] as well as glutamine and lysine [17].

Nonetheless, studies on the effect of PL concentration and composition, comparing preterm and term milk over the course of lactation, are scarce and total quantification of major PL species has not been performed. Thus, the objective of this study was to elucidate how age at gestation affects the PL composition and concentration of human colostrum and milk during the first 15 weeks postpartum, utilizing liquid chromatography tandem-mass spectrometry (LC-MS/MS).

2. Materials and Methods

2.1. Ethical Approval

All subjects gave their informed consent for inclusion before they participated in the study. The study was conducted in accordance with the Declaration of Helsinki and the protocol was approved by The Clinical Research Ethics Committee of the Cork Teaching Hospitals, Cork, Ireland (clinical number reference ECM 4(s) 06/08/13).

2.2. Chemicals and Reagents

The standards L-alpha-phosphatidylcholine (from egg yolk, ≥99%) and 3-sn-phosphatidylethanolamine (from bovine brain, ≥98%) were purchased from Sigma-Aldrich (St. Louis, MO, USA); The sphingomyelin (bovine buttermilk, >98%) standard was purchased from Larodan (Solna, Sweden). The SPLASH®LipidoMix™ deuterated internal standard (containing deuterated PC C15:0/C18:1, PE C15:0/18:1 and SM C15:0/18:1) was purchased from Avanti Polar Lipids Inc. (Birmingham, AL, USA). Mass spectrometry-grade ammonium formate (99.0%) and HPLC-grade chloroform (≥ 99.8%), acetonitrile and methanol were purchased from Sigma-Aldrich (St. Louis, MO, USA). LC-MS grade formic acid (98–100%) was purchased from Merck (Darmstadt, Germany). Preparation of ultrapure water was carried out using a water purification system from Holm & Halby (Brøndby, Denmark).

2.3. Preparation of Standards

Stock solutions of PC (100 mg/100 mL), PE (100 mg/100 mL) and SM (125 mg/100 mL) standards were prepared in chloroform and stored at −20 °C. For the calibration curve, a mixed solution containing all three standards was prepared in ice-cold methanol. The mixture solution was serial diluted in methanol to obtain eight dilutions of the following standard concentrations; PC 0.2–25 mg/100 mL; PE 0.1–13 mg/100 mL; and SM 0.1–15.8 mg/100 mL. To each of the eight dilutions was added 1 µL internal standard mix. High and low quality controls (QC) were constructed based on the mixture solution, to control drift and deterioration of the system. High QC were constructed with 6.3 mg/100 mL PC, 3.3 mg/100 mL PE and 4.0 mg/100 mL SM; Low QC were constructed with 0.2 mg/100 mL PC, 0.1 mg/100 mL PE and 0.1 mg/100 mL SM.

2.4. Milk Sample Collection

Preterm and term colostrum and milk samples of known gestational age and lactational stage were collected from non-smoking, non-diabetic mothers as described earlier [15]. Briefly, preterm HM samples were provided by the neonatal intensive care unit of Cork University Maternity Hospital (Wilton, Co., Cork, Ireland). Colostrum and HM samples ($n = 57$) spanning lactation from 1 day to 15 weeks postpartum were donated by 14 mothers who had delivered healthy, preterm (delivered before the 37th week of gestation) infants. The mean age of the mothers was 34 ± 5.3 years, mean gestational age was 30.8 ± 3.4 weeks and mean infant weight at birth was 2064 ± 847.2 grams.

Pasteurized and microbiologically screened term (delivered from the 37th week of gestation) milk samples were provided by The Western Trust Milk Bank (Irvinestown, Co., Fermanagh, Ireland). These samples ($n = 22$) were donated by 22 mothers, spanning stages of lactation from <5 days postpartum to 15 weeks postpartum. Preterm and term samples were kept at $-80\ °C$ until analysis.

2.5. Liquid-Liquid Extraction of Phospholipids

Extraction of polar lipids from technical replicates was carried out in random order, creating $n = 158$ HM samples. A Modified Bligh and Dyer protocol was used for PL extraction [18]. Milk samples were thawed on ice prior to extraction, followed by 40 seconds of homogenization by vortexing. In a centrifuge tube, 70 µL of milk was mixed with 1 µL of deuterated internal standard and vortexed briefly. To the centrifuge tube, 1 mL of ice-cold chloroform:methanol:water (1:2:0.8) was added and vortexed for 20 seconds, followed by centrifugation at $4\ °C$ for 10 minutes at 20,000 g. In a new centrifuge tube, the organic and aqueous phases were transferred and the solvents evaporated at $30\ °C$. One mL ice cold methanol was added for resuspension of the dried lipids and proteins were precipitated by incubation at $-20\ °C$ for 30 minutes. Centrifugation at 20,000 g for 10 minutes at $4\ °C$ was followed by transfer of the supernatant to a new centrifuge tube. Until analysis, the lipid extracts were stored at $-80\ °C$. Prior to analysis, centrifugation of the lipid extract was carried out at 20,000 g for 10 minutes at $4\ °C$ to precipitate residual proteins, followed by transfer of 200 µL of lipid extract to glass vials.

2.6. Analysis of Phospholipid Species

An established LC-MS method for quantification of PL species (mg/100 mL) was based on a method described by Zhao et al. [19]. Sample analysis was performed on an Agilent 1290 Ultra Performance Liquid Chromatography system coupled to an Agilent 6495 Triple Quadrupole Mass Spectrometer (Agilent Technologies, Palo Alto, CA, USA). The LC was equipped with an autosampler (kept at $4\ °C$) and the MS with an ESI source. Chromatographic separation was conducted on a Waters HILIC column (2.1 × 100 mm with particle size 1.7 µm) protected by a Waters HILIC guard column (2.1 mm × 5 mm) (Waters Corporation, Milford, MA, USA). The column oven temperature was set to $25\ °C$. The LC conditions consisted of water with 10 mM ammonium formate (eluent A) at pH 3 (adjusted with formic acid) and acetonitrile (eluent B). The column was equilibrated prior to chromatographic separation. The following gradient was applied for separation of analytes: the mobile phase was held constant at 95% eluent B from 0 to 0.10 min; from 0.10 to 20 min the gradient decreased to 80% eluent B; from 20 to 21 min the gradient decreased to 70% eluent B, after which the mobile phase was held constant from 21 to 23 min; from 23 to 24 min the mobile phase was increased to 95% eluent B, followed by equilibration of the column at 95% eluent B from 24 to 26 min. An injection volume of 5 µL was used. The ESI ion source (operated in positive mode) and MS were operated under the following conditions: capillary voltage 4500 V, nebulizer 1000 V, temperature of the source $210\ °C$, cone gas (ultrahigh-purity nitrogen) 17 L/min, desolvation temperature and sheath gas temperatures of $230\ °C$ and $400\ °C$, respectively, both with gas flow rates of 11 L/h. High and low funnel radio frequencies were 200 eV and 160 eV, respectively. PE, PC and SM PL standards were used to develop a dynamic multiple reaction monitoring (MRM) method for the MS. Internal standards added during sample preparation were used for the absolute quantification of PLs.

2.7. Preprocessing of Data and Statistical Analysis

For retrieval and pre-processing of LC-MS data, the Agilent MassHunter Workstation Software -Data Acquisition for 6400 Series Triple Quadrupole program version B.07.00 was used (Agilent Technologies Inc., Santa Clara, CA, USA). Chromatograms were manually inspected and fitted when needed. Calibration curves were constructed containing at minimum of five points. A regression coefficient cut-off of ≥ 0.95 was used. Multivariate exploratory data analysis was carried out using SIMCA version 15.0.2 (Sartorius Stedim Biotech, Umeå, Sweden) to identify clustering as well as used for outlier detection. Unit variance scaling was conducted prior to analysis. Minitab version 18 (Minitab Inc., Pennsylvania, United States) was used for analysis of variance (ANOVA) utilizing a two-way factorial comparison of PL classes and -species as the dependent variable(s), with gestational age and lactational stage as independent variables. A p-value of $p < 0.05$ was considered significant for testing of PL classes. For the multiple testing of PL species, a corrected significance level of $p < 0.002$ was calculated using Bonferroni correction. Tukey's test was used as a post-hoc test for multiple comparisons of means, using upper- and lower-case letters to illustrate significance related to lactational stage and gestational age, respectively. In case of statistical interaction, that is, if either of the main factors were dependent on the other, the superscript "i" was used for annotation purposes.

3. Results

3.1. Percentage Distribution of Phospholipids

Identification and quantification of 31 PL species were achieved by LC-MS/MS. Milk samples were divided into three lactational stages depending on days postpartum, that is, colostrum (<5 days), transitional milk (≥ 5 days and ≤ 2 weeks) and mature milk (> 2 weeks and ≤ 15 weeks), as well as divided according to gestational age, that is, preterm milk and term milk. Figure 1 illustrates the percentage distribution of the three PL classes. The majority of the PLs were constituted by PE, followed by PC and SM, in both preterm and term colostrum, transitional and mature milk.

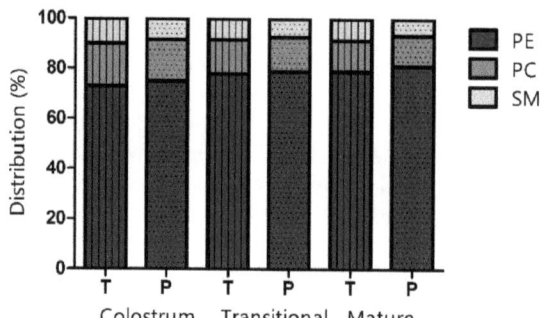

Figure 1. Percentage distribution of phosphatidylethanolamine (PE), phosphatidylcholine (PC) and sphingomyelin (SM) in term (T; vertically striped bars) and preterm (P; dotted bars) colostrum, transitional and mature milk.

In both preterm and term milk, interaction ($p < 0.05$) was observed between gestational age and lactational stage for all three PL classes; nevertheless, the percentage of PE was observed to increase from colostrum to mature milk ($p < 0.05$), whereas the percentage of PC decreased ($p < 0.05$). The percentage constituted by SM in preterm milk decreased from colostrum to mature milk, whereas the percentage of SM in term milk showed a decrease from colostrum to transitional milk, followed by a slight increase ($p < 0.05$).

3.2. Total Concentration of Phospholipids

The concentration (mg/100 mL) of total PLs in preterm and term milk at three lactational stages (colostrum, transitional and mature milk) is shown in Table 1. The concentration of total PLs in both preterm and term milk was found to decrease significantly ($p < 0.05$) with progression from colostrum to transitional milk and from transitional milk to mature milk. Furthermore, the concentration of total PLs was significantly higher in preterm milk compared to term milk throughout lactation ($p < 0.05$). The same pattern of significance was observed for the PL classes SM and PC. No significant differences in the concentration of PE were observed between colostrum and transitional milk in both preterm and term milk.

3.3. Concentrations of Individual Phospholipid Species

A total of 31 individual PL species from three different PL classes were quantified in preterm and term milk at three lactational stages (Table 2). The concentrations of the majority of the PL species were found to be significantly affected by both gestational age of the neonate and lactational stage postpartum. In general, preterm milk exhibited higher concentrations of individual PL species compared to term milk. The concentration of the individual PL species also decreased with time postpartum.

Significant differences ($p < 0.002$) in the concentration of PL species were found in the earlier stages of lactation. A decrease in the concentration of the SM species C18:1/C20:0, C18:0/C22:1 and C18:0/C22:0 as well as the PC species C16:0/C18:2 and C16:0/C18:1 was observed between colostrum and transitional milk in both preterm and term milk, with higher concentrations in preterm milk compared to term milk.

Effects throughout the entire lactational period were observed related to the SM species C18:0/C20:0, the PC species C16:0/C16:0 and C16:0/C16:1 and the PE species C18:1/C16:0 and C18:1/C18:0. A significant decrease ($p < 0.002$) in their concentrations were observed from colostrum to transitional milk, as well as from transitional to mature milk, in both preterm and term milk. No significant differences were observed between colostrum and transitional milk with regards to the concentration of PE species C18:0/C22:6 and C18:0/C16:0; however, a significant decrease ($p < 0.002$) was observed in mature milk compared to colostrum and transitional milk. Moreover, for the PE species C18:0/C18:1, C16:0/C22:6 and C16:0/C20:4 and the PC species C18:0/C18:2, C16:0/C20:5 and C16:0/C20:3, no significant differences were observed between colostrum and transitional milk, as well as between transitional milk and colostrum. However, the species decreased significantly ($p < 0.002$) from colostrum to mature milk. Finally, preterm milk contained significantly higher ($p < 0.002$) concentrations of all of the aforementioned PL species compared to term milk throughout lactation.

Neonate age at parturition alone also showed to affect the concentration of several PLs, as the concentration of the SM species C18:1/C14:0, C18:0/C24:0 and C18:0/C14:0 and the PE species C16:0/C18:1, C16:0/C18:2, C18:0/C18:2 and C18:1/C18:1 were all significantly higher ($p < 0.002$) in preterm milk compared to term milk.

Table 1. Concentration of phospholipids (mg/100 mL) in colostrum, transitional and mature milk from mothers with preterm and term gestations.

Phospholipid Class	Preterm			Term		
	Colostrum	Transitional	Mature	Colostrum	Transitional	Mature
PE	87.85 (27.30) A,a	75.25 (24.85) A,a	46.64 (19.81) B,a	49.40 (13.68) A,a	37.86 (14.00) A,b	29.15 (13.04) B,a
PC	19.31 (11.46) A,a	13.09 (3.94) B,a	6.92 (3.23) C,a	11.44 (2.64) A,b	6.56 (3.26) B,b	4.50 (1.97) C,b
SM	10.13 (6.24) A,a	7.21 (1.95) B,a	3.93 (1.63) C,a	6.90 (1.26) A,b	4.23 (1.88) B,b	3.29 (1.73) C,b
Total PL	117.30 (42.08) A,a	95.49 (29.43) B,a	57.49 (23.77) C,a	67.74 (14.47) A,b	48.65 (18.11) B,b	36.94 (16.41) C,b

Two-way ANOVA with Tukey's test of concentration means and standard deviations (in brackets). PE: Phosphatidylethanolamine, PC: Phosphatidylcholine, SM: Sphingomyelin, PL: phospholipid. Values with different superscript letters are significantly different ($p < 0.05$) within a row (upper case letters A, B, C indicate significant differences related to lactational stage; lower case letters a, b indicate significant differences related to gestational age).

Table 2. Concentration (mg/100 mL) of phospholipid species in colostrum, transitional and mature milk from mothers with preterm and term gestations.

#	Phospholipid Species	Preterm			Term		
		Colostrum	Transitional	Mature	Colostrum	Transitional	Mature
1	PE (18:1/18:1)	0.78 (0.47) a	1.09 (0.63) a	0.92 (0.54) a	0.43 (0.26) b	0.38 (0.15) b	0.52 (0.39) b
2	PE (18:1/18:0)	>12.00 (5.80) A,a	>12.00 (5.53) B,a	6.66 (3.28) C,a	12.00 (4.67) A,b	6.58 (3.02) B,b	3.52 (2.10) C,b
3	PE (18:1/16:0)	7.55 (2.88) A,a	4.63 (2.03) B,a	2.79 (1.58) C,a	4.46 (2.36) A,b	2.76 (0.58) B,b	1.56 (0.85) C,b
4	PE (18:0/22:6)	7.45 (3.64) A,a	7.94 (3.55) A,a	3.57 (2.44) B,a	3.60 (1.21) A,b	3.05 (1.64) A,b	1.81 (1.34) B,b
5	PE (18:0/20:4)	6.43 (1.97) i	4.09 (1.19) ii	2.79 (1.13) iii	3.05 (1.04) ii,iii	2.58 (1.64) ii,iii	2.13 (0.96) iii
6	PE (18:0/18:2)	6.88 (3.13) a	7.20 (2.54) a	5.80 (2.56) a	3.86 (1.42) b	4.24 (2.50) b	4.18 (1.82) b
7	PE (18:0/18:1)	2.51 (1.11) A,a	2.49 (0.85) AB,a	1.50 (0.66) B,a	1.46 (0.36) A,b	1.18 (0.58) AB,b	1.19 (0.64) B,b
8	PE (18:0/16:0)	10.16 (3.50) A,a	7.69 (3.57) A,a	4.40 (2.07) B,a	5.43 (2.25) A,b	4.71 (1.17) A,b	2.86 (1.26) B,b
9	PE (16:0/22:6)	4.99 (2.15) A,a	4.34 (1.87) AB,a	2.42 (1.64) B,a	2.24 (0.63) A,b	1.57 (0.77) AB,b	1.02 (0.63) B,b
10	PE (16:0/20:4)	>12.00 (6.97) A,a	11.78 (4.07) AB,a	8.43 (3.97) B,a	6.57 (2.37) A,b	5.31 (2.46) AB,b	4.94 (1.83) B,b
11	PE (16:0/18:2)	9.00 (3.58) a	9.11 (3.32) a	6.12 (3.14) a	5.07 (1.31) b	4.50 (2.24) b	4.52 (2.21) b
12	PE (16:0/18:1)	1.99 (1.13) a	2.30 (1.14) a	1.25 (0.91) a	1.23 (0.31) b	1.01 (0.55) b	0.90 (0.57) b
13	PC (18:1/20:4)	0.41 (0.36) i	0.22 (0.06) ii	0.12 (0.06) iii	0.15 (0.02) ii,iii	0.11 (0.06) ii,iii	0.07 (0.04) iii
14	PC (18:0/18:2)	1.25 (0.68) A,a	1.01 (0.27) AB,a	0.66 (0.25) B,a	0.67 (0.13) A,b	0.63 (0.41) AB,b	0.57 (0.29) B,b
15	PC (18:0/18:1)	0.60 (0.38) i	0.40 (0.11) ii	0.21 (0.08) iii	0.33 (0.05) ii	0.20 (0.12) ii,iii	0.19 (0.10) iii
16	PC (16:0/20:5)	1.97 (1.46) A,a	1.42 (0.47) AB,a	0.80 (0.49) B,a	0.76 (0.13) A,b	0.64 (0.33) AB,b	0.46 (0.26) B,b
17	PC (16:0/20:4)	0.54 (0.37) i	0.35 (0.09) ii	0.20 (0.11) iii	0.20 (0.02) ii,iii	0.17 (0.10) ii,iii	0.11 (0.05) iii
18	PC (16:0/20:3)	1.61 (1.05) A,a	1.30 (0.38) B,a	0.79 (0.40) B,a	0.77 (0.15) A,b	0.61 (0.35) B,b	0.50 (0.20) B,b
19	PC (16:0/18:2)	0.61 (0.34) A,a	0.42 (0.12) B,a	0.25 (0.10) B,a	0.33 (0.05) A,b	0.20 (0.11) B,b	0.18 (0.06) B,b
20	PC (16:0/18:1)	0.55 (0.37) A,a	0.34 (0.10) B,a	0.18 (0.08) B,a	0.34 (0.10) A,b	0.17 (0.08) B,b	0.13 (0.05) B,b

Table 2. Cont.

#	Phospholipid Species	Preterm						Term					
		Colostrum		Transitional		Mature		Colostrum		Transitional		Mature	
21	PC (16:0/16:1)	1.09	(0.59) A,a	0.79	(0.30) B,a	0.37	(0.18) C,a	0.81	(0.25) A,b	0.38	(0.24) B,b	0.23	(0.11) C,b
22	PC (16:0/16:0)	10.68	(6.06) A,a	6.78	(2.38) B,a	3.36	(1.66) C,a	7.06	(2.29) A,b	3.44	(1.72) B,b	2.06	(0.94) C,b
23	SM (18:1/23:0)	0.22	(0.16)	0.19	(0.13)	0.15	(0.14)	0.13	(0.09)	0.12	(0.07)	0.09	(0.06)
24	SM (18:1/20:0)	1.33	(0.98) A,a	0.85	(0.25) B,a	0.48	(0.22) B,a	0.96	(0.26) A,b	0.47	(0.21) B,b	0.40	(0.17) B,b
25	SM (18:1/14:0)	0.52	(0.28) a	0.56	(0.27) a	0.31	(0.16) a	0.36	(0.09) b	0.21	(0.02) b	0.23	(0.14) b
26	SM (18:0/24:0)	0.08	(0.08) a	0.04	(0.04) a	0.03	(0.03) a	0.02	(0.01) b	0.01	(0.01) b	0.02	(0.02) b
27	SM (18:0/23:0)	0.02	(0.05)	0.01	(0.01)	0.01	(0.01)	0.02	(0.01)	0.01	(0.004)	0.01	(0.005)
28	SM (18:0/22:1)	1.35	(0.96) A,a	0.90	(0.24) B,a	0.52	(0.19) B,a	0.85	(0.11) A,b	0.52	(0.28) B,b	0.51	(0.28) B,b
29	SM (18:0/22:0)	1.29	(0.88) A,a	0.89	(0.28) B,a	0.45	(0.22) B,a	0.81	(0.14) A,b	0.47	(0.30) B,b	0.40	(0.23) B,b
30	SM (18:0/20:0)	5.02	(2.99) A,a	3.39	(0.93) B,a	1.78	(0.76) C,a	3.56	(0.71) A,b	2.27	(1.10) B,b	1.47	(0.83) C,b
31	SM (18:0/14:0)	0.29	(0.17) a	0.38	(0.20) a	0.22	(0.11) a	0.19	(0.05) b	0.15	(0.02) b	0.17	(0.09) b

Two-way ANOVA with Tukey's test of concentration means and standard deviations (in brackets). #: Numbering of phospholipid species. PE: Phosphatidylethanolamine, PC: Phosphatidylcholine, SM: Sphingomyelin. Fatty acid moieties are indicated in parentheses. Values with different superscript letters are significantly different ($p < 0.002$) within a row (upper case letters A, B, C indicate significant differences related to lactational stage; lower case letters a, b indicate significant differences related to gestational age; roman letters i, ii, iii indicate significant differences related to interaction). The significance level has been adjusted using Bonferroni correction.

3.4. Exploratory Analysis of Preterm and Term Milk

Exploratory multivariate data analysis was used to evaluate the changes in PL composition with progression of lactation in preterm and term milk. A Principal Component Analysis (PCA) model was constructed and principal components 1 and 2 collectively explained 76.8% of the variance. The accompanying score scatter plot (Figure 2A) showed a clear progression with lactation along the first principal component, whereas the separation of preterm and term colostrum was evident on both first and second principal components. Furthermore, grouping of preterm and term milk became more pronounced with progression from transitional milk to mature milk. These findings suggest that the PL composition of both preterm and term milk changes with progression of lactation. In addition, differences in the composition of PLs between preterm and term milk became less apparent as lactation progressed. The loadings scatter plot (Figure 2B) furthermore substantiated that the observed variance was associated with a higher content of PLs in early lactation; thereafter, differences in PL content in preterm and term milk subsided with progression of lactation. The PL species that contributed the most to the variance are presented in Figure 3.

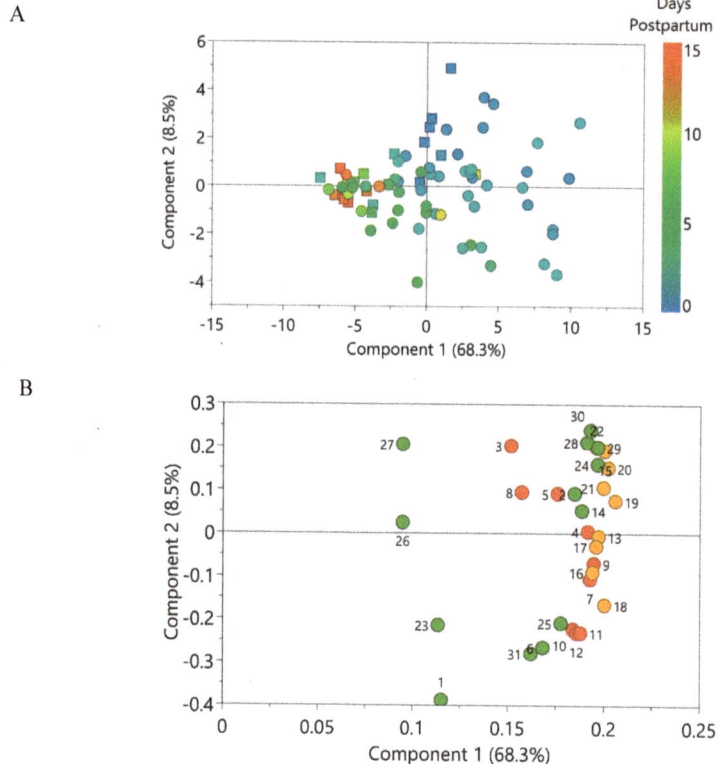

Figure 2. (**A**) PCA score scatter plot of preterm milk (squares) and term milk (circles) expressed between day 1 and 15 weeks post-partum. Score colours indicate day postpartum at which the samples were expressed. (**B**) Corresponding loading scatter plot containing the 31 phospholipid species quantified in human milk belonging to the phospholipid classes phosphatidylethanolamine (red circles), phosphatidylcholine (orange circles) and sphingomyelin (green circles). Details about the numbering of phospholipid species can be found in Table 2.

Based on the exploratory analysis (Figure 2), PL species from the PC class was found to strongly contribute to the differences observed between preterm and term milk throughout lactation. There was large variability in preterm and term colostrum with regards to the concentration of two SM species and several PC species (Figure 3), nearly all of which contained long-chain polyunsaturated fatty acids (LC-PUFA). These differences diminished with time postpartum as the concentration decreased in both preterm and term milk (albeit to a higher extend in the former), resulting in discrete differences in the concentrations of these PLs in mature preterm and term milk.

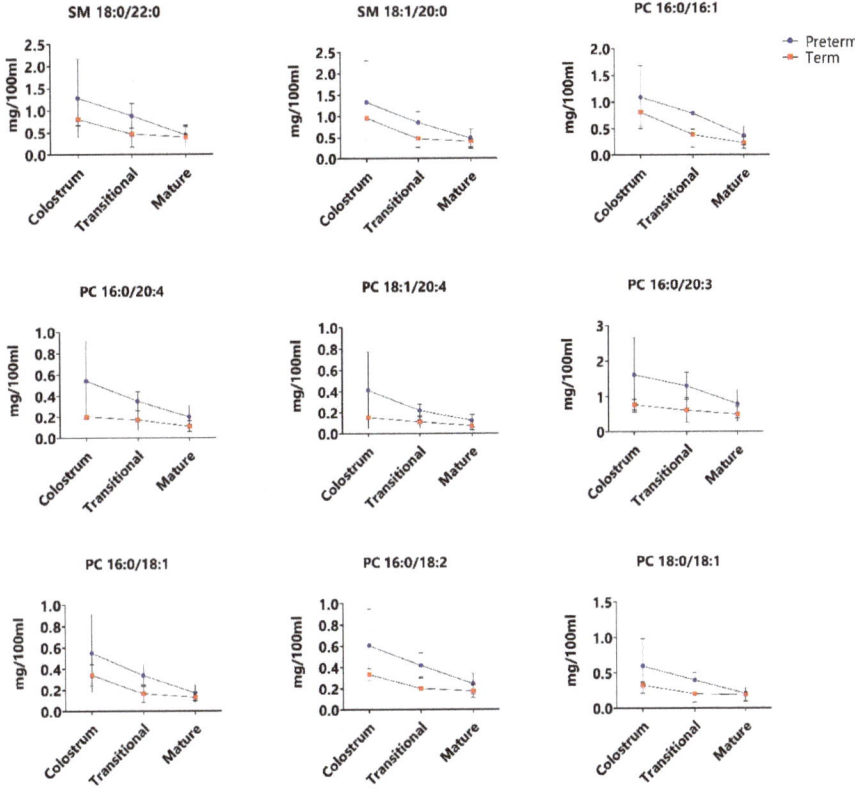

Figure 3. Concentration (mg/100 mL) of phosphatidylcholine (PC) and sphingomyelin (SM) species in preterm (blue circle) and term (red square) milk in colostrum (<5 days), transitional milk ($\geq$5 days and $\leq$2 weeks) and mature milk (>2 weeks and $\leq$15 weeks).

4. Discussion

In general, the PL content of HM was found to be influenced by the age of the infant at parturition as well as days postpartum. Throughout lactation, preterm colostrum and milk were found to contain significantly higher concentrations of total PLs, as well as several PL species and -classes, compared to term colostrum and milk. The differences observed in PL composition of preterm and term colostrum were particularly influenced by differences in the concentration of PC species containing LC-PUFA moieties. The observed differences between preterm and term colostrum were less prominent in mature milk.

The percentage distribution of PLs in term HM have been studied by several authors [9,20–22] but there are significant discrepancies between studies, as previously reported [12]. In the present study, regardless of gestational age, PE represented the largest proportion of PLs throughout lactation, followed by PC and SM. These findings deviate from findings by Bitman et al. [21] and Benoit et al. [22], who reported SM to constitute the largest proportion of PLs in HM (36% [21]; 43.3% [22]), followed by PE (20% [21], 21.3% [22]) and PC (28% [21], 19.0% [22]). In contrast, a study by Sala-Vila et al. [20] reported even larger differences in the PL distribution of term milk, ascribing SM and PC to contribute the vast majority of PLs (39.2–41.0% and 31.3–38.4%, respectively), while PE only contributed with a minor proportion (5.9–12.76%). Nonetheless, Wang et al. [9] reported PE to be the major PL contributor (~36.1%) in term milk, followed by SM (~30.7%) and PC (~23.1%). Our findings suggest that the proportion of PE increased with progression of lactation in both preterm and term milk; however, this main effect was dependent on gestational age. Nonetheless, findings by Sala-Vila et al. [20] showed that the proportion of PE increased significantly from colostrum to transitional milk and from transitional milk to mature milk.

To our knowledge, absolute quantification of individual PL species in preterm milk over the course of lactation has not been published previously. Independent of gestational age, the concentration of total PLs quantified was found to be the highest in colostrum. In addition, the concentration of total PLs decreased with progressing lactation for both preterm and term milk. This is in accordance with the report of Bitman et al. [21] of a decrease in relative PL content from colostrum to mature milk. Additionally, higher contents of PLs throughout lactation in very preterm milk (delivered between the 26–30th gestational week) and preterm milk compared to term milk, were reported [21]. The measured concentrations of PC and SM in term milk over the course of lactation are in agreement with Claumarchirant et al. [23] and Ma et al. [24]. Conversely, the concentration of PE contributed heavily to the discrepancies between this and other studies. The concentrations found in this study was 49.64 ± 13.68 mg/100 mL in term colostrum and 29.15 ± 13.04 mg/100 mL in term mature milk (≤ 15 weeks postpartum). In contrast, Claumarchirant et al. [23] found the concentration of PE to range between 15.9-9.5 mg/100 mL, while the concentrations of PE measured by Ma et al. [24] was 9.0 ± 2.6 mg/100 mL in colostrum and 3.9 ± 1.6 mg/100 mL in mature milk.

Discrepancies between the present study and other studies might stem from the diverse nature of PLs. GPL consist of a glycerol backbone, to which fatty acyls are bound to the first and/or second carbon and a phosphate group with a polar headgroup attached to the third carbon; accordingly, the SL SM consists of a sphingosine backbone with phosphocholine and a fatty acyl attached [25]. The structural composition of both GPL and SL enables an array of individual PL species to be found within a PL class, depending on the attached fatty acid moieties. Hence, the discrepancies between studies might originate from the fact that measurements are conducted on different PL species. Furthermore, the majority of studies on the PL composition of HM do not report on the individual PL species but rather the distribution of individual fatty acyl moieties contributing to the various PL classes, thus precluding direct comparisons between studies. Discrepancies might also be influenced by the use of an internal standard during sample preparation. In the present study, the method used for quantification of PLs included the addition of deuterated internal standards prior to PL extraction. Addition of an internal standard is carried out to correct for potential analyte losses or analytical errors during sample preparation [26] and similarly during MS acquisition, which, along with the use of dynamic MRM, strengthens the quantification performed in this study. This approach might also explain the higher concentrations of PE observed in this study. In contrast, the study by Ma et al. [24] employed external standards for method validation.

Besides quantification of the three major PL classes in HM, total lipid content of the preterm and term milk samples from the present study was furthermore analysed by Guerra et al. [16] using GC-MS. To summarize, the mean concentration of total lipids in the preterm milk samples was significantly higher than the term milk samples [16]. Lactational stage also affected total lipid content, as the lipid concentration increased with maturation of the milk [16]. This inverse relationship between total lipid content and PLs is in agreement with finding by others [27]. As the milk matures, TAG synthesis in the ER of the alveolar cells increases, along with the demand for membrane material for the formation of cytoplasmic lipid droplets (CLD). As a result of the limited availability of PLs, coalescence of CLD can relieve the demand for PLs, subsequently leading to the release of larger MFGs into the lumen.

In preterm milk, both the total PL and total lipid concentration decreased with progression of lactation, which is in contrast to the findings in term milk [16]. These findings might indicate an altered MFG size distribution and/or concentration of MFGs in preterm milk during early lactation. Differences in the size distribution of fat globules in preterm and term milk was also observed by Simonin et al [28], who found a higher content of small milk fat globules (SMFG) and large milk fat globules (LMFG) in preterm milk, as well as differences in the concentration of MFGs. These differences were observed to be more pronounced in colostrum and diminished in mature milk [28].

For hydrolysis and digestion of the globules in the infant gastrointestinal tract, an increased surface area for anchoring of gastric lipase [29] would be the result of a higher SMFG fraction in preterm colostrum and milk, which might subsequently be of benefit due to the immature gastrointestinal tract and reduced lipolytic capacity of premature infants. A higher PL content have previously been related to the SMFG fraction in bovine milk [30].

Independent of gestation, PLs containing the LC-PUFA AA and DHA were especially prevalent in PE, which is in accordance with previous studies, that reported that DHA and EPA (an intermediate of DHA synthesis) were more abundant in PE than PC and SM [9]. DHA and AA are often found as FA moieties of PE and PC and contribute to myelination, as well as neurological and visual development of the neonate [31,32].

The concentration of 25 out of 31 individual PL species were found to be higher in preterm milk compared to term milk throughout lactation. Interestingly, multivariate analysis showed that the PL species which contributed the most to the differences in preterm and term milk early postpartum belonged to the PC and SM class. PC and SM are both important sources of choline, as they account for approximately 17% of choline in HM [12]. Choline is also found in HM as free choline, phosphocholine and glycerophosphocholine. Accretion of choline begins in utero, as choline is transferred from the placenta to the foetus, playing crucial roles in organ development of, for example, the brain and neural tube [33], in membrane biosynthesis [12] and in normal cell function [34]. The importance of choline for infant development continues postpartum and includes involvement in the structural integrity of cell membranes [35], as a precursor for acetylcholine in cholinergic neurotransmission [11] and as a methyl-donor for the synthesis of methionine [34]. Studies have furthermore suggested increased choline requirements of preterm infants [34].

Besides palmitic acid (C16:0), the LC-PUFA AA (C20:4), DGLA (C20:3) and linoleic acid (C18:2) from the N6-series of FAs were the principal FA that were attached to the PC species that contributed to the observed differences in preterm and term PLs early postpartum. The accretion of these LC-PUFA is likewise involved in neural and visual development and they are accumulated in the foetus during the last trimester [20].

The altered PL composition of preterm colostrum and milk subsided with maturation and our findings indicate that the composition of mature (3–15 weeks) preterm milk might be similar to that of mature term milk. A similar trend was observed by Sundekilde et al. [15] for the metabolite profiles of the preterm and term colostrum and milk samples analysed in the present study. By proton NMR spectroscopy, the metabolome of preterm colostrum was revealed to differ from that of term colostrum, after which preterm milk changed with progression of lactation, which resulted in the composition of preterm milk resembling term milk at 5–7 weeks postpartum [15]. Other milk constituents such as carbohydrates, fat [36], protein and minerals [37] have similarly been reported to be affected by gestational age. Preterm infants are delivered when the in utero growth-rate is at its highest [38]; thus, the growth-rate of preterm infants is much higher than for infants delivered at term [39].

The composition of HM, especially the lipid fraction, is known to be subject to both inter- and intraindividual variability [3,40]. For the concentration of PL in both preterm and term samples, large standard deviations were calculated. Diurnal variation, variation during a feed (foremilk or hindmilk), as well as maternal diet, are factors known to affect the lipid fraction [41]. The mentioned factors were uncontrolled in this study, which could possibly have contributed to the observed variations. The small sample size is a substantial limitation to this study and corroboration of the work presented is therefore necessary.

5. Conclusions

In this study, the effect of time postpartum and gestational age on the composition of the three major PL classes in milk (PE, PC and SM) constituting a total of 31 PL species, was evaluated. The concentration of PC and SM as well as total PL was significantly higher in milk from preterm mothers throughout lactation. Accordingly, the majority of PL species from all three PL classes were significantly higher in preterm milk. The observed differences between preterm and term PL composition were most pronounced in colostrum and attenuated in mature milk. However, due to the small sample size, confirmation of these findings is needed.

The underlying mechanisms behind the alterations of early lactation PL composition of preterm milk, are unknown. However, immaturity of the mammary and/or a compensatory mechanism to meet the higher metabolic demands in preterm infants gland might influence the observed phenotype [42]. Nevertheless, advancements must be made in the understanding of how to adequately meet the nutritional needs of preterm infants, to ensure preterm growth and development on par with infants delivered at term.

Author Contributions: Conceptualization, U.K.S., E.D., A.L.K.; methodology, I.E.I.L., V.M.A.; formal analysis, I.E.I.L., U.K.S.; investigation, E.D., C.-A.O., I.E.I.L., U.K.S., V.M.A.; resources, I.E.I.L., V.M.A., E.D., J.A.O., C.-A.O., C.A.R., A.L.K., H.C.B., U.K.S.; writing—original draft preparation, I.E.I.L.; writing—review and editing, I.E.I.L., V.M.A., E.D., J.A.O., C.-A.O., C.A.R., A.L.K., H.C.B., U.K.S.; supervision, U.K.S.; funding acquisition, U.K.S.

Funding: This research was funded by Aarhus University, grant number 2017-524.

Acknowledgments: The authors would like to acknowledge the Western Trust Milk Bank, Irvinestown, Co. Fermanagh, Ireland, the staff of the neonatal intensive care unit of Cork University Maternity Hospital and all of the mothers and infants who kindly donated their milk.

Conflicts of Interest: The authors declare no conflict of interest. The funders had no role in the design of the study; in the collection, analyses or interpretation of data; in the writing of the manuscript or in the decision to publish the results.

References

1. Cai, X.; Wardlaw, T.; Brown, D.W. Global Trends in Exclusive Breastfeeding. *Int. Breastfeed. J.* **2012**, *7*, 1–5. [CrossRef] [PubMed]
2. Bridgman, S.L.; Konya, T.; Azad, M.B.; Sears, M.R.; Becker, A.B.; Turvey, S.E.; Mandhane, P.J.; Subbarao, P.; CHILD Study Investigators; Scott, J.A.; et al. Infant Gut Immunity: A Preliminary Study of IgA Associations with Breastfeeding. *J. Dev. Orig. Health Dis.* **2016**, *7*, 68–72. [CrossRef] [PubMed]
3. Delplanque, B.; Gibson, R.; Koletzko, B.; Lapillonne, A.; Strandvik, B. Lipid Quality in Infant Nutrition: Current Knowledge and Future Opportunities. *J. Pediatr. Gastroenterol. Nutr.* **2015**, *61*, 8–17. [CrossRef] [PubMed]
4. EFSA Panel on Dietetic Products, Nutrition and Allergies (NDA). Scientific Opinion on the Essential Composition of Infant and Follow-on Formulae. *EFSA J.* **2014**, *12*, 3760. [CrossRef]
5. Shoji, H.; Shimizu, T.; Kaneko, N.; Shinohara, K.; Shiga, S.; Saito, M.; Oshida, K.; Shimizu, T.; Takase, M.; Yamashiro, Y. Comparison of the Phospholipid Classes in Human Milk in Japanese Mothers of Term and Preterm Infants. *Acta Paediatr.* **2006**, *95*, 996–1000. [CrossRef] [PubMed]
6. Gallier, S.; Gragson, D.; Cabral, C.; Jiménez-Flores, R.; Everett, D.W. Composition and Fatty Acid Distribution of Bovine Milk Phospholipids From Processed Milk Products. *J. Agric. Food Chem.* **2010**, *58*, 10503–10511. [CrossRef] [PubMed]
7. Blans, K.; Hansen, M.S.; Sørensen, L.V.; Hvam, M.L.; Howard, K.A.; Möller, A.; Wiking, L.; Larsen, L.B.; Rasmussen, J.T. Pellet-free Isolation of Human and Bovine Milk Extracellular Vesicles by Size-Exclusion Chromatography. *J. Extracell. Vesicles* **2017**, *6*, 1–16. [CrossRef]

8. Sæle, Ø.; Rød, K.E.L.; Quinlivan, V.H.; Li, S.; Farber, S.A. A Novel System to Quantify Intestinal Lipid Digestion and Transport. *BBA Mol. Cell Biol. Lipids* **2018**, *1863*, 948–957. [CrossRef]
9. Wang, L.; Shimizu, Y.; Kaneko, S.; Hanaka, S.; Abe, T.; Shimasaki, H.; Hisaki, H.; Nakajima, H. Comparison of the Fatty Acid Composition of Total Lipids and Phospholipids in Breast Milk from Japanese Women. *Pediatr. Int.* **2000**, *42*, 14–20. [CrossRef]
10. Moukarzel, S.; Dyer, R.A.; Keller, B.O.; Elango, R.; Innis, S.M. Human Milk Plasmalogens Are Highly Enriched in Long-Chain PUFAs. *J. Nutr.* **2016**, *146*, 2412–2417. [CrossRef]
11. Maas, C.; Franz, A.R.; Shunova, A.; Mathes, M.; Bleeker, C.; Poets, C.F.; Schleicher, E.; Bernhard, W. Choline and Polyunsaturated Fatty Acids in Preterm Infants' Maternal Milk. *Eur. J. Nutr.* **2017**, *56*, 1733–1742. [CrossRef] [PubMed]
12. Cilla, A.; Quintaes, K.D.; Barberá, R.; Alegría, A. Phospholipids in Human Milk and Infant Formulas: Benefits and Needs for Correct Infant Nutrition. *Crit. Rev. Food Sci. Nutr.* **2016**, *56*, 1880–1892. [CrossRef] [PubMed]
13. Gidrewicz, D.A.; Fenton, T.R. A Systematic Review and Meta-Analysis of the Nutrient Content of Preterm and Term Breast Milk. *BMC Pediatr.* **2014**, *14*, 216. [CrossRef] [PubMed]
14. Lemons, J.A.; Moye, L.; Hall, D.; Simmons, M. Differences in the Composition of Preterm and Term Human Milk During Early Lactation. *Pediatr. Res.* **1982**, *16*, 113–117. [CrossRef] [PubMed]
15. Sundekilde, U.K.; Downey, E.; O'Mahony, J.; O'Shea, C.; Ryan, C.A.; Kelly, A.; Bertram, H.C. The Effect of Gestational and Lactational Age on the Human Milk Metabolome. *Nutrients* **2016**, *8*, 304. [CrossRef] [PubMed]
16. Guerra, E.; Downey, E.; O'Mahony, J.A.; Caboni, M.F.; O'Shea, C.; Ryan, A.C.; Kelly, A.L. Influence of Duration of Gestation on Fatty Acid Profiles of Human Milk. *Eur. J. Lipid Sci. Technol.* **2016**, *118*, 1775–1787. [CrossRef]
17. Spevacek, A.R.; Smilowitz, J.T.; Chin, E.L.; Underwood, M.A.; German, J.B.; Slupsky, C.M. Infant Maturity at Birth Reveals Minor Differences in the Maternal Milk Metabolome in the First Month of Lactation. *J. Nutr.* **2015**, *145*, 1698–1708. [CrossRef]
18. Bligh, E.G.; Dyer, W.J. A Rapid Method of Total Lipid Extraction and Purification. *Can. J. Biochem. Physiol.* **1959**, *37*, 911–917. [CrossRef]
19. Zhao, Y.Y.; Xiong, Y.; Curtis, J.M. Measurement of Phospholipids by Hydrophilic Interaction Liquid Chromatography Coupled to Tandem Mass Spectrometry: The Determination of Choline Containing Compounds in Foods. *J. Chromatogr. A* **2011**, *1218*, 5470–5479. [CrossRef]
20. Sala-Vila, A.; Castellote, A.I.; Rodriguez-Palmero, M.; Campoy, C.; López-Sabater, M.C. Lipid Composition in Human Breast Milk from Granada (Spain): Changes During Lactation. *Nutrition* **2005**, *21*, 467–473. [CrossRef]
21. Bitman, J.; Wood, L.; Hamosh, M.; Mehta, N.R. Comparison of the Phospholipid Composition of Breast Milk from Mothers of Term and Preterm Infants During Lactation. *Am. J. Clin. Nutr.* **1984**, *40*, 1103–1119. [CrossRef] [PubMed]
22. Benoit, B.; Fauquant, C.; Daira, P.; Peretti, N.; Guichardant, M.; Michalski, M.C. Phospholipid Species and Minor Sterols in French Human Milks. *Food Chem.* **2010**, *120*, 684–691. [CrossRef]
23. Claumarchirant, L.; Cilla, A.; Matencio, E.; Sanchez-Siles, L.M.; Castro-Gomez, P.; Fontecha, J.; Alegría, A.; Lagarda, M.J. Addition of Milk Fat Globule Membrane as an Ingredient of Infant Formulas for Resembling the Polar Lipids of Human Milk. *Int. Dairy J.* **2016**, *61*, 228–238. [CrossRef]
24. Ma, L.; MacGibbon, A.K.H.; Jan Mohamed, H.J.B.; Loy, S.L.; Rowan, A.; McJarrow, P.; Fong, B.Y. Determination of phospholipid concentrations in breast milk and serum using a high performance liquid chromatography–mass spectrometry–multiple reaction monitoring method. *Int. Dairy J.* **2017**, *71*, 50–59. [CrossRef]
25. Smoczynski, M. Role of Phospholipid Flux during Milk Secretion in the Mammary Gland. *J. Mammary Gland Biol. Neoplasia* **2017**, *22*, 117–129. [CrossRef] [PubMed]
26. Wang, M.; Wang, C.; Han, X. Selection of Internal Standards for Accurate Quantification of Complex Lipid Species in Biological Extracts by Electrospray Ionization Mass Spectrometry—What, How and Why? *Mass Spectrom. Rev.* **2017**, *36*, 693–714. [CrossRef] [PubMed]
27. Koletzko, B.; Rodriguez-palmero, M. Polyunsaturated Fatty Acids in Human Milk and Their Role in Early Infant Development. *J. Mammary Gland Boil. Neoplasia* **1999**, *4*, 269–284. [CrossRef]

28. Simonin, C.; Rüegg, M.; Sidiropoulos, D. Comparison of the fat content and fat globule size distribution of breast milk from mothers delivering term and preterm. *Am. J. Clin. Nutr.* **1984**, *40*, 820–826. [CrossRef] [PubMed]
29. Singh, H.; Gallier, S. Nature's Complex Emulsion: The Fat Globules of Milk. *Food Hydrocoll.* **2017**, *68*, 81–89. [CrossRef]
30. Lopez, C.; Briard-Bion, V.; Ménard, O.; Beaucher, E.; Rousseau, F.; Fauquant, J.; Leconte, N.; Robert, B. Fat Globules Selected from Whole Milk According to Their Size: Different Compositions and Structure of the Biomembrane, Revealing Sphingomyelin-rich Domains. *Food Chem.* **2011**, *125*, 355–368. [CrossRef]
31. González, H.F.; Visentin, S. Nutrients and Neurodevelopment: Lipids. Update. *Arch. Argent. Pediatr.* **2016**, *114*, 472–476. [CrossRef] [PubMed]
32. Campoy, C.; Escolano-Margarit, V.; Anjos, T.; Szajewska, H.; Uauy, R. Omega 3 Fatty Acids on Child Growth, Visual Acuity and Neurodevelopment. *Br. J. Nutr.* **2012**, *107*, S85–S106. [CrossRef] [PubMed]
33. Fisher, M.C.; Zeisel, S.H.; Mar, M.; Sadler, T.W. Inhibitors of Choline Uptake and Metabolism Cause Developmental Abnormalities in Neurulating Mouse Embryos. *Teratology* **2001**, *64*, 114–122. [CrossRef] [PubMed]
34. Bernhard, W.; Full, A.; Arand, J.; Maas, C.; Poets, C.F.; Franz, A.R. Choline Supply of Preterm Infants: Assessment of Dietary Intake and Pathophysiological Considerations. *Eur. J. Nutr.* **2013**, *52*, 1269–1278. [CrossRef] [PubMed]
35. Bernhard, W.; Raith, M.; Kunze, R.; Koch, V.; Heni, M.; Maas, C.; Abele, H.; Poets, C.F.; Franz, A.R. Choline Concentrations are Lower in Postnatal Plasma of Preterm Infants Than in Cord Plasma. *Eur. J. Nutr.* **2015**, *54*, 733–741. [CrossRef] [PubMed]
36. Bauer, J.; Gerss, J. Longitudinal Analysis of Macronutrients and Minerals in Human Milk Produced by Mothers of Preterm Infants. *Clin. Nutr.* **2011**, *30*, 215–220. [CrossRef]
37. Tudehope, D.I. Human milk and the nutritional needs of preterm infants. *J. Pediatr.* **2013**, *162*, S17–S25. [CrossRef] [PubMed]
38. Robinson, D.T.; Martin, C.R. Fatty acid requirements for the preterm infant. *Semin. Fetal Neonatal Med.* **2017**, *22*, 8–14. [CrossRef] [PubMed]
39. Caudill, M.A. Pre- and Postnatal Health: Evidence of Increased Choline Needs. *J. Am. Diet. Assoc.* **2010**, *110*, 1198–1206. [CrossRef]
40. Sabel, K.G.; Lundqvist-Persson, C.; Bona, E.; Petzold, M.; Strandvik, B. Fatty acid patterns early after premature birth, simultaneously analysed in mothers' food, breast milk and serum phospholipids of mothers and infants. *Lipids Health Dis.* **2009**, *8*, 1–15. [CrossRef]
41. Ballard, O.; Morrow, A.L. Human Milk Composition: Nutrients and Bioactive Factors. *Pediatr. Clin. N. Am.* **2013**, *60*, 49–74. [CrossRef] [PubMed]
42. Nilsson, A.K.; Löfqvist, C.; Najm, S.; Hellgren, G.; Sävman, K.; Andersson, M.X.; Smith, L.E.H.; Hellström, A. Long-chain polyunsaturated fatty acids decline rapidly in milk from mothers delivering extremely preterm indicating the need for supplementation. *Acta Paediatr. Int. J. Paediatr.* **2018**, 1–8. [CrossRef] [PubMed]

© 2019 by the authors. Licensee MDPI, Basel, Switzerland. This article is an open access article distributed under the terms and conditions of the Creative Commons Attribution (CC BY) license (http://creativecommons.org/licenses/by/4.0/).

Article

Carotenoid Content in Breastmilk in the 3rd and 6th Month of Lactation and Its Associations with Maternal Dietary Intake and Anthropometric Characteristics

Monika A. Zielinska [1], Jadwiga Hamulka [1,*] and Aleksandra Wesolowska [2]

1. Department of Human Nutrition, Faculty of Human Nutrition and Consumer Sciences, Warsaw University of Life Sciences—SGGW, 159 Nowoursynowska St., 02-776 Warsaw, Poland; monika_zielinska@sggw.pl
2. Laboratory of Human Milk and Lactation Research at Regional Human Milk Bank in Holy Family Hospital, Department of Neonatology, Faculty of Health Sciences, Medical University of Warsaw, 63A Zwirki i Wigury St., 02-091 Warsaw, Poland; aleksandra.wesolowska@wum.edu.pl
* Correspondence: jadwiga_hamulka@sggw.pl; Tel.: +48-22-593-71-12

Received: 20 December 2018; Accepted: 14 January 2019; Published: 18 January 2019

Abstract: Carotenoids are diet-dependent milk components that are important for the visual and cognitive development of an infant. This study determined β-carotene, lycopene and lutein + zeaxanthin in breastmilk and its associations with dietary intake from healthy Polish mothers in the first six months of lactation. Concentrations of carotenoids in breastmilk were measured by HPLC (high-performance liquid chromatography) (first, third, sixth month of lactation) and dietary intake was assessed based on a three-day dietary record (third and sixth month of lactation). The average age of participants (n = 53) was 31.4 ± 3.8 years. The breastmilk concentrations of carotenoids were not changed over the progress of lactation. Lycopene was a carotenoid with the highest content in breastmilk (first month 112.2 (95% CI 106.1–118.3)—sixth month 110.1 (103.9–116.3) nmol/L) and maternal diet (third month 7897.3 (5465.2–10329.5) and sixth month 7255.8 (5037.5–9474.1) µg/day). There was a positive correlation between carotenoids in breastmilk and dietary intake (lycopene r = 0.374, r = 0.338; lutein + zeaxanthin r = 0.711, r = 0.726, 3rd and 6th month, respectively) and an inverse correlation with maternal BMI in the third month of lactation (β-carotene: r = −0.248, lycopene: r = −0.286, lutein + zeaxanthin: r = −0.355). Adjusted multivariate regression models confirmed an association between lutein + zeaxanthin intake and its concentration in breastmilk (third month: $β$ = 0.730 (0.516–0.943); 6th: $β$ = 0.644 (0.448–0.840)). Due to the positive associations between dietary intake and breastmilk concentrations, breastfeeding mothers should have a diet that is abundant in carotenoids.

Keywords: bioactive factors; carotenoids; dietary intake; high-performance liquid chromatography (HPLC); human milk; lactation; maternal diet; prospective study

1. Introduction

Breastfeeding is the best feeding method for newborns, infants and toddlers. International organizations (e.g., WHO, UNICEF, AAP, ESPGHN) recommend exclusive breastfeeding during the first six months of life and further breastfeeding along with complementary feeding up to two years or more, as long as the infant and mother desire [1–3]. Exclusive breastfeeding at this time provides sufficient energy and macronutrients as well as most of the micronutrients (except vitamin D and K) to meet the mean requirements of healthy term-born infants [3,4]. During this time, breastmilk provides not only necessary nutrients, however also a variety of bioactive factors, such as immunoglobins, stem cells, cytokines, hormones, growth factors and phytochemicals (e.g., carotenoids and flavonoids) [5–7]. These compounds are crucial for optimal development and further health, including decreased risk of infectious diseases during childhood as well as chronic non-communicable diseases throughout the lifespan [2,8]. Previous studies have shown that breastmilk composition is influenced by many determinants, including maternal factors (e.g., age, nutritional status, dietary intake, tobacco use or passive smoking), infant factors (gestational age, age, gender) and physiological factors (lactation stage, nursing stage, diurnal variation) [9–14]. The content of some nutrients in breast milk is dependent on maternal dietary intake, including fatty acid profile [15], iodine [16] and vitamins B_1, B_2, B_6 and B_{12} [17]. However, maternal dietary intake has a limited influence on breastmilk macronutrient composition [14] and some micronutrients, e.g., folic acid [17] and zinc [18]. Some studies have indicated that maternal nutritional status measured as body mass index (BMI) may influence the milk fat concentrations [14] as well as carotenoid concentrations [19].

Carotenoids (β-carotene, lycopene, lutein, zeaxanthin, β-cryptoxanthin, astaxanthin) are plant bioactive compounds that cannot be synthesized by mammals however are supplied with a high dietary intake of vegetables and fruits (especially green, yellow, orange or red) or some animal products (e.g., yolk, salmon or rainbow trout) [20–22]. Numerous studies in recent years have shown their positive impact on human health, mainly in adults [23–25], however also during pregnancy and early childhood, as noted previously [6,26]. Recent studies focusing on breastmilk carotenoids have shown that their concentration depends on maternal dietary intake [27] and varies among populations [19,28–30], especially due to different dietary habits. Other factors may also affect the carotenoid concentration in breast milk, such as education status or anthropometric parameters of mothers [19]. Giordano and Quadro [26], in a recent review, emphasize the importance of monitoring maternal and infant lutein and zeaxanthin status due to its importance during pregnancy and the neonatal period [26]. An analysis of breastmilk carotenoids may be a useful, non-invasive method due to the correlations between maternal carotenoid status, breastmilk carotenoids and infant carotenoid status [27,30–33]. To date, few studies have considered the impact of dietary intake of carotenoids and maternal factors on breastmilk carotenoid concentration and there is a lack of long-term studies on the subject.

The aims of this study were (1) to investigate the concentrations of selected carotenoids (β-carotene, lycopene and lutein + zeaxanthin) in breastmilk from healthy mothers living in an urban area of Poland in the first six months of lactation (2) to determine the dietary intake of carotenoids in the third and sixth months of lactation and (3) to explore the associations between breastmilk carotenoids and dietary intake and other maternal factors.

2. Materials and Methods

2.1. Ethical Approval

The study was approved by the Ethics Committee of the Medical University of Warsaw in 2015, Resolution No. AKBE/139/15. Written consent was obtained from all participants and the study was conducted in compliance with the Helsinki Declaration.

2.2. Study Design

The study consisted of three study sessions at the first, third and sixth month of lactation. All mothers completed a questionnaire which included detailed questions about lifestyle, sociodemographic, health factors and nutrition (e.g., maternal education, household income, dietary supplement use) during the pre-conceptional, prenatal and postnatal periods. At each study visit, breastmilk samples were also collected and anthropometric measurements were assessed in infants and mothers, maternal psychological status was evaluated, as well as the infant psychomotor development in the sixth month. Detailed information about the study design is shown in Table 1. The data used in this paper are part of the data that were obtained during this study.

Table 1. Study design characteristics.

Study Visit		Assessment			
		Breastmilk	Nutritional	Anthropometric	Psychological
1st (1st month)	Mother	- macronutrients - carotenoids - fatty acid profile	- Food Frequency Questionnaire FFQ-6 [1] [34] - dietary supplement use	- pre-pregnancy, 14 Hbd [2], 27 Hbd body mass - current body mass	- EPDS [3] [35] - PSS-10 [4] [36,37]
	Infant	-	- breastfeeding frequency - dietary supplement use	- birth parameters - body mass - body length - head circumference	-
2nd (3rd month)	Mother	- macronutrients - carotenoids - fatty acid profile	- 3-day dietary record - carotenoid intake - dietary supplement use	- current body mass	- EPDS [3] [35] - PSS-10 [4] [36,37]
	Infant	-	- breastfeeding frequency - dietary supplement use	- body mass - body length - head circumference	-
3rd (6th month)	Mother	- macronutrients - carotenoids - fatty acid profile	- 3-day dietary record - carotenoid intake - dietary supplement use	- current body mass	- EPDS [3] [35] - PSS-10 [4] [36,37]
	Infant	-	- breastfeeding frequency - dietary supplement use - introduced to other foods or drinks	- body mass - body length - head circumference	- DSR [5] Scale [38].

[1] FFQ-6—Food Frequency Questionnaire; [2] Hbd—hebdomas (weeks of gestation); [3] EPDS—Edinburgh Postpartum Depression Scale; [4] PSS-10—Perceived Stress Scale; [5] DSR—Children Development Scale.

2.3. Study Participants

Recruitment was conducted between April 2015 and July 2017 in the Holy Family Hospital in Warsaw among patients of Obstetrics Clinic as well as breastfeeding women from the local community using social media groups. Eligibility criteria included women between 18 and 45 years of age giving birth to a single, healthy infant. The exclusion criteria included maternal chronic disease (kidney, liver, gastrointestinal diseases influencing nutrient absorption, hypertension, diabetes type I or II) and pregnancy complications (including preeclampsia or eclampsia, pregnancy induced hypertension). The study also included 53 mother-infant pairs (Figure 1). The mean age of mothers (n = 53) was 31.4 ± 3.8 years. Most of the mothers had a university education and a high average income. Before pregnancy, most of them had a normal BMI (n = 46; 87%); at the first month of lactation, 78% were classified as having normal body weight (n = 40), 86% (n = 42) at the third month and 85% (n = 40) at the sixth month. Further, 25 participants (47%) were primiparas and 53% of the infants were females (n = 28; detailed study group characteristics are shown in Table S1). Most of the study visits took place during the spring-summer period (no statistically significant differences between the number of visits in particular seasons, p = 0.482).

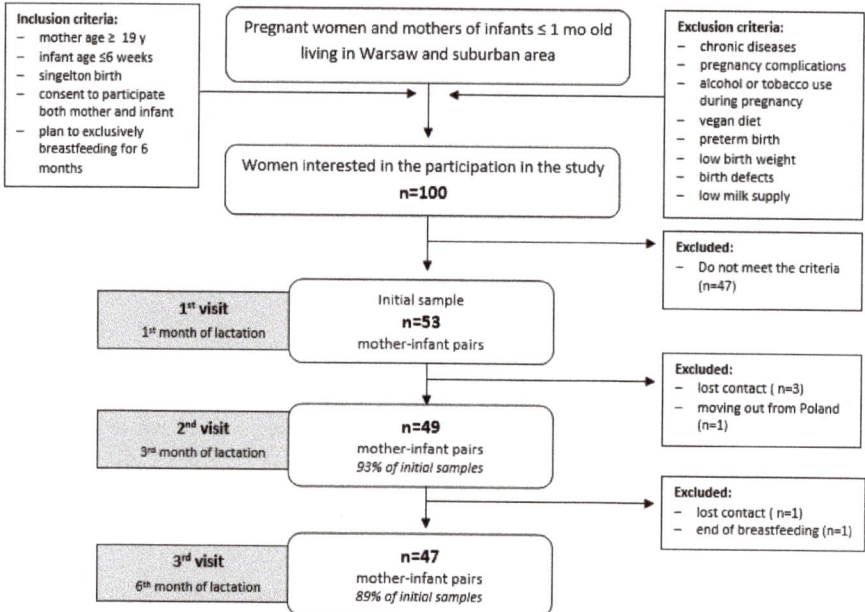

Figure 1. Flowchart: study design and study sample collection.

2.4. Breastmilk Collection

Breastmilk samples were collected by participants at home after being given detailed instructions. The same amount (5–10 mL) of pre-feeding and post-feeding breastmilk was collected 24 hours prior to each study visit at four time periods (06:00–12:00; 12:00–18:00; 18:00–24:00; 24:00–06:00) to minimize the daily differences in milk composition due to the 24-hour variation in carotenoid concentrations that is described in literature [39,40]. The samples were obtained by a breast pump or manually from the breast(s) that the infant fed from. Breastmilk samples were transported to the Holy Family Hospital in Warsaw under cooling conditions. Next, the same amount from all four samples was collected and mixed in a Vortex shaker IKA MS2 (IKA Works Inc., Wilmington, North Carolina, USA) for one minute. The pooled sample was distributed in 2 mL clear polypropylene tubes, labelled and stored at −80 °C for later analysis within six months of collection. Precautions were taken throughout the breastmilk sample collection and procedures to minimize the exposure of samples to temperature, light and air.

2.5. Breastmilk Composition Analysis

Breastmilk energy value and fat content per 100 mL of the raw sample of breastmilk were analysed using a MIRIS human milk analyser HMA (Miris, Uppsala, Sweden). Prior to the analysis, each sample (n = 149) was warmed to 40 °C and homogenized (1.5 s/1 mL of sample) using a sonicator (milk homogenizer, Miris, Uppsala, Sweden) according to the method described previously [14]. From each raw pooled sample, macronutrient composition was analysed three times (~2 mL per analysis) and the average of the three measurements was used for further statistical analysis.

2.6. Breastmilk Carotenoid Analysis

Carotenoid concentration (β-carotene, lycopene and lutein + zeaxanthin) in milk samples was assessed using high-performance liquid chromatography (HPLC). Milk samples for analysis were prepared based on the modified method published by Macias and Schweigert [41]. To 2 mL of milk, 500 μL of a 12% solution of pyrogallol, 50 μL, 1% ascorbic acid in 0.1 M HCl, 1.5 mL 30% KOH solution and 2.5 mL ethanol were successively added. The mixture was shaken for 30 seconds with a Vortex shaker and was then incubated in a 50 °C water bath for 40 minutes. Subsequently, 1 mL of saturated NaCl solution and 1 mL of n-hexane were added to the ice-cooled samples, followed by vigorous shaking for three minutes. The sample was then centrifuged for 10 minutes at 4 °C (8000 rpm). Immediately after centrifugation, the upper hexane layer was transferred to a new tube. The extraction was repeated two more times. The combined hexane extracts were evaporated in a vacuum evaporator (30 minutes at 30 °C). The formed precipitate was dissolved in 0.5 mL of hexane and was transferred to dark glass vials.

The content of individual carotenoids in the prepared milk samples was determined using the Shimadzu HPLC system (Japan: 2 LC-20AD pumps, CMB-20A controller system, SIL-20AC autosampler, UV/IS SPD-20AV detector, CTD-20AC controller) using C18 Synergi Fusion-RP 80i columns (250 × 4.60 mm, Phenomenex, CA, USA). The determination of the studied carotenoids was carried out at a wavelength of 471 nm for lycopene, 450 nm for β-carotene and 445 nm for lutein + zeaxanthin. The mobile phase consisted of two phases—phase A: acetonitrile/methanol mixed in proportions of 90:10 (90/10; v/v) and phase B: methanol/ethyl acetate, in a 34:16 ratio (34/16; v/v). The flow rate of the developing mixture was 1.0 cm^3/min and the sample injection was 100 μL. Individual carotenoid concentrations in milk were calculated by comparing them with a corresponding standards curve (standards of catalogue numbers: β-carotene C4582, lutein + zeaxanthin X6250, lycopene L9879 from Sigma-Aldrich Inc., Merck KGaA, Darmstadt, Germany). The standard curves for all carotenoids are shown in supplementary materials (Figure S1). The concentrations of the studied carotenoids were expressed in nmol/L. Since lutein and zeaxanthin could not be completely resolved and were summed, all references to milk lutein concentrations refer to lutein + zeaxanthin.

2.7. Dietary Intake Analysis

The maternal diet at the 3rd and 6th month of lactation was assessed using the 3-day dietary record conducted by respondents before the study visits (on three typical days, two weekdays and one weekend day). Respondents received all necessary information and instructions on conducting a dietary record, including the importance of scrupulosity recording all consumed foods and beverages during the period covered by the 3-day dietary records. During the recording, respondents used a kitchen scale to measure the weight of the serving, or if they did not have kitchen scales, they estimated the servings using typical household measures. All sizes of portions were verified using the "Album of Photographs of Food Products and Dishes" created by the National Food and Nutrition Institute [42]. All nutritional data were collected and analysed by the qualified dietician.

Based on the data from the 3-day record of daily energy value and dietary intake of selected macronutrients (protein, carbohydrates, fat) and micronutrients (vitamin A, E, D, B_1, B_2, B_3, B_6, B_{12}, C, folic acid; iodine, calcium, potassium, phosphor, magnesium, iron, zinc, copper, manganese), the dietary fibre and fatty acid profile were calculated using Dieta 5.0 Software (National Food and Nutrition Institute, Warsaw, Poland) based on Polish food-composition tables [43]. The total dietary intake of selected carotenoids (lutein + zeaxanthin, lycopene and β-carotene) were estimated using data from the USDA Database [22]. In addition, data on dietary supplement use in the preconceptional prenatal and postnatal periods were collected including the used dose, name and the brand of dietary supplement. The data were used to calculate the average daily nutrient intake with dietary supplements. The intake of nutrients was calculated as a mean from the three recorded days to obtain the mean daily intake from diet, dietary supplements and the sum of both for each participant. For the purpose of this study, we used only energy value, fat intake and intake of antioxidant fat-soluble vitamins (A and E) and carotenoids (lutein + zeaxanthin, lycopene and β-carotene) for analysis.

2.8. Anthropometric Measurements

The body weight (kg) and height (cm) of mothers were measured according to the International Standards for Anthropometric Assessment [44] to the nearest 0.1 kg or 0.1 cm, respectively, using a professional stadiometer (Seca 799, Hamburg, Germany). All measurements were taken in light clothing and without shoes. Body mass index (BMI, kg/m^2) was calculated and classified according to the World Health Organization (WHO) [45]. BMI and body mass changes between visits were calculated.

2.9. Statistical Analysis

The parameters that were analysed in this study were presented as: means and standard deviation or 95% coefficient intervals (CI), minimum and maximum values. To assess the normality of distributions, the Shapiro–Wilk test was applied. The differences between groups were analysed using the Wilcoxon matched pairs test (for two repeated measurements) or the Friedman ANOVA test (for three repeated measurements). For qualitative variables, the Cochrane Q test was used to check the differences between three repeated measurements and the Chi^2 McNemar test was used for the two repeated measurements.

The correlations between breastmilk carotenoids and the maternal anthropometric and sociodemographic characteristics, as well as dietary intake, were estimated with Spearman's rank or Kendall's *tau* correlations (if one of the variables was measured on an ordinal scale) coefficient. Linear regression was used to investigate the relationship between breastmilk carotenoids and the dietary intake of its carotenoids. The one univariate and final three multivariate models were specifically adjusted for season, maternal age, BMI, education, mode of delivery, fat, vitamin E and vitamin A intake. All of the analyses were performed using Statistica 13.1 software (Dell Inc., Round Rock, TX, USA). For all of the tests, $p \leq 0.05$ was considered as significant.

3. Results

3.1. Breastmilk Composition and Carotenoid Concentration

The average energy value, fat content and carotenoid concentrations in breastmilk at the 1st, 3rd and 6th month of lactation are shown in Table 2. The energy value and fat content in breastmilk were, on average, around 68–69 kcal/100mL and 3.8–3.9 g/100mL, respectively, and this level was similar regardless of the lactation month. The highest concentration of determined carotenoids was observed for lycopene (minimum–maximum values 77.2–176.9 nmol/L). The concentrations of β-carotene and lutein + zeaxanthin were at similar levels—mean values around 33 nmol/L (despite the insignificant 12% higher concentration of lutein + zeaxanthin at the sixth month due to the higher dietary intake at this month) and also did not differ between the months of lactation.

Table 2. Energy value, fat and carotenoid content in breastmilk in the first, third and sixth month of lactation.

Nutrient (Unit)	Breastmilk Composition Mean Value (95% CI [1]) Min–Max			p-Value [2]
	First Month (n = 53)	Third Month (n = 49)	Sixth Month (n = 47)	
Energy (kcal/100 mL)	69.5 (67.5–71.6) 57.0–89.7	68.2 (65.3–71.0) 49.3–95.5	68.4 (66.5–70.4) 53.7–90.3	0.777
Total fat (g/100 mL)	3.8 (3.6–4.1) 2.2–6.0	3.8 (3.5–4.1) 1.7–6.9	3.9 (3.7–4.1) 2.2–6.4	0.595
β-carotene (nmol/L)	33.2 (33.0–33.5) 31.9–36.3	33.1 (32.9–33.3) 31.8–35.9	33.3 (33.1–33.6) 32.3–35.4	0.436
Lycopene (nmol/L)	112.2 (106.1–118.3) 77.2–169.0	111.2 (105.0–117.3) 75.4–176.9	110.1 (103.9–116.3) 78.3–176.7	0.457
Lutein + zeaxanthin (nmol/L)	33.0 (26.3–39.7) 2.7–123.2	33.0 (24.1–41.8) 3.2–139.9	37.1 (26.5–47.8) 1.8–169.0	0.640

[1] CI—coefficient interval; [2] Friedman ANOVA test.

3.2. Dietary Intake

Table 3 shows the energy value and dietary intake of selected nutrients at the third and sixth month of lactation. There was no statistical difference between the third and sixth month for energy value and fat intake. We observed statistically significant differences between total vitamin E intake (dietary supplements and diet; $p = 0.000$) and for vitamin A, both dietary and dietary with supplements ($p = 0.010$ and $p = 0.009$, respectively). These differences are associated with the higher use of dietary supplements in the third month of lactation compared with the sixth month (90% vs. 81%; $p < 0.001$). The highest consumption was recorded for lycopene, followed by β-carotene and lutein + zeaxanthin in the lowest amount. We observed no significant differences in carotenoid intake between the months of lactation.

Table 3. Energy value and selected nutrients and carotenoid intake by participants at the third and sixth month of lactation.

Nutrient (Unit)	Dietary Intake Mean Value ± SD [1] or (95% CI [2]) Min–Max		p-Value [3]
	Third Month (n = 49)	Sixth Month (n = 47)	
Energy (kcal/day)	2193.7 ± 631.17 1186.5–3914.0	2046.2 ± 502.9 1051.4–3317.5	0.083
Total fat (g/day)	84.3 ± 28.5 37.2–185.6	76.4 ± 26.1 18.9–135.7	0.085
Vitamin E, dietary (mg α-tocopherol Equivalent/day)	12.6 ± 5.9 4.3–32.7	11.4 ± 5.1 4.7–24.5	0.258
Vitamin E, dietary & supplements (mg α-tocopherol Equivalent/day)	21.5 ± 13.9 4.3–57.7	13.1 ± 6.7 4.7–27.5	0.000
Vitamin A, dietary (μg Retinol Equivalent/day)	1289.5 ± 591.4 226.4–2447.3	1030.1 ± 500.0 213.1–2745.2	0.010
Vitamin A, dietary & supplements (μg Retinol Equivalent/day)	1295.0 ± 588.3 226.4–2447.3	1030.1 ± 500.0 213.1–2745.2	0.009

Table 3. Cont.

Nutrient (Unit)	Dietary Intake Mean Value ± SD [1] or (95% CI [2]) Min–Max		p-Value [3]
	Third Month (n = 49)	Sixth Month (n = 47)	
β-carotene (μg/day)	4480.8 (3575.0–5386.7) 319.6–16461.0	3441.9 (5037.5–9474.1) 716.2–9552.9	0.232
Lycopene (μg/day)	7897.3 (5465.2–10329.5) 477.2–30472.7	7255.8 (5037.5–9474.1) 339.2–38852.9	0.422
Lutein + zeaxanthin (μg/day)	2945.2 (1910.8–3979.6) 263.7–16678.9	3739.3 (2834.9–4643.7) 128.9–12207.3	0.054

[1] SD—standard deviation; [2] CI—coefficient interval; [3] Wilcoxon matched pairs test.

3.3. Associations between Carotenoid Concentrations, Dietary Intake and Maternal Characteristics

The associations between carotenoid concentrations, maternal factors and dietary intake are shown in Table 4. The intake of all of the analysed carotenoids positively correlated with its carotenoid breastmilk concentrations at the third and sixth months. For lutein, we observed a strong correlation (third month: $r = 0.711$, sixth month: $r = 0.726$; $p \leq 0.001$), for β-carotene—a weak and moderate correlation (third month: $r = 0.442$, sixth month: $r = 0.532$; $p \leq 0.001$), whereas for the lycopene, only a weak correlation (third month: $r = 0.374$; $p \leq 0.01$, sixth month: $r = 0.338$; $p \leq 0.05$). We found statistically significant adverse correlations between maternal BMI (kg/m^2) and all three carotenoid concentrations in breastmilk at the third month of lactation (β-carotene $r = -0.337$; lycopene $r = -0.286$; lutein + zeaxanthin $r = -0.355$; $p \leq 0.05$) and also β-carotene at the sixth month ($r = -0.337$; $p \leq 0.05$). However, the analysis based on BMI categories found significant inverse associations only for β-carotene at the sixth month ($r = -0.453$; $p \leq 0.05$).

Table 4. Correlations between breastmilk carotenoids and maternal characteristics and dietary intake.

Variables	Breastmilk β-carotene r Coefficient (p-Value)		Breastmilk Lycopene r Coefficient (p-Value)		Breastmilk Lutein + Zeaxanthin r Coefficient (p-Value)	
	Third Month (n = 49)	Sixth Month (n = 47)	Third Month (n = 49)	Sixth Month (n = 47)	Third Month (n = 49)	Sixth Month (n = 47)
Maternal characteristic						
Maternal age (years) [1]	0.015	−0.197	−0.136	0.113	−0.048	−0.009
Maternal education [2]	0.022	−0.091	0.026	0.046	−0.029	−0.020
Mode of delivery [2]	−0.009	−0.001	−0.094	−0.050	−0.175	−0.115
Number of children [2]	0.079	0.061	−0.003	0.105	0.057	0.037
BMI at third or sixth month (kg/m^2) [1]	−0.248 *	−0.337 *	−0.286 *	−0.119	−0.355*	−0.205
Maternal dietary intake						
Carotenoid intake (μg) [1,3]	0.442 ***	0.532 ***	0.374 **	0.338 *	0.711 ***	0.726 ***
Energy intake (kcal/day) [1]	0.041	0.182	−0.097	−0.005	−0.078	0.169
Fat intake (g/day) [1]	0.049	0.302 *	−0.197	−0.155	−0.105	0.167
Vitamin E intake (mg/day) [1]	0.270	0.062	0.082	0.013	0.123	−0.069
Vitamin A intake (μg/day) [1]	0.008	0.136	0.175	0.241	−0.027	−0.001
Breastmilk composition						
Breastmilk fat (g/100 mL) [1]	0.235	0.167	0.274	0.064	0.399 *	0.237

[1] Spearman rank correlation coefficient; [2] tau Kendall coefficient; [3] intake of respective carotenoid; * $p \leq 0.05$; ** $p \leq 0.01$; *** $p \leq 0.001$.

Table 5 presents the results of a linear regression analysis of carotenoid concentrations in breastmilk and its dietary intake. The univariate analysis reveals significant associations between dietary individual carotenoids intake in all analysed compounds. After the adjustments (models 2–4), we also found significant associations between dietary intake and breastmilk lycopene and β-carotene at the third month of lactation, however the models were not significant and explained only 3–14% of the variance in breastmilk carotenoids. However, even after adjustment for maternal age, BMI, education and mode of delivery, fat and vitamin A and E intake (model 4), dietary intake of lutein + zeaxanthin explained 51–68% of the variation at both times (third month, β = 0.730 (95% CI 0.516–0.943), p = 0.000; sixth month β = 0.644 (95% CI 0.448–0.840)) and β-carotene explained 35% of the variation in breastmilk β-carotene at the sixth month (β = 0.428, 95% CI 0.180–0.676).

Table 5. Univariate and multivariate linear regression models between breastmilk carotenoids and dietary intake and third and sixth months of lactation.

Model	Dietary Intake	Breastmilk Carotenoids at Third Month of Lactation			Breastmilk Carotenoids at Sixth Month of Lactation		
		β-carotene (n = 49)	Lycopene (n = 49)	Lutein + Zeaxanthin (n = 49)	β-carotene (n = 47)	Lycopene (n = 47)	Lutein + Zeaxanthin (n = 47)
1	β (95% CI) p-value	0.342 (0.066–0.618) 0.016	0.364 (0.087–0.640) 0.011	0.711 (0.504–0.917) 0.000	0.397 (0.121–0.672) 0.016	0.364 (0.084–0.643) 0.012	0.779 (0.591–0.967) 0.000
	Model parameters	$R^2 = 0.10$	$R^2 = 0.11$	$R^2 = 0.49$	$R^2 = 0.14$	$R^2 = 0.11$	$R^2 = 0.60$
2	β (95% CI) p-value	0.325 (0.054–0.596) 0.020	0.369 (0.088–0.650) 0.011	0.680 (0.468–0.891) 0.000	0.391 (0.116–0.665) 0.006	0.379 (0.092–0.665) 0.011	0.785 (0.593–0.977) 0.000
	Model parameters	$R^2 = 0.14$ 0.013	$R^2 = 0.10$ 0.039	$R^2 = 0.50$ 0.000	$R^2 = 0.07$ 0.037	$R^2 = 0.10$ 0.037	$R^2 = 0.60$ 0.000
3	β (95% CI) p-value	0.325 (0.054–0.596) 0.020	0.369 (0.088–0.650) 0.011	0.680 (0.468–0.891) 0.000	0.391 (0.116–0.665) 0.006	0.379 (0.092–0.665) 0.011	0.785 (0.593–0.977) 0.000
	Model parameters	$R^2 = 0.06$ 0.238	$R^2 = 0.11$ 0.126	$R^2 = 0.62$ 0.000	$R^2 = 0.29$ 0.005	$R^2 = 0.07$ 0.191	$R^2 = 0.62$ 0.000
4	β (95% CI) p-value	0.407 (0.094–0.721) 0.012	0.415 (0.104–0.726) 0.010	0.730 (0.516–0.943) 0.000	0.428 (0.180–0.676) 0.001	0.401 (0.089–0.713) 0.013	0.644 (0.448–0.840) 0.000
	Model parameters	$R^2 = 0.04$ 0.337	$R^2 = 0.06$ 0.262	$R^2 = 0.51$ 0.000	$R^2 = 0.35$ 0.003	$R^2 = 0.03$ 0.351	$R^2 = 0.68$ 0.000

Model 1: univariate analysis; Model 2: multivariate analysis adjusted for season; Model 3: multivariate analysis adjusted for maternal age, BMI, education and mode of delivery; Model 4: multivariate analysis adjusted for maternal age, BMI, education and mode of delivery, fat and vitamin A and E intake.

4. Discussion

In this study, we found that the breastmilk carotenoid concentrations were unchanged through the first six months of lactation. The carotenoid with the highest concentration in breastmilk was lycopene, whereas the β-carotene and lutein + zeaxanthin contents were similar. In addition, lycopene was the main dietary carotenoid. We observed a strong positive relationship between lutein + zeaxanthin dietary intake and its concentration in breastmilk, even after adjustment for confounders. In addition, we noted adverse associations between the breastmilk carotenoids in the third month of lactation and maternal BMI.

Previous studies investigated the breastmilk carotenoid changes during lactation and found that their concentrations decreased with the duration of lactation [19,31,41,46]. Colostrum contains significantly higher concentrations of carotenoids compared to mature milk, regardless of dietary intake or plasma carotenoid concentration [19,30,31,41,46,47]. However, in our study, we did not find any differences in carotenoid concentration between the first and sixth month of lactation which may be explained by the fact that we analysed only mature milk. According to other studies, the decrease occurs only between colostrum and mature milk and the mature milk then has a stable level of carotenoids [30,31,47]. This difference in carotenoid concentration between colostrum and mature milk suggests the occurrence of a special mechanism of transporting carotenoids into the mammary gland during the first day postpartum [30,41,46,48]. This may be associated with the importance of carotenoids, especially lutein, to the health and development of newborns from the first days of life, including decreasing oxidative stress, as well as protection of the retina and participation in its proper development [6,24,26]. Other studies reported not only longitudinal changes in carotenoid concentration, however also diurnal changes, and even within a feeding session, there were changes as a result of changes in breastmilk fat concentrations [39,40,49].

Previous research has demonstrated that the primary carotenoids in breastmilk are β-carotene, lutein and zeaxanthin, lycopene, α-carotene and β-cryptoxanthin, although differences in their proportions were observed [19,28–31,39,41,46,50]. In the current study, lycopene was the main breastmilk carotenoid with a concentration of 112.2 at the first month and 110.1 nmol/L at the sixth month. However, studies from other populations reported much lower concentrations of lycopene in mature milk, from 14.0 to 59.8 nmol/L (German [31,41], Brazil [50], China [19,28,30] Australia [28], Canada [28], Chile [28], Japan [28], Mexico [28,30], Philippines [28], UK [28], USA [28,30]). In addition, some studies found much higher concentrations of lutein compared to our results (33.0 and 37.1 nmol/L in the first and sixth month of lactation, respectively), from 44.0–114.4 nmol/L (studies conducted in Chile [28], China [28,30,33], Japan [28], Mexico [28,30]), and some reported lower 6.0–29.0 nmol/L (studies conducted in Germany [41], Brazil [50,51], Australia [28], Canada [28], USA [28], UK [28], China [19]). Similarly, other authors observed higher concentrations of β-carotene 36.2–78.2 nmol/L (Germany [31,41], Australia [28], Canada [28], Chile [28], China [28,30], Japan [28], Mexico [28], UK [28], USA [28,30]), although some studies showed a lower concentration of 16.0–22.0 nmol/L (Philippines [28], Brazil [50,51], China [19]). These differences can be explained by variations in dietary habits between populations, including the availability of fruit and vegetables, as well as preferred dish preparation methods. It is also important to note that other studies use different methods of breastmilk sample collection (for example, total volume of one breast or just 5–12 millilitres of foremilk which had 25% lower concentrations of carotenoids compared to the hindmilk [40]). Breastmilk carotenoids are the only source of carotenoids for newborns and infants because infant formula are not fortified with them [6]. β-carotene has pro-vitamin A properties, so it can be converted, if necessary, to contribute to meeting the nutritional requirement for vitamin A [52]. A growing body of studies highlight the importance of lutein during the neonatal and infancy period due to its role during neuro and visual development, as well as reducing oxidative stress and the risk of disorders associated with prematurity [6,7,24,26]. However, lycopene may also decrease the risk of adverse neonatal outcomes (birth parameters, Respiratory Distress Syndrome, Newborn Intensive Care Unit admission), as shown in a recently published study which assessed the maternal serum lycopene during pregnancy in 180 maternal-infant pairs from the USA [53].

As carotenoids cannot be synthesized by mammals, all breastmilk carotenoids are diet-derived. The source of carotenoids are vegetables and fruits, especially tomatoes and its products for lycopene, green, yellow and orange for lutein and yellow and orange for β-carotene [24]. In our study, lycopene was not only the main carotenoid in breastmilk, however it was also the carotenoid with the highest intake in our study group (7897.3, 7255.8 µg/d, third and sixth month of lactation, respectively), followed by β-carotene (4480.0, 3441.9 µg/day) and lutein + zeaxanthin (2945.2, 3739.3 µg/day). These amounts were similar to the carotenoid intake during the first and second trimester that was reported in a study that was conducted in the USA [54]. β-carotene intake was also similar to the results of a study that was conducted among pregnant Polish women, however the authors reported a lower intake of lycopene and lutein + zeaxanthin [55], although other studies of pregnant women reported a much lower intake of these carotenoids [56–59]. The NHANES study also showed a much lower lutein intake by women who were at reproductive age [60], although a recent study that was conducted in Canada found a similar intake of lycopene and lutein + zeaxanthin and higher intake of β-carotene [61]. Detailed data on the dietary intake of carotenoids during lactation are limited. Cena et al. [27] calculated lutein consumption in 15 Italian women at the third and 30th day postpartum, which was 1209 ± 157 and 1258 ± 102 µg/day, which is also lower compared to our study. However, this data was obtained almost 10 years ago and dietary habits may have changed since then due to the increasing availability of fruits and vegetables. A study that was conducted among Chinese populations found an intake of lutein + zeaxanthin at the level 3.3 ± 0.41 mg/day [33], which was similar to our study.

Several studies have found that a dietary intake of carotenoids determines the serum carotenoids [24,27,62] and, hence, it determines the breastmilk concentrations [27,30–33,50,51]. Other studies analysed serum carotenoids and found strong correlations between maternal plasma and breastmilk, where breastmilk concentrations were 10–120 times lower compared to the serum concentrations [30,51]. Cena et al. [27] described the strong associations between the dietary intake of lutein and its concentration in serum ($r = 0.94$) and breastmilk ($r = 0.84$), however a more recent study by Xu et al. [33] did not find any associations between maternal dietary lutein + zeaxanthin and serum or breastmilk concentrations. Furthermore, a Brazilian study found no association between pro-vitamin A carotenoids and breastmilk β-carotene [51]. Despite the inconsistency of these results, several interventional studies clearly indicated that the concentration of breastmilk carotenoids increases after the consumption of high-carotenoid food products (e.g., carrot or tomato paste, Chlorella) or dietary supplements [62–65]. A multinational study by Canfield et al. [28] also found that the major milk carotenoids are consistent with major dietary carotenoids. The current study confirmed the associations between dietary carotenoid intake and the breastmilk concentrations that were observed by other authors. Nonetheless, the strength of the association noted in the current study differed depending on the carotenoid—it was weaker for lycopene ($r = 0.374$, $r = 0.338$, third and sixth month, respectively) and β-carotene at the third month of lactation ($r = 0.442$), moderate at the sixth month ($r = 0.532$) and was the strongest for lutein ($r = 0.711$, $r = 0.726$). After adjustment for potential confounders, we confirmed this observation only for lutein and β-carotene. We hypothesized that stronger associations of lutein compared to β-carotene or lycopene were the result of better nutritional economy of lutein and more efficient uptake to the mammary gland and breastmilk due to its important role during early postnatal development [6,26]. Lutein and zeaxanthin, as xanthophylls, are also more polar compared to carotenes and in vitro studies suggest that xanthophylls may have higher bioavailability [30,48]. Due to their end hydroxyl groups, they have a higher polarity and are easily transferred into lipoproteins, which are responsible for carotenoid transport, and further into milk fat globules [30,48]. Milk fat globules are covered by a membrane trilayer that is derived from the endoplasmic reticulum and apical plasma membrane of mammary epithelial cells [30,48,66]. Lutein and zeaxanthin have a higher membrane solubility and a perpendicular orientation within the membrane bilayer which allows them to be more stably bound within the membrane layer compared to carotenes [67]. A recent study comparing dietary and serum carotenoids in men and non-lactating women showed that women have significantly higher concentrations of serum carotenoids than men despite a lower dietary intake of carotenoids, which indicates a sex-difference in the nutritional economy of carotenoids [61].

Maternal dietary intake is not the only factor influencing the breastmilk carotenoid status. In the current study, we found adverse associations between all breastmilk carotenoids at the third month of lactation and β-carotene in the sixth month and maternal BMI. A recent study by Xue et al. [19] found that overweight women had lower concentrations of β-carotene compared to women who were of normal weight, however they did not find any associations for lutein, similar to Meneses and Trugo [51] who also did not observe any associations between breastmilk carotenoids and maternal BMI. However, studies analysing the impact of body mass on serum carotenoids found adverse associations between them [61,68]. The explanation of this association could be twofold. Firstly, both being overweight and obese are associated with greater body fat storage and the carotenoids may be uptaken and stored in adipose tissue [48]. A previous study by Johnson et al. [69] showed that women have a higher lutein concentration in adipose tissue compared to men despite a lower dietary intake and similar serum concentration, which also indicates more effective carotenoid accumulation in adipose tissue in women. Previous studies found that circulating concentrations of other micronutrients, e.g., folate and vitamin B_{12}, may also be altered in women who are overweight and obese, independent of other maternal factors [70,71]. This observation indicated the possibility of a modification in micronutrient metabolism, including carotenoids, which resulted in reduced plasma levels and increased uptake in other tissues, especially adipose tissue [70,71]. Secondly, being overweight and obese coincide with elevated oxidative stress levels which reduces the carotenoid level, one of the antioxidant compounds [72]. On the other hand, we observed adverse associations only between the BMI category and β-carotene at the sixth month of lactation. However, more research is needed to explain this association during the lactation period as well as the impact of pre-pregnancy and pregnancy diet on the breastmilk carotenoids due to physiological weight loss during the lactation period.

Strengths and Limitations

The strength of this study is that we used twice repeated three-day records to collect the data on maternal nutrition which helped to decrease the risk of underestimating or overestimating long-term dietary habits. Second, we collected breastmilk samples from a 24-hour period, as well as both foremilk and hindmilk which minimized possible errors due to changes in breastmilk composition. Third, we had a low drop-out rate (11%), despite a relatively long follow-up period.

Finally, a number of potential limitations need to be considered. First, we had a sample with a very good socio-economic status and education level that were higher than the national average. Our group also had a moderate number of participants which was generally larger than some studies, mainly prospective ($n = 21$ [27,31], $n = 23$ [40], $n = 46$ [51]) and lower compared to other mainly cross-sectional studies ($n = 56$ [33]; $n = 140$ [46]; $n = 365$ [30]; $n = 456$ [28]; $n = 509$ [19]). Second, we used convenience sampling. Due to this factor, precautions must be taken when extrapolating the results. Third, we used three-day records only at the second and third visit, so we cannot analyse the associations between dietary intake and breastmilk carotenoids at the first month of lactation. Fourth, we did not collect maternal serum, so we cannot analyse the relationship between the maternal dietary intake of carotenoids and plasma correlations.

5. Conclusions

The concentrations of β-carotene, lycopene and lutein + zeaxanthin in breastmilk samples, as well as its dietary intake, were studied in the first six months of lactation in healthy women from an urban area of Poland. Overall, our findings reveal that maternal dietary intake of carotenoids positively correlates with breastmilk concentrations, especially for lutein. Furthermore, breastmilk carotenoids in the studied population were moderate (β-carotene and lutein + zeaxanthin) or relatively high (lycopene) compared to other populations and remained stable over time. It was also found that an increase in maternal BMI depletes breastmilk carotenoid concentrations. Our results indicate that the maternal dietary intake of carotenoids is an important factor that influences breastmilk carotenoid concentrations and it is easily modifiable through nutritional intervention. Research should continue to explore the biological impact of such results and improve the knowledge of the unique composition of human milk. These findings may help to determine the nutritional recommendations for the dietary intake of carotenoids for breastfeeding mothers and infants and may also form the basis for the development of nutritional programs or dietary supplements. Moreover, an analysis of breastmilk carotenoids may be useful as a non-invasive method to monitor maternal carotenoid status, as well as its intake by infants.

Supplementary Materials: The following are available online at http://www.mdpi.com/2072-6643/11/1/193/s1, Table S1: Characteristics of the study group, Figure S1: Standard curves for all carotenoids that were used in the presented manuscript.

Author Contributions: M.A.Z., J.H. and A.W. were responsible for the conception and design of this study. J.H. and M.A.Z. were involved in the funding acquisition in respect to the project and managing of the project. M.A.Z. and A.W. were responsible for the data collection of this study. M.A.Z. was responsible for the data cleaning and statistical analysis for this particular paper. M.A.Z. was responsible for writing the manuscript. J.H. and A.W. were responsible for revising the manuscript critically for important intellectual content. The manuscript has been revised by all co-authors.

Funding: This research was partially funded by the Polish Ministry of Science and Higher Education. This research was partially funded by the Faculty of Human Nutrition and Consumer Sciences, Warsaw University of Life Sciences, grant number 505-10-100200-N00322-99.

Acknowledgments: We thank all of the mothers and infants for participating in the study. Thanks are also extended to Holy Family Hospital in Warsaw, especially to Elżbieta Łodykowska for her help in participant recruitment.

Conflicts of Interest: The authors declare no conflict of interest.

References

1. World Health Organization/United Nations Children's Fund. *Global Strategy for Infant and Young Child Feeding*; World Health Organization: Geneva, Switzerland, 2003.
2. Eidelman, A.I.; Schanler, R.J.; Johnston, M.; Landers, S.; Noble, L.; Szucs, K.; Viehmann, L. Breastfeeding and the use of human milk. *Pediatrics* **2012**, *129*, e827–e841. [CrossRef]
3. Fewtrell, M.; Bronsky, J.; Campoy, C.; Domellöf, M.; Embleton, N.; Fidler Mis, N.; Hojsak, I.; Hulst, J.M.; Indrio, F.; Lapillonne, A.; et al. Complementary Feeding: A Position Paper by the European Society for Paediatric Gastroenterology, Hepatology, and Nutrition (ESPGHAN) Committee on Nutrition. *J. Pediatr. Gastroenterol. Nutr.* **2017**, *64*, 119–132. [CrossRef] [PubMed]
4. Butte, N.F.; Lopez-Alarcon, M.G.; Garza, C. *Nutrient Adequacy of Exclusive Breastfeeding for the Term Infant During the First 6 Months of Life*; World Health Organization: Geneva, Switzerland, 2001; Available online: http://www.who.int/nutrition/publications/infantfeeding/nut_adequacy_of_exc_bfeeding_eng.pdf/ (accessed on 20 November 2018).
5. Ballard, O.; Morrow, A.L. Human milk composition: Nutrients and bioactive factors. *Pediatr. Clin. N. Am.* **2013**, *60*, 49–74. [CrossRef] [PubMed]
6. Zielińska, M.A.; Wesołowska, A.; Pawlus, B.; Hamułka, J. Health Effects of Carotenoids during Pregnancy and Lactation. *Nutrients* **2017**, *9*, 838. [CrossRef] [PubMed]
7. Tsopmo, A. Phytochemicals in human milk and their potential antioxidative protection. *Antioxidants* **2018**, *7*, 32. [CrossRef] [PubMed]
8. Victora, C.G.; Bahl, R.; Barros, A.J.; França, G.V.; Horton, S.; Krasevec, J.; Murch, S.; Sankar, M.J.; Walker, N.; Rollins, N.C.; et al. Breastfeeding in the 21st century: Epidemiology, mechanisms, and lifelong effect. *Lancet* **2016**, *387*, 475–490. [CrossRef]
9. Mitoulas, L.R.; Kent, J.C.; Cox, D.B.; Owens, R.; Sherriff, J.L.; Hartmann, P.E. Variation in fat, lactose and protein in human milk over 24 h and throughout the first year of lactation. *Br. J. Nutr.* **2002**, *88*, 29–37. [CrossRef] [PubMed]
10. Kent, J.C.; Mitoulas, L.R.; Cregan, M.D.; Ramsay, D.T.; Doherty, D.A. Volume and frequency of breastfeedings and fat content of breast milk throughout the day. *Pediatrics* **2006**, *117*, e387. [CrossRef]
11. Gidrewicz, D.A.; Fenton, T.R. A systematic review and meta-analysis of the nutrient content of preterm and term breast milk. *BMC Pediatr.* **2014**, *14*, 216. [CrossRef]
12. Napierala, M.; Mazela, J.; Merritt, T.A.; Florek, E. Tobacco smoking and breastfeeding: Effect on the lactation process, breast milk composition and infant development. A critical review. *Environ. Res.* **2016**, *151*, 321–338. [CrossRef]
13. Grote, V.; Verduci, E.; Scaglioni, S.; Vecchi, F.; Contarini, G.; Giovannini, M.; Koletzko, B.; Agostoni, C.; European Childhood Obesity Project. Breast milk composition and infant nutrient intakes during the first 12 months of life. *Eur. J. Clin. Nutr.* **2016**, *70*, 250–256. [CrossRef] [PubMed]
14. Bzikowska-Jura, A.; Czerwonogrodzka-Senczyna, A.; Olędzka, G.; Szostak-Węgierek, D.; Weker, H.; Wesołowska, A. Maternal Nutrition and Body Composition During Breastfeeding: Association with Human Milk Composition. *Nutrients* **2018**, *10*, 1379. [CrossRef] [PubMed]
15. Barrera, C.; Valenzuela, R.; Chamorro, R.; Bascuñán, K.; Sandoval, J.; Sabag, N.; Valenzuela, F.; Valencia, M.P.; Puigrredon, C.; Valenzuela, A. The Impact of Maternal Diet during Pregnancy and Lactation on the Fatty Acid Composition of Erythrocytes and Breast Milk of Chilean Women. *Nutrients* **2018**, *10*, 839. [CrossRef] [PubMed]
16. Henjum, S.; Lilleengen, A.M.; Aakre, I.; Dudareva, A.; Gjengedal, E.L.F.; Meltzer, H.M.; Brantsæter, A.L. Suboptimal Iodine Concentration in Breastmilk and Inadequate Iodine Intake among Lactating Women in Norway. *Nutrients* **2017**, *9*, 643. [CrossRef] [PubMed]
17. Allen, L.H. B vitamins in breast milk: Relative importance of maternal status and intake, and effects on infant status and function. *Adv. Nutr.* **2012**, *3*, 362–369. [CrossRef]
18. Aumeistere, L.; Ciproviča, I.; Zavadska, D.; Bavrins, K.; Borisova, A. Zinc Content in Breast Milk and Its Association with Maternal Diet. *Nutrients* **2018**, *10*, 1438. [CrossRef] [PubMed]
19. Xue, Y.; Campos-Giménez, E.; Redeuil, K.M.; Lévèques, A.; Actis-Goretta, L.; Vinyes-Pares, G.; Zhang, Y.; Wang, P.; Thakkar, S.K. Concentrations of Carotenoids and Tocopherols in Breast Milk from Urban Chinese Mothers and Their Associations with Maternal Characteristics: A Cross-Sectional Study. *Nutrients* **2017**, *9*, 1229. [CrossRef] [PubMed]

20. Ambati, R.R.; Phang, S.M.; Ravi, S.; Aswathanarayana, R.G. Astaxanthin: Sources, extraction, stability, biological activities and its commercial applications—A review. *Mar. Drugs* **2014**, *12*, 128–152. [CrossRef]
21. Saini, R.K.; Nile, S.H.; Park, S.W. Carotenoids from fruits and vegetables: Chemistry, analysis, occurrence, bioavailability and biological activities. *Food Res. Int.* **2015**, *76*, 735–750. [CrossRef]
22. US Department of Agriculture, Agricultural Research Service, Nutrient Data Laboratory. USDA National Nutrient Database for Standard Reference, Legacy. Available online: https://ndb.nal.usda.gov/ndb/ (accessed on 20 September 2018).
23. Fiedor, J.; Burda, K. Potential role of carotenoids as antioxidants in human health and disease. *Nutrients* **2014**, *6*, 466–488. [CrossRef]
24. Johnson, E.J. Role of lutein and zeaxanthin in visual and cognitive function throughout the lifespan. *Nutr. Rev.* **2014**, *72*, 605–612. [CrossRef] [PubMed]
25. Eggersdorfer, M.; Wyss, A. Carotenoids in human nutrition and health. *Arch. Biochem. Biophys.* **2018**, *652*, 18–26. [CrossRef] [PubMed]
26. Giordano, E.; Quadro, L. Lutein, zeaxanthin and mammalian development: Metabolism, functions and implications for health. *Arch. Biochem. Biophys.* **2018**, *647*, 33–40. [CrossRef] [PubMed]
27. Cena, H.; Castellazzi, A.M.; Pietri, A.; Roggi, C.; Turconi, G. Lutein concentration in human milk during early lactation and its relationship with dietary lutein intake. *Public Health Nutr.* **2009**, *12*, 1878–1884. [CrossRef]
28. Canfield, L.M.; Clandinin, M.T.; Davies, D.P.; Fernandez, M.C.; Jackson, J.; Hawkes, J.; Goldman, W.J.; Pramuk, K.; Reyes, H.; Sablan, B.; et al. Multinational study of major breast milk carotenoids of healthy mothers. *Eur. J. Nutr.* **2003**, *42*, 133–141.
29. Jackson, J.G.; Zimmer, J.P. Lutein and zeaxanthin in human milk independently and significantly differ among women from Japan, Mexico, and the United Kingdom. *Nutr. Res.* **2007**, *27*, 449–453. [CrossRef]
30. Lipkie, T.E.; Morrow, A.L.; Jouni, Z.E.; McMahon, R.J.; Ferruzzi, M.G. Longitudinal survey of carotenoids in human milk from urban cohorts in China, Mexico, and the USA. *PLoS ONE* **2015**, *10*, e0127729. [CrossRef]
31. Schweigert, F.J.; Bathe, K.; Chen, F.; Büscher, U.; Dudenhausen, J.W. Effect of the stage of lactation in humans on carotenoid levels in milk, blood plasma and plasma lipoprotein fractions. *Eur. J. Nutr.* **2004**, *43*, 39–44. [CrossRef] [PubMed]
32. Henriksen, B.S.; Chan, G.; Hoffman, R.O.; Sharifzadeh, M.; Ermakov, I.V.; Gellermann, W.; Bernstein, P.S. Interrelationships between maternal carotenoid status and newborn infant macular pigment optical density and carotenoid status. *Investig. Ophthalmol. Vis. Sci.* **2013**, *54*, 5568–5578. [CrossRef]
33. Xu, X.; Zhao, X.; Berde, Y.; Low, Y.; Kuchan, M. Milk and Plasma Lutein and Zeaxanthin Concentrations in Chinese Breast-Feeding Mother-Infant Dyads with Healthy Maternal Fruit and Vegetable Intake. *J. Am. Coll. Nutr.* **2018**, 1–6. [CrossRef] [PubMed]
34. Wądołowska, L.; Niedźwiedzka, E. Food Frequency Questionnaire FFQ-6. Available online: www.uwm.edu.pl/edu/lidiawadolowska (accessed on 4 November 2018). (In Polish)
35. Cox, J.L.; Holden, J.M.; Sagovsky, R. Detection of postnatal depression. Development of the 10-item Edinburgh Postnatal Depression Scale. *Br. J. Psychiatry* **1987**, *150*, 782–786. [CrossRef] [PubMed]
36. Cohen, S.; Kamarck, T.; Mermelstein, R. A global measure of perceived stress. *J. Health Soc. Behav.* **1983**, *24*, 385–396. [CrossRef] [PubMed]
37. Juczyński, Z.; Ogińska-Bulik, N. *PSS-10–Perceived Stress Scale*; Psychological Test Laboratory of the Polish Psychological Association: Warsaw, Poland, 2009. (In Polish)
38. Matczak, A.; Jaworowska, A.; Ciechanowicz, A.; Fecenec, D.; Stańczak, J.; Zalewska, E. *DSR—Children Development Scale DSR*; Psychological Test Laboratory of the Polish Psychological Association: Warsaw, Poland, 2007. (In Polish)
39. Giuliano, A.R.; Neilson, E.M.; Yap, H.H.; Baier, M.; Canfield, L.M. Methods of nutritional biochemistry quantitation of and inter/intra-individual variability in major carotenoids of mature human milk. *J. Nutr. Biochem.* **1994**, *5*, 551–556. [CrossRef]
40. Jackson, J.G.; Lien, E.L.; White, S.J.; Bruns, N.J.; Kuhlman, C.F. Major carotenoids in mature human milk: Longitudinal and diurnal patterns. *J. Nutr. Biochem.* **1998**, *9*, 2–7. [CrossRef]
41. Macias, C.; Schweigert, F.J. Changes in the concentration of carotenoids, vitamin A, alpha-tocopherol and total lipids in human milk throughout early lactation. *Ann. Nutr. Metab.* **2001**, *45*, 82–85. [CrossRef] [PubMed]

42. Szponar, L.; Wolnicka, K.; Rychlik, E. *Album of Photographs of Food Products and Dishes*; National Food and Nutrition Institute: Warsaw, Poland, 2000. (In Polish)
43. Kunachowicz, H.; Nadolna, I.; Przygoda, B.; Iwanow, K. (Eds.) *Food Composition Tables*; PZWL: Warsaw, Poland, 2017. (In Polish)
44. ISAK. *International Standards for Anthropometric Assessment*; International Society for the Advancement of Kinanthropometry: Potchefstroom, South Africa, 2001; Available online: http://www.ceap.br/material/MAT17032011184632.pdf (accessed on 18 November 2018).
45. World Health Organization. *Obesity: Preventing and Managing the Global Epidemic*; Report of WHO Consultation, WHO Technical Report Series 894; World Health Organization: Geneva, Switzerland, 2000.
46. Xavier, A.A.O.; Díaz-Salido, E.; Arenilla-Vélez, I.; Aguayo-Maldonado, J.; Garrido-Fernández, J.; Fontecha, J.; Sánchez-García, A.; Pérez-Gálvez, A. Carotenoid Content in Human Colostrum is Associated to Preterm/Full-Term Birth Condition. *Nutrients* **2018**, *10*, 1654. [CrossRef] [PubMed]
47. Song, B.J.; Jouni, Z.E.; Ferruzzi, M.G. Assessment of phytochemical content in human milk during different stages of lactation. *Nutrition* **2013**, *29*, 195–202. [CrossRef]
48. Bohn, T.; Desmarchelier, C.; Dragsted, L.O.; Nielsen, C.S.; Stahl, W.; Rühl, R.; Keijer, J.; Borel, P. Host-related factors explaining interindividual variability of carotenoid bioavailability and tissue concentrations in humans. *Mol. Nutr. Food Res.* **2017**, *61*, 1600685. [CrossRef]
49. Khan, S.; Prime, D.K.; Hepworth, A.R.; Lai, C.T.; Trengove, N.J.; Hartmann, P.E. Investigation of short-term variations in term breast milk composition during repeated breast expression sessions. *J. Hum. Lact.* **2013**, *29*, 196–204. [CrossRef]
50. De Azeredo, V.B.; Trugo, N.M. Retinol, carotenoids, and tocopherols in the milk of lactating adolescents and relationships with plasma concentrations. *Nutrition* **2008**, *24*, 133–139. [CrossRef]
51. Meneses, F.; Trugo, N.M.F. Retinol, β-carotene, and lutein + zeaxanthin in the milk of Brazilian nursing women: Associations with plasma concentrations and influences of maternal characteristics. *Nutr. Res.* **2005**, *25*, 443–451. [CrossRef]
52. Strobel, M.; Tinz, J.; Biesalski, H.K. The importance of beta-carotene as a source of vitamin A with special regard to pregnant and breastfeeding women. *Eur. J. Nutr.* **2007**, *46*, 1–20. [CrossRef] [PubMed]
53. Hanson, C.; Lyden, E.; Furtado, J.; Van Ormer, M.; White, K.; Overby, N.; Anderson-Berry, A. Serum Lycopene Concentrations and Associations with Clinical Outcomes in a Cohort of Maternal-Infant Dyads. *Nutrients* **2018**, *10*, 204. [CrossRef] [PubMed]
54. Litonjua, A.A.; Rifas-Shiman, S.L.; Ly, N.P.; Tantisira, K.G.; Rich-Edwards, J.W.; Camargo, C.A., Jr.; Weiss, S.T.; Gillman, M.W.; Gold, D.R. Maternal antioxidant intake in pregnancy and wheezing illnesses in children at 2 y of age. *Am. J. Clin. Nutr.* **2006**, *84*, 903–911. [CrossRef] [PubMed]
55. Hamułka, J.; Sulich, A.; Zielińska, M.; Wawrzyniak, A. Assessment of carotenoid intake in a selected group of pregnant women. *Probl. Hig. Epidemiol.* **2015**, *96*, 763–768.
56. Mathews, F.; Yudkin, P.; Smith, R.F.; Neil, A. Nutrient intakes during pregnancy: The influence of smoking status and age. *J. Epidemiol. Community Health* **2000**, *54*, 17–23. [CrossRef]
57. Scaife, A.R.; McNeill, G.; Campbell, D.M.; Martindale, S.; Devereux, G.; Seaton, A. Maternal intake of antioxidant vitamins in pregnancy in relation to maternal and fetal plasma levels at delivery. *Br. J. Nutr.* **2006**, *95*, 771–778. [CrossRef]
58. Watson, P.E.; McDonald, B.W. Seasonal variation of nutrient intake in pregnancy: Effects on infant measures and possible influence on diseases related to season of birth. *Eur. J. Clin. Nutr.* **2007**, *61*, 1271–1280. [CrossRef]
59. Miyake, Y.; Sasaki, S.; Tanaka, K.; Hirota, Y. Consumption of vegetables, fruit, and antioxidants during pregnancy and wheeze and eczema in infants. *Allergy* **2010**, *65*, 758–765. [CrossRef]
60. Johnson, E.J.; Maras, J.E.; Rasmussen, H.M.; Tucker, K.L. Intake of lutein and zeaxanthin differ with age, sex, and ethnicity. *J. Am. Diet. Assoc.* **2010**, *110*, 1357–1362. [CrossRef]
61. Allore, T.; Lemieux, S.; Vohl, M.C.; Couture, P.; Lamarche, B.; Couillard, C. Correlates of the difference in plasma carotenoid concentrations between men and women. *Br. J. Nutr.* **2018**, 1–10. [CrossRef] [PubMed]
62. Sherry, C.L.; Oliver, J.S.; Renzi, L.M.; Marriage, B.J. Lutein supplementation increases breast milk and plasma lutein concentrations in lactating women and infant plasma concentrations but does not affect other carotenoids. *J. Nutr.* **2014**, *144*, 1256–1263. [CrossRef] [PubMed]
63. Lietz, G.; Mulokozi, G.; Henry, J.C.; Tomkins, A.M. Xanthophyll and hydrocarbon carotenoid patterns differ in plasma and breast milk of women supplemented with red palm oil during pregnancy and lactation. *J. Nutr.* **2006**, *136*, 1821–1827. [CrossRef] [PubMed]

64. Nagayama, J.; Noda, K.; Uchikawa, T.; Maruyama, I.; Shimomura, H.; Miyahara, M. Effect of maternal Chlorella supplementation on carotenoid concentration in breast milk at early lactation. *Int. J. Food Sci. Nutr.* **2014**, *65*, 573–576. [CrossRef] [PubMed]
65. Haftel, L.; Berkovich, Z.; Reifen, R. Elevated milk β-carotene and lycopene after carrot and tomato paste supplementation. *Nutrition* **2015**, *31*, 443–445. [CrossRef] [PubMed]
66. Koletzko, B. Human Milk Lipids. *Ann. Nutr. Metab.* **2016**, *69*, 28–40. [CrossRef]
67. Widomska, J.; Subczynski, W.K. Why has nature chosen lutein and zeaxanthin to protect the retina? *J. Clin. Exp. Ophthalmol.* **2014**, *5*, 326. [CrossRef]
68. Wang, L.; Gaziano, J.M.; Norkus, E.P.; Buring, J.E.; Sesso, H.D. Associations of plasma carotenoids with risk factors and biomarkers related to cardiovascular disease in middle-aged and older women. *Am. J. Clin. Nutr.* **2008**, *88*, 747–754. [CrossRef]
69. Johnson, E.J.; Hammond, B.R.; Yeum, K.J.; Qin, J.; Wang, X.D.; Castaneda, C.; Snodderly, D.M.; Russell, R.M. Relation among serum and tissue concentrations of lutein and zeaxanthin and macular pigment density. *Am. J. Clin. Nutr.* **2000**, *71*, 1555–1562. [CrossRef]
70. Knight, B.A.; Shields, B.M.; Brook, A.; Hill, A.; Bhat, D.S.; Hattersley, A.T.; Yajnik, C.S. Lower Circulating B12 Is Associated with Higher Obesity and Insulin Resistance during Pregnancy in a Non-Diabetic White British Population. *PLoS ONE* **2015**, *10*, e0135268. [CrossRef]
71. Maffoni, S.; De Giuseppe, R.; Stanford, F.C.; Cena, H. Folate status in women of childbearing age with obesity: A review. *Nutr. Res. Rev.* **2017**, *30*, 265–271. [CrossRef] [PubMed]
72. Keaney, J.F., Jr.; Larson, M.G.; Vasan, R.S.; Wilson, P.W.; Lipinska, I.; Corey, D.; Massaro, J.M.; Sutherland, P.; Vita, J.A.; Benjamin, E.J.; et al. Obesity and systemic oxidative stress: Clinical correlates of oxidative stress in the Framingham Study. *Arterioscler. Thromb. Vasc. Biol.* **2003**, *23*, 434–439. [CrossRef] [PubMed]

© 2019 by the authors. Licensee MDPI, Basel, Switzerland. This article is an open access article distributed under the terms and conditions of the Creative Commons Attribution (CC BY) license (http://creativecommons.org/licenses/by/4.0/).

Article

Temporal Progression of Fatty Acids in Preterm and Term Human Milk of Mothers from Switzerland

Sagar K. Thakkar [1], Carlos Antonio De Castro [2], Lydie Beauport [3], Jean-François Tolsa [3], Céline J. Fischer Fumeaux [3], Michael Affolter [4] and Francesca Giuffrida [4,*]

1. Nestlé Institute of Health Sciences, Nestlé Research, Lausanne 1000, Switzerland; Sagar.Thakkar@rd.nestle.com
2. Clinical Development Unit, Nestlé Research Asia, Singapore 138567, Singapore; CarlosAntonio.DeCastro@rdsg.nestle.com
3. Clinic of Neonatology, Department Woman Mother Child, University Hospital of Lausanne, Lausanne 1011, Switzerland; Lydie.Beauport@chuv.ch (L.B.); Jean-Francois.Tolsa@chuv.ch (J.-F.T.); Celine-Julie.Fischer@chuv.ch (C.J.F.F.)
4. Nestlé Institute of Food Safety & Analytical Science, Nestlé Research, Lausanne 1000, Switzerland; Michael.Affolter@rdls.nestle.com
* Correspondence: Francesca.Giuffrida@rdls.nestle.com; Tel.: +41-21-785-8084

Received: 22 November 2018; Accepted: 25 December 2018; Published: 8 January 2019

Abstract: We longitudinally compared fatty acids (FA) from human milk (HM) of mothers delivering term and preterm infants. HM was collected for 4 months postpartum at 12 time points for preterm and for 2 months postpartum at 8 time points for term group. Samples were collected from the first feed of the morning, and single breast was fully expressed. FA were analyzed by gas chromatography coupled with flame ionization detector. Oleic, palmitic and linoleic acids were the most abundant FA across lactation and in both groups. Preterm colostrum contained significantly ($p < 0.05$) higher 8:0, 10:0, 12:0, sum medium chain fatty acids (MCFA), 18:3 n-3 FA compared to term counterparts. Preterm mature milk contained significantly higher 12:0, 14:0, 18:2 n-6, sum saturated fatty acids (SFA), and sum MCFA. We did not observe any significant differences between the preterm and term groups for docosahexaenoic acid, arachidonic acid and eicosapentaenoic acid at any stage of lactation. Overall, preterm milk was higher for SFA with a major contribution from MCFA and higher in 18:2 n-6. These observational differences needs to be studied further for their implications on preterm developmental outcomes and on fortification strategies of either mothers' own milk or donor human milk.

Keywords: human milk; preterm; term; infants; lipids; fatty acids; human milk fortification; docosahexaenoic acid (DHA); mothers' own milk; donor human milk; arachidonic acid (ARA); eicosapentaenoic acid (EPA)

1. Introduction

Given the documented short- and long-term advantages of breastfeeding for both the mother and the infant, it is no surprise that breastfeeding and human milk (HM) feeding are considered as normative standards by health care professionals and organizations such as the World Health Organization (WHO), American Academy of Pediatrics and European Commission [1]. WHO recommends exclusive breastfeeding for the first six months of infant life followed by introduction of complementary foods and continued breastfeeding for up to two years of life, and even beyond until mutually agreeable by the mother-infant dyad [2]. Indeed, it has been stated that HM feeding during this early stage of life is able to meet the nutritional demands of not only growth and development, but also imparts the immune factors and protects from later in life metabolic abnormalities of apparently healthy term infants [3]. For preterm infants, HM confers additional benefits, while reducing morbidity and mortality and enhancing neurodevelopment of this vulnerable population [4]. However, feeding a preterm infant requires special considerations to meet the nutritional demands to allow for mimicking growth that would otherwise take place in-utero.

Understanding the roles of nutrition in general and of lipids in particular, developmental outcomes of preterm infants has been a subject of much research in an effort to define their nutritional needs [5–9]. Since HM feeding is well tolerated and reduces the risk of co-morbidities such as necrotizing enterocolitis and sepsis in preterm infants, it is important that feeding HM to preterm infants has become the primary choice of nutritional source [10]. Furthermore, strategies have also been developed to supplement mothers' own milk (MOM) with either multi-nutrient or single nutrient human milk fortifiers (HMF) based on bed-side analyses of MOM [11]. Nevertheless, the bedside analyses of MOM has been focused on macronutrients (total proteins, total lipids, total lactose and calculated total energy) often using mid-infrared spectroscopic methods and not on their detailed profile or other micronutrients that may also contribute to growth and development [12].

Lipids in unfortified HM/MOM provide approximately 50% of the total energy to its consumers and the majority (90–95%) of lipids are present in the form of triacylglycerol (TAG), a glycerol molecule bound to three fatty acids (FA). These FA range from medium chain to very long chain and may be saturated, mono-unsaturated or polyunsaturated. Additionally, essential FA as well as non-nutritive bioactive FA are also part of the FA pool supplied via HM. Conventional data suggests that FA in HM may be modified by the maternal dietary or nutritional supplement intake but not the quantity or concentration of total lipids [13]. The concentration of HM lipids increases with advancing stages of lactation [14–16]. In fact, HM composition may be impacted by a multitude of factors ranging from maternal to infant parameters and even including the physiological and behavioral aspects observed in the mother-infant dyad. These parameters have been recently reviewed and summarized by Fields and colleagues [17]. Certainly one of those parameters that may influence the HM composition is the gestational age at the birth of infants. Mothers of preterm infants have a higher risk of delayed onset of lactogenesis II and potentially the mammary tissues may not be developed to the extent of their term mother counterparts [18]. Undoubtedly, more research is needed in this area to understand the impact of delayed milk production on the mother and the infant.

A handful of reports already exists on comparison of HM from mothers of term and preterm infants. However, they have either focused on selective FA such as arachidonic acid (ARA) or docosahexaenoic acid (DHA) [19,20], or focused only on transitional and mature milk [21]. Therefore, in this study we aimed to explore and compare the composition of FA in HM produced by mothers delivering a preterm infant to that of a term infant from colostrum, transitional and mature milk.

2. Materials and Methods

2.1. Ethical and Legal Considerations

This study was conducted according to the guidelines of the Declaration of Helsinki. The study protocol with all procedures involving human subjects was approved by the Ethical Board (Commission cantonal d'éthique de la recherché sur l'être humain) of the Canton de Vaud, Switzerland (Protocol 69/13, clinical study 11.39.NRC; April 9, 2013). Written consent was obtained from all participating subjects of the study. The study was registered at ClinicalTrials.gov with the identifier NCT02052245.

2.2. Study Settings and Subjects

The study was conducted between October 2013 and July 2014 at the neonatal intensive care unit (NICU) of the University Hospital (CHUV) in Lausanne, Switzerland. A longitudinal HM sampling from lactating mothers was performed for HM characterization. Subjects were recruited at the hospital within 2–3 days after giving birth (preterm and term births) by a single dedicated lactation nurse who managed all interactions with subjects from start to finish of the study. A total of 61 mothers were recruited for the study, out of which 27 had preterm deliveries and 34 had term deliveries.

2.3. Inclusion and Exclusion Criteria

Eligibility criteria for this study included women older than 18 years of age giving (1) preterm birth (between gestational ages 28 0/7 and 32 6/7 weeks) or (2) term birth (>37 0/7 and not above 41 6/7 weeks) with mothers' intention to exclusively or partially breastfeed at least until 4 month post-partum. Exclusion criteria included gestational and pre-gestational diabetes (type I or II), alcohol or illicit drug consumption and insufficient skills to understand study questionnaire. Availability of refrigerator/freezer at home for storage of collected human milk samples was required.

2.4. Data Collection

After the subjects were enrolled and their signature obtained on the informed consent form, the following information was collected for (a) mothers: age, height, weight before pregnancy and at delivery, and (b) for the infants: date of birth, sex, gestational age, delivery method, weight, and head circumference and sibling related data. The dedicated study nurse conducted data collection in face-to-face interviews with the subjects.

2.5. Human Milk Sampling, Handling, and Storage

HM from mothers of preterm infants were sampled weekly for eight weeks post-partum and then once every two weeks for eight more weeks totaling 12 longitudinal samples. HM from mothers of term infants were sampled weekly for eight weeks post-partum totaling eight longitudinal samples. Various aspects of HM sampling were standardized for all subjects. Milk was collected between 06h00 and 12h00 using an electric breast pump (Symphony®, Medela, Baar, Switzerland) allowing the mothers flexibility to express at home or at the NICU. The side of the breast selected by the mother was kept the same during the entire study and the mothers were requested to empty the breast in the previous feed or the pumping session. Single full breast was sampled and an aliquot of 10 mL HM for each time point (or 1–3 mL for the first two sampling time points in the preterm group) was reserved for biochemical characterization. The remainder of the HM was returned to the mother for feeding to the infant at a later time point, if so required. Each sample was transferred to freezing tubes, labelled with subject number and collection information, stored at −18 °C in the home freezer, transferred to the hospital (storage at −80 °C) and then shipped to the Nestlé Research Centre (Lausanne, Switzerland) where it was stored at −80 °C until analysis. The frozen HM samples were thawed once for aliquoting into 15 individual small volume fractions (min 0.2 mL to max 2 mL) in separate polypropylene Eppendorf tubes dedicated to the different analyses. The aliquoting approach was implemented to avoid repeated thawing-freezing cycles and to adapt the required volumes to the specific needs of the individual analytical methods.

2.6. Quantification of Total Lipids in Human Milk

Total lipid content was measured in HM samples by a human milk analyzer (HMA). The device employed for analyses was a HMA generation 3 (Miris AB, Uppsala, Sweden) using the XMA-SW software version 2.87 (Miris AB, Uppsala, Sweden). This HMA is based on semi-solid middle infrared (MIR) transmission spectroscopy. The wavebands used are specific for the functional carbonyl groups (5.7 µm) for fat determination, amide groups (6.5 µm) for protein determination, and hydroxyl groups (9.6 µm) for carbohydrate determination. Prior to analysis, a daily calibration check was performed using the calibration solution provided by the supplier. All samples were homogenized for 3×10 s using the MIRIS sonicator (MIRIS AB, Uppsala, Sweden) as recommended by MIRIS and were kept in a water bath at 40 °C prior to measurement. Homogenized samples (1 mL) were injected into the flow cell and measured within a minute. Once the analysis was completed, the built-in cell and all lines were rinsed with deionized water. After five milk samples, the system was cleaned with the recommended MIRIS detergent. An in-house control sample as well as a calibration standard provided by the manufacturer were analyzed after every tenth measurement for quality control purposes.

2.7. Direct Method Procedure to Prepare Fatty Acid Methyl Esters (FAME) from Human Milk

Fatty acids were quantified in HM as described by Cruz-Hernandez et al. [22]. Acid FAME were prepared using HCl/Methanol (3N) as a catalyst. The methylation procedure was as follows: In a 15 mL test tube equipped with Teflon-lined screw caps, 250 µL of HM was added followed by 300 µL of internal standard FAME 11:0 and 300 µL of internal standard TAG 13:0, 2 mL of methanol, 2 mL of methanol/HCL (3N) and 1 mL of *n*-hexane. Test tubes were firmly capped and shaken vigorously and heated at 100 °C for 60 min, with occasional shaking. Care was taken to fit the cap tightly with cap liner to avoid leaks when tubes are heated at 100 °C. After cooling down to room temperature, 2 mL water was added and shaken vigorously for centrifugation at $1200\times g$ for 5 min followed by transfer of the upper phase (hexane) into gas chromatography vials. Analyses were performed on a 7890A gas-chromatograph (Agilent Technologies, Palo Alto, CA, USA) equipped with a fused-silica CP-Sil 88 capillary column (100% cyanopropylpolysiloxane; 100 m, 0.25 mm id, 0.25 µm film thickness; Agilent, Palo Alto, CA, USA) have been used with a split injector (1:25 ratio) heated at 250 °C and a flame-ionization detector operated at 300 °C. Oven temperature programming used was 60 °C isothermal for 5 min, increased to 165 °C at 15 °C/min, isothermal for 1 min at this temperature, and then increased to 195 °C at 2 °C/min and held isothermal for 14 min and then increased to 215 °C at 5 °C/min and held isothermal for 8 min at 215 °C. Hydrogen was used as carrier gas under constant flow mode at 1.5 mL/min.

2.8. Statistics

The scarcity of quantitative data on the fatty acid content in preterm HM precluded a power calculation in this exploratory study. Sample size was initially set at $n = 20$ subjects per group (preterm and term groups), according to the estimated recruitment feasibility at the study center within a one year period. The longitudinal evolution of fatty acid content was compared in preterm and term HM postpartum age (categorized by lactation stages: colostrum ($\leq$1 week postpartum), transitional (>1 week and $\leq$2 weeks post-partum) and mature (>2 weeks and $\leq$16 weeks)). No aggregation was done for observations from the same participant within each lactation stage (i.e., colostrum has 1 observation, transitional milk has 1 observation and mature milk has 4 observations per participant). Mixed linear models were used to estimate the differences between preterm and term HM. The models used age (colostrum, transitional milk and mature milk stages), term/preterm birth status, and interaction between age and term/preterm status and delivery mode. Within subject variability was accounted by declaring the subject ID as a random effect. Logarithmic transformation was applied to the FA as they are generally skewed. Only the ratios (n-6 to n-3 ratio and ARA to DHA ratio) were assumed to be normally distributed, therefore no transformation was applied. Contrast estimates of the model were calculated by comparing preterm and term HM groups at each time point. No imputation method was applied for missing data (both in between visits and loss to follow up) as the method used does not require a complete data set. A conventional 2-sided 5% error rate was used without adjusting for multiplicity. A similar analysis was done separating term and preterm infants but looking at the age differences (colostrum vs. transitional milk, transitional milk vs. mature milk and colostrum vs. mature milk). Statistical analyses were done with SAS 9.3 (SAS Institute Inc., Cary, NC, USA) and R 3.2.1. (R Foundation, Vienna, Austria) Differences were considered statistically significant when *p* values were <0.05.

3. Results

3.1. Subject Characteristics

This study included convenient sampling of 27 mothers who delivered 33 preterm infants and 34 mothers who delivered 34 term infants at CHUV neonatal unit in Lausanne, Switzerland. Multiple deliveries (twins) were frequent (36%) in the preterm group, but absent in the term group. Figure 1 displays the study flow chart. Two out of 27 (7.4%) preterm infant mothers and 6 out of 34 (17.6%) term infant mothers were lost to follow-up. No adverse events were reported along the study period. In total, 498 HM samples, 279 from preterm and 219 from full term infant mothers, were available for fatty acid analyses. Table 1 reports mother, infant demographic, and baseline anthropometric data. Maternal characteristics were comparable among groups. Caesarean delivery was more frequent in the preterm group. Preterm and full term infants significantly differed in all parameters except for gender distribution.

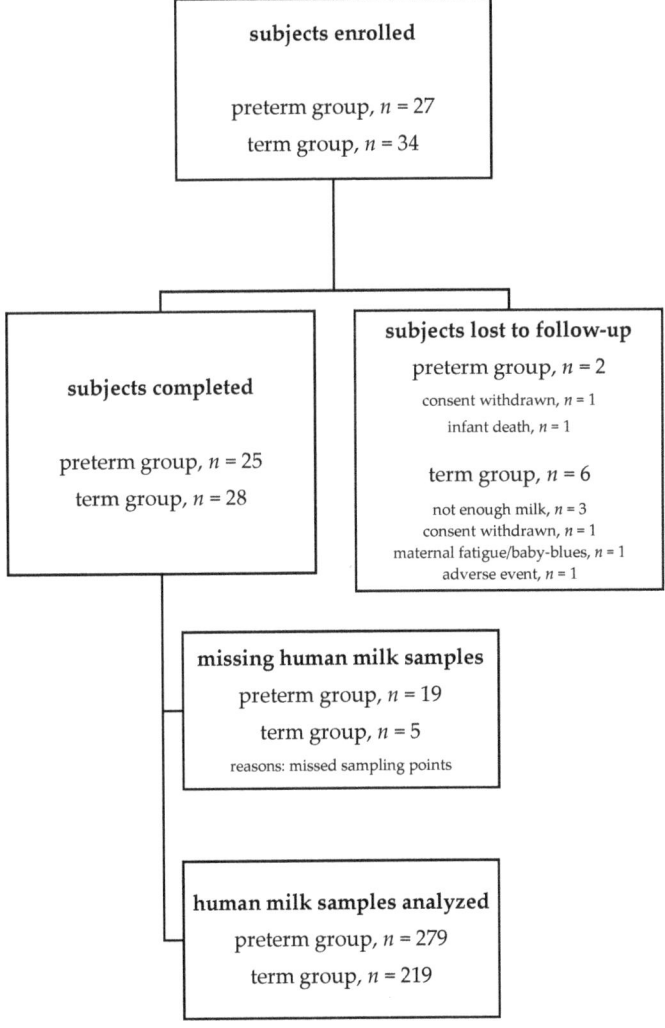

Figure 1. Study flow chart.

Table 1. Maternal and infant characteristics of the study population.

Study Population	Preterm	Term	p-value
Maternal	n = 27	n = 34	
Age (years), mean ± SD	32.4 ± 5.6	31.2 ± 4.2	0.3173
Height (cm), mean ± SD	165.2 ± 7.1	166.8 ± 6.6	0.3601
Weight before pregnancy (kg), mean ± SD	62.1 ± 9.5	64.3 ± 12.0	0.4479
Weight at birth (kg), mean ± SD	70.3 ± 10.6	74.5 ± 11.3	0.1426
BMI before pregnancy (kg/m^2), mean ± SD	22.8 ± 3.4	23.2 ± 4.9	0.6990
BMI at birth (kg/m^2), mean ± SD	25.8 ± 3.7	26.9 ± 4.7	0.3141
Caesarean delivery, %	63.0	23.5	0.0019
Infant	n = 33	n = 34	
Gestational age at birth (weeks), mean ± SD	30.8 ± 1.4	39.5 ± 1.0	<0.0001
Males, %	54.5	52.9	0.8952
Twins, %	36.4	0.0	0.0001
Height (cm), mean ± SD	40.4 ± 3.2	49.4 ± 1.7	<0.0001
Weight (g), mean ± SD	1421.4 ± 372.8	3277.6 ± 353.6	<0.0001
Head circumference (cm), mean ± SD	27.8 ± 2.1	34.4 ± 1.5	<0.0001

3.2. Total Lipids in Preterm and Term Human Milk

Total lipids was measured by MIRIS® HMA as previously described by Giuffrida et al. [23] and results are listed in Table 2. Total lipids content increased from colostrum (2.4 and 1.7 g/100 mL preterm and term HM, respectively) to mature milk (3.1 and 3.6 g/100 mL in preterm and term HM, respectively). Significant differences in total lipid content were observed between preterm and term groups but only in mature milk (Figure 2).

Table 2. Fatty acid (FA) composition of human milk (HM) expressed by mothers who delivered either preterm or term infant(s).

Fatty Acids	Colostrum		Transitional Milk		Mature Milk	
	Preterm	Term	Preterm	Term	Preterm	Term
total lipids (g/100 mL)	2.40 (1.25)	1.70 (1.35)	3.05 (1.35)	3.10 (0.95)	3.10 [c] (1.40)	3.60 [c] (1.95)
8:0 (caprylic acid)	0.08 [a] (0.08)	0.03 [a] (0.05)	0.20 (0.09)	0.22 (0.04)	0.21 (0.08)	0.22 (0.06)
10:0 (capric acid)	0.60 [a] (0.44)	0.29 [a] (0.3)	1.58 (0.58)	1.64 (0.47)	1.48 (0.47)	1.46 (0.38)
12:0 (lauric acid)	3.14 [a] (2.22)	2.24 [a] (1.28)	6.76 (2.41)	6.33 (2.02)	5.91 [c] (2.44)	5.26 [c] (2.10)
14:0 (myristic acid)	6.20 (1.52)	5.83 (1.77)	7.86 (3.35)	7.62 (1.8)	7.36 [c] (2.9)	6.27 [c] (1.93)
16:0 (palmitic acid)	24.02 (1.97)	25.68 (2.83)	22.75 (4.86)	23.49 (3.11)	23.10 (3.46)	23.29 (3.31)
16:1 n-7 (palmitoleic acid)	2.34 (0.88)	2.18 (0.51)	1.96 (1.24)	2.34 (0.74)	2.17 [c] (0.82)	2.44 [c] (0.77)
18:0 (stearic acid)	6.30 (1.55)	6.79 (1.51)	6.27 (1.86)	6.23 (1.03)	7.03 [c] (2.06)	6.75 [c] (1.69)
18:1 n-9 (oleic acid)	37.64 (2.73)	39.36 (3.02)	34.62 (5.77)	35.85 (4.06)	35.22 [c] (5.16)	37.67 [c] (4.82)
18:1 n-7 (vaccenic acid)	2.62 (0.62)	2.67 (0.48)	2.06 (0.44)	2.13 (0.4)	1.83 [c] (0.45)	1.96 [c] (0.38)
18:1 trans fatty acids	0.70 (0.28)	0.75 (0.19)	0.68 (0.3)	0.75 (0.31)	0.71 [c] (0.42)	0.82 [c] (0.36)
18:2 n-6 (linoleic acid)	9.61 (2.19)	7.92 (1.17)	9.55 [b] (2.98)	8.70 [b] (2.23)	10.21 [c] (3.64)	9.35 [c] (2.90)
18:3 n-3 (α-linolenic acid)	0.77 [a] (0.25)	0.51 [a] (0.15)	0.72 (0.28)	0.67 (0.26)	0.75 (0.43)	0.74 (0.30)
18:3 n-6 (γ-linolenic acid)	0.03 (0.02)	0.03 (0.02)	0.05 (0.03)	0.08 (0.04)	0.09 (0.04)	0.10 (0.06)
20:0 (arachidic acid)	0.21 [a] (0.05)	0.27 [a] (0.09)	0.20 (0.06)	0.21 (0.04)	0.20 (0.07)	0.20 (0.04)
20:1 n-9 (eicosenoic acid)	0.76 (0.23)	0.99 (0.20)	0.60 [b] (0.17)	0.54 [b] (0.09)	0.47 (0.14)	0.45 (0.12)
20:2 n-6 (eicosadienoic acid)	0.52 (0.22)	0.58 (0.16)	0.42 [b] (0.12)	0.34 [b] (0.06)	0.29 (0.12)	0.26 (0.07)
20:3 n-6 (dihomo-γ-linolenic acid)	0.51 [a] (0.17)	0.66 [a] (0.35)	0.41 (0.10)	0.48 (0.18)	0.35 (0.12)	0.38 (0.13)
20:5 n-3 (EPA)	0.07 (0.07)	0.07 (0.03)	0.06 (0.05)	0.07 (0.03)	0.06 (0.04)	0.06 (0.04)
22:1 n-9 (erucic acid)	0.19 [a] (0.06)	0.25 [a] (0.06)	0.12 (0.06)	0.12 (0.02)	0.09 (0.03)	0.08 (0.03)
20:4 n-6 (ARA)	0.71 (0.36)	0.78 (0.32)	0.55 (0.19)	0.53 (0.11)	0.40 (0.12)	0.42 (0.12)
24:0 (lignoceric acid)	0.19 (0.05)	0.23 (0.1)	0.14 (0.05)	0.12 (0.04)	0.08 (0.04)	0.08 (0.03)
24:1 n-9 (nervonic acid)	0.30 [a] (0.09)	0.39 [a] (0.14)	0.13 (0.09)	0.13 (0.03)	0.07 (0.04)	0.07 (0.03)
22:6 n-3 (DHA)	0.61 (0.41)	0.64 (0.28)	0.35 (0.19)	0.42 (0.15)	0.27 (0.16)	0.28 (0.17)
Sum SFA	43.49 (7.03)	42.03 (2.84)	47.14 (7.76)	46.08 (6.37)	45.88 [c] (7.45)	43.86 [c] (5.93)
Sum MUFA	44.59 (2.90)	46.51 (3.72)	40.65 (6.19)	42.12 (4.88)	40.44 [c] (5.6)	43.84 [c] (4.96)
Sum MCFA (< 14:0)	3.95 [a] (2.52)	2.51 [a] (1.65)	8.52 (2.67)	7.93 (2.66)	7.63 [c] (2.89)	6.94 [c] (2.38)
Sum PUFA	12.90 (2.95)	11.33 (1.95)	12.19 (3.05)	11.29 (2.03)	12.76 (4.00)	11.77 (3.43)
Sum PUFA n-3	1.48 (0.91)	1.16 (0.36)	1.24 (0.54)	1.15 (0.23)	1.15 (0.62)	1.11 (0.36)
Sum PUFA n-6	10.16 (2.68)	9.92 (1.66)	10.88 (2.96)	10.20 (2.22)	11.37 (3.82)	10.61 (3.17)
n-6 to n-3 ratio	7.29 (2.49)	8.67 (2.24)	8.68 (4.56)	8.58 (1.93)	9.72 (4.19)	9.58 (3.31)
ARA to DHA ratio	1.36 [a] (0.71)	1.40 [a] (0.53)	1.52 (0.59)	1.40 (0.41)	1.77 (1.15)	1.51 (0.68)

The data expressed in this table are medians (and interquartile range in parentheses) expressed as g/100 g FA except total lipids which is expressed in g/100 mL of human milk and ratios. Values within a row with a letter ([a, b, c]) indicate statistically significant differences (p < 0.05) between preterm and term HM for colostrum, transitional and mature milk, respectively. SFA—Saturated Fatty Acids, MUFA—Mono-Unsaturated Fatty Acids, PUFA—Poly-Unsaturated Fatty Acids, MCFA—Medium Chain Fatty Acids, ARA—Arachidonic Acid, DHA—Docosahexaenoic Acid, EPA—Eicosapentaenoic Acid.

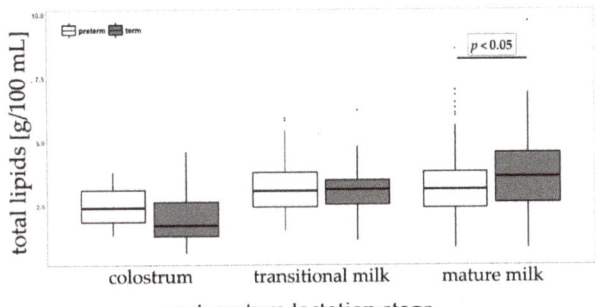

Figure 2. Total lipids content (g/100 mL) in colostrum, transitional and mature milk in preterm and term groups.

3.3. Fatty Acids in Preterm and Term Human Milk

FA were measured by gas chromatography—flame ionization detection (GC-FID), and the results are enumerated in Table 2.

The sum of saturated fatty acid (SFA) content increased significantly from colostrum (43.49 and 42.03% sum of FA in preterm and term HM, respectively) to mature milk (45.88 and 43.86% sum of FA in preterm and term HM, respectively). The sum of SFA content was significantly higher in preterm than in term mature milk. Palmitic acid (16:0) was the most abundant SFA and overall second most abundant FA in preterm and term HM in this study. In spite of increases in total SFA, palmitic acid content significantly decreased from colostrum (24.02 and 25.68% sum of FA in preterm and term HM, respectively) to mature milk (23.10 and 23.29% sum of FA in preterm and term HM, respectively). Stearic acid (18:0) increased significantly only in preterm milk (6.30% in colostrum and 7.03% in mature milk). Furthermore, short (SC; 8:0) and medium chain (MC; 10:0 and 12:0) FA content was significantly low in colostrum (3.95 and 2.51% sum of FA in preterm and term HM, respectively) compared to transitional (8.52 and 7.93% sum of FA in preterm and term HM, respectively) and mature milk (7.63 and 6.94% sum of FA in preterm and term HM, respectively). Significant differences were observed between preterm and term colostrum for caprylic (8:0), capric (10:0) and lauric (12:0) being higher in preterm than in term. In mature HM, lauric and myristic (14:0) acids were significantly higher in preterm than in term. Figure 3 shows capric and lauric acid concentration in colostrum, transitional and mature milk and differences between preterm and term. Arachidic acid (20:0) was lower in preterm milk but only for the colostrum stage of lactation.

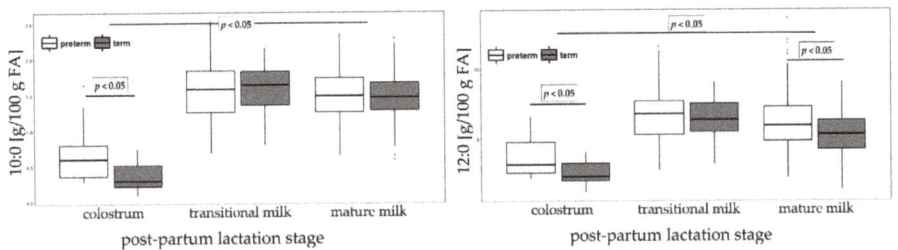

Figure 3. Capric (10:0, left panel) and lauric (12:0, right panel) acids in colostrum, transitional and mature milk in preterm and term groups. The results are expressed as percentages (g/100 g sum of FA).

The sum of MUFA in HM decreased significantly from colostrum (44.59 and 46.51% of sum of FA in preterm and term HM, respectively) to mature milk (40.44 and 43.84% of sum of FA in preterm and term HM, respectively). Oleic acid (18:1 n-9), the most abundant FA, decreased significantly along the lactation (from 37.64 to 35.22 % of total FA in preterm HM and from 39.36 to 37.67% of total FA in term HM) (Figure 4). Other MUFA (i.e., 18:1 n-7, 20:1 n-9, 22:1 n-9 and 24:1 n-9) also decreased significantly over lactation. The only exception was 16:1 n-7 for which no significant trend was observed. The total MUFA content was significantly higher in term than in preterm mature milk. However, significant differences were observed between preterm and term colostrum with 22:1 n-9 and 24:1 n-9 being significantly lower in preterm and in mature milk for 18:1 n-9 being significantly lower in preterm HM.

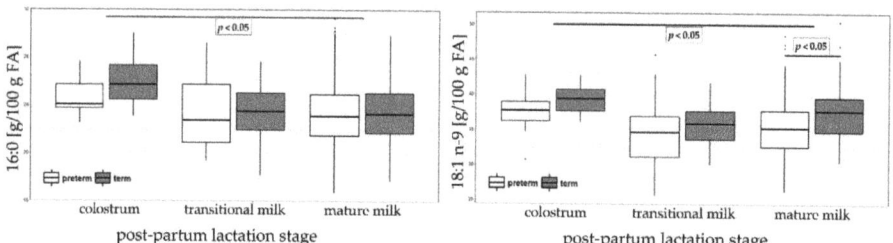

Figure 4. Palmitic (16:0, left panel) and oleic (18:1 n-9, right panel) acids in colostrum, transitional and mature milk in preterm and term. The results are expressed as percentages (g/100 g sum of FA).

Among poly-unsaturated fatty acids (PUFA), n-6 linoleic acid (LA; 18:2 n-6) was the most abundant FA and it increased significantly from colostrum (9.61 and 7.92% sum of FA in preterm and term HM, respectively) to mature milk (10.21 and 9.35% sum of FA in preterm and term HM, respectively). ARA (20:4 n-6) content decreased significantly from colostrum to mature milk in both preterm and term HM. No significant differences on PUFA n-6 content were observed between preterm and term in colostrum, transitional milk and mature milk (Figure 5). Among PUFA, n-3 alpha-linolenic acid (ALA; 18:3 n-3) was the most abundant FA and it increased significantly in term from colostrum (0.51% of total FA) to mature milk (0.74% of total FA) but it was stable at about 0.7% of sum of FA in preterm group. DHA (22:6 n-3) decreased significantly over the lactation period from 0.6% in colostrum to 0.3% of total FA in mature milk, in both preterm and term HM. Eicosapentaenoic acid (EPA; 20:5 n-3) was present in minute quantities (0.07% of the sum of FA in colostrum, transitional and mature milk). Between preterm and term, the only significant differences were observed for ALA in colostrum, being higher in preterm than in term HM.

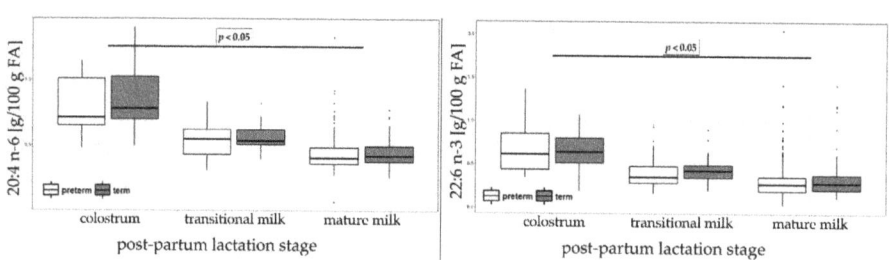

Figure 5. Arachidonic acid (ARA) (20:4 n-6, top panel) and docosahexaenoic acid (DHA) (22:6 n-3, bottom panel) acids in colostrum, transitional and mature milk in preterm and term. The results are expressed as percentages (g/100 g sum of FA).

4. Discussion

There are different classes of fatty acids (FA) in HM with putative biological functions. Most widely studied are long chain polyunsaturated FA (LCPUFA) with potential roles in the development of visual and cognitive functions in early life [24,25]. HM also contains essential FA, such as linoleic (18:2 n-6) and alpha-linolenic (18:3 n-3) acid that must be supplied orally as *de novo* synthesis is low to non-existent [7]. Furthermore, there is also the presence of short to medium chain and saturated FA (SFA) along with monounsaturated FA (MUFA). In this study, we longitudinally characterized total lipids and FA from milk of mothers who delivered either preterm or term infants.

Total lipid content increased from colostrum to mature milk in agreement with multiple previous reports [16,26–33]. However, our observation that term HM had higher lipids than preterm milk when the milk was mature is not corroborated by a systematic review and meta-analysis of Gidrewicz and Fenton [34]. This could be due to inclusion of multiple studies in meta-analysis with varied sampling and analytical procedures as compared to our study.

The saturated FA we characterized in this study ranges from short chain FA (8:0), medium chain FA (10:0, 12:0, 14:0), to long chain FA (16:0, 18:0, 20:0) and very long chain FA (24:0). Overall, the sum of all SFA in colostrum was lower than transitional and mature milk for both preterm and term groups. In colostrum, short and medium chain FA were higher in preterm than in term HM, in agreement with previous works [35,36]. Benefits of medium chain FA in preterm infant nutrition has been a topic of research for past few decades. It has been demonstrated [21] that, after triacylglycerol hydrolysis of 10:0 and 12:0, FA are absorbed directly in the blood circulation without being incorporated in chylomicrons, thus may be more bioavailable and/or readily available sources of energy in the immature preterm digestive system that longer chain FA. Additionally, entering the cells these FA get into mitochondria without the assistance of a carnitine transporter [21], therefore sparing adenosine triphosphate (ATP) for other cellular process. Palmitic acid (16:0) was the most abundant saturated FA and accounted for approximately 60% of sum of SFA and it also represented the second most abundant FA in HM. Palmitic acid did not show any major temporal changes in either of the groups, suggesting minor variations amongst different populations, a phenomenon that has also been observed in other studies [36–38].

Amongst all FA characterized in this study, oleic acid (OA, 18:1 n-9) was the most abundant FA in both term and preterm groups. While it did not show any statistically significant difference between groups in colostrum and transitional milk, the content was higher in term mature milk. This observation is in line with findings of Rueda et al., [39] which not only agreed with the ranges of OA present in both term and preterm milk but also demonstrated that term HM contained higher proportions of OA in comparison to their preterm counterparts.

Linoleic acid (LA) was the most abundant n-6 FA in both term and preterm milk. Both groups showed slight increases in the concentrations from colostrum to mature milk. Not only is linoleic acid an essential fatty acid that is a required precursor for production of ARA, but also downstream products of ARA yields leukotrienes, prostaglandins and thromboxane that are physiologically active and provide diversified functions of signaling. The relative percentage of linoleic acid in our study agreed with previous reports of Luukkainen [37], Genzel-Boroviczény [38], Rueda [39] and Sabel [40]. However, Kovacs [36] reported to have 40 to 50% more LA in milk of both term and preterm infants. However, since n-6 FA can be modulated by maternal intakes, the groups can attribute the observation to differential intake. Alpha-linolenic acid (ALA), the precursor to EPA and DHA, represented between 40 and 60% of all n-3 FA characterized in this study for both term and preterm infants. In our study, colostrum of preterm HM contained statistically significant higher proportions of ALA than their term counterparts. However, this significance did not sustain over time and the relative percentages of ALA were comparable in both term and preterm groups for transitional and mature milk. The mature milk of other studies also reportedly has no differences between the groups [36,37]. A limitation of this study is that neither dietary intake, nor supplement intake was recorded preventing us from associating HM FA to these factors.

The concentration of DHA decreased over stages of lactation for both groups, preterm and term HM. This trend has been described by numerous studies reported in the literature [19,31,35,37–39]. However, consistency is not observed when comparing the concentrations of DHA in milk within term and preterm groups. In our study we did not observe significant differences, yet it has been reported by Kovács et al. [36] that there is significantly higher DHA in preterm milk than in term milk for first 21 day post-partum. On the other hand Rueda et al. [39], reported to have higher DHA in term milk over preterm milk for first week post-partum. Since DHA content of mothers milk is sensitive to maternal intake of food rich in sources of DHA [41], it may explain the differences observed in different studies. It also may be prudent to note that preterm offspring may benefit from fortifying mothers' own milk in populations where dietary/supplementary intake of sources of DHA may be lower than ideal.

5. Conclusions

In summary, oleic, palmitic and linoleic acids were the most abundant FA across lactation and in both groups. Preterm colostrum contained significantly higher 8:0, 10:0, 12:0, sum medium chain fatty acids (MCFA), 18:3 n-3 FA compared to term counterparts. Preterm mature milk contained significantly higher 12:0, 14:0, 18:2 n-6, sum SFA, and sum MCFA. Preterm colostrum contained significantly lower 20:0, 20:3 n-6, 22:1 n-9 and 24:1 n-9. Preterm mature milk contained significantly lower total lipids, 16:1 n-7 and 18:1 n-9. We did not observe any significant differences between the preterm and term groups for DHA, ARA and EPA at any stage of lactation. Overall, preterm milk was higher for SFA with major contributions from MCFA and higher in 18:2 n-6. These observational differences need to be studied further for their implications on preterm developmental outcomes and on fortification strategies of either mothers' own milk or donor human milk.

Author Contributions: S.K.T. and F.G. contributed to the interpretation of the results and drafted the manuscript. M.A. designed and conducted the research, contributed to the interpretation of the results and to manuscript writing and approved the final version. J.-F.T., C.J.F.F. contributed to the study design and to the development of overall research plan, conducted the research, reviewed the manuscript and approved the final version. C.A.D.C. performed the statistical analysis, participated in manuscript writing and approved the final version. L.B. conducted the research, reviewed the manuscript and approved the final version.

Funding: This study was funded by Nestlé Research, Lausanne (Nestec Ltd.). Nestlé Research sponsored the study, participated to the protocol building, and provided the infrastructure and the material and human resources to analyze the HM samples and lipids data. Employees of Nestlé Research further participated in the interpretation of the results and the manuscript writing.

Acknowledgments: First and foremost, a sincere note of gratitude to all mothers who enthusiastically participated in this study. Furthermore, sincere thanks to Nassima Grari, the study lactation nurse who admirably cared about all mothers during the full study duration, to Céline Romagny and Emilie Darcillon who managed the organizational and operational part of the clinical study and the data management. Sincere thanks also to Clara L. Garcia-Rodenas for critical inputs on this manuscript.

Conflicts of Interest: S.K.T., M.A., C.A.D.C., and F.G. are all employees of Nestec Ltd. L.B., J.-F.T., and C.J.F.F. declare no conflict of interest.

References

1. Gartner, L.M.; Morton, J.; Lawrence, R.A.; Naylor, A.J.; O'Hare, D.; Schanler, R.J.; Eidelman, A.I. Breastfeeding and the use of human milk. *Pediatrics* **2005**, *115*, 496–506. [PubMed]
2. Kramer, M.S.; Kakuma, R. Optimal duration of exclusive breastfeeding. *Cochrane Database Syst. Rev.* **2012**, CD003517. [CrossRef] [PubMed]
3. Victora, C.G.; Bahl, R.; Barros, A.J.; França, G.V.; Horton, S.; Krasevec, J.; Murch, S.; Sankar, M.J.; Walker, N.; Rollins, N.C.; et al. Breastfeeding in the 21st century: Epidemiology, mechanisms, and lifelong effect. *Lancet* **2016**, *387*, 475–490. [CrossRef]
4. Quigley, M.; Embleton, N.D.; McGuire, W. Formula versus donor breast milk for feeding preterm or low birth weight infants. *Cochrane Database Syst. Rev.* **2018**, *6*, CD002971. [CrossRef] [PubMed]
5. Ziegler, E.E. 3.14 Preterm and low-birth-weight infants. *World Rev. Nutr. Diet* **2015**, *113*, 214–217.
6. Kumar, R.K.; Singhal, A.; Vaidya, U.; Banerjee, S.; Anwar, F.; Rao, S. Optimizing nutrition in preterm low birth weight infants—Consensus summary. *Front. Nutr.* **2017**, *4*, 20. [CrossRef]
7. Heird, W.C.; Lapillonne, A. The role of essential fatty acids in development. *Annu. Rev. Nutr.* **2005**, *25*, 549–571. [CrossRef]
8. Molloy, C.; Doyle, L.W.; Makrides, M.; Anderson, P.J. Docosahexaenoic acid and visual functioning in preterm infants: A review. *Neuropsychol. Rev.* **2012**, *22*, 425–437. [CrossRef]

9. Schneider, N.; Garcia-Rodenas, C. Early nutritional interventions for brain and cognitive development in preterm infants: A review of the literature. *Nutrients* **2017**, *9*, 187. [CrossRef]
10. Ziegler, E.E. Human milk and human milk fortifiers. *World Rev. Nutr. Diet.* **2014**, *110*, 215–227.
11. Arslanoglu, S.; Moro, G.E.; Ziegler, E.E. Optimization of human milk fortification for preterm infants: New concepts and recommendations. *J. Perinat. Med.* **2010**, *38*, 233–238. [CrossRef] [PubMed]
12. Rochow, N.; Fusch, G.; Choi, A.; Chessell, L.; Elliott, L.; McDonald, K.; Kuiper, E.; Purcha, M.; Turner, S.; Chan, E.; et al. Target fortification of breast milk with fat, protein, and carbohydrates for preterm infants. *J. Pediatr.* **2013**, *163*, 1001–1007. [CrossRef] [PubMed]
13. Dror, D.K.; Allen, L.H. Overview of Nutrients in Human Milk. *Adv. Nutr.* **2018**, *9* (Suppl.1), 278S–294S. [CrossRef] [PubMed]
14. Thakkar, S.K.; Giuffrida, F.; Cristina, C.H.; De Castro, C.A.; Mukherjee, R.; Tran, L.A.; Steenhout, P.; Lee, L.Y.; Destaillats, F. Dynamics of human milk nutrient composition of women from Singapore with a special focus on lipids. *Am. J. Hum. Biol.* **2013**, *25*, 770–779. [CrossRef] [PubMed]
15. Giuffrida, F.; Cruz-Hernandez, C.; Bertschy, E.; Fontannaz, P.; Masserey Elmelegy, I.; Tavazzi, I.; Marmet, C.; Sanchez-Bridge, B.; Thakkar, S.K.; De Castro, C.A.; et al. Temporal changes of human breast milk lipids of chinese mothers. *Nutrients* **2016**, *8*, 715. [CrossRef] [PubMed]
16. Mimouni, F.B.; Lubetzky, R.; Yochpaz, S.; Mandel, D. Preterm human milk macronutrient and energy composition: A systematic review and meta-analysis. *Clin. Perinatol.* **2017**, *44*, 165–172. [CrossRef]
17. Fields, D.A.; Schneider, C.R.; Pavela, G. A narrative review of the associations between six bioactive components in breast milk and infant adiposity. *Obesity* **2016**, *24*, 1213–1221. [CrossRef]
18. Meier, P.; Patel, A.L.; Wright, K.; Engstrom, J.L. Management of breastfeeding during and after the maternity hospitalization for late preterm infants. *Clin. Perinatol.* **2013**, *40*, 689–705. [CrossRef]
19. Berenhauser, A.C.; Pinheiro do Prado, A.C.; da Silva, R.C.; Gioielli, L.A.; Block, J.M. Fatty acid composition in preterm and term breast milk. *Int. J. Food Sci. Nutr.* **2012**, *63*, 318–325. [CrossRef]
20. Granot, E.; Ishay-Gigi, K.; Malaach, L.; Flidel-Rimon, O. Is there a difference in breast milk fatty acid composition of mothers of preterm and term infants? *J. Matern. Fetal Neonatal Med.* **2016**, *29*, 832–835. [CrossRef]
21. Bobinski, R.; Mikulska, M.; Mojska, H.; Simon, M. Comparison of the fatty acid composition of transitional and mature milk of mothers who delivered healthy full-term babies, preterm babies and full-term small for gestational age infants. *Eur. J. Clin. Nutr.* **2013**, *67*, 966–971. [CrossRef] [PubMed]
22. Cruz-Hernandez, C.; Goeuriot, S.; Giuffrida, F.; Thakkar, S.K.; Destaillats, F. Direct quantification of fatty acids in human milk by gas chromatography. *J. Chromatogr. A* **2013**, *1284*, 174–179. [CrossRef] [PubMed]
23. Giuffrida, F.; Austin, S.; Cuany, D.; Sanchez-Bridge, B.; Longet, K.; Bertschy, E.; Sauser, J.; Thakkar, S.K.; Lee, L.Y.; Affolter, M. Comparison of macronutrient content in human milk measured by mid-infrared human milk analyzer and reference methods. *J. Perinatol.* **2018**. [CrossRef] [PubMed]
24. Koletzko, B.; Rodriguez-Palmero, M. Polyunsaturated fatty acids in human milk and their role in early infant development. *J. Mammary Gland Biol. Neoplasia* **1999**, *4*, 269–284. [CrossRef] [PubMed]
25. Lapillonne, A. Enteral and parenteral lipid requirements of preterm infants. *World Rev. Nutr. Diet.* **2014**, *110*, 82–98. [PubMed]
26. Bauer, J.; Gerss, J. Longitudinal analysis of macronutrients and minerals in human milk produced by mothers of preterm infants. *Clin. Nutr.* **2011**, *30*, 215–220. [CrossRef] [PubMed]
27. Narang, A.P.; Bains, H.S.; Kansal, S.; Singh, D. Serial composition of human milk in preterm and term mothers. *Indian J. Clin. Biochem.* **2006**, *21*, 89–94. [CrossRef]
28. Anderson, G.H.; Atkinson, S.A.; Bryan, M.H. Energy and macronutrient content of human milk during early lactation from mothers giving birth prematurely and at term. *Am. J. Clin. Nutr.* **1981**, *34*, 258–265. [CrossRef]
29. Corvaglia, L.; Martini, S.; Aceti, A.; Capretti, M.G.; Galletti, S.; Faldella, G. Cardiorespiratory events with bolus versus continuous enteral feeding in healthy preterm infants. *J. Pediatr.* **2014**, *165*, 1255–1257. [CrossRef]
30. Gross, S.J.; David, R.J.; Bauman, L.; Tomarelli, R.M. Nutritional composition of milk produced by mothers delivering preterm. *J. Pediatr.* **1980**, *96*, 641–644. [CrossRef]
31. Molto-Puigmarti, C.; Castellote, A.I.; Carbonell-Estrany, X.; López-Sabater, M.C. Differences in fat content and fatty acid proportions among colostrum, transitional, and mature milk from women delivering very preterm, preterm, and term infants. *Clin. Nutr.* **2011**, *30*, 116–123. [CrossRef] [PubMed]

32. Mahajan, S.; Chawla, D.; Kaur, J.; Jain, S. Macronutrients in breastmilk of mothers of preterm infants. *Indian Pediatr.* **2017**, *54*, 635–637. [CrossRef] [PubMed]
33. Saint, L.; Maggiore, P.; Hartmann, P.E. Yield and nutrient content of milk in eight women breast-feeding twins and one woman breast-feeding triplets. *Br. J. Nutr.* **1986**, *56*, 49–58. [CrossRef] [PubMed]
34. Gidrewicz, D.A.; Fenton, T.R. A systematic review and meta-analysis of the nutrient content of preterm and term breast milk. *BMC Pediatr.* **2014**, *14*, 216. [CrossRef] [PubMed]
35. Kuipers, R.S.; Luxwolda, M.F.; Dijck-Brouwer, D.J.; Muskiet, F.A. Fatty acid compositions of preterm and term colostrum, transitional and mature milks in a sub-Saharan population with high fish intakes. *Prostaglandins Leukot. Essent. Fatty Acids* **2012**, *86*, 201–207. [CrossRef]
36. Kovacs, A.; Funke, S.; Marosvölgyi, T.; Burus, I.; Decsi, T. Fatty acids in early human milk after preterm and full-term delivery. *J. Pediatr. Gastroenterol. Nutr.* **2005**, *41*, 454–459. [CrossRef] [PubMed]
37. Luukkainen, P.; Salo, M.K.; Nikkari, T. The fatty acid composition of banked human milk and infant formulas: The choices of milk for feeding preterm infants. *Eur. J. Pediatr.* **1995**, *154*, 316–319. [CrossRef]
38. Genzel-Boroviczeny, O.; Wahle, J.; Koletzko, B. Fatty acid composition of human milk during the 1st month after term and preterm delivery. *Eur. J. Pediatr.* **1997**, *156*, 142–147. [CrossRef]
39. Rueda, R.; Ramírez, M.; García-Salmerón, J.L.; Maldonado, J.; Gil, A. Gestational age and origin of human milk influence total lipid and fatty acid contents. *Ann. Nutr. Metab.* **1998**, *42*, 12–22. [CrossRef]
40. Sabel, K.G.; Lundqvist-Persson, C.; Bona, E.; Petzold, M.; Strandvik, B. Fatty acid patterns early after premature birth, simultaneously analysed in mothers' food, breast milk and serum phospholipids of mothers and infants. *Lipids Health Dis.* **2009**, *8*, 20. [CrossRef]
41. Francois, C.A.; Connor, S.L.; Wander, R.C.; Connor, W.E. Acute effects of dietary fatty acids on the fatty acids of human milk. *Am. J. Clin. Nutr.* **1998**, *67*, 301–308. [CrossRef] [PubMed]

© 2019 by the authors. Licensee MDPI, Basel, Switzerland. This article is an open access article distributed under the terms and conditions of the Creative Commons Attribution (CC BY) license (http://creativecommons.org/licenses/by/4.0/).

Review

The Functional Power of the Human Milk Proteome

Jing Zhu [1,2] and Kelly A. Dingess [1,2,*]

1. Biomolecular Mass Spectrometry and Proteomics, Bijvoet Center for Biomolecular Research and Utrecht Institute for Pharmaceutical Sciences, University of Utrecht, Padualaan 8, 3584 CH Utrecht, The Netherlands
2. Netherlands Proteomics Center, Padualaan 8, 3584 CH Utrecht, The Netherlands
* Correspondence: k.a.dingess@uu.nl

Received: 12 July 2019; Accepted: 6 August 2019; Published: 8 August 2019

Abstract: Human milk is the most complete and ideal form of nutrition for the developing infant. The composition of human milk consistently changes throughout lactation to meet the changing functional needs of the infant. The human milk proteome is an essential milk component consisting of proteins, including enzymes/proteases, glycoproteins, and endogenous peptides. These compounds may contribute to the healthy development in a synergistic way by affecting growth, maturation of the immune system, from innate to adaptive immunity, and the gut. A comprehensive overview of the human milk proteome, covering all of its components, is lacking, even though numerous analyses of human milk proteins have been reported. Such data could substantially aid in our understanding of the functionality of each constituent of the proteome. This review will highlight each of the aforementioned components of human milk and emphasize the functionality of the proteome throughout lactation, including nutrient delivery and enhanced bioavailability of nutrients for growth, cognitive development, immune defense, and gut maturation.

Keywords: human milk; protein; glycoprotein; endogenous peptide; breastfeeding

1. Introduction

Human milk via breastfeeding is the gold standard for infant feeding, as it provides not only nutritional excellence, but also protective effects during a time of unmatched antigenic and pathogenic challenges. Both short and long term health benefits have been attributed to breastfeeding by clinical and epidemiological studies [1]. Some short and long term benefits include, but are not limited to, lower incidences of diarrhea, respiratory and urinary tract infections, and otitis media; reduction of disease risk, such as, asthma, allergy, and type I and II *diabetes mellitus* [2–5]. Additionally, it has been shown to have protective effects against the development of noncommunicable diseases commonly associated with inflammatory pathogenesis, such as obesity and cardiovascular disease [2–5]. The act of breastfeeding has also shown to be beneficial for maternal health, including reduced risk for development of rheumatoid arthritis, cardiovascular disease, diabetes, and breast and ovarian cancer [1]. For all these reasons and more, exclusive breastfeeding for six months, and in conjunction with complementary food feeding through one year of life or longer, as desired by the mother and infant, are recommended by the American Academy of Pediatrics and the World Health Organization [1].

The human milk proteome is comprised of not only proteins, which are highly glycosylated, but also endogenous peptides that are derived from proteins within the mammary gland maintaining their own distinct functions. One must consider the overall composition of milk, as it is one of the most complex and dynamic biofluids of the human body, to understand the function of the human milk proteome. Human milk is comprised of three major macronutrients, carbohydrates, fats, and proteins, listed by order of their relative abundance [6]. These macronutrients are continuously changing in their composition and concentration throughout lactation to meet the demanding needs of the infant during growth and development.

In this review, we aimed to provide a comprehensive overview of the human milk proteome, which is obtained by applying multiple omics approaches, including proteomics, peptidomics, and glycoproteomics. We believe that combined analysis of multiple omics approaches will help in providing an increased understanding of the functionality of each constituent of the proteome. This review will highlight each component of the human milk proteome, proteins, including enzymes/proteases, glycoproteins, and endogenous peptides. Additionally, we will emphasize the functionality of the proteome throughout lactation, including nutrient delivery and enhanced bioavailability of nutrients for growth, cognitive development, immune defense, and gut maturation.

2. Factors that Affect Milk Composition

The lactational stage is one of the key designators for determining the composition of human milk, which is traditionally based on day postpartum and further categorized as colostrum (< 72 h postpartum), transitional (> 3–15 days), and mature (> 16 days) milk [1,6,7]. These stages of lactation are more accurately categorized by the maturity of the mammary gland and the functions of the proteome it is producing. Colostrum is produced in the lactational stage when the protein concentration is highest, especially regarding immune modulatory proteins, as the mammary gland is still maturing. Typically, the maturity of the mammary gland involves tight junction closures of the mammary epithelium and regulation of the $Na^+:K^+$ efflux [7]. Colostrum is then followed by transitional milk, in which there is an upregulation of protein synthesis and an inhibition of protein degradation. Last is mature milk in which there is a switch from protein production to fatty acid synthesis.

Reports indicate that breast milk composition becomes relatively stable between 2–12 weeks of lactations and that as the mammary gland matures from a transitional to a mature stage there is less protein variance [8]. Overall, while considering the entire lactational period, the protein concentration in human milk is typically in the range of 10–20 g/L [9], starting high in colostrum and then steadily declining until it reaches a plateau and then becomes designated as mature milk. Table 1 illustrates detailed protein concentrations and functionality. For generating Table 1, individual proteins were searched, articles were considered first for term milk and mass spectrometry (MS) derived data. If this was not available, then literature values based on preterm milk and or immune assays were used. Overall, we aimed to use the most current literature possible. The overall shift in milk composition is important, because it shows the adaptability of milk to meet the changing functional needs of the infant. Shifting from innate immune initiation to adaptive immune learning and becoming more energy dense to better meet the caloric needs of the growing infant [10] and their diversifying gut microbiome.

Recent research has focused on exploring the human milk proteome longitudinally and determined that the protein content of human milk changes throughout the course of lactation based on functionality, rather than by differing mothers or differing populations [11–13]. This change in functionality meets the changing needs of the developing infant where an early lactational stage was characterized by greater concentrations of the immunoglobulins IgA and IgM and a switch at later lactational stage was characterized by the enhanced abundance of IgG [11,12]. This data suggests that human milk transitions from a defense mechanism of the newborn and direct pathogen-killing, to a more mature milk supporting an independent immune system [11,12]. Overall, there is ample evidence that human milk helps to establish both innate and adaptive immune responses of the infant as the infant matures.

There are many factors that affect the composition of human milk, independent of the lactational stage. These factors can be maternal, such as diet, smoking status, exercise, BMI, and infant factors, such as gestational age and sex of the infant [9,10]. Additionally, the breast milk composition can fluctuate within feeds and diurnally, and it is influenced by maternal diet [10]. Recent evidence has shown that foreign proteins, from bovine milk, can be detected in human milk, and that the most likely source for these foreign proteins is via the maternal diet [14]. Diurnal changes have been identified in the fat composition of milk, but such changes are not well investigated or reported for proteins, protein post-translational modifications (PTMs), and/or endogenous peptides. Additionally, the overall proteome can be impacted by inhibitors and activators, present within the mammary gland, and throughout the body allowing for constituents of the proteome to be translocated to the mammary gland via circulation, as described in more detail in Section 2.1. Future research should aim to explore the diurnal changes of the human milk proteome and all of its components.

Human milk cannot be artificially mass-produced in all details of its natural complexity, instead it is tailor made to meet the precise and unique needs of each mother-infant pair. Moreover, it has undergone more than 300 million years [15] of evolutionary selective pressure to become this intricate, dynamic, and essential element of human nutrition. The proteome is an exemplary model of this, as it can reveal the complex differences between individuals while meeting the functional needs of the infant. Wherein the biological functionality of human milk is determined by a series of interacting and synergistic factors that are conserved, regardless of the considerable qualitative and quantitative differences in the human milk proteome between individuals.

Table 1. Overview of reported human milk protein concentrations over lactation and functionality.

	Protein Name	Total	Colostrum	Early	Transitional	Mature	References	Function
	Total protein	203-1752	360-1690	606-1675	203-1752	362-1632	[16]	
	Total caseins	19-591	42-507	103-355	87-591	19-743	[16]	
	Ratio whey/casein		90:10	78:22	72:28	60:40	[9,16]	
	Whey Proteins							
Proteins	α-Lactalbumin	275-372	300-560	NA	420	275-372	[17,18]	Lactose synthesis
	Lactoferrin	97-291	291	NA	180	97	[17,19]	Antimicrobial; Gut development
	Osteopontin	6-149	149	NA	NA	6-22	[3,20]	Cell adhesion
	sIgA	22-545	545	NA	150	22-130	[17,18]	Adaptive immunity
	IgG	2-7	NA	NA	5	2-7	[18,21]	Adaptive immunity
	sIgM	1-3	NA	NA	12	1-3	[18,21]	Adaptive immunity
	Lysozyme	3-110	32	NA	30	3-110	[17,18]	Antimicrobial
	α1-Antitrypsin	2-5	NA	NA	NA	2-5	[21]	Protease inhibitor
	Serum albumin	35-69	35	NA	62	37-69	[18]	Transport
	Lactoperoxidase	70 *	NA	NA	NA	70 *,#	[22]	Antimicrobial
	Haptocorrin	70-700 *	NA	NA	NA	70-700 *	[3]	Vitamin B12 transport
	Complement C3	11-12	NA	11	NA	12	[23,24]	Innate immunity
	Complement C4	5	NA	5	NA	5	[23,24]	Innate immunity
	Complement factor B	2	NA	2	NA	NA	[23]	Innate immunity
	Casein Proteins							
	β-casein	4-442	4-364	18-204	6-414	5-442	[17]	Calcium transport
	α-S1-casein	4-168	12-58	15-85	9-110	4-168	[16]	Calcium transport
	κ-casein	10-172	25-150	47-134	10-172	10-134	[16]	Calcium transport
	MFGM Proteins							
	Mucin 1	13-294 §	NA	NA	13-250 §	35-294 §	[25]	Growth promoter
	Lactadherin	3-33 §	NA	NA	4-33 §	3-13 §	[25]	Cell adhesion
	Butyrophilin subfamily 1	500-10,000 *,§	NA	NA	800-8200 *,§	500-10,000 *,§	[25]	Regulation of immune response
	Bile salt-activated lipase	10-20	NA	NA	NA	NA	[3]	Lipid digestion
Enzymes	Total protease activity	0.76-1.38 †	1.38 †	NA	NA	0.76 †	[26]	Coagulation
	Thrombin	7100 **,§	NA	NA	NA	7100 **,§,#	[27]	Proteolysis
	Plasmin	14600 **,§	NA	NA	NA	14,600 **,§,#	[27]	Proteolysis
	Elastase	200 **,§	NA	NA	NA	200 **,§,#	[27]	Proteolysis
Hormone peptides	Total endogenous peptides	1-2	NA	NA	NA	NA	[3]	
	Ghrelin	7-16 **	6-9 **	NA	7-10 **	13-16 **	[28,29]	Appetite stimulator
	Leptin	16-194 **	16-700 **	NA	20-84 **	165-194 **	[28,30,31]	Energy regulator
	Epidermal growth factor	4-5 *	NA	NA	NA	4-5 **	[28]	Stimulates magnesium reabsorption
	Insulin-like growth factor-1	6-12 *	NA	NA	NA	6-12 *	[28]	Insulin regulator and growth-promoting

Table 1. Cont.

Adiponectin	420–8790 **	NA	NA	661–2156 **	420–8790 **	[31,32]	Glucose and fat regulator
Parathyroid	1029–5840 ‡	1029 ‡	4584 ‡	5840 ‡	NA	[33]	Epidermis development

Notes: Values were obtained from studies analyzing term milk samples unless otherwise noted. MS data from literature was considered first and values are indicated in blue, if no MS data was available, immune assays were used and are indicated in black. Total values were derived from the ranges observed across lactation from all referenced literature. Lactational stages designated as, colostrum: ≤2 days, early: 3–5 days, transitional: 6–15 days, mature: ≥16 days. # wide range of lactational days reported and values were designated as mature. Values were reported as ranges unless only a single value was designated. All values were rounded to the nearest whole number and are in mg/100 mL, unless designated as * ug/100 mL. ** ng/100mL. † Units for total protease activity are μmol tyrosine/1000 mL/min. ‡ Units for parathyroid hormone related protein were in units of pmol/L, only mean values from reference were reported. Most articles do not distinguish between sIgA and IgA or IgA1 vs. IgA2, we therefore choose to represent all values as sIgA. § Enzymes and MFGM proteins were from preterm human milk samples, MFGM values from paper designated as <15 days were considered transitional. Values from hormone peptides were reported only from normal weight infants. Functions were determined by UniProt assignments based on Protein ID.

2.1. Mammary Gland Physiology

The lactating mammary gland, which is comprised of branching networks of ducts formed from epithelial cells making up extensive lobulo-alveolar clusters, is the site of milk production and secretion. These lactocytes are responsible for 80–90% of the protein content of human milk, with the remaining percentage coming from maternal circulation [10]. Proteins and endogenous peptides enter the inner luminal space of the mammary gland through four general pathways (Figure 1), belonging to either transcellular or paracellular routes that were extensively summarized by other reviews [26,34–37]. In brief, the exocytotic pathway (pathway 1) is the dominant way for the secretion of endogenously generated proteins, including the major milk proteins, e.g., caseins, α-lactalbumin, and lactoferrin. This secretion mechanism is similar to exocytotic pathways that were found in other cell types [37]. Lipid-associated proteins existing in milk fat globule membrane (MFGM) are secreted by a process that is unique to mammary epithelial cells (pathway 2) [38], including mucins and enzymes [39]. Another transcellular pathway, the transcytosis pathway (pathway 3), is responsible for the transporting of proteins from serum or stromal cells, e.g., secretory immunoglobulin A (sIgA) [40], albumin [41] and transferrin [42]. A benefit of this pathway is that proteins can be released into the lumen in their intact and active forms [26]. Finally, one paracellular pathway (pathway 4), directly transports serum substances and cells [43]. This paracellular pathway is only available on special occasions, such as inflammation or preterm birth [44], and it is normally closed by the tight-junctions between epithelial cells [44,45].

Human milk glycoproteins follow these same four general pathways, but glycosylation can be locally impacted by the glycan biosynthesis in the mammary gland or globally throughout the body. Maternal genetic factors may also contribute to the protein glycosylation profile that is produced. Two fucosyltransferases, including fucosyltransferases 2 (FUT2) and fucosyltransferases 3 (FUT3), which are encoded by the secretor and Lewis genes, respectively, are responsible for adding the terminal fucose in an $\alpha 1,2$ linkage and subterminal fucose in $\alpha 1,3$ and $\alpha 1,4$ linkage [46]. Some glycans that are bound to proteins then carry distinct elongated glycan epitopes that are determined by polymorphisms [47].

The secretion and transport of endogenous peptides is impacted by protein associations and conformational availability to be digested, or not, to shorter peptides by proteases and protease inhibitors. The transport of proteases from circulation to MECs is not fully understood and much remains unanswered, such as, whether proteases pass MECs by transcellular or paracellular pathways or whether specific receptors and transports exist for proteases [26]. Literature referring to bovine milk indicates that proteases in milk are not the source of somatic cells and that these proteases are therefore derived from blood [26,48]. Maternal health also plays a role in protein degradation to peptides as infections, especially those within the mammary gland, can result in an increase in proteolysis [26]. The health of the infant can influence this same process, as there is feedback communication between the infant and the mother during breastfeeding. When the infant is sick salivary proteins can make their way up the mammary ducts via the nipple during suckling, which then signals the transcription of immune modulating proteins within the mammary gland that are needed by the infant [49,50]. This opens up an exciting and unexplored avenue of research to better elucidate the maternal infant communication mechanism that is taking place at the mammary-oral interface. Furthermore, this could then lead to potentially novel human milk protein and or endogenous peptide biomarkers in relation to infant health.

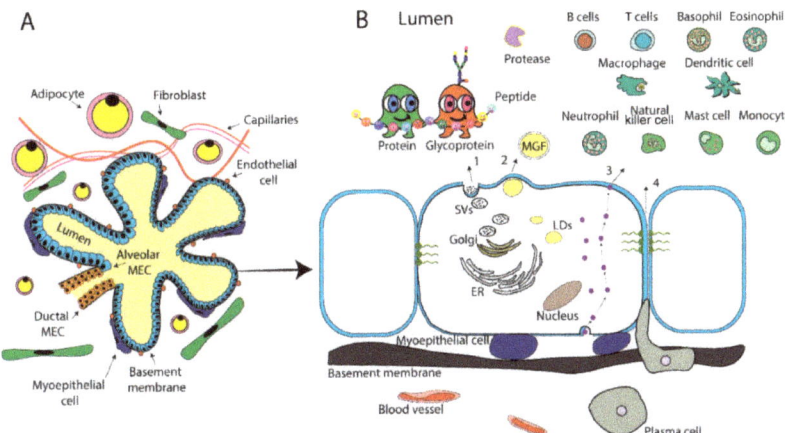

Figure 1. Physiology and transport of the proteome in a lactating mammary gland. (**A**) One lobulo-alveolar cluster connected to a lactiferous duct is depicted. The alveolus cluster, made up of a monolayer of polarized alveolar mammary epithelial cells (MEC) surrounding the lumen, and is connected to the lactiferous duct that is surrounded by a bilayer of ductal MEC. The alveolar MEC are surrounded by basement membrane and a single layer of polarized myoepithelial cells that contract to stimulate milk ejection from the lumen. The alveoli are embedded in a stoma of vascularized connective-tissue that contains lipid-depleted adipocytes, fibroblasts, endothelial cells and capillaries. (**B**) A zoomed in representation of the alveolar MEC is depicted to show the four key transport pathways of the milk proteome. Pathway (1) The exocytotic pathway is the dominant way for the secretion of endogenously generated proteins. These proteins and other aqueous components of milk are transported in secretory vesicles (SVs). Pathway (2) Secretion of lipid-associated proteins by the formation of lipid droplets (LDs), formed in the endoplasmic reticulum (ER), that move to the apical membrane to be secreted as milk fat globule (MFG). MFG are excreted by budding and are enwrapped by the apical plasma membrane of the MEC and become MFGM. Pathway (3) The vesicular transcytosis of proteins from serum or stromal cells. Pathway (4) Direct transport via the paracellular pathway for serum substances and cells, such as immune cells and stem cells. This route of transport is only available during pregnancy, early lactation prior to tight junction closure of MEC, involution, during times of inflammation or preterm birth.

3. Proteins

Human milk, as the preferred source of infant nutrition, provides essential and non-essential amino acids via its protein and free peptide fraction, which in turn are used by the infant for protein synthesis required for growth. Therefore, the analysis of human milk proteins is important for the determination of protein requirements to meet the infant's needs. Moreover, these needs are individual specific and they are based on the infant's ability to utilize and break down dietary proteins.

Historically milk proteins have been classified into two groups, whey and caseins, and they have recently expanded to include a third group, proteins from the MFGM [6,16]. Within these groups, the proportions of whey and casein change throughout lactation, with the ratio of whey to casein being the highest during colostrum and then reaching a steady state in mature milk [9,11,12,16,51], see Table 1 for details. This effect is also attributed to the increase in milk volume throughout lactation [51]. Mucins are found within the MFGM, and therefore are not designated as a percentage of protein, since they are associated within the fat layer and typically make up less than one percent of total protein [6]. Proteins making up the whey fraction of milk predominantly include: α-lactalbumin, lactoferrin, secretory IgA (sIgA), albumin, and lysozyme; and, the casein fraction of milk include: α-S1-, β-, and κ-casein. Unlike bovine milk, human milk does not contain β-lactoglobulin or α-S2-casein, which can trigger immunogenic responses to cow's milk proteins and then potentially other subsequent food allergies later in life when consumed by susceptible infants [52–55].

It is often difficult to make comparisons of the protein content in human milk across differing literature sources due to the variability in protein concentrations reported. These variations arise because of the expanse of laboratory methods that are used to determine protein concentration. Examples of differing methods of protein assessment include, in order of low to increasing complexity and accuracy: colorimetric assays, like bicinchoninic acid (BCA); nitrogen determination methods, like Kjeldahl; and, chromatographic methods, like matrix-assisted laser desorption/ionization (MALDI) and nano-spray liquid chromatography tandem mass spectrometry (LC-MS/MS). Mass spectrometry-based methods for the analysis of the human milk proteome are the most ideal, as they allow for both quantitative and qualitative analysis of the entire proteome, proteins with PTMs, like glycosylation, and the endogenous peptidome in one systematic approach.

Clear objectives need to be determined prior to protein analysis to define whether protein assessment is to be done, as true protein measurement vs total nitrogen estimates. This distinction is important, because measurements between the two can differ anywhere from 1–37% [8]. Protein estimated from the measurement of nitrogen often over-estimates the true protein content of human milk because it is assumed that all nitrogen is protein and does not consider the non-protein nitrogen (NPN) compounds [8]. In human milk, protein accounts for approximately 75% of nitrogen-containing compounds [56]. The remaining 25% of the nitrogen in milk is accounted for by NPN compounds, such as urea, nucleotides, endogenous peptides, free amino acids, DNA, and RNA [57,58]. From this, it is estimated that endogenous peptides make up 10–15% of NPN [58]. However, nitrogen studies have not considered glycosylation, and therefore what percentage of nitrogen is coming from this PTM has not be determined.

4. Glycoproteins

Protein glycosylation, the process of adding sugar units to a protein, is one of the most prominent PTMs on human milk proteins. It has been estimated that up to 70% of human milk proteins are glycosylated [59]. Unlike protein sequences, the biosynthesis of glycosylation cannot be directly predicted from the gene. However, the enzymes, including glycosyltransferases, glycosidases, and transporters, which are involved in the process, are directly encoded in the genome [60,61]. A glycoprotein carries one or more glycans covalently attached to a polypeptide backbone, usually via N- or O-linkages [62,63]. The N-glycosylation in human biology is designated as a sugar chain, starting with a N-acetylglucosamine (GlcNAc) residue that is linked to an asparagine residue of a polypeptide chain, normally with consensus peptide sequence: Asn-X-Ser/Thr (X is any amino acid except Pro) [64,65]. Additionally, some human milk proteins have N-glycans that occur at Asn-X-Cys, e.g., alpha-lactalbumin [66]. Alternatively, O-glycosylation is frequently linked glycans via N-acetylgalactosamine (GalNAc) to a serine or threonine residue of the polypeptide [67]. The expression level and glycosylation level of glycoproteins in human milk can vary throughout lactation and or biological situations [68]. Protein glycosylation in milk is of special interest, since it is relevant to proteolytic susceptibility, and it functions as competitive inhibitors of pathogen binding and immunomodulators in the gut. As a result, the glycosylated protein in human milk helps to shape the developing gut and immune system of the growing infant [47].

The quantitative variation in expression for some major glycoproteins has been reported, (Table 1). The level of glycosylation, especially the site-specific information throughout lactation, is less detailed in the literature, with the exception of a few proteins e.g., lactoferrin, bile salt activated lipase (BSSL), sIgA, secretory IgM (sIgM), and α-antitrypsin [69]. The less characterized site-specific glycosylation is mainly due to two analytical challenges: (1) the dynamic range of human milk proteins and (2) the complexity of glycosylation.

The global human milk glycosylation analyses is not often carried out on the intact glycopeptide level with respective site-specific glycosylation information due to these analytical challenges. Instead, analysis has mainly focused on the mapping of N-glycosylation sites of enzymatic deglycosylated peptides or on the enzymatically released N-glycans, as these analyses circumvent the analytical challenges. Interestingly, the analysis of the released N-glycans has revealed that a specific feature of human milk glycosylation is the multiple and abundant N-glycan fucosylation, when compared with bovine milk or human serum [70]. In humans, mono-fucosylation tends to be core fucosylation [71], while additional fucosylation is likely terminal fucosylation. Core fucosylation is important, because it can prevent the enzymatic cleavage of proteins from membrane surfaces [72]. The feature of having terminal fucosylation on glycoproteins represents a structural homology to human milk oligosaccharides (HMOs). This is of interest, because, if these two components have similar structures, then perhaps they share similar functionalities, which is not well characterized for terminal fucosylation on glycoproteins. However, it is well described that the fucosylated HMOs are correlated with the increasing diversity of the neonates gut microbiota. Future research should aim to further investigate the glycoproteome and the development of the infant gut.

5. Endogenous Peptides

To date, the literature on the endogenous peptidome in human milk remains limited and there are no studies investigating PTMs of the peptidome, to our knowledge. This is truly astounding when considering that human milk peptides are more diverse than proteins [73]. Additionally, peptides make ideal signaling molecules and they have extreme potential for biomarkers, as they serve as messengers encoding the status of specific regions of the body and potentially the entire organisms as they are less restricted in the movement in relation to proteins.

Several groups have developed similar MS-based methods to investigate endogenous peptides in human milk [74–77]. Recently, we established and validated an MS-based method, optimizing peptide extraction, MS fragmentation, and database searching, for a robust and reproducible analysis of endogenous peptides in human milk [78]. Within this method development, we determined that the most optimal workflow for endogenous peptides in human milk fundamentally depended on the depth of analysis desired and the time to be invested.

An understanding of naturally occurring endogenous human milk peptides verses peptides that are the digestion products of proteins in the infant's intestinal tract is important. This distinction is required for understanding functionality. Human milk peptidome studies have consistently found peptides abundantly from the casein fraction of milk proteins, where approximately 50% are derived from β-casein [74–77,79]. In addition to this, other studies have reported no endogenous peptides identified from major whey proteins, such as α-lactalbumin, lactoferrin, or immunoglobulins, [74–77,79]. Further studies suggested that the proteases in human milk are responsible for the observed proteolytic activity and that this activity is specific and conserved [76]. However, since the whey portion of the proteome is heavily glycosylated, peptides from these proteins are either not degraded by proteases or are not detected due to the binding glycan. Future analysis of endogenous peptides in human milk should aim to address this gap in knowledge.

Human milk peptides have been shown to have functional properties beyond the sources of amino acids; for instance, they are involved in immunomodulation, opioid-like activity, antioxidant, antimicrobial, and antiviral action, and probiotic action [74,77,80–83]. The functionality of human milk peptides require that they are hydrolyzed from their parent protein, in which case proteins can be seen as the carriers of functional peptides. Previous studies have suggested that the peptide sequences of human milk are not due to random proteolytic digestion but rater selective proteolysis of the mammary gland [77]. The driving forces for this selectivity may be for protective factors for both the mother and the infant [77]. The benefits of intact peptides for the mother and infant must have served an evolutionary advantage as the energy cost of protein production is high, and therefore it would not be opportune for biological systems to break down these proteins if they did not serve some importance. The proposed health benefits for the mother and infant include the prevention of bacterially induced mastitis and protection against gastrointestinal infections, respectively [77]. Additionally, it has been shown that proteolysis within the mammary gland is increased during times of inflammation or infection [26], leading to altered milk compositions, which can include increased serum proteins and potentially serum derived peptides in milk. Thus, the analysis of the human milk peptidome could provide insights into the maternal-infant health dynamic.

To date, there is huge interest in determining the functionality of human milk endogenous peptides, which are unique and diverse from that of their parent protein. Particularly, endogenous peptides with a wide range of bioactivities, including antimicrobial, antihypertensive, antithrombotic, and immunomodulatory [84]. While some peptides have established functionalities, other peptides only have proposed functionalities that are based on sequence motif. Additionally, biological insights of endogenous peptides can be achieved by different strategies, such as site visualization, mapping the peptide to the precursor protein back bone; enzymatic mapping, assessing which enzymes released the peptide from the precursor protein; peptide structure predictions, predicting the three-dimensional structure of the peptide from the amino acid sequences and PTMs; and, predicted functionality by homology searches against databases of known functional sequences. Dallas et al. previously reviewed these topics in detail [84]. A major limitation in the analysis of bioactive functionality is that it is reliant on predictive bioinformatics, which is contingent on the quality and completeness of databases currently available. One group has sought to bridge this gap in knowledge by deriving a database specific to milk peptides from various mammalian species with known functionality, Milk Bioactive Peptide Database (MBPDB) [83].

6. Enzymes

Human milk is comprised of a mixture of proteolytic enzymes, zymogens, protease activators, and protease inhibitors, and therefore the net proteolytic activity is dependent on the quantitative interaction of these components. Additionally, these numerous enzymes all have differing functionalities. The main proteases in human milk include plasmin, trypsin-2, cathepsin-D, neutrophil elastase, thrombin, kallikrein, and several amino- and carboxypeptidases [26]. These proteases are secreted in their inactive form and are then activated by protease activators, such as tissue-type plasminogen activator (t-PA) and urokinase-type activator (u-PA) [26]. The system maintains balance with protease inhibitors, such as α1-antichymotrypsin and α1-antitrypsin [85,86]. The mammary gland is an ideal environment for these proteolytic enzymes, as most of them are capable of functioning at body temperature and they can function at milk's neutral pH [82].

In general, it is well established that human milk contains higher levels of enzymes than bovine milk [87], and further that the enzymes in human milk are inherently different than the same enzyme being expressed in a different body fluid. This difference arises from human milk enzymes having a more highly organized tertiary structure, resulting in greater hydrophobicity, which in turn may account for their resistance to proteolysis and denaturation in the infant's gastrointestinal system [26,85–90]. Studies have shown that the neutral pH of human milk provides a buffering capacity in the infants stomach, which increases the pH and limits proteolysis from pepsin [91]. This increase in the pH of the infant's stomach is thought to be important for facilitating bacterial colonization of the gut, and that proteolysis is facilitated by milk proteases, which remain active in the infants stomach, such as cathepsin_D and plamin [91]. There are extensive reviews on the enzymes in the mammary gland, human milk, and throughout infant digestion [26,85–91], for this reason we choose to highlight enzymes that function within the mammary gland to establish the proteome, especially highlighting plasmin, and enzymes that act in secreted milk, which function later on in the infant's digestive tract.

6.1. Enzymes that Function in the Mammary Gland

Within the mammary gland, enzymatic proteolysis specifically and reproducibly acts as a controlled proteolytic system [88]. Evidence for this can be observed by endogenous peptides, which are characterized by overlapping ladder peptide products that originate from a specific region of the parent protein. These characteristics are due to the nature of the parent protein, PTMs, secondary and tertiary structure, and the abundance in relation to available proteases. It has been identified by Timmer et al. that four key factors determine the specificity of proteolysis: (1) spaciotemporal co-localization, (2) exosite interactions, (3) sub-site specificity, and (4) structural presentation [92]. Guerrero et al. considered these aspects of proteolysis in their analysis of endogenous peptides in human milk and investigated the intrinsic disorder of a protein in relation to the identified peptides [76]. Wherein, the intrinsic disorder is where a part of the protein is "natively unfolded" [93], containing regions that lack stable tertiary structures, which allow for easier proteolytic access and degradation. Overall, in this analysis, it was determined that the observed specificity of endogenous peptides could be explained by the protease activity within the mammary gland. This activity was explained by three factors: (1) the intrinsic disorder of proteins; (2) the advantage of a single cleavage for the generation of peptides at the N and/or C-termini; and, (3) the participation of enzymes with high specificity for peptides being observed at internal regions of proteins [76].

It has been previously demonstrated that peptides from major whey proteins, such as lactoferrin, α-lactalbumin, and secretory immunoglobulins, remain intact and they do not contribute to the milk peptidome [74,76,77,84], under healthy conditions. These studies demonstrated that the majority of endogenous peptides were derived from caseins, osteopontin, and polymeric immunoglobulin receptor, and that these peptides represented a minority of the total protein content. Further, these studies hypothesized that this was due to the proteases that were present in human milk, originating in the mammary gland, such as plasmin, cathepsin D, elastase, cytosol aminopeptidase, and carboxypeptidase B2, and that they were active throughout lactation. It is also thought that major whey proteins are not digested in the mammary gland due to a lack of association with micelles and inhibited protease activity, potentially due to PTMs [77].

Plasmin is one of the main proteases that hydrolyzes human milk proteins in the mammary gland [77]. It is thought to be an important protease in human milk, as it helps to facilitate the capacity for protein digestion of the infant. Plasmin is a trypsin-like, serine-type protease and it is highly specific for cleavage of peptide bonds of lysine, and to a lesser extent, arginine, at the N-terminal domain. Plasmin activity in milk is controlled by a heterogeneous system of inhibitors and activators. The inactivated zymogen, plasminogen, is converted to active plasmin by two serine proteases, t-PA and u-PA [26]. In addition to these activators are inhibitors of plasmin, including type-1 plasminogen activator inhibitor (PAI-1), inter-α-trypsin inhibitor, and α1-antitrypsin [26]. The inactive form of plasmin, plasminogen is known to associate with the casein micelles in milk [94], as does the activator t-PA. Therefore, the majority of active plasmin in milk is associated with the casein micelle structure. This has been demonstrated in studies that showed β-casein derived peptides that were made up the greatest abundance and relative peptide count, even though it is not the most abundant protein in milk [26,77]. Subsequently, it is also reasonable to expect low amounts of endogenous peptides from whey derived proteins in human milk, due to a lack of association with plasminogen and activator t-PA, and an association with plasmin inhibitors [26,77].

6.2. Enzymes Present in Milk

The incorporation of proteases in human milk may be an inherent evolutionary design, which is meant to aid in infant digestion. This is important, as infants are born with developmentally naïve digestion systems, starting in the mouth with little salivary proteases and extended to the gastrointestinal system, in which they produce relatively little gastric acid and express low protease activity in comparison to older children and adults. Additionally, the consumption of milk makes the pH of the stomach more neutral than that of an adult. However, even with these known developmental disadvantages, we know that infants are capable of digesting and absorbing milk proteins. It has been hypothesized that this is because proteases are simultaneously delivered via milk to aid in infant digestion [26], which makes sense, taken together with the infant's low digestive capacity. Evidence for supporting this has been recently reported via peptidomic and bioinformatics analysis that showed milk proteases were actively breaking down protein within the human infant's gastrointestinal tract [95–97]. The concentrations of proteases, their inhibitors, and activators have also been reported to change across lactation. Wherein, the mean protease activity, in milk, is reported to decline throughout lactation, reducing in concentration as the infant becomes more capable of utilizing its own digestive system. It is reported that protease activity increases during involution, and it is thought that this is to aid in returning the mammary gland to pre-gestation status. Dallas et al. recently reviewed these topics extensively [26].

The digestibility of the human milk proteome in the infant's gastrointestinal tract is of importance, because of the consequences of a lack of digestion. For instance, if proteins are not properly digested in the small intestine than this can result in an incomplete amino acid hydrolysis required for growth and would, therefore, increase the infant's total protein requirements. Further on, if proteins reach the colon, and they are not excreted in feces, they could contribute to over growth of bacteria that are capable of protein-fermentation, such as *Clostridium perfringens* and various *Bacteroides* species [26]. This, in turn, could result in lower populations of beneficial carbohydrate consuming bacteria in the infant gut, such as bifidobacteria [26].

7. Functionality

The human milk proteome contributes to a wide range of functionalities, including: growth as sources of amino acids; enhanced bioavailability of micronutrients, such as vitamins, minerals, and trace elements; improved cognitive development; immunogenic training as innate and adaptive immunity; and, promoting intestinal growth and maturation via interactions with the microbiome. These functionalities are achieved by varying proteins, endogenous peptides, and their PTMs. Importantly, some proteins are capable of exerting multiple functionalities. One example of this is lactoferrin, which can act as a nutrient transporter, defend against pathogens, stimulator of commensal microbiota, and more [2–4,9]. Moreover, the individual components of the human milk proteome are able to synergistically work to drive functionality. Several examples of this exist, including the interaction of lactoferrin and lysozyme on antibacterial functionality and the interplay between sIgA and endogenous peptides from pIgR to sequester bacteria at the mucosal interface in the gut. We break down these functional topics over the next several sections and discuss the major parts of the proteome that contribute to specific functions.

7.1. Growth

Dietary reference intake values for infant protein are based on the protein content of human milk, as this is optimal for meeting the requirements for growth and maintenance. Across the ages of 0 to 4 months, the average protein intake for infants consuming human milk is 8 g/day [98]. The composition of the human milk proteome is one of the major contributing factors to infant growth trajectories. For example, an overall high percentage of whey vs casein proteins corresponds to slowing growth rates in human infants [10]. This is due to an even lower level of casein protein in human milk that already contains one of the lowest concentration of caseins relative to other studied species [51]. Human milk proteins drive infant growth, as they are important sources of essential amino acids after digestion. Moreover, proteins and their derived peptides support infant growth by the enhancement of nutrient absorption and digestion.

One way digestive enhancement is achieved is by increasing the solubility and bioavailability of nutrients. A well-known example is β-casein, which forms casein phosphopeptides (CPPs) in the mammary gland and during infant digestion and it functions to chelate with minerals and promote the absorption of calcium, zinc, and iron [99–101]. Another example is the most prevalent whey protein α-lactalbumin, which has binding sites for calcium and zinc [102,103], which help to increase mineral absorption. The glycoprotein haptocorrin helps to protect the acid-sensitive vitamin B_{12} pass the infants' stomach, for later absorption in the small intestine [104]. Another mechanism of enhanced nutrient uptake is via protein receptors. For instance, iron that is bound to lactoferrin is not released in the intestine due to the high binding affinity, but for infants, the uptake of iron is increased via the lactoferrin receptor [105]. The lactoferrin receptor is suggested to be the principal iron transport pathway in early life [106], which indicates the nutritional importance of lactoferrin for iron absorption for neonates. Additionally, iron that is provided by lactoferrin is also well utilized in adults [107].

Enzymes in human milk are essential as they contribute to digestion in neonates, as they have rather immature digestive systems and are incapable of producing sufficient quantities of enzymes to fully facilitate digestion [88]. One well-accepted example is bile salt-stimulated lipase (BSSL) which has a wide substrate specificity to hydrolyze mono-, di-, and triglycerides, cholesterol esters, fat-soluble vitamin esters, phospholipids, galactolipids as well as ceramides [108,109], thus aiding in the digestion of milk lipids. [110]. The newborn normally has the ability to digest lactose, the main carbohydrate in human milk, by lactase present in the small intestine [111]. However, the main enzyme for complex carbohydrates, α-amylase, is low at birth [112], but it is supplemented by human milk [113] and it may aid in the digestion of complex carbohydrates when complementary foods are introduced to breastfed infants [114].

Cognitive development is another important aspect of growth, however the proteome is often overlooked in this regard and more attention is placed on human milk oligosaccharides and fatty acids. Human milk lactoferrin and proteins of the MFGM have been shown to be important in the cognitive development of the infant [9]. In piglet models, it has been shown that feeding bovine lactoferrin at human milk concentrations was associated with the differential expression of 10 genes in the brain-derived neurotrophin factor (BDNF) signaling pathway and later downstream target proteins of this pathway that are important for neurodevelopment and cognition [115]. Other studies using piglet models have shown that feeding lactoferrin resulted in upregulated intestinal gene expression of BDNF and improved gut maturation, linking the gut-brain-microbe axis in a way not previously reported [116]. While intriguing, these results have not been investigated in the human infant. The only randomized controlled trial (RCT) investigating formula with supplemented MFGM containing 4% protein vs standard formula found that the MFGM supplemented group had significantly higher mean cognitive scores in the Bayley Scales of Infant and Toddler Development when compared to the standard formula group [117]. Moreover, the cognitive scores of the MFGM supplemented group were not different as compared to that of the breast-fed reference group at 12 months of age [117].

7.2. Immune

The specificity of human milk composition to meet the needs of an individual infant can be exemplified in the immune modulating components, which aim to protect the infant and help to drive the immune development of the infant's naive immune system. There are a multitude of immune modulating components in human milk, including: antigens, cytokines, immunoglobulins (Ig), polyunsaturated fatty acids, and chemokines [118]; leucocytes, including macrophages, neutrophilic granulocytes, and lymphocytes [119]; immune stimulating proteins and glycoproteins, such as lactoferrin and sIgA [54]; and, additional components, such as hormones, growth factors, and endogenous peptides [120]. The immunogenic activities of the milk proteome are critical for establishing innate immunity and developing adaptive immunity. Additionally, it may play a role to some extent in allergy prevention or development, depending on whether it is acting to build tolerance or potentially cause sensitization. However, many characteristics of this are unknown and current research is aiming to better understand these mechanisms.

7.2.1. Innate Immunity

The innate immune system is the first line of host defense against pathogens [121] and it is extremely important for infants, since they are lacking mature adaptive immunity. The human milk proteome compensates neonatal innate immunity through many different ways: (1) the inhibition of growth of pathogens; (2) the inhibition of the binding of a pathogen to its receptor; and, (3) regulation of immune response and inflammation.

One way to inhibit the growth of pathogens is to have competition for the resources required for growth. For example, lactoferrin has a high affinity to bind free ions, which are essential for bacterial growth. By making the free ions unavailable, lactoferrin has a broad bacteriostatic effect. Same holds true for haptocorrin, as the major vitamin B_{12} binding protein, normally unsaturated in human milk, to withhold vitamin B_{12} and inhibit the growth of bacteria [122]. However, haptocorrin does not have a general antibacterial activity, rather its activity was found for a single enteropathogenic *Escherichia coli* (*E. coli*) O127 strain (EPEC) [123] and *Bifidobacterium breve* [124].

Another way to inhibit the growth of pathogens is the disruption of membrane structure. Human milk contains a high concentration of lysozyme, which is an enzyme that is capable of hydrolyzing β-1,4 linkages of *N*-acetylmuramic acid (NAM) and *N*-acetylglucosamine (NAG) in bacterial cell walls, leading to the instability of cell wall and bacterial cell death [125]. Lysozyme has been shown to synergistically work with lactoferrin to kill gram-negative bacteria in vitro [126]. With the help of lactoferrin to bind and remove lipopolysaccharide from the outer cell membrane of gram-negative bacteria, the inner proteoglycan matrix is left accessible for lysozyme. Human milk also contains defensins and cathelicidins [127], which have shown to have synergistic antibacterial effects with lysozyme [128]. In addition to the enzymatic activity, human lysozyme is cationic [129] and it can insert into and form pores in negatively charged bacteria membranes [130,131].

The membrane attack complex (MAC) as a result of the activation of the complement system is capable of forming pores in the lipid bilayers of bacteria and it leads to cell lysis and death [132]. The complement system is an important part of innate immunity and it also plays a role in adaptive responses [133]. Human milk contains many proteins in the complement system, yet in low levels, although systematic research of the whole system is still lacking. The assessment of individual complement proteins indicate that the complement system in human milk might provide additional immunological and non-immunological protection for infants [134].

The membrane structure of pathogens could be damaged and the pathogen growth could be inhibited by a so-called "lactoperoxidase system (LPS)", which consists of lactoperoxidase and thiocyanate (SCN^-) which occur naturally in human milk, and H_2O_2, which is generated by bacteria [135]. When the H_2O_2 is present, lactoperoxidase uses it to oxidase SCN^- to hypothiocyanite ($OSCN^-$), which has broad antimicrobial activity [136]. Moreover, the overexpressed $OSCN^-$ could be eliminated by antioxidants presented in milk to limit the local tissue damage by oxidative stress [5]. Lactoperoxidase is a glycoprotein and it is resistant to proteolysis [137], thus playing a role in infant host defense.

When pathogens enter the infant's gastrointestinal tract, the first stage of the infection is colonization by adhesion to host epithelial cells. The intestinal epithelial cells are heavily covered with glycans. Cell surface glycan epitope recognition is the first step of enteric pathogens in their pathogenesis [138]. Milk glycans, in the free form of HMOs or conjugated form in glycoprotein and glycolipid, might have epitopes as part of their structures, and competitively recognize and bind to either the lectin receptors of epithelial cells or pathogen lectin receptors [139]. Thereby, inhibiting the adhesion of a pathogen and infection afterward. Many glycoproteins in human milk have demonstrated the capability to block the interaction of epithelial cells and enteric pathogens, e.g., Tenascin-C for HIV-1 [140]; lactoferrin for *Escherichia coli* O157:H7 and *Salmonella enterica* [141]; BSSL for Norwalk virus [142]; mucins for rotavirus [143], Norwalk virus [142]; and, *Salmonella enterica serovar Typhimurium* SL1344 [144]. One critical aspect in exploring the roles of milk glycoproteins is to characterize the attached glycans and their specificities since the epitope recognition is structure-based. Lactoferrin has a lower ability to interrupt the adhesion of bacteria to epithelial cells when fucose moieties are enzymatically removed [141]. Sialylated glycans of secretory immunoglobulin A (sIgA) were able to bind the S-fimbriated *E. coli* strains and inhibit adhesion [145]. Moreover, mothers have different abilities to produce specific glycan structures [146], as discussed in the glycoproteomics section, as well as attached glycans of glycoproteins [71]. The individual-specific properties add another level of complexity and lead to unique selective inhibition for endemic pathogens.

The immature intestinal mucosa of the neonates is overly sensitive to infection, because of overexpressed inflammatory genes and under-expressed negative feedback regulator genes [147]. Human milk helps to regulate this immunologic balance in neonates, not only by reducing exposure to pathogens and the prevention of infection, but also via modulating the immune response to minimize inflammation pathology for the breastfed infant [148]. Many components in human milk, including proteins/glycoproteins and peptides, have immunomodulatory functions. For example, soluble isoforms of toll-like receptor (TLR)-2 serves as a decoy receptor, suppressing TLR2 activation and decreasing interleukin (IL)-8 and tumor necrosis factor (TNF)-α release, thus reducing inflammation [149]. Glycoproteins, such as CD14 [150], lactoferrin [151], and lactadherin [152], can regulate TLR4 signaling, whereas the endogenous peptide, β-defensin 2, suppresses TLR7 expression [153]. In a recent RCT, bovine osteopontin was added to infant formula and the infants were fed either regular formula or osteopontin supplemented formula [154]. The results from this RCT showed that infants consuming osteopontin supplemented formula had significantly lower serum concentrations of pro-inflammatory cytokine transforming growth factor-α as compared to non-supplemented infants, and moreover that these infants had cytokine profiles more similar to that of breastfed infants [154].

7.2.2. Adaptive Immunity

While their own adaptive immune system is maturing, neonates may get protection from the products of the adaptive immune system of the mother, which was first reported in 1892 [155]. One primary protective product is a protein family, called immunoglobulins or antibodies, which come from several sources and they are transported into milk by receptor-mediated processes, reflecting the antigenic stimulation of immunity of the mother [156]. Immunoglobulins include several classes, like IgM, IgA, IgG, IgE, and IgD [157]. In human milk, the most abundant immunoglobulin is sIgA, followed by sIgM and IgG [158].

Both sIgA and sIgM are polymeric, typically dimeric IgA and pentameric IgM, which are formed through the covalent interaction with a joining (J) chain and secretory component (SC) from the endoproteolytic cleavage of the polymeric immunoglobulin receptor (pIgR) [159]. sIgA and sIgM are produced in a similar manner and represent the history of the antigenic stimulation of mucosal immunity in the mother [119]. When a pathogen enters the mother's gut or upper airways, the Peyer's patch acquires the pathogen and its antigens are presented by M cells to circulating B cells, which migrates to the serosal (basolateral) side of the mammary epithelial cell and it produces IgA and IgM. As the IgA and IgM move from the serosal to the luminal (apical) side of the mammary epithelial cell, they are glycosylated and complexed to form sIgA and sIgM, which are secreted into milk [160]. When the infant consumes milk, the sIgA and sIgM associate with free SC and they are resistant to digestion [161]. Additionally, free SC is responsible for intracellular neutralization of some viruses and it is capable of binding bacterial components, like *Streptococcus pneumonia*-derived SpsA protein and prevent the invasion of epithelial cells [161]. sIgA and sIgM both carry the memory of pathogens faced by the mother, which allows for them to provide the same protection to the infant by binding to recognized pathogens and inhibiting their ability to infect the infant [162]. This mechanism is called the enteromammary link, where maternal immunity is transferred to the breastfed infant [163]. Milk sIgA is considered to be the dominant immunoglobulin to protect mucosal surfaces of infants from pathogens and enteric toxins, through intracellular neutralization, virus excretion, and immune exclusion [158,164]. Additionally, sIgM has been demonstrated to activate the complement cascade in vitro [165].

IgG is required to provide systemic immunity and it is transferred before birth. During pregnancy, IgG is transferred from mother to the fetus via transplacental passage and it provides crucial protection to the neonate in the first weeks of life after birth [166]. The main portion of human milk IgG is transported from serum through neonatal Fc receptor (FcRn) [167], and its four subclasses are ordered in cord serum level as IgG1, IgG2, IgG4, and IgG3 [168]. The concentration of IgG in human milk is much less than sIgA since the human mammary gland has considerably less FcRn than pIgR. Additionally, IgG is less resistant to digestion than sIgA. The remaining intact IgG in the intestinal lumen might be involved in immune surveillance by binding antigens and enhancing the local mucosal immune response by transporting the IgG-antigen complexes into *lamina propria* for the subsequent induction of immune activation or tolerance [169]. On the other hand, maternal antibodies may inhibit infant vaccine response depending on the ratio of maternal antibodies to the vaccine antigen [170], although maternal vaccination, such as pertussis and influenza, showed protective effects for infants [171,172].

7.2.3. Potential Allergens in Human Milk and Immunity

The presence of non-human proteins in human milk was reported many decades ago [173], especially some potential food allergens, including proteins from cow's milk, eggs, peanuts, and wheat, in the form of degraded peptides, intact proteins, and/or an immune complex with antibodies [14,174–179]. Besides food proteins, other sources are also reported to be present in human milk, e.g., house dust mite [180]. It is still not quite clear how foreign proteins enter human milk and the consequences for infant health. It is difficult to elucidate the exact roles of foreign proteins in a complicated mixture as human milk, where pro-inflammatory and anti-inflammatory factors are both present and change constantly [53]. However, some critical reviews have revealed that breastfeeding positively promotes the development of tolerance in the infant and protects against allergic diseases, asthma, and atopic dermatitis [53,181,182]. Moreover, studies have shown that avoidance of food allergens in maternal diet during lactation, or postponed introduction of risky foods in children have not shown a clear benefit [183]. Instead, introducing a food antigen in early life with low levels of continuous exposure might reduce the risk of developing related allergies [184]. The potential allergens introduced in human milk by months of breastfeeding might be the ideal way to promote tolerance [185]. Tolerance to harmless antigens plays a crucial role in immune homeostasis by preventing potentially dangerous hypersensitivity reactions [186].

7.3. Gut Development

The infant's immune system and gut develop concurrently. As the infant's immune system is immature at birth, so is the mucosal lining of the gastrointestinal tract and the gut with limited microbiota. The maturation of the gut during infancy relies, extensively, on adequate nutritional support, and it is shaped by the delivery of human milk bioactive constituents. The human milk proteome plays a large role in this maturation by providing support to intestinal epithelia, the mucosal lining and gut microbiota for developmental programming [187]. Moreover, intestinal tropism is primarily derived from human milk protein, rather than by lipid or carbohydrate [187]. The bacteriostatic properties of the milk proteome help to create an environment within the infants' gastrointestinal tract, where unrestrained bacterial growth is prohibited and the microorganisms are removed from the small intestine without causing inflammation [3]. In turn, this contributes to the development of a healthy microbiome for the infant [3]. All together, the milk proteome results in enhanced responsiveness of the intestine to dietary, physiological, and pathological challenges [187].

The infants gut physiology consists of intestinal epithelial cells (IECs) covered by a mucus layer, and it is at this interface that innate and adaptive immunities must cooperatively function to protect the infant from the vast assault of stimuli [161]. Intestinal homeostasis is maintained by IECs, which act as a physical barrier driving synergistic immunity against invading pathogens. The main player of the IECs is the polymeric immunoglobulins (pIgs). As described in the adaptive immunity section, pIgR is complexed to dimeric IgA (dIgA) by the J-chain; this is a critical step in the translocation of IgA across the IECs and secreted into the lumen, where sIgA can exert its protective functions. This complex formation and translocation occurs both in mammary epithelial cells and in IECs. Once this dIgA-pIgR complex is expressed on the apical surface of IECs, it is released by proteolytic cleavage. Research studies have shown that pIgR cleavage occurs on the cell surface and not inside the cells, which indicates that this is a highly localized process [161]. However, the enzyme responsible for the cleavage of pIgR has not been identified, although some research suggests that the enzyme is a serine proteinase [161]. Maternal sIgA in breast milk is supplied to the infant in sufficient quantities to promote gut health and maturation, as the infant is incapable of producing sufficient quantities of sIgA until approximately two years of age. After sIgA is translocated from the IECs it enters the intestinal mucus layer where it helps to maintain spatial segregation between the microbiota and epithelial surface [188]. It has been postulated that the production of sIgA and pIgR is regulated by the gut microbiota by products that are shed from the microbial community [188]. This, in turn, helps to stimulate a commensal relationship between the hosts gut and microbiome, in which the human milk proteome is a key driver for the developing infant.

Mucosal development is a major determinant of infant's gut health and homeostasis. Mucus producing goblet cells, Paneth and M cells in the intestine, function beyond nutrient absorption and have a major role in bacterial mucosal cross talk and innate barrier function [187]. These cells are responsible for the secretion of mucus and mucin proteins, which are important in the nonspecific protection of the gut lining [187]. Mucus in the gut lining is a complex gel that is primarily composed of water and electrolytes, but it also contains many proteome components, such as mucins, glycoproteins, immunoglobulins, albumin [187], and bioactive peptides. It is at the mucus interface that a commensal microbiome can be established, while pathogenic bacteria can become trapped and acted upon by the functional proteome.

Many human milk proteins play a part in this, for instance, the sIgA-pIgR complex, as just described. Another protein, κ-casein, which is heavily glycosylated with glycan structures that are similar to exposed glycans on mucosal surfaces, helps to prevent pathogen binding. This protective functionality is most likely occurring from the proteolytic product of κ-casein, glycomacropeptide (GMP), which binds to mucosal surfaces, thereby preventing the attachment and infection of pathogens [4]. Similarly, haptocorrin may act in the fashion as κ-casein in inhibiting pathogen interaction at the mucosal surface.

Proteins that are associated with the MFGM are important for gut development, such as mucin-1, xanthine oxidoreductase, butyrophilin, CD36, adipophilin, lactadherin, and fatty acid-binding protein [3,4,9]. These proteins help to stimulate intestinal epithelial health, promote gastric stability, and contribute to the antiviral and antibacterial activities in the infant gastrointestinal tract [3,4,9]. Specifically, mucin-1 has been shown to inhibit the invasion of *Salmonella typhimurium*, in a model of fetal intestinal cells, at concentrations that are similar to that of human milk [144].

Recent research has investigated the proteome as a cipher for functionally encrypted peptides that are released upon infant digestion. As this review aimed to assess the human milk proteome and endogenous peptides and not the products of digestion, we did not address these current topics. However, we wanted to point readers in the right direction, as these functional peptides are described to exert their bioactivities directly in the gastrointestinal lumen and at peripheral organs after being absorbed at the intestinal mucosa. Please see a recent review by Y. Wada and B Lonnerdal [80], and a few experimental studies [95–97].

Some proteins have been hypothesized to be more beneficial for the infants gut as intact, such as lactoferrin, lysozyme, α-lactalbumin, and immunoglobulins due to their important functions as antimicrobial and immunologic, and for the presence of the glycan groups. An overall lack of digestion of these proteins may then be functionally beneficial for helping to drive the establishment of commensal gut microbiome, as it is commonly observed that these proteins can be found back intact in infant feces [2,4]. The glycans that are released from these undigested glycoproteins by microbial glycosidases are bifidogenic [189]. Further, it has been suggested that the degree of glycan diversity and rate of glycan site occupancy of these glycoproteins changes to meet the developing gut microbial diversity [141].

8. Conclusions

Human milk has historically been known to prevent high morbidity and mortality rates of infants in the first months of life. This is, in large part, due to the human milk proteome, which helps the immature infant fight against infectious diseases, such as otitis media, respiratory tract infections, gastrointestinal infections, and more. Additionally, breastfeeding, in general, has been shown to reduce the risk of development of maternal cancer and the proteome in particular has been shown to reduce the risk of mastitis. The power of the human milk proteome is derived from the enteromammary transfer of maternal immunity and its broad and synergistic functionality, Figure 2. Moreover, the protective and developmental proteome is unique to each individual mother-infant pair.

There is an overall need for a more comprehensive characterization and quantification of the human milk proteome throughout lactation. Preferred methodologies, such as modern MS, should be considered, especially for human milk, where the complex matrix can cause unwanted complications in other types of analysis, to fully characterize the proteome and derive deeper understandings of its functionality. These current gaps in knowledge could be overcome by some of the topics covered throughout this review, such as quantitative MS to derive concentrations of individual proteins and endogenous peptides, and advanced MS methodologies to better characterize PTMs, like glycans. The complexity of the human milk proteome and its corresponding functions is a challenging but demanding topic for personalized nutrition, as it is responsible for driving infant growth and development, and it even has the potential for the discovery of novel biomarkers and therapeutics.

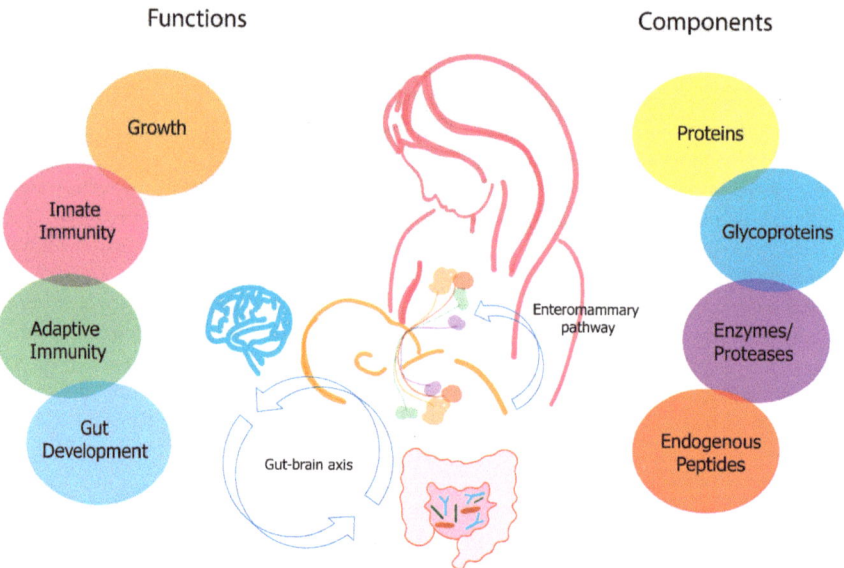

Figure 2. The power of the human milk proteome stems from an evolutionary advantage of cross communication between the mother-infant dyad, via the enteromammary pathway, which in turn stimulated the rise of the uniquely functioning proteome and all of its components. These components work both individually and synergistically to provided protective immunity and overall development to the infant while maintaining health benefits for the mother as well.

Author Contributions: Conceptualization, investigation, data curation, writing—Original draft preparation, writing—Review and editing, and visualization: K.A.D. and J.Z.

Funding: We acknowledge Albert Heck for supporting us to write this review independently. Both K.A.D. and J.Z. received support from Utrecht University and Danone Nutricia Research. J.Z. additionally acknowledges support from the Chinese Scholarship Council (CSC).

Acknowledgments: We thank Zorah Vogel for her contribution to the literature searching for this project and for her contributions in helping to put together Table 1. Additionally, we thank Marko Mank (Danone Nutricia Research), Bernd Stahl (Danone Nutricia Research), Vojtech Franc (Utrecht University), Lynette K. Roggers (Nationwide Children's Hospital) and Albert J. R. Heck (Utrecht University) for proof-reading parts of this review.

Conflicts of Interest: Kelly A. Dingess and Jing Zhu were enrolled as Ph.D. students at Utrecht University during this study, partly financially supported by Danone Nutricia Research.

References

1. Eidelman, A.I.; Schanler, R.J. Breastfeeding and the use of human milk. *Pediatrics* **2012**, *129*, 827–841. [CrossRef]
2. Donovan, S.M. The Role of Lactoferrin in Gastrointestinal and Immune Development and Function: A Preclinical Perspective. *J. Pediatr.* **2016**, *173*, S16–S28. [CrossRef] [PubMed]
3. Demmelmair, H.; Prell, C.; Timby, N.; Lonnerdal, B. Benefits of Lactoferrin, Osteopontin and Milk Fat Globule Membranes for Infants. *Nutrients* **2017**, *9*, 817. [CrossRef] [PubMed]
4. Lonnerdal, B. Bioactive Proteins in Human Milk: Health, Nutrition, and Implications for Infant Formulas. *J. Pediatr.* **2016**, *173*, S4–S9. [CrossRef] [PubMed]
5. Palmeira, P.; Carneiro-Sampaio, M. Immunology of breast milk. *Revista da Associação Médica Brasileira* **2016**, *62*, 584–593. [CrossRef] [PubMed]
6. Ballard, O.; Morrow, A.L. Human milk composition: Nutrients and bioactive factors. *Pediatr. Clin.* **2013**, *60*, 49–74. [CrossRef]
7. Lemay, D.G.; Ballard, O.A.; Hughes, M.A.; Morrow, A.L.; Horseman, N.D.; Nommsen-Rivers, L.A. RNA sequencing of the human milk fat layer transcriptome reveals distinct gene expression profiles at three stages of lactation. *PLoS ONE* **2013**, *8*, e67531. [CrossRef]
8. Gidrewicz, D.A.; Fenton, T.R. A systematic review and meta-analysis of the nutrient content of preterm and term breast milk. *BMC Pediatr.* **2014**, *14*, 216. [CrossRef]
9. Donovan, S.M. *Human Milk Proteins: Composition and Physiological Significance*; Karger Publishers: Basel, Switzerland, 2019.
10. Andreas, N.J.; Kampmann, B.; Mehring Le-Doare, K. Human breast milk: A review on its composition and bioactivity. *Early Hum. Dev.* **2015**, *91*, 629–635. [CrossRef]
11. Xinliu, G.; Robert, J.M.; Jessica, G.W.; Barbara, S.D.; Ardythe, L.M.; Qiang, Z. Temporal changes in milk proteomes reveal developing milk functions. *J. Proteome Res.* **2012**, *11*, 3897–3907. [CrossRef]
12. Zhang, Q.; Cundiff, J.; Maria, S.; McMahon, R.; Woo, J.; Davidson, B.; Morrow, A. Quantitative Analysis of the Human Milk Whey Proteome Reveals Developing Milk and Mammary-Gland Functions across the First Year of Lactation. *Proteomes* **2013**, *1*, 128–158. [CrossRef] [PubMed]
13. Zhang, L.; de Waard, M.; Verheijen, H.; Boeren, S.; Hageman, J.A.; van Hooijdonk, T.; Vervoort, J.; van Goudoever, J.B.; Hettinga, K. Changes over lactation in breast milk serum proteins involved in the maturation of immune and digestive system of the infant. *J. Proteom.* **2016**, *147*, 40–47. [CrossRef] [PubMed]
14. Zhu, J.; Garrigues, L.; Van den Toorn, H.; Stahl, B.; Heck, A.J.R. Discovery and quantification of non-human proteins in human milk. *J. Proteom. Res.* **2018**, *18*, 225–238. [CrossRef]
15. Oftedal, O.T. The evolution of milk secretion and its ancient origins. *Animal* **2012**, *6*, 355–368. [CrossRef] [PubMed]
16. Liao, Y.; Weber, D.; Xu, W.; Durbin-Johnson, B.P.; Phinney, B.S.; Lonnerdal, B. Absolute Quantification of Human Milk Caseins and the Whey/Casein Ratio during the First Year of Lactation. *J. Proteome Res.* **2017**, *16*, 4113–4121. [CrossRef] [PubMed]
17. Gan, J.; Robinson, R.C.; Wang, J.; Krishnakumar, N.; Manning, C.J.; Lor, Y.; Breck, M.; Barile, D.; German, J.B. Peptidomic profiling of human milk with LC-MS/MS reveals pH-specific proteolysis of milk proteins. *Food Chem.* **2019**, *274*, 766–774. [CrossRef] [PubMed]
18. Lonnerdal, B.; Erdmann, P.; Thakkar, S.K.; Sauser, J.; Destaillats, F. Longitudinal evolution of true protein, amino acids and bioactive proteins in breast milk: A developmental perspective. *J. Nutr. Biochem.* **2017**, *41*, 1–11. [CrossRef] [PubMed]

19. Ke, X.; Chen, Q.; Pan, X.; Zhang, J.; Mo, W.; Ren, Y. Quantification of lactoferrin in breast milk by ultra-high performance liquid chromatography-tandem mass spectrometry with isotopic dilution. *RSC Adv.* **2016**, *6*, 12280–12285. [CrossRef]
20. Chatterton, D.E.; Nguyen, D.N.; Bering, S.B.; Sangild, P.T. Anti-inflammatory mechanisms of bioactive milk proteins in the intestine of newborns. *Int. J. Biochem. Cell Biol.* **2013**, *45*, 1730–1747. [CrossRef]
21. Huang, J.; Kailemia, M.J.; Goonatilleke, E.; Parker, E.A.; Hong, Q.; Sabia, R.; Smilowitz, J.T.; German, J.B.; Lebrilla, C.B. Quantitation of human milk proteins and their glycoforms using multiple reaction monitoring (MRM). *Anal. Bioanal. Chem.* **2017**, *409*, 589–606. [CrossRef]
22. Shin, K.; Hayasawa, H.; Lonnerdal, B. Purification and quantification of lactoperoxidase in human milk with use of immunoadsorbents with antibodies against recombinant human lactoperoxidase. *Am. J. Clin. Nutr.* **2001**, *73*, 984–989. [CrossRef] [PubMed]
23. Cole, F.S.; Schneeberger, E.E.; Lichtenberg, N.A.; Colten, H.R. Complement biosynthesis in human breast-milk macrophages and blood monocytes. *Immunology* **1982**, *46*, 429–441. [PubMed]
24. Kassim, O.O.; Afolabi, O.; Ako-Nai, K.A.; Torimiro, S.E.; Littleton, G.K.; Mbogua, C.N.; Oke, O.; Turner, W.; Grissom, F. Immunoprotective factors in breast milk and sera of mother-infant pairs. *Trop. Geogr. Med.* **1986**, *38*, 362–366. [PubMed]
25. Peterson, J.A.; Hamosh, M.; Scallan, C.D.; Ceriani, R.L.; Henderson, T.R.; Mehta, N.R.; Armand, M.; Hamosh, P. Milk Fat Globule Glycoproteins in Human Milk and in Gastric Aspirates of Mother's Milk-Fed Preterm Infants. *Pediatr. Res.* **1998**, *44*, 499–506. [CrossRef] [PubMed]
26. Dallas, D.C.; Murray, N.M.; Gan, J. Proteolytic Systems in Milk: Perspectives on the Evolutionary Function within the Mammary Gland and the Infant. *J. Mammary Gland Biol. Neoplasia* **2015**, *20*, 133–147. [CrossRef]
27. Demers-Mathieu, V.; Nielsen, S.D.; Underwood, M.A.; Borghese, R.; Dallas, D.C. Changes in Proteases, Antiproteases, and Bioactive Proteins From Mother's Breast Milk to the Premature Infant Stomach. *J. Pediatr. Gastroenterol. Nutr.* **2018**, *66*, 318–324. [CrossRef]
28. Khodabakhshi, A.; Ghayour-Mobarhan, M.; Rooki, H.; Vakili, R.; Hashemy, S.I.; Mirhafez, S.R.; Shakeri, M.T.; Kashanifar, R.; Pourbafarani, R.; Mirzaei, H.; et al. Comparative measurement of ghrelin, leptin, adiponectin, EGF and IGF-1 in breast milk of mothers with overweight/obese and normal-weight infants. *Eur. J. Clin. Nutr.* **2015**, *69*, 614–618. [CrossRef]
29. Aydin, S.; Aydin, S.; Ozkan, Y.; Kumru, S. Ghrelin is present in human colostrum, transitional and mature milk. *Peptides* **2006**, *27*, 878–882. [CrossRef]
30. Ilcol, Y.O.; Hizli, Z.B.; Ozkan, T. Leptin concentration in breast milk and its relationship to duration of lactation and hormonal status. *Int. Breastfeed. J.* **2006**, *1*, 21. [CrossRef]
31. Gridneva, Z.; Kugananthan, S.; Rea, A.; Lai, C.T.; Ward, L.C.; Murray, K.; Hartmann, P.E.; Geddes, D.T. Human Milk Adiponectin and Leptin and Infant Body Composition over the First 12 Months of Lactation. *Nutrients* **2018**, *10*, 1125. [CrossRef]
32. Martin, L.J.; Woo, J.G.; Geraghty, S.R.; Altaye, M.; Davidson, B.S.; Banach, W.; Dolan, L.M.; Ruiz-Palacios, G.M.; Morrow, A.L. Adiponectin is present in human milk and is associated with maternal factors. *Am. J. Clin. Nutr.* **2006**, *83*, 1106–1111. [CrossRef] [PubMed]
33. Lubetzky, R.; Weisman, Y.; Dollberg, S.; Herman, L.; Mandel, D. Parathyroid hormone-related protein in preterm human milk. *Breastfeed. Med.* **2010**, *5*, 67–69. [CrossRef] [PubMed]
34. Truchet, S.; Honvo-Houeto, E. Physiology of milk secretion. *Best Pract. Res. Clin. Endocrinol. Metab.* **2017**, *31*, 367–384. [CrossRef] [PubMed]
35. McManaman, J.L.; Neville, M.C. Mammary physiology and milk secretion. *Adv. Drug Deliv. Rev.* **2003**, *55*, 629–641. [CrossRef]
36. Truchet, S.; Chat, S.; Ollivier-Bousquet, M. Milk secretion: The role of SNARE proteins. *J. Mammary Gland Biol. Neoplasia* **2014**, *19*, 119–130. [CrossRef] [PubMed]
37. Linzell, J.L.; Peaker, M. Mechanism of milk secretion. *Physiol. Rev.* **1971**, *51*, 564–597. [CrossRef] [PubMed]
38. Wooding, F.B. The mechanism of secretion of the milk fat globule. *J. Cell Sci.* **1971**, *9*, 805–821.
39. Lee, H.; Padhi, E.; Hasegawa, Y.; Larke, J.; Parenti, M.; Wang, A.; Hernell, O.; Lonnerdal, B.; Slupsky, C. Compositional Dynamics of the Milk Fat Globule and Its Role in Infant Development. *Front. Pediatr.* **2018**, *6*, 313. [CrossRef]
40. Hunziker, W.; Kraehenbuhl, J.P. Epithelial transcytosis of immunoglobulins. *J. Mammary Gland Biol. Neoplasia* **1998**, *3*, 287–302. [CrossRef]

41. Monks, J.; Neville, M.C. Albumin transcytosis across the epithelium of the lactating mouse mammary gland. *J. Physiol.* **2004**, *560*, 267–280. [CrossRef]
42. Ollivier-Bousquet, M. Transferrin and prolactin transcytosis in the lactating mammary epithelial cell. *J. Mammary Gland Biol. Neoplasia* **1998**, *3*, 303–313. [CrossRef] [PubMed]
43. Ninkina, N.; Kukharsky, M.S.; Hewitt, M.V.; Lysikova, E.A.; Skuratovska, L.N.; Deykin, A.V.; Buchman, V.L. Stem cells in human breast milk. *Hum. Cell* **2019**, *32*, 223–230. [CrossRef] [PubMed]
44. Verd, S.; Ginovart, G.; Calvo, J.; Ponce-Taylor, J.; Gaya, A. Variation in the Protein Composition of Human Milk during Extended Lactation: A Narrative Review. *Nutrients* **2018**, *10*, 1124. [CrossRef] [PubMed]
45. Shennan, D.B.; Peaker, M. Transport of milk constituents by the mammary gland. *Physiol. Rev.* **2000**, *80*, 925–951. [CrossRef] [PubMed]
46. Ma, B.; Simala-Grant, J.L.; Taylor, D.E. Fucosylation in prokaryotes and eukaryotes. *Glycobiology* **2006**, *16*, 158R–184R. [CrossRef] [PubMed]
47. Georgi, G.; Bartke, N.; Wiens, F.; Stahl, B. Functional glycans and glycoconjugates in human milk. *Am. J. Clin. Nutr.* **2013**, *98*, 578S–585S. [CrossRef] [PubMed]
48. Wickramasinghe, S.; Rincon, G.; Islas-Trejo, A.; Medrano, J.F. Transcriptional profiling of bovine milk using RNA sequencing. *BMC Genom.* **2012**, *13*, 45. [CrossRef] [PubMed]
49. Riskin, A.; Almog, M.; Peri, R.; Halasz, K.; Srugo, I.; Kessel, A. Changes in immunomodulatory constituents of human milk in response to active infection in the nursing infant. *Pediatr. Res.* **2012**, *71*, 220–225. [CrossRef] [PubMed]
50. Hassiotou, F.; Hepworth, A.R.; Metzger, P.; Tat Lai, C.; Trengove, N.; Hartmann, P.E.; Filgueira, L. Maternal and infant infections stimulate a rapid leukocyte response in breastmilk. *Clin. Transl. Immunol.* **2013**, *2*, e3. [CrossRef]
51. Lonnerdal, B. Nutritional and physiologic significance of human milk proteins. *Am. J. Clin. Nutr.* **2003**, *77*, 1537s–1543s. [CrossRef]
52. Kramer, M.S. Breastfeeding and allergy: The evidence. *Ann. Nutr. Metab.* **2011**, *59* (Suppl. 1), 20–26. [CrossRef] [PubMed]
53. Iyengar, S.R.; Walker, W.A. Immune factors in breast milk and the development of atopic disease. *J. Pediatr. Gastroenterol. Nutr.* **2012**, *55*, 641–647. [CrossRef] [PubMed]
54. Matheson, M.C.; Allen, K.J.; Tang, M.L. Understanding the evidence for and against the role of breastfeeding in allergy prevention. *Clin. Exp. Allergy* **2012**, *42*, 827–851. [CrossRef] [PubMed]
55. Matheson, M.C.; Erbas, B.; Balasuriya, A.; Jenkins, M.A.; Wharton, C.L.; Tang, M.L.; Abramson, M.J.; Walters, E.H.; Hopper, J.L.; Dharmage, S.C. Breast-feeding and atopic disease: A cohort study from childhood to middle age. *J. Allergy Clin. Immunol.* **2007**, *120*, 1051–1057. [CrossRef] [PubMed]
56. LÖNnerdal, B.O.; Atkinson, S. CHAPTER 5—Nitrogenous Components of Milk: A. Human Milk Proteins. In *Handbook of Milk Composition*; Jensen, R.G., Ed.; Academic Press: San Diego, CA, USA, 1995.
57. Carlson, S.E. Human milk nonprotein nitrogen: Occurrence and possible functions. *Adv. Pediatr.* **1985**, *32*, 43–70. [PubMed]
58. Atkinson, S.A.; LÖNnerdal, B.O. B-Nonprotein Nitrogen Fractions of Human Milk. In *Handbook of Milk Composition*; Jensen, R.G., Ed.; Academic Press: San Diego, CA, USA, 1995.
59. Froehlich, J.W.; Dodds, E.D.; Barboza, M.; McJimpsey, E.L.; Seipert, R.R.; Francis, J.; An, H.J.; Freeman, S.; German, J.B.; Lebrilla, C.B. Glycoprotein expression in human milk during lactation. *J. Agric. Food Chem.* **2010**, *58*, 6440–6448. [CrossRef] [PubMed]
60. Lairson, L.L.; Henrissat, B.; Davies, G.J.; Withers, S.G. Glycosyltransferases: Structures, functions, and mechanisms. *Annu. Rev. Biochem.* **2008**, *77*, 521–555. [CrossRef]
61. Henrissat, B.; Surolia, A.; Stanley, P. A Genomic View of Glycobiology. In *Essentials of Glycobiology*, 3rd ed.; Varki, A., Cummings, R.D., Esko, J.D., Stanley, P., Hart, G.W., Aebi, M., Darvill, A.G., Kinoshita, T., Packer, N.H., Eds.; Cold Spring Harbor Press: Cold Spring Harbor, NY, USA, 2015; pp. 89–97.
62. Chung, C.Y.; Majewska, N.I.; Wang, Q.; Paul, J.T.; Betenbaugh, M.J. SnapShot: N-Glycosylation Processing Pathways across Kingdoms. *Cell* **2017**, *171*, 258. [CrossRef]
63. Joshi, H.J.; Narimatsu, Y.; Schjoldager, K.T.; Tytgat, H.L.P.; Aebi, M.; Clausen, H.; Halim, A. SnapShot: O-Glycosylation Pathways across Kingdoms. *Cell* **2018**, *172*, 632. [CrossRef]
64. Schwarz, F.; Aebi, M. Mechanisms and principles of N-linked protein glycosylation. *Curr. Opin. Struct. Biol.* **2011**, *21*, 576–582. [CrossRef]

65. Aebi, M. N-linked protein glycosylation in the ER. *Biochim. Biophys. Acta* **2013**, *1833*, 2430–2437. [CrossRef] [PubMed]
66. Giuffrida, M.G.; Cavaletto, M.; Giunta, C.; Neuteboom, B.; Cantisani, A.; Napolitano, L.; Calderone, V.; Godovac-Zimmermann, J.; Conti, A. The unusual amino acid triplet Asn-Ile-Cys is a glycosylation consensus site in human alpha-lactalbumin. *J. Protein Chem.* **1997**, *16*, 747–753. [CrossRef] [PubMed]
67. Levery, S.B.; Steentoft, C.; Halim, A.; Narimatsu, Y.; Clausen, H.; Vakhrushev, S.Y. Advances in mass spectrometry driven O-glycoproteomics. *Biochim. Biophys. Acta* **2015**, *1850*, 33–42. [CrossRef] [PubMed]
68. Orczyk-Pawilowicz, M.; Hirnle, L.; Berghausen-Mazur, M.; Katnik-Prastowska, I.M. Lactation stage-related expression of sialylated and fucosylated glycotopes of human milk alpha-1-acid glycoprotein. *Breastfeed. Med.* **2014**, *9*, 313–319. [CrossRef] [PubMed]
69. Goonatilleke, E.; Huang, J.; Xu, G.; Wu, L.; Smilowitz, J.T.; German, J.B.; Lebrilla, C.B. Human Milk Proteins and Their Glycosylation Exhibit Quantitative Dynamic Variations during Lactation. *J. Nutr.* **2019**. [CrossRef] [PubMed]
70. Yoshida, Y.; Furukawa, J.I.; Naito, S.; Higashino, K.; Numata, Y.; Shinohara, Y. Quantitative analysis of total serum glycome in human and mouse. *Proteomics* **2016**, *16*, 2747–2758. [CrossRef] [PubMed]
71. Schneider, M.; Al-Shareffi, E.; Haltiwanger, R.S. Biological functions of fucose in mammals. *Glycobiology* **2017**, *27*, 601–618. [CrossRef] [PubMed]
72. Sumer-Bayraktar, Z.; Grant, O.C.; Venkatakrishnan, V.; Woods, R.J.; Packer, N.H.; Thaysen-Andersen, M. Asn347 glycosylation of corticosteroid-binding globulin fine-tunes the host Immune response by modulating proteolysis by pseudomonas aeruginosa and neutrophil elastase. *J. Biol. Chem.* **2016**, *291*, 17727–17742. [CrossRef]
73. Schrader, M. Origins, Technological Development, and Applications of Peptidomics. *Methods Mol. Biol* **2018**, *1719*, 3–39. [CrossRef]
74. Dallas, D.C.; Smink, C.J.; Robinson, R.C.; Tian, T.; Guerrero, A.; Parker, E.A.; Smilowitz, J.T.; Hettinga, K.A.; Underwood, M.A.; Lebrilla, C.B.; et al. Endogenous human milk peptide release is greater after preterm birth than term birth. *J. Nutr.* **2015**, *145*, 425–433. [CrossRef]
75. Wan, J.; Cui, X.W.; Zhang, J.; Fu, Z.Y.; Guo, X.R.; Sun, L.Z.; Ji, C.B. Peptidome analysis of human skim milk in term and preterm milk. *Biochem. Biophys. Res. Commun.* **2013**, *438*, 236–241. [CrossRef] [PubMed]
76. Guerrero, A.; Dallas, D.C.; Contreras, S.; Chee, S.; Parker, E.A.; Sun, X.; Dimapasoc, L.; Barile, D.; German, J.B.; Lebrilla, C.B. Mechanistic peptidomics: Factors that dictate specificity in the formation of endogenous peptides in human milk. *Mol. Cell. Proteom.* **2014**, *13*, 3343–3351. [CrossRef] [PubMed]
77. Dallas, D.C.; Guerrero, A.; Khaldi, N.; Castillo, P.A.; Martin, W.F.; Smilowitz, J.T.; Bevins, C.L.; Barile, D.; German, J.B.; Lebrilla, C.B. Extensive in vivo human milk peptidomics reveals specific proteolysis yielding protective antimicrobial peptides. *J. Proteome Res.* **2013**, *12*, 2295–2304. [CrossRef] [PubMed]
78. Dingess, K.A.; van den Toorn, H.W.P.; Mank, M.; Stahl, B.; Heck, A.J.R. Toward an efficient workflow for the analysis of the human milk peptidome. *Anal. Bioanal. Chem.* **2019**, *411*, 1351–1363. [CrossRef] [PubMed]
79. Dingess, K.A.; de Waard, M.; Boeren, S.; Vervoort, J.; Lambers, T.T.; van Goudoever, J.B.; Hettinga, K. Human milk peptides differentiate between the preterm and term infant and across varying lactational stages. *Food Funct.* **2017**, *8*, 3769–3782. [CrossRef] [PubMed]
80. Wada, Y.; Lonnerdal, B. Bioactive peptides derived from human milk proteins-mechanisms of action. *J. Nutr. Biochem.* **2014**, *25*, 503–514. [CrossRef]
81. Dallas, D.C.; Weinborn, V.; de Moura Bell, J.M.; Wang, M.; Parker, E.A.; Guerrero, A.; Hettinga, K.A.; Lebrilla, C.B.; German, J.B.; Barile, D. Comprehensive peptidomic and glycomic evaluation reveals that sweet whey permeate from colostrum is a source of milk protein-derived peptides and oligosaccharides. *Food Res. Int.* **2014**, *63*, 203–209. [CrossRef]
82. Nielsen, S.D.; Beverly, R.L.; Dallas, D.C. Milk Proteins Are Predigested Within the Human Mammary Gland. *J. Mammary Gland Biol. Neoplasia* **2017**, *22*, 251–261. [CrossRef]
83. Nielsen, S.D.; Beverly, R.L.; Qu, Y.; Dallas, D.C. Milk bioactive peptide database: A comprehensive database of milk protein-derived bioactive peptides and novel visualization. *Food Chem.* **2017**, *232*, 673–682. [CrossRef]
84. Dallas, D.C.; Guerrero, A.; Parker, E.A.; Robinson, R.C.; Gan, J.; German, J.B.; Barile, D.; Lebrilla, C.B. Current peptidomics: Applications, purification, identification, quantification, and functional analysis. *Proteomics* **2015**, *15*, 1026–1038. [CrossRef]
85. Hamosh, M. Bioactive factors in human milk. *Pediatr. Clin.* **2001**, *48*, 69–86. [CrossRef]

86. Hamosh, M. C—Enzymes in Human Milk. In *Handbook of Milk Composition*; Jensen, R.G., Ed.; Academic Press: San Diego, CA, USA, 1995; pp. 388–427.
87. Shahani, K.M.; Kwan, A.J.; Friend, B.A. Role and significance of enzymes in human milk. *Am. J. Clin. Nutr.* **1980**, *33*, 1861–1868. [CrossRef] [PubMed]
88. Dallas, D.C.; German, J.B. *Enzymes in Human Milk*; Karger Publishers: Basel, Switzerland, 2017.
89. Shahani, K.M. Milk enzymes: Their role and significance. *J. Dairy Sci.* **1966**, *49*, 907–920. [CrossRef]
90. Jenness, R. The composition of human milk. *Semin. Perinatol.* **1979**, *3*, 225–239. [PubMed]
91. Dallas, D.C.; Underwood, M.A.; Zivkovic, A.M.; German, J.B. Digestion of Protein in Premature and Term Infants. *J. Nutr. Disord.* **2012**, *2*, 112. [CrossRef]
92. Timmer, J.C.; Zhu, W.; Pop, C.; Regan, T.; Snipas, S.J.; Eroshkin, A.M.; Riedl, S.J.; Salvesen, G.S. Structural and kinetic determinants of protease substrates. *Nat. Struct. Mol. Biol.* **2009**, *16*, 1101–1108. [CrossRef] [PubMed]
93. Tompa, P. Intrinsically unstructured proteins. *Trends Biochem. Sci.* **2002**, *27*, 527–533. [CrossRef]
94. Heegaard, C.W.; Larsen, L.B.; Rasmussen, L.K.; Hojberg, K.E.; Petersen, T.E.; Andreasen, P.A. Plasminogen activation system in human milk. *J. Pediatr. Gastroenterol. Nutr.* **1997**, *25*, 159–166. [CrossRef]
95. Beverly, R.L.; Underwood, M.A.; Dallas, D.C. Peptidomics Analysis of Milk Protein-Derived Peptides Released over Time in the Preterm Infant Stomach. *J. Proteome Res.* **2019**, *18*, 912–922. [CrossRef]
96. Dallas, D.C.; Guerrero, A.; Khaldi, N.; Borghese, R.; Bhandari, A.; Underwood, M.A.; Lebrilla, C.B.; German, J.B.; Barile, D. A peptidomic analysis of human milk digestion in the infant stomach reveals protein-specific degradation patterns. *J. Nutr.* **2014**, *144*, 815–820. [CrossRef]
97. Holton, T.A.; Vijayakumar, V.; Dallas, D.C.; Guerrero, A.; Borghese, R.A.; Lebrilla, C.B.; German, J.B.; Barile, D.; Underwood, M.A.; Shields, D.C.; et al. Following the digestion of milk proteins from mother to baby. *J. Proteome Res.* **2014**, *13*, 5777–5783. [CrossRef] [PubMed]
98. Richter, M.; Baerlocher, K.; Bauer, J.M.; Elmadfa, I.; Heseker, H.; Leschik-Bonnet, E.; Stangl, G.; Volkert, D.; Stehle, P. Revised Reference Values for the Intake of Protein. *Ann. Nutr. Metab.* **2019**, *74*, 242–250. [CrossRef] [PubMed]
99. Kibangou, I.; Bouhallab, S.; Bureau, F.; Allouche, S.; Thouvenin, G.; Bougle, D. Caseinophosphopeptide-bound iron: Protective effect against gut peroxidation. *Ann. Nutr. Metab.* **2008**, *52*, 177–180. [CrossRef] [PubMed]
100. Garcia-Nebot, M.J.; Barbera, R.; Alegria, A. Iron and zinc bioavailability in Caco-2 cells: Influence of caseinophosphopeptides. *Food Chem.* **2013**, *138*, 1298–1303. [CrossRef] [PubMed]
101. Hansen, M.; Sandstrom, B.; Lonnerdal, B. The effect of casein phosphopeptides on zinc and calcium absorption from high phytate infant diets assessed in rat pups and Caco-2 cells. *Pediatr. Res.* **1996**, *40*, 547–552. [CrossRef] [PubMed]
102. Lonnerdal, B.; Glazier, C. Calcium binding by alpha-lactalbumin in human milk and bovine milk. *J. Nutr.* **1985**, *115*, 1209–1216. [CrossRef] [PubMed]
103. Ren, J.; Stuart, D.I.; Acharya, K.R. Alpha-lactalbumin possesses a distinct zinc binding site. *J. Biol. Chem.* **1993**, *268*, 19292–19298. [PubMed]
104. Adkins, Y.; Lonnerdal, B. Mechanisms of vitamin B(12) absorption in breast-fed infants. *J. Pediatr. Gastroenterol. Nutr.* **2002**, *35*, 192–198. [CrossRef]
105. Suzuki, Y.A.; Shin, K.; Lonnerdal, B. Molecular cloning and functional expression of a human intestinal lactoferrin receptor. *Biochemistry* **2001**, *40*, 15771–15779. [CrossRef]
106. Lopez, V.; Suzuki, Y.A.; Lonnerdal, B. Ontogenic changes in lactoferrin receptor and DMT1 in mouse small intestine: Implications for iron absorption during early life. *Biochem. Cell Biol.* **2006**, *84*, 337–344. [CrossRef]
107. Lonnerdal, B.; Bryant, A. Absorption of iron from recombinant human lactoferrin in young US women. *Am. J. Clin. Nutr.* **2006**, *83*, 305–309. [CrossRef] [PubMed]
108. Blackberg, L.; Hernell, O. Further characterization of the bile salt-stimulated lipase in human milk. *FEBS Lett.* **1983**, *157*, 337–341. [CrossRef]
109. Li, X.; Lindquist, S.; Lowe, M.; Noppa, L.; Hernell, O. Bile salt-stimulated lipase and pancreatic lipase-related protein 2 are the dominating lipases in neonatal fat digestion in mice and rats. *Pediatr. Res.* **2007**, *62*, 537–541. [CrossRef] [PubMed]
110. Hamosh, M. Lipid metabolism in premature infants. *Neonatology* **1987**, *52* (Suppl. 1), 50–64. [CrossRef]
111. Segurel, L.; Bon, C. On the Evolution of Lactase Persistence in Humans. *Annu. Rev. Genom. Hum. Genet.* **2017**, *18*, 297–319. [CrossRef]

112. Sevenhuysen, G.P.; Holodinsky, C.; Dawes, C. Development of salivary alpha-amylase in infants from birth to 5 months. *Am. J. Clin. Nutr.* **1984**, *39*, 584–588. [CrossRef]
113. Hamosh, M. Enzymes in milk: Their function in the mammary gland, in milk, and in the infant. *Biol. Hum. Milk* **1988**, *15*, 45–61.
114. Heitlinger, L.A.; Lee, P.C.; Dillon, W.P.; Lebenthal, E. Mammary amylase: A possible alternate pathway of carbohydrate digestion in infancy. *Pediatr. Res.* **1983**, *17*, 15–18. [CrossRef]
115. Chen, Y.; Zheng, Z.; Zhu, X.; Shi, Y.; Tian, D.; Zhao, F.; Liu, N.; Hüppi, P.S.; Troy, F.A.; Wang, B. Lactoferrin Promotes Early Neurodevelopment and Cognition in Postnatal Piglets by Upregulating the BDNF Signaling Pathway and Polysialylation. *Mol. Neurobiol.* **2015**, *52*, 256–269. [CrossRef]
116. Yang, C.; Zhu, X.; Liu, N.; Chen, Y.; Gan, H.; Troy, F.A.; Wang, B. Lactoferrin up-regulates intestinal gene expression of brain-derived neurotrophic factors BDNF, UCHL1 and alkaline phosphatase activity to alleviate early weaning diarrhea in postnatal piglets. *J. Nutr. Biochem.* **2014**, *25*, 834–842. [CrossRef]
117. Timby, N.; Domellof, E.; Hernell, O.; Lonnerdal, B.; Domellof, M. Neurodevelopment, nutrition, and growth until 12 mo of age in infants fed a low-energy, low-protein formula supplemented with bovine milk fat globule membranes: A randomized controlled trial. *Am. J. Clin. Nutr.* **2014**, *99*, 860–868. [CrossRef] [PubMed]
118. Friedman, N.J.; Zeiger, R.S. The role of breast-feeding in the development of allergies and asthma. *J. Allergy Clin. Immunol.* **2005**, *115*, 1238–1248. [CrossRef] [PubMed]
119. Brandtzaeg, P. The mucosal immune system and its integration with the mammary glands. *J. Pediatr.* **2010**, *156*, S8–S15. [CrossRef] [PubMed]
120. Field, C.J. The immunological components of human milk and their effect on immune development in infants. *J. Nutr.* **2005**, *135*, 1–4. [CrossRef] [PubMed]
121. Akira, S.; Uematsu, S.; Takeuchi, O. Pathogen recognition and innate immunity. *Cell* **2006**, *124*, 783–801. [CrossRef] [PubMed]
122. Morkbak, A.L.; Poulsen, S.S.; Nexo, E. Haptocorrin in humans. *Clin. Chem. Lab. Med.* **2007**, *45*, 1751–1759. [CrossRef]
123. Adkins, Y.; Lonnerdal, B. Potential host-defense role of a human milk vitamin B-12-binding protein, haptocorrin, in the gastrointestinal tract of breastfed infants, as assessed with porcine haptocorrin in vitro. *Am. J. Clin. Nutr.* **2003**, *77*, 1234–1240. [CrossRef]
124. Jensen, H.R.; Laursen, M.F.; Lildballe, D.L.; Andersen, J.B.; Nexo, E.; Licht, T.R. Effect of the vitamin B12-binding protein haptocorrin present in human milk on a panel of commensal and pathogenic bacteria. *BMC Res. Notes* **2011**, *4*, 208. [CrossRef]
125. Callewaert, L.; Michiels, C.W. Lysozymes in the animal kingdom. *J. Biosci.* **2010**, *35*, 127–160. [CrossRef]
126. Ellison, R.T.; Giehl, T.J. Killing of gram-negative bacteria by lactoferrin and lysozyme. *J. Clin. Investig.* **1991**, *88*, 1080–1091. [CrossRef]
127. Grapov, D.; Lemay, D.G.; Weber, D.; Phinney, B.S.; Azulay Chertok, I.R.; Gho, D.S.; German, J.B.; Smilowitz, J.T. The human colostrum whey proteome is altered in gestational diabetes mellitus. *J. Proteome Res.* **2015**, *14*, 512–520. [CrossRef] [PubMed]
128. Chen, X.; Niyonsaba, F.; Ushio, H.; Okuda, D.; Nagaoka, I.; Ikeda, S.; Okumura, K.; Ogawa, H. Synergistic effect of antibacterial agents human beta-defensins, cathelicidin LL-37 and lysozyme against Staphylococcus aureus and Escherichia coli. *J. Derm. Sci.* **2005**, *40*, 123–132. [CrossRef] [PubMed]
129. Nash, J.A.; Ballard, T.N.; Weaver, T.E.; Akinbi, H.T. The peptidoglycan-degrading property of lysozyme is not required for bactericidal activity in vivo. *J. Immunol.* **2006**, *177*, 519–526. [CrossRef] [PubMed]
130. Zhang, X.; Jiang, A.; Yu, H.; Xiong, Y.; Zhou, G.; Qin, M.; Dou, J.; Wang, J. Human Lysozyme Synergistically Enhances Bactericidal Dynamics and Lowers the Resistant Mutant Prevention Concentration for Metronidazole to Helicobacter pylori by Increasing Cell Permeability. *Molecules* **2016**, *21*, 1435. [CrossRef] [PubMed]
131. Ragland, S.A.; Criss, A.K. From bacterial killing to immune modulation: Recent insights into the functions of lysozyme. *PLoS Pathog.* **2017**, *13*, e1006512. [CrossRef] [PubMed]
132. Serna, M.; Giles, J.L.; Morgan, B.P.; Bubeck, D. Structural basis of complement membrane attack complex formation. *Nat. Commun.* **2016**, *7*, 10587. [CrossRef] [PubMed]
133. Dunkelberger, J.R.; Song, W.C. Complement and its role in innate and adaptive immune responses. *Cell Res.* **2010**, *20*, 34–50. [CrossRef]

134. Ogundele, M. Role and significance of the complement system in mucosal immunity: Particular reference to the human breast milk complement. *Immunol. Cell Biol.* **2001**, *79*, 1–10. [CrossRef]
135. Sharma, S.; Singh, A.K.; Kaushik, S.; Sinha, M.; Singh, R.P.; Sharma, P.; Sirohi, H.; Kaur, P.; Singh, T.P. Lactoperoxidase: Structural insights into the function, ligand binding and inhibition. *Int. J. Biochem. Mol. Biol.* **2013**, *4*, 108–128.
136. Sarr, D.; Toth, E.; Gingerich, A.; Rada, B. Antimicrobial actions of dual oxidases and lactoperoxidase. *J. Microbiol.* **2018**, *56*, 373–386. [CrossRef]
137. Kussendrager, K.D.; van Hooijdonk, A.C. Lactoperoxidase: Physico-chemical properties, occurrence, mechanism of action and applications. *Br. J. Nutr* **2000**, *84* (Suppl. 1), S19–S25. [CrossRef] [PubMed]
138. Newburg, D.S. Neonatal protection by an innate immune system of human milk consisting of oligosaccharides and glycans. *J. Anim. Sci.* **2009**, *87*, 26–34. [CrossRef] [PubMed]
139. Lis-Kuberka, J.; Orczyk-Pawilowicz, M. Sialylated Oligosaccharides and Glycoconjugates of Human Milk. The Impact on Infant and Newborn Protection, Development and Well-Being. *Nutrients* **2019**, *11*, 306. [CrossRef] [PubMed]
140. Fouda, G.G.; Jaeger, F.H.; Amos, J.D.; Ho, C.; Kunz, E.L.; Anasti, K.; Stamper, L.W.; Liebl, B.E.; Barbas, K.H.; Ohashi, T.; et al. Tenascin-C is an innate broad-spectrum, HIV-1-neutralizing protein in breast milk. *Proc. Natl. Acad. Sci. USA* **2013**, *110*, 18220–18225. [CrossRef] [PubMed]
141. Barboza, M.; Pinzon, J.; Wickramasinghe, S.; Froehlich, J.W.; Moeller, I.; Smilowitz, J.T.; Ruhaak, L.R.; Huang, J.; Lonnerdal, B.; German, J.B.; et al. Glycosylation of human milk lactoferrin exhibits dynamic changes during early lactation enhancing its role in pathogenic bacteria-host interactions. *Mol. Cell. Proteom.* **2012**, *11*. [CrossRef] [PubMed]
142. Ruvoen-Clouet, N.; Mas, E.; Marionneau, S.; Guillon, P.; Lombardo, D.; Le Pendu, J. Bile-salt-stimulated lipase and mucins from milk of 'secretor' mothers inhibit the binding of Norwalk virus capsids to their carbohydrate ligands. *Biochem. J.* **2006**, *393*, 627–634. [CrossRef] [PubMed]
143. Yolken, R.H.; Peterson, J.A.; Vonderfecht, S.L.; Fouts, E.T.; Midthun, K.; Newburg, D.S. Human milk mucin inhibits rotavirus replication and prevents experimental gastroenteritis. *J. Clin. Investig.* **1992**, *90*, 1984–1991. [CrossRef]
144. Liu, B.; Yu, Z.; Chen, C.; Kling, D.E.; Newburg, D.S. Human milk mucin 1 and mucin 4 inhibit Salmonella enterica serovar Typhimurium invasion of human intestinal epithelial cells in vitro. *J. Nutr.* **2012**, *142*, 1504–1509. [CrossRef]
145. Schroten, H.; Stapper, C.; Plogmann, R.; Kohler, H.; Hacker, J.; Hanisch, F.G. Fab-independent antiadhesion effects of secretory immunoglobulin A on S-fimbriated Escherichia coli are mediated by sialyloligosaccharides. *Infect. Immun.* **1998**, *66*, 3971–3973.
146. Bode, L. Human milk oligosaccharides: Every baby needs a sugar mama. *Glycobiology* **2012**, *22*, 1147–1162. [CrossRef]
147. Nanthakumar, N.; Meng, D.; Goldstein, A.M.; Zhu, W.; Lu, L.; Uauy, R.; Llanos, A.; Claud, E.C.; Walker, W.A. The mechanism of excessive intestinal inflammation in necrotizing enterocolitis: An immature innate immune response. *PLoS ONE* **2011**, *6*, e17776. [CrossRef] [PubMed]
148. Buescher, E.S. Anti-inflammatory characteristics of human milk: How, where, why. *Adv. Exp. Med. Biol.* **2001**, *501*, 207–222. [CrossRef] [PubMed]
149. Langjahr, P.; Diaz-Jimenez, D.; De la Fuente, M.; Rubio, E.; Golenbock, D.; Bronfman, F.C.; Quera, R.; Gonzalez, M.J.; Hermoso, M.A. Metalloproteinase-dependent TLR2 ectodomain shedding is involved in soluble toll-like receptor 2 (sTLR2) production. *PLoS ONE* **2014**, *9*, e104624. [CrossRef] [PubMed]
150. Kitchens, R.L.; Thompson, P.A. Modulatory effects of sCD14 and LBP on LPS-host cell interactions. *J. Endotoxin Res.* **2005**, *11*, 225–229. [CrossRef] [PubMed]
151. Ando, K.; Hasegawa, K.; Shindo, K.; Furusawa, T.; Fujino, T.; Kikugawa, K.; Nakano, H.; Takeuchi, O.; Akira, S.; Akiyama, T.; et al. Human lactoferrin activates NF-kappaB through the Toll-like receptor 4 pathway while it interferes with the lipopolysaccharide-stimulated TLR4 signaling. *FEBS J.* **2010**, *277*, 2051–2066. [CrossRef] [PubMed]
152. Aziz, M.; Jacob, A.; Matsuda, A.; Wang, P. Review: Milk fat globule-EGF factor 8 expression, function and plausible signal transduction in resolving inflammation. *Apoptosis* **2011**, *16*, 1077–1086. [CrossRef] [PubMed]
153. Stroinigg, N.; Srivastava, M.D. Modulation of toll-like receptor 7 and LL-37 expression in colon and breast epithelial cells by human beta-defensin-2. *Allergy Asthma Proc.* **2005**, *26*, 299–309. [PubMed]

154. Lonnerdal, B.; Kvistgaard, A.S.; Peerson, J.M.; Donovan, S.M.; Peng, Y.M. Growth, Nutrition, and Cytokine Response of Breast-fed Infants and Infants Fed Formula With Added Bovine Osteopontin. *J. Pediatr. Gastroenterol. Nutr.* **2016**, *62*, 650–657. [CrossRef]
155. Famulener, L. On the transmission of immunity from mother to offspring: A study upon serum hemolysins in goats [with discussion]. *J. Infect. Dis.* **1912**, *10*, 332–368. [CrossRef]
156. Hurley, W.L.; Theil, P.K. Perspectives on immunoglobulins in colostrum and milk. *Nutrients* **2011**, *3*, 442–474. [CrossRef]
157. Mix, E.; Goertsches, R.; Zett, U.K. Immunoglobulins-basic considerations. *J. Neurol.* **2006**, *253* (Suppl. 5), V9–V17. [CrossRef] [PubMed]
158. Van de Perre, P. Transfer of antibody via mother's milk. *Vaccine* **2003**, *21*, 3374–3376. [CrossRef]
159. Johansen, F.E.; Braathen, R.; Brandtzaeg, P. Role of J chain in secretory immunoglobulin formation. *Scand. J. Immunol.* **2000**, *52*, 240–248. [CrossRef] [PubMed]
160. Newburg, D.S. Innate immunity and human milk. *J. Nutr.* **2005**, *135*, 1308–1312. [CrossRef] [PubMed]
161. Asano, M.; Komiyama, K. Polymeric immunoglobulin receptor. *J. Oral Sci.* **2011**, *53*, 147–156. [CrossRef] [PubMed]
162. Hanson, L.; Silfverdal, S.A.; Stromback, L.; Erling, V.; Zaman, S.; Olcen, P.; Telemo, E. The immunological role of breast feeding. *Pediatr. Allergy Immunol.* **2001**, *12* (Suppl. 14), 15–19. [CrossRef]
163. Goldman, A.S. Evolution of the mammary gland defense system and the ontogeny of the immune system. *J. Mammary Gland Biol. Neoplasia* **2002**, *7*, 277–289. [CrossRef]
164. Mantis, N.J.; Rol, N.; Corthesy, B. Secretory IgA's complex roles in immunity and mucosal homeostasis in the gut. *Mucosal Immunol.* **2011**, *4*, 603–611. [CrossRef]
165. Michaelsen, T.E.; Emilsen, S.; Sandin, R.H.; Granerud, B.K.; Bratlie, D.; Ihle, O.; Sandlie, I. Human Secretory IgM Antibodies Activate Human Complement and Offer Protection at Mucosal Surface. *Scand. J. Immunol.* **2017**, *85*, 43–50. [CrossRef]
166. Wilcox, C.R.; Holder, B.; Jones, C.E. Factors Affecting the FcRn-Mediated Transplacental Transfer of Antibodies and Implications for Vaccination in Pregnancy. *Front. Immunol.* **2017**, *8*, 1294. [CrossRef]
167. Jiang, X.; Hu, J.; Thirumalai, D.; Zhang, X. Immunoglobulin Transporting Receptors Are Potential Targets for the Immunity Enhancement and Generation of Mammary Gland Bioreactor. *Front. Immunol.* **2016**, *7*, 214. [CrossRef] [PubMed]
168. Gasparoni, A.; Avanzini, A.; Ravagni Probizer, F.; Chirico, G.; Rondini, G.; Severi, F. IgG subclasses compared in maternal and cord serum and breast milk. *Arch. Dis. Child.* **1992**, *67*, 41–43. [CrossRef] [PubMed]
169. Rojas, R.; Apodaca, G. Immunoglobulin transport across polarized epithelial cells. *Nat. Rev. Mol. Cell Biol.* **2002**, *3*, 944–955. [CrossRef] [PubMed]
170. Siegrist, C.A. Mechanisms by which maternal antibodies influence infant vaccine responses: Review of hypotheses and definition of main determinants. *Vaccine* **2003**, *21*, 3406–3412. [CrossRef]
171. Munoz, F.M. Current Challenges and Achievements in Maternal Immunization Research. *Front. Immunol.* **2018**, *9*, 436. [CrossRef] [PubMed]
172. Blanchard-Rohner, G.; Eberhardt, C. Review of maternal immunisation during pregnancy: Focus on pertussis and influenza. *Swiss Med. Wkly.* **2017**, *147*, w14526. [CrossRef] [PubMed]
173. Donnally, H.H. The question of the elimination of foreign protein (egg-white) in woman's milk. *J. Immunol.* **1930**, *19*, 15–40.
174. Kilshaw, P.J.; Cant, A.J. The passage of maternal dietary proteins into human breast milk. *Int. Arch. Allergy Appl. Immunol.* **1984**, *75*, 8–15. [CrossRef] [PubMed]
175. Chirdo, F.G.; Rumbo, M.; Anon, M.C.; Fossati, C.A. Presence of high levels of non-degraded gliadin in breast milk from healthy mothers. *Scand. J. Gastroenterol.* **1998**, *33*, 1186–1192. [PubMed]
176. Palmer, D.J.; Gold, M.S.; Makrides, M. Effect of maternal egg consumption on breast milk ovalbumin concentration. *Clin. Exp. Allergy* **2008**, *38*, 1186–1191. [CrossRef] [PubMed]
177. Coscia, A.; Orru, S.; Di Nicola, P.; Giuliani, F.; Varalda, A.; Peila, C.; Fabris, C.; Conti, A.; Bertino, E. Detection of cow's milk proteins and minor components in human milk using proteomics techniques. *J. Matern. Fetal Neonatal Med.* **2012**, *25* (Suppl. 4), 54–56. [CrossRef] [PubMed]
178. Bernard, H.; Ah-Leung, S.; Drumare, M.F.; Feraudet-Tarisse, C.; Verhasselt, V.; Wal, J.M.; Creminon, C.; Adel-Patient, K. Peanut allergens are rapidly transferred in human breast milk and can prevent sensitization in mice. *Allergy* **2014**, *69*, 888–897. [CrossRef] [PubMed]

179. Metcalfe, J.R.; Marsh, J.A.; D'Vaz, N.; Geddes, D.T.; Lai, C.T.; Prescott, S.L.; Palmer, D.J. Effects of maternal dietary egg intake during early lactation on human milk ovalbumin concentration: A randomized controlled trial. *Clin. Exp. Allergy* **2016**, *46*, 1605–1613. [CrossRef] [PubMed]
180. Baiz, N.; Macchiaverni, P.; Tulic, M.K.; Rekima, A.; Annesi-Maesano, I.; Verhasselt, V. Early oral exposure to house dust mite allergen through breast milk: A potential risk factor for allergic sensitization and respiratory allergies in children. *J. Allergy Clin. Immunol.* **2017**, *139*, 369–372. [CrossRef] [PubMed]
181. van Odijk, J.; Kull, I.; Borres, M.P.; Brandtzaeg, P.; Edberg, U.; Hanson, L.A.; Host, A.; Kuitunen, M.; Olsen, S.F.; Skerfving, S.; et al. Breastfeeding and allergic disease: A multidisciplinary review of the literature (1966–2001) on the mode of early feeding in infancy and its impact on later atopic manifestations. *Allergy* **2003**, *58*, 833–843. [CrossRef] [PubMed]
182. Dogaru, C.M.; Nyffenegger, D.; Pescatore, A.M.; Spycher, B.D.; Kuehni, C.E. Breastfeeding and childhood asthma: Systematic review and meta-analysis. *Am. J. Epidemiol.* **2014**, *179*, 1153–1167. [CrossRef] [PubMed]
183. Mavroudi, A.; Xinias, I. Dietary interventions for primary allergy prevention in infants. *Hippokratia* **2011**, *15*, 216–222. [PubMed]
184. Ierodiakonou, D.; Garcia-Larsen, V.; Logan, A.; Groome, A.; Cunha, S.; Chivinge, J.; Robinson, Z.; Geoghegan, N.; Jarrold, K.; Reeves, T.; et al. Timing of Allergenic Food Introduction to the Infant Diet and Risk of Allergic or Autoimmune Disease: A Systematic Review and Meta-analysis. *JAMA* **2016**, *316*, 1181–1192. [CrossRef] [PubMed]
185. Jeurink, P.V.; Knipping, K.; Wiens, F.; Baranska, K.; Stahl, B.; Garssen, J.; Krolak-Olejnik, B. Importance of maternal diet in the training of the infant's immune system during gestation and lactation. *Crit. Rev. Food Sci. Nutr.* **2019**, *59*, 1311–1319. [CrossRef] [PubMed]
186. Pabst, O.; Mowat, A.M. Oral tolerance to food protein. *Mucosal Immunol.* **2012**, *5*, 232–239. [CrossRef] [PubMed]
187. Neu, J. Gastrointestinal maturation and implications for infant feeding. *Early Hum. Dev.* **2007**, *83*, 767–775. [CrossRef] [PubMed]
188. Kaetzel, C.S. Cooperativity among secretory IgA, the polymeric immunoglobulin receptor, and the gut microbiota promotes host-microbial mutualism. *Immunol. Lett.* **2014**, *162*, 10–21. [CrossRef] [PubMed]
189. Karav, S.; Le Parc, A.; Leite Nobrega de Moura Bell, J.M.; Frese, S.A.; Kirmiz, N.; Block, D.E.; Barile, D.; Mills, D.A. Oligosaccharides Released from Milk Glycoproteins Are Selective Growth Substrates for Infant-Associated Bifidobacteria. *Appl. Environ. Microbiol.* **2016**, *82*, 3622–3630. [CrossRef] [PubMed]

© 2019 by the authors. Licensee MDPI, Basel, Switzerland. This article is an open access article distributed under the terms and conditions of the Creative Commons Attribution (CC BY) license (http://creativecommons.org/licenses/by/4.0/).

Review

A Review of Bioactive Factors in Human Breastmilk: A Focus on Prematurity

Andrea Gila-Diaz [1], Silvia M. Arribas [1], Alba Algara [1], María A. Martín-Cabrejas [2], Ángel Luis López de Pablo [1], Miguel Sáenz de Pipaón [3,4] and David Ramiro-Cortijo [1,*]

1. Department of Physiology, Faculty of Medicine, Universidad Autónoma de Madrid, 28029 Madrid, Spain; andrea.gila@estudiante.uam.es (A.G.-D.); silvia.arribas@uam.es (S.M.A.); alba.algara@estudiante.uam.es (A.A.); angel.lopezdepablo@uam.es (Á.L.L.d.P.)
2. Department of Agricultural Chemistry and Food Science, Universidad Autónoma de Madrid, Institute of Food Science Research, CIAL (UAM-CSIC), 28049 Madrid, Spain; maria.martin@uam.es
3. Neonatology Service, La Paz Hospital-Universidad Autónoma de Madrid, 28046 Madrid, Spain; miguel.saenz@salud.madrid.org
4. Carlos III Health Institute, Maternal and Child Health and Development Research Network, 28029 Madrid, Spain
* Correspondence: david.ramiro@uam.es; Tel.: +34-914-975-416

Received: 10 April 2019; Accepted: 4 June 2019; Published: 10 June 2019

Abstract: Preterm birth is an increasing worldwide problem. Prematurity is the second most common cause of death in children under 5 years of age. It is associated with a higher risk of several pathologies in the perinatal period and adulthood. Maternal milk, a complex fluid with several bioactive factors, is the best option for the newborn. Its dynamic composition is influenced by diverse factors such as maternal age, lactation period, and health status. The aim of the present review is to summarize the current knowledge regarding some bioactive factors present in breastmilk, namely antioxidants, growth factors, adipokines, and cytokines, paying specific attention to prematurity. The revised literature reveals that the highest levels of these bioactive factors are found in the colostrum and they decrease along the lactation period; bioactive factors are found in higher levels in preterm as compared to full-term milk, they are lacking in formula milk, and decreased in donated milk. However, there are still some gaps and inconclusive data, and further research in this field is needed. Given the fact that many preterm mothers are unable to complete breastfeeding, new information could be important to develop infant supplements that best match preterm human milk.

Keywords: adipokines; antioxidants; breastfeeding; cytokines; growth factors

1. Introduction

Preterm infants are those born before 37 weeks of gestation [1]. They are considered extremely preterm at less than 28 weeks, very preterm between 28 and 32 weeks, and moderate to late preterm between 32 and 37 weeks [2].

Prematurity is a serious health problem worldwide, with increasing rates in both low- and high-income countries [2,3]. The World Health Organization (WHO) estimated that the global incidence of preterm delivery was around 11.1% in 2010. The frequency of preterm births (PTBs) in low-income settings is between 12% and 18.1% [2], with Asia and Africa being the areas with the highest rates [4]. The frequency of PTB can vary between 5% and 13% in high-income areas such as the United States of America [5].

PTB is a multifactorial syndrome and has multiple etiologies. The main factors that may increase women's risk are maternal malnutrition, poor pregnancy weight gain, infections and maternity at extreme ages, short interpregnancy intervals, or/and obstetric complications. Infertility and the subsequent need for assisted reproduction techniques resulting in twin pregnancies are also important factors. Finally, exposure to lifestyle-related toxic substances and environmental pollutants are also considered as potential risk factors [3–8].

Preterm infants, in comparison with term infants, have higher morbimortality rates. Throughout the world, prematurity is the second most common cause of death among children under 5, and is the leading cause in high-income countries [4,9]. Late onset sepsis [10], necrotizing enterocolitis (NEC) [11], retinopathy of prematurity (ROP) [12], bronchopulmonary dysplasia (BPD) [13], and neurodevelopmental problems [14] are among the most frequent morbidities. Besides, growing evidence is proving that individuals born preterm are at higher risk of cardiovascular and metabolic diseases in adulthood [15–17].

Nutrition during the earliest stage of life is of great importance for the growth and maturation of tissues and organs, especially for those infants with PTB. Breastmilk (BM) is tailored to cover the needs of the newborns, providing the adequate amount of macro and micronutrients for infant growth and development. In addition, it contains several bioactive compounds, which contribute to the maturation of their immune system, among other important aspects. The BM composition is dynamic and varies according to the maternal age, number of pregnancies [18], body mass index (BMI) [19], maternal diet, time of the day, lactation period, and other environmental factors [20]. Therefore, the WHO recommends that infants be exclusively breastfed for the first six month of life, and lactation, together with nutritionally adequate foods, can continue up to two years of age or beyond [21,22]. This recommendation has been endorsed by several Pediatric and Nutrition Associations [23–25].

Among the most important bioactive compounds found in BM, the most outstanding are antioxidants, which may help counteracting the negative effects of oxidative stress in newborns [26]. Other bioactive compounds to bear in mind are growth factors and hormones, which regulate the energy intake and maturation of organs. Also, cytokines are present in BM, which may help to protect against infections or reduce inflammatory processes [20]. These bioactive compounds may contribute to the long-term protection of preterm infants against the development of cardiometabolic diseases. Premature infants fed with BM exhibit lower rates of metabolic syndrome, hypertension, or insulin resistance in adolescence, compared with the newborns who are not breastfed [24,27].

BM is the ideal food for neonates during the first days of life and cannot be equaled by artificial substitutes [28]. However, there are mothers who, for several reasons, do not breastfeed their newborns. In 2018, 7.6 million babies were not breastfed [29]. In these cases, commercial formulas and donated BM, generally from mothers with term deliveries [30], are used. The human milk is recognized as a better scavenger of free radicals than the infant formula due to its bioactive compounds, which are lacking in commercial formulas [31,32].

The aim of this review is to summarize the role of some bioactive factors present in BM, namely antioxidants, growth factors, adipokines, and cytokines, with specific attention to the differences between preterm and full-term human milk; we also summarize their role on the development and the potential beneficial actions on neonatal and long-term health of premature infants.

2. Bioactive Compounds in Breastmilk

There is wide evidence that BM prevents many of the perinatal complications associated with preterm labor [33], and BM is recognized as a protective factor against morbidity and mortality. Some of the beneficial factors against preterm complications may be related to the bioactive compounds and interactive elements present in the milk, such as antioxidants, growth factors, adipokines, cytokines, or antimicrobial compounds [34].

2.1. Antioxidants in Human Milk

Reactive oxygen species (ROS) are physiologically relevant molecules that participate in cellular signaling processes. However, ROS are highly oxidizing and, in excess, can cause damage to cellular structures. To counteract the oxidative effects of ROS, there are a wide variety of antioxidant systems.

There is a delicate balance between ROS and other reactive species and antioxidants. If this balance is lost, the result is oxidative stress. This can be due to an excessive ROS production that exceeds the antioxidant capacity, or insufficient antioxidant systems.

Birth represents a significant oxidative challenge because of the switch from the relatively low-oxygen intrauterine environment to the high-oxygen extrauterine atmosphere [35]. Thus, newborns are exposed to an increase in ROS during labor and the transition to neonatal life [36]. PTB disrupts the normal developmental upregulation of antioxidant systems. Increased oxidative stress is observed in preterm neonates compared with full-term infants [37], being a critical factor that exacerbates perinatal morbidities of prematurity [38].

Endogenous antioxidants can be classified as enzymatic (i.e., superoxide dismutase (SOD), catalase, or glutathione peroxidase (GPx)), small non-enzymatic molecules (like glutathione (GSH)), or hormones with antioxidant capacity (such as melatonin) [37,39]. In addition to endogenous antioxidants, several foods, mainly vegetables and fruits, contain antioxidants such as vitamins, carotenoids, and polyphenols, among others.

2.1.1. Antioxidant Properties of Breastmilk

BM has a powerful antioxidant composition and all the above-mentioned compounds have been found in it [18,19,26,40,41]. Antioxidants are important for newborn protection against disease [28], and may be critical for infants with PTB. There is evidence that premature neonates nourished with BM have less oxidative stress, evidenced by the lower levels of oxidative damage biomarkers compared with the infants who were formula fed [40,42]. It has been proposed that the reduction in oxidative damage by BM may be related to the reduction in ROS synthesis—evidenced for hydroxyl radicals—or to an increase in the antioxidant defense systems [42]. Some reports show that preterm infants fed with BM plus a fortifier had higher antioxidant urinary levels compared to both infants exclusively fed BM and with those who were formula-fed; the mechanisms remain unclear [43] and this aspect should be further analyzed.

The total antioxidant capacity of BM seems to be higher in colostrum compared to mature milk [44,45] and its radical scavenging activity decreases along the lactation period. Pasteurization processes and storage conditions of BM may reduce antioxidants [46], as well as some immunological and nutritional properties. For example, refrigeration decreases the concentration of vitamins, lactoferrin, and lipases. Therefore, it is not recommended for BM. However, despite the fact that these treatments alter the bioactive compounds, donated milk should be pasteurized and frozen. As stated by the Spanish Association of Pediatrics, the development of methods to improve the preservation of the antioxidant capacity of donated BM, or any factors improving maternal antioxidant status, deserve further investigation [47]. BM treatment is important, particularly in the context of prematurity. Certain viruses, such as cytomegalovirus, may infect the newborn [48] through raw BM transmission [49]. The cytomegalovirus infection is problematic in preterm infants, particularly in those with very low birth weight [48].

Differences in antioxidant capacity of term and preterm BM remain controversial. Some data show that preterm BM has more antioxidants and equal resistance to oxidative stress as compared to term human milk [43]. On the other hand, other studies show a lower total antioxidant capacity in preterm BM compared to full-term BM [44,50]. It has been suggested that this is due to the fact that antioxidants tend to accumulate over the last three months of gestation. Thus, it is proposed that a mother with a PTB would synthetize a less antioxidants [50,51]. Discrepancies in the antioxidant activity between term and preterm BM may be related to differences in the ethnic group studied or the maternal nutritional habits. Another possible explanation to this controversy could be the time point when the studies were conducted. In this sense, it has been observed that the total antioxidant capacity of premature BM does not decline after one week, and preterm neonates benefit more from the antioxidant capacity of colostrum and transitional milk [50]. Term milk may differ in this respect. Taken together, these reports suggest that more research is needed in this area. Aspects such as ethnicity, maternal nutritional status, or breastfeeding period should be taken into account in future studies to clarify differences in antioxidant properties in BM and its relationship with infant needs.

2.1.2. Antioxidants Present in Breastmilk

The antioxidant properties of human BM are related to the combination of different compounds, both exogenous and endogenous molecules.

Several food-derived antioxidants, including polyphenols, carotenoids, and vitamins, have been reported in BM, and there is evidence that they are more abundant in preterm than in full-term BM. The levels of these bioactive components also differ in formula milk. As stated in Table 1, the levels of antioxidants included in formula milk do not always match the reports in BM, and in some instances are clearly much higher. An excess in antioxidants may be deleterious, since some antioxidants may turn into pro-oxidant molecules under certain circumstances. Therefore, research on the adequate requirements of antioxidants in both term and preterm infants and adjustment of the antioxidant content in formula are needed.

Table 1. Exogenous antioxidants in breastmilk.

Antioxidant Compounds	Preterm Infants	Term Infants	Formula Feeding	
			Preterm	Term
α-carotene	7.7	3.6	0.51	1.40
β-carotene	49.1	13.7	71.1	63.9
Lycopene	66.1	11.9	1.5	5.8
Retinol	401.6	185.8	3086.2	911.8
α-tocopherol	5880.8	1381.9	20,109.1	13,360.2
γ-tocopherol	1207.1	622.8	6787.1	6561.6

Adapted from Hanson et al. (2016) [31]. The data are expressed as µg/L.

Enzymatic antioxidants are also important to counteract oxidative processes. The main enzymatic systems which detoxify ROS are SOD, catalase, and GPx. SOD catalyzes the dismutation of superoxide anion to hydrogen peroxide, which must be removed by catalase. Three SOD isoforms (copper–zinc, manganese, and extracellular SOD) are present in mammals, with different subcellular locations and tissue distribution [52]. Catalase consists of four protein subunits, which make it very resistant to pH changes, thermal denaturation and proteolysis [53]. This enzyme eliminates the hydrogen peroxide generated by SOD and completes the reaction to eliminate ROS [52]. Finally, GPx participates, together with catalase, in the detoxification of hydrogen peroxide among other organic hydroperoxides [54,55].

There is evidence of some of these enzymatic systems and their concentrations have been reported. Catalase has been found in human BM, with some differences between term and preterm BM [19]. In milk from term mothers, the catalase concentration ranges between 0.43 and 0.84 U/mg protein, and in preterm BM the concentration has been found to be 0.5–0.97 U/mg protein [40]. GPx has also been found in BM, with a maximum value of 31.2 mM/min/L [56].

Regarding non-enzymatic systems, GSH is a three amino acid peptide, which is the main intracellular low molecular weight antioxidant. It participates in the regeneration of other antioxidants, such as vitamin C and E to their active forms [57,58]. GSH has been found in BM in concentrations ranging from 10.4 to 43.1 nmol/mg.

Melatonin is the major endocrine product of the pineal gland, which plays a physiological role in neuroimmunomodulation. It is synthetized from tryptophan via serotonin in pinealocytes and many other cells, with a circadian regulation [59–64]. Melatonin is interesting in the context of BM bioactive compounds due to its pleiotropic actions. It has been demonstrated to exhibit protective effects against cellular aging due to its antioxidant effects, both as a direct scavenger, and stimulating the expression of SOD, catalase, and GPx [60,65–68]. On the other hand, melatonin appears to be one of the most promising molecules for neuroprotection in preterm infants due to its effects on the modulation of neuroinflammatory pathways [69].

Melatonin is present in BM in much higher concentrations during night time, being almost undetectable during the day. So far, the studies have not been able to detect differences between preterm and term BM mothers. Some studies have observed higher levels in colostrum, which has been reported to increase the phagocytic activity of cells against bacteria. A longer sleep time was observed in newborns who were breastfed than in those who were formula-fed [70]. Table 2 describes several endogenous antioxidants and their concentrations in BM.

Table 2. Endogenous antioxidants in breastmilk.

Antioxidant	Activity	Range
Superoxide dismutase	Eliminates superoxide anion	2.01–6.26 nmol/min/mL
Catalase	Eliminates hydrogen peroxide	1.84–26.1 nmol/min/mL
Glutathione peroxidase	Eliminates hydrogen peroxide	6.6–17.7 mM/min/L
Glutathione	Regeneration of other antioxidants	10.4–43.1 nmol/mg of protein
Melatonin	Free radical scavenger, antioxidant expression	<10–23 ng/L

Adapted from [18,26,41,54–57,60–64,66].

2.2. Growth Factors in Human Milk

Preterm infants are immature neonates who usually exhibit growth retardation, poor development and physical and neurological deficits. Postnatal growth retardation is likely the result of inadequate nutrition support after delivery, also contributing to poor neonatal health. In this context, in addition to a macronutrient supply, growth factors provided by breastfeeding may be of great importance. Growth factors play a role in the growth, maturation, and integrity of several organs, particularly for the neonatal gastrointestinal tract [71]. They help with the maturation of gut immunity and have anti-inflammatory effects [72,73]. Hirai et al. described the trophic effects of growth factors on fetal and neonatal gastrointestinal tract by promoting the proliferation and differentiation on their immature cells [74]. The highest concentrations of growth factors are provided by colostrum, as the first milk released after birth [75].

The main growth factors present in BM and their trophic effects on neonatal organs and systems are summarized in Figure 1.

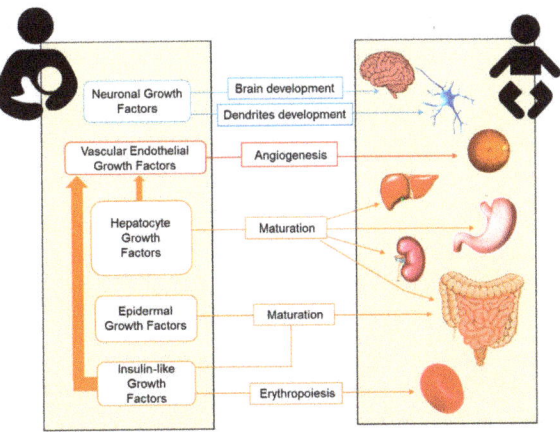

Figure 1. Transference of maternal growth factors through breastmilk and their trophic effects on the growth and maturation of neonatal organs and systems.

2.2.1. Hepatocyte Growth Factor (HGF)

HGF was first identified as a potent mitogen of primary cultured hepatocytes. The essential role is promoting organogenesis. It is also involved in the formation of the kidney, lung, mammary gland, teeth, muscle, and neuronal tissues [76]. To preserve proliferation, angiogenesis, and intestinal tissue development through paracrine and endocrine signaling, high levels of HGF are required in BM. This factor is released into BM by multipotent mesenchymal stem cells [77]. In addition to direct proliferative properties, HGF may also regulate the vascular endothelial growth factor (VEGF) synthesis [78].

2.2.2. Epidermal Growth Factor (EGF)

EGF is recognized as a critical trophic factor for the normal intestinal cell development (Table 3). The members of the EGF family are first synthetized as transmembrane precursors, eventually undergoing proteolysis into the mature, secreted form of the growth factor [71].

Both the amniotic fluid and the BM contain EGF [79,80]. A member of the EGF family is the heparin-binding growth factor (HB-EGF). Its exogenous administration protects from intestinal ischemia-reperfusion injury, hemorrhagic shock, and NEC by enhancing the healing of intestinal anastomoses and reducing anastomotic complications [81,82]. In BM, EGF levels are higher at the beginning of the lactation period and decrease over time. Furthermore, preterm BM contains higher levels of EGF than term BM, which may help in the reduction of NEC incidence [83]. However, very preterm BM contains lower levels than preterm milk, although still higher compared to term milk. In preterm colostrum EGF content has been found in the range of 22.8–373 µg/L and in term colostrum between 27.7 and 209 µg/L [84].

2.2.3. Neuronal Growth Factors

Brain-derived neurotrophic factor (BDNF) is a small neurotrophic protein, which is widely expressed in the mammalian adult brain [85,86]. BDNF, together with S100B protein and glial cell line-derived neurotrophic factor (GDNF), plays a critical role in the development and maintenance of the nervous system, and in neuronal survival and proliferation [87]. BDNF, S100B, and GDNF are present in human milk. S100B protein and GDNF levels increase within the lactation period [86].

2.2.4. Insulin-Like Growth Factor (IGF) Superfamily

Human BM contains IGFs such as IGF-I and IGF-II [88]. IGF-I synthesis is regulated by the availability of amino acids and the overall energy intake, and is a marker of the nutritional status. IGF-I levels in BM are higher during the first days after delivery, decreasing as the milk matures [89]. No significant differences were found between preterm and term milk in other growth factors from this family, except for IGF-I and IGF-II, which is higher in preterm milk [88,90]. IGF-I could be important in the protection of enterocytes after intestinal damage caused by ROS [91]. Enteral IGF-I administration enhances erythropoiesis and augments hematocrit, but its function is still not clearly known [20].

2.2.5. Vascular Endothelial Growth Factor (VEGF)

VEGF mediates vascularization, which is also controlled by IGF-I. In preterm infants, the relative hyperoxia found in the extrauterine environment inhibits the expression of VEGF, interrupting the growth of retinal blood vessels. Pulmonary immaturity and the subsequent need for oxygen therapy contribute to the susceptibility of the retinal tissue to the oxidative injury and the subsequent development of ROP [92].

VEGF levels in BM are higher at the beginning of the lactation period, which helps reducing ROP's burden during the first days of life. There are some controversies regarding the differences in this growth factor between preterm and term BM. While no differences have been reported in some studies [93], other authors demonstrate both lower and higher levels of VEGF in preterm compared to full-term milk (Table 3) [73,90].

2.2.6. CD14 Protein

CD14 acts as a co-receptor for the detection of bacterial lipopolysaccharide (LPS) [94]. CD14 is a protein with two forms: one is a soluble form (sCD14) and the other is anchored to the cellular membrane (mCD14). This latter membrane-bound form is primarily expressed on the surface of monocytes, macrophages and neutrophils [95,96]. CD14 may have a major implication due to the protection provided against subsequent allergy manifestations [97]. There is some evidence that low levels of sCD14 in BM are associated with eczema development [98]. In addition, other factors such as the newborn genotype and the interaction with the bioactive factor present in BM may also participate in the development of allergies [96].

Table 3. Growth factors in breastmilk.

Growth Factors	Main Tissue Synthesized	Range (µg/mL)	Main Neonatal Functions
Epidermal-GF	Submandibular salivary gland	24–37	Intestinal mucosa maturation and healing, nutrient absorption, protein synthesis
Neuronal-GF	Cerebral cortex and hippocampus	2.8–934	Nervous system maturation, learning, and memory
Insulin-like-GF	Placenta and digestive system	5–35	Retinal vascularization, brain maturation
Vascular Endothelial-GF		505–650	Angiogenesis

Adapted from [20,71,80,85–87,90,99]; GF, growth factors

2.3. Adipokines in Human Milk

In addition to growth factors, adipokines constitute another group of compounds present in BM which is important for metabolism and infant growth. These cytokines derived from adipocytes have been demonstrated to modify weight gain and fat and lean body mass in infants in the early postpartum period [100] and have long-term effects on metabolic programming. They are also involved in the regulation of food intake and energy balance [101]. Several adipokines, with opposing actions on food intake and energy expenditure, have been found in BM (summarized in Table 4). Thus, it can be hypothesized that the programming of food intake and body composition may be influenced by the relative concentration of these compounds in BM. Likewise, BM adipokines may also modulate the development of metabolic diseases in adulthood such as obesity, type 2 diabetes mellitus, or insulin resistance [102].

2.3.1. Leptin

Leptin is an anorexigenic hormone encoded in the *ob gene* and mostly synthesized by white adipose tissue, which acts through the arcuate nucleus of the hypothalamus. Leptin minimizes the energy intake and increases the energy expenditure [101], and it plays a role in fetal and neonatal growth [103,104].

Several studies have reported the presence of leptin in BM, which may be produced by various cell types in the mammary tissue. The *ob gene* is expressed in the epithelium of the mammary gland of lactating women and it produces leptin [105]. Leptin is also transported from maternal circulation to BM [106] and it has been reported that plasma and BM levels are directly associated with the fat content and the body mass index [100,107].

Leptin levels in human milk vary over the lactation period [108]. Differences in BM leptin concentrations are also reported comparing whole and skimmed milk. In mothers with term infants, leptin concentration in whole milk ranged between 0.2 to 10.1 ng/mL [109,110], being lower in skimmed milk at between 0.1 to 3.4 ng/mL [111,112]. Leptin levels seem to be higher in BM from mothers with preterm newborns, with concentrations ranging between 0.6 and 5.3 ng/mL in skimmed milk [108,113].

Table 4. Adipokines in human breastmilk.

Adipokines	Tissue Synthesized	Range (ng/mL)	Preterm Infants	Main Neonatal Functions
Leptin	White adipose Placenta Mammary	0.2–10.1	↑/?	Anorexigenic T-lymphocyte responses
Adiponectin	Adipocytes	4.2–87.9	≈/?	Orexigenic Regulation of lipid/glucose metabolism Improvement of insulin sensitivity Anti-inflammatory actions
Resistin	Immune cells Epithelial cells	0.2–1.8	↑/?	Regulation of glucose homeostasis Inhibition of adipocyte differentiation Inflammatory response
Ghrelin	Stomach Pituitary Other	0.07–6	?	Orexigenic Gastric motility and secretion Adipogenesis Anti-inflammatory actions
Obestatin	Stomach Small intestine	0.4–1.3	?	Anorexigenic Body weight regulation
Nesfatin	Neurons Pancreas Other	0.008–0.01	?	Anorexigenic Production of body fat
Apelin	Heart Lung Other	43–81	?	Regulation of cardiovascular system Fluid homeostasis Angiogenesis Regulation of insulin secretion

Adapted from Catli et al. (2014) [102]. Arrows indicate the comparison with term infants (↑, higher; ↓, lower; ≈, similar); ? indicates unavailable or inconclusive data and the need for research.

Several reports indicate that leptin levels in BM can play an important role in infant growth, and its production by human mammary epithelial cells might be regulated physiologically according to the necessity and the state of the infant [114]. Maternal milk of small, appropriate, and large gestational age infants (SGA, AGA, and LGA) has different leptin levels, especially during the first month of life. Human milk leptin levels are significantly reduced in the SGA neonates compared to AGA and LGA infants, together with a rapid growth during the first postnatal 15 days. It has also been reported that leptin levels in milk of infants with accelerated postnatal growth were lower than in infants with normal growth [115]. These findings suggest that the presence of leptin in BM might play an important role in growth, appetite and regulation of nutrition in infancy, especially during the early lactation period.

2.3.2. Adiponectin

Adiponectin is an orexigenic hormone which regulates lipid and glucose metabolism. Adiponectin enhances insulin sensitivity and stimulates fatty acid oxidation through the activation of AMP-activated protein kinase (AMPK) in peripheral tissues, and inhibits hepatic glucose production [101]. In addition, adiponectin has powerful anti-inflammatory effects, influencing the vascular endothelium. Adiponectin stimulates the food intake through the hypothalamus and reduces the energy expenditure through its central activity [116]. The production of adiponectin is regulated by the peroxisomal proliferator-activated receptor-γ (PPAR-γ), a nuclear receptor expressed in the liver and muscle, which protects against obesity-related insulin resistance [117].

Adiponectin is present in BM, in a range between 4.2 and 87.9 ng/mL [118], being more than 40 times greater than the concentrations of leptin [119]. In the colostrum higher levels have been reported, with concentrations in the range between 2.9 and 317 ng/mL [118], in accordance with studies showing that BM adiponectin levels are negatively associated with the lactation period [120]. Other factors that may influence BM adiponectin levels are the maternal nutritional status and body composition, although this aspect is still controversial. Some studies report a positive association between adiponectin and maternal body fat mass [120]; others demonstrate that the adiponectin concentration in colostrum is markedly dependent on maternal diet and nutritional status during pregnancy, and there are also reports that failed to observe an association with maternal BMI or infant birth weight [121]. The maternal hormonal and inflammatory profile may also alter the BM adiponectin levels [122].

BM adiponectin also seems to have an influence on infant growth. In term infants who were breastfed for at least six months, high levels of adiponectin are associated with overweight [123]. These data support that BM adiponectin in the first stages of life could have an important implication in the regulation of infant growth [124], which deserves further consideration.

2.3.3. Resistin

Resistin is an adipocyte-derived hormone, which regulates glucose homeostasis and counteracts the action of insulin in peripheral tissues, inhibits adipocyte differentiation and may function as a regulator of adipogenesis [101]. Resistin has been identified in BM in a range between 0.2 and 1.8 ng/mL [125] and its levels decrease along the lactation period [126]. In the perinatal period, it seems that resistin is not directly involved in the regulation of insulin sensitivity or adipogenesis [127]. It has been suggested that resistin could have a role in controlling fetal growth together with other BM hormones, and could be involved in the appetite regulation and in the metabolic development of infants [104]. Resistin could also have an important role in the regulation of the energy metabolism and adiposity in utero. Higher serum resistin levels have been found in term infants compared to preterm infants [102], and it has been suggested that newborns could benefit from these higher concentrations of circulating resistin, facilitating the production of hepatic glucose and preventing hypoglycemia after delivery [128].

2.3.4. Ghrelin

Ghrelin is an amino acid synthesized in several organs from the digestive and nervous system, heart and lungs, the stomach being the main site of production [129]. It is one of the most important orexigenic peptides. However, in addition to the functions related to the regulation of food intake and metabolism, ghrelin has other physiological actions, such as in gastric motility and acid secretion, reduction of insulin secretion, adipogenesis, and cardiovascular function, as well as anti-inflammatory effects. Ghrelin secretion declines in situations of positive energy balance, such as after food intake or in obesity, while increasing when fasting or during weight loss.

The presence of ghrelin in BM may be a strong factor influencing the feeding behavior, and body composition later in life, through its effects on short-term food intake and long-term body weight [104]. Ghrelin is found in both term and preterm human BM [130], with concentrations in the range of 73–6000 pg/mL [131,132]. A gradual and parallel increase has been observed between post-partum ghrelin plasma levels and the concentration in BM [130]. However, this physiological increase is impaired in women with gestational diabetes and pre-gestational diabetes mellitus [133]. These alterations may alter the infant's feeding behavior.

A positive correlation between ghrelin concentrations in mature BM and infant weight gain has been found, suggesting that this hormone is involved in postnatal growth [134]. Savino et al observed that infants who were fed with formula milk had higher serum ghrelin levels than those who were breastfed. They propose that formula-fed infants have a better feeding stimulus, which makes them eat more, and consequently, boosts their growth and weight gain. This can explain the protective effect of BM against the development of obesity in childhood and adulthood [135].

2.3.5. Obestatin

Obestatin drifts from ghrelin and its synthesis is mainly produced by digestive system cells, especially from the stomach and small intestine [136]. Obestatin is an anorexigenic hormone which reduces food intake, regulates weight gain and gastric emptying by suppressing the intestinal motility [125]. It was proposed that, in addition to its effects on energy balance regulation, it opposes the actions of ghrelin [130]. Some data have identified higher levels of obestatin in BM than in maternal blood [130]. The reported range of obestatin in BM is from 0.4 to 1.3 ng/mL [125]. Although it is not completely confirmed, some authors have suggested that colostrum have higher levels of obestatin to prepare the digestive system to receive milk by reducing newborn appetite. [130].

2.3.6. Nesfatin

Nesfatin is an anorexigenic [93] neuropeptide related to the melanocortin signaling pathway in the hypothalamus. It is mainly manifested in nervous system cells and peripheral tissues. Nesfatin acts as an appetite regulator and a body fat producer. [137]. Nesfatin has been found in BM in a range between 8 and 14 pg/mL [138].

2.3.7. Apelin

Apelin, an endogenous ligand for the G-protein-receptors [139], participates by maintaining the cardiovascular and fluid homeostasis, regulates appetite, cell proliferation and angiogenesis [140]. Although apelin has been found to regulate food intake, most of the studies have been made on rats, and the results are controversial. More studies in humans are needed to clarify its effect on food intake [140–143]. Apelin concentrations in BM have been ranged between 43 and 81 pg/mL, being lower in women who developed gestational diabetes [133].

2.4. Cytokines in Human Milk

Cytokines are small proteins synthesized by nearly all the nucleated cells [144]. They are signaling molecules involved in the communication between cells [82]. BM is a cytokine-rich food [145]. According to their role in the inflammatory response, cytokines can be divided into those that promote inflammation or protects against infection, and those that decrease inflammation.

A possible role of some cytokines present in BM in preterm infants may be to compensate the delay of the immune system development [146]. It has been hypothesized that immune mediators in human milk may have a powerful role to play in maturing the infant intestine and stimulating the immune system [97]. The stimulation of immune activity by BM cytokines may be related to their capacity to make connections with cells by crossing the intestinal barrier.

There is large variability between cytokine concentrations among breastfeeding women, being pro-inflammatory cytokines generally low [147]. BM cytokine content can be affected by different factors, such as gestational age, maternal diet, infections, smoking, maternal ethnicity, or exercise [147]. In addition, the cytokine profile of BM fluctuates along the different phases of breastfeeding and the clinical significance of cytokine concentrations in neonatal health outcomes is still under debate [148,149].

2.4.1. Anti-Inflammatory Cytokines in Human Breastmilk

The anti-inflammatory cytokines found in BM include transforming growth factor-β (TGF-β), interleukin 7 (IL-7), and IL-10 (Figure 2). IL-7 BM is known to cross the intestinal barrier and contribute to thymus and T-lymphocyte development [150]. In colostrum, IL-10 ranges from 5.9 to 7.3 ng/L in term and from 1.1 to 8.8 ng/L in preterm infants. It has also been reported that these concentrations decrease over lactation, with an average of 0.7 to 1.3 ng/L and 0.5 to 0.9 ng/L for term or preterm mature milk, respectively [81]. The possible differences in IL-10 concentrations between preterm and term milk are still controversial. While some authors have reported lower levels in preterm milk [151], others fail to demonstrate significant differences [152]. An interesting aspect is the finding of a lower IL-10 in BM of infants with increased risk of NEC [153]. These results suggest that the accessibility of this cytokine from BM may be important to modulate the neonatal inflammatory response; however, this is an aspect which deserves further research.

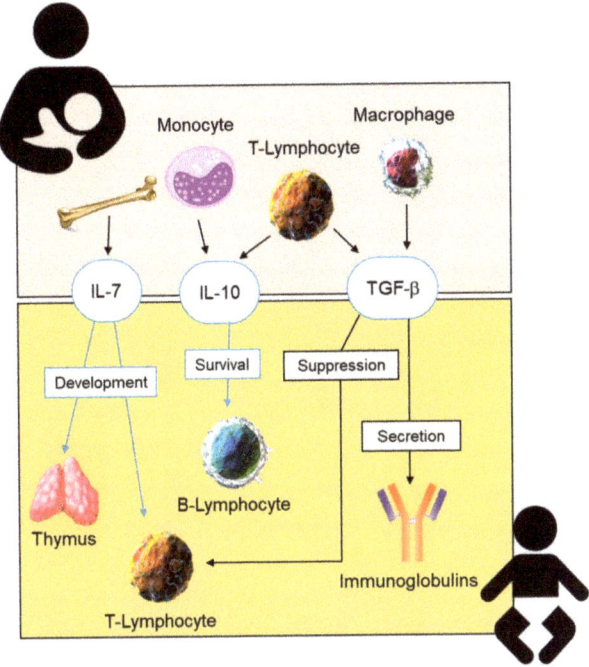

Figure 2. Site of synthesis of anti-inflammatory cytokines present in human breastmilk, and their effects on the neonatal immune system. IL-6, interleukin-6; IL-10, interleukin-10; TGF-β, transforming growth factor-β.

Transforming Growth Factor- β (TGF- β)

TGF-β is an anti-inflammatory cytokine mainly produced by parenchymal cells and infiltrating cells such as lymphocytes, macrophages and platelets [154]. It is found in human milk, with higher levels in colostrum. During breastfeeding, all three isoforms of the TGF-β superfamily are produced, the majority being TGF-β2 [75]. The levels of this cytokine range between 0.1 and 13.3 µg/L in term colostrum and between 1.4 and 43 µg/L in preterm colostrum. These levels decrease along the lactation period, with concentrations in the order of 0.4–2.8 µg/L in term and 0.9–6.3 µg/L in preterm mature milk [84].

The TGF-β found in BM is of great importance for the newborn as it suppresses the neonatal T-lymphocytes activity [155], enabling oral and intestinal tolerance. It could also reduce atopic sensitization by controlling the inflammatory processes involved [156]. TGF-β regulates the production of secretory immunoglobulin A (IgA). IgA in human milk confers passive immune protection to the infant [157]. IgA levels are higher in preterm colostrum (1.8–16.4 g/L) than in term colostrum (1.2–11.6 g/L). Levels greatly decrease along the lactation period to an average of 0.2–0.8 g/L [84].

Therefore, TGF-β-mediated tolerance and IgA antibody synthesis play an essential role in the immunological development of newborns. Some studies have shown how maternal exposure to highly microbial environments increased their TGF-β breastmilk concentrations [147], and it has been proposed that the maternal immune mediators are transferred to the infant by the BM. Thus, the increased TGF-β and IgA levels in the very first milk reflect their purpose of targeting microbial antigens and improve the mucosal barrier function [158]. It has been proposed that, during the gestational and lactation period, there is an increased synthesis of TGF-β, due to the infiltration of immune cells into the breast tissue [159]. These studies propose that the local mammary gland production of TGF-β may be responsible for the elevated concentrations observed in human milk.

2.4.2. Inflammatory Cytokines in Human Breastmilk

Most of the inflammatory cytokines, such as tumor necrosis factor alpha (TNF-α), IL-1β, IL-6, IL-8, and interferon gamma (IFNγ), are present at lower concentrations compared to anti-inflammatory cytokines (Figure 3). The levels of these cytokines decrease over lactation [20] and are associated with the timing of delivery [160], being higher in preterm milk than in term BM [82].

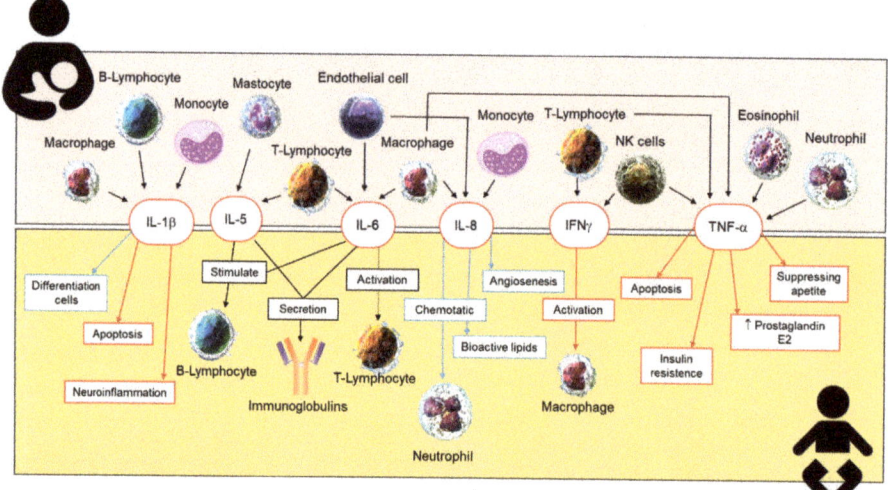

Figure 3. Synthesis of the inflammatory cytokines present in human breastmilk and their effect on the neonate. IL-1β, interleukin-1β; IL-5, interleukin-5; IL-6, interleukin-6; IL-8, interleukin-8; IFNγ, interferon gamma; TNF-α, tumor necrosis factor alpha.

IL-6 levels range between 4.4 and 340 ng/L in term colostrum, between 15.3 and 362 ng/L in preterm colostrum, and between 9.3 and 67.9 ng/L in very preterm colostrum.

IL-8 concentrations vary between 0.04 and 26.3 μg/L in term colostrum, between 0.13 and 14.7 μg/L in preterm colostrum, and between 0.1 and 3.0 μg/L in very preterm colostrum.

TNF-α levels decrease from an average of 11.4 ng/L in term colostrum to 1.6 ng/L in term mature milk, from 18.2 ng/L in preterm colostrum to 3.2 ng/L in preterm mature milk, and from 4.2 ng/L in very preterm colostrum to 1.8 ng/L in very preterm mature milk [84].

There is a controversy about the effect of these cytokines on the health of offspring [149]. Some reports demonstrate that preterm gestation does not substantially influence the cytokine content of BM during the first month of lactation compared to full-term gestation, which can be beneficial for the regional and systemic immune response of the very preterm infant [152].

IL-1β is a mediator of the inflammatory response, and is involved in cell proliferation, differentiation, and apoptosis. Several studies have obtained a range of between 2 and 2500 pg/mL for

IL-1β in human milk [161]. It has been reported that the higher the levels of IL-1β in human milk, the greater the protection against eczema for infants [96].

IFNγ improves the Th1/inflammation response while abolishing the Th2/allergic reaction [145]. The colostrum of allergic mothers contains less IFNγ but more Th2, IL-4 and IL-13 when compared to non-allergic mothers [162].

Other studies have demonstrated that IL-8 and TNFα levels are slightly elevated in advanced BM of mothers who suffered from preeclampsia [163]. However, not all studies reveal a negative effect of the inflammatory cytokines. IL-8 also protects intestinal cells against chemical injury, since it has a trophic function in the developing human intestine [151]. Other authors have shown the relevance of cytokines as risk factors for immunological disease in infants. Thus, the presence of IL-5 and IL-13 in human milk, although extremely low, are risk factors for asthma at the age of one [164].

More research is needed on the role of cytokines in BM and how they influence neonatal health. The contribution of cytokines, anti and pro-inflammatory in BM, is different, and this variability could make the difference between health and disease in preterm infants.

3. Human Milk Cells

3.1. Stem Cells in Human Milk

Recent data suggest that up to 6% of the cells in human milk are stem cells, and mesenchymal stem cells isolated from BM are potentially reprogrammable to multiple tissue types [165,166]. These cells may be involved in the development of immune cells including the regulatory T cell, which may produce tolerance to non-inherited maternal antigens and suppress anti-maternal immunity. It induces pregnancy microchimerism, leading to intestinal tissue repair and immune development and protection against infectious diseases [166].

3.2. Leukocytes in Human Milk

Leukocytes are highly present in colostrum, which means that breastfed infants are exposed to up to 1010 maternal leukocytes/day [167]; however the contribution of this exposure in infants' immune development is not yet clear [148]. A study carried out with 61 mothers and neonates showed a significantly smaller proportion of macrophages in BM of mothers with infants that developed an allergy to cow's milk. In contrast, a higher content of neutrophils, eosinophils, or lymphocytes was associated with lower allergies to cow's milk [168].

The role of these cells is still far from understood; for example, how these cells pass the stomach and intestinal barriers and access the infant, or their mechanisms of action, remain unclear. Therefore, further research to clarify these aspects of the inflammation and immunity development is required.

4. Human Milk Microbiota

Early microbial colonization is essential for the infant's metabolic and immunological maturation. Its development begins at birth, and the most important changes occur during the first year of life [169,170]. The microbiome is constantly changing, and it is influenced by hormones, cytokines, and chemokines. After birth, the transference continues along breastfeeding, and it is considered the main cause of variability between exclusively breastfed and formula-fed infants during the first months of life [165,171]. Raw BM is not a sterile food [167] and several reports confirm more than 100 types of viable bacteria/mL in human milk [171], including 65% of the phyla Proteobacteria and 34% of the phyla Firmicutes [82]. Regarding the genera, the most common are *Staphylococcus, Streptococcus, Lactobacillus, Enterococcus, Lactococcus, Weissella, Veillonella*, and *Bifidobacterium* [172,173]. It has been described that human milk microbiota at an early age is strongly linked with other perinatal factors such as place of residence, delivery mode, or maternal food intake [96]. With respect to the comparison between preterm and term milk, few studies have reported a difference in the BM microbiome. Some trends include more *Bifidobacterium* in term milk and more *Enterococcus* in preterm milk [97,167].

The association between the microbiome and various disorders, such as visceral pain, autism spectrum disorder, cardiovascular risk, obesity, depression, or multiple sclerosis, has been well demonstrated. Besides, the microbiota exerts immune-modulating effects that influence allergic reactions [169]. Thus, the gut microbiome from allergic children differs from non-allergic ones in composition and diversity [174]. It has been hypothesized that early gut colonizers could help in developing and maturing the immune system in infants [175]. Probiotic intake throughout gestation and lactation also leads to specific changes in the neonate [176]. A clinical trial reports the beneficial effect of oral supplementation with *Lactobacillus rhamnosus* in women during pregnancy and breastfeeding to reduce the allergy risk in infants. Modulatory effects were observed both on milk composition and the infant gut [177]. Variations in microbiota of preterm infants have been associated with a higher predisposition for developing NEC [178]. Taking together the information available, BM could be considered a probiotic food for infants. However, the potential protective effect of the BM microbiome is not fully understood, and additional research in this area is necessary to understand its role in the newborn's health.

5. Conclusions and Proposal for Future Research

Breastmilk is a complex food and is the gold standard for infant nutrition. The aim of the present review was to summarize current knowledge regarding some important bioactive molecules which participate in growth, defense against oxidative damage, and infections, and are particularly important in prematurity, which is a global rising problem.

Breastmilk is a dynamic fluid; it is known that its macronutrient composition changes along lactation to match infant needs. The present review provides evidence that bioactive factors also change over time and, in general, the highest levels are found in colostrum. Another relevant conclusion is that bioactive factors are higher in preterm compared to full-term milk. However, the review of the literature evidences some controversy. Regarding antioxidants, there is no consensus in relation to the content and the requirements in preterm and full-term infants, and there is insufficient information on the changes in donated milk due to thermal treatments. The growth factors and adipokines present in breastmilk, together with macronutrients, contribute to organ maturation and development. The issues that deserve further investigation include how maternal dietary habits or metabolic status influence the content of these bioactive compounds. Accessibility of cytokines from breastmilk is important to modulate the immediate and long-term inflammatory response of the infant. There is no consensus on the concentrations of different cytokines in breastmilk and their relative role for neonatal and adult health, an aspect that deserves further investigation. Finally, the immune-modulating effects of breastmilk leucocytes, stem cells, and the microbiota are still far from being understood, and represent an expanding field of research.

Another conclusion is the insufficient clinical evidence to demonstrate that raw breastmilk is better than donated milk for infant growth. However, due to the better content of bioactive factors, it is possible that it may improve other clinical outcomes such neurodevelopment or prevention of adult diseases.

Among aspects which require further attention, we suggest several research lines. Firstly, it is necessary to improve and systematize methods to quantify bioactive factors in breastmilk. It would also be of interest to gain insight on how body fat, nutritional status, and health of the mother influence bioactive breastmilk factors. The results obtained from these studies may help to improve clinical practice, particularly in preterm infants, in the progress of treatments to preserve bioactive factors in donated milk, and to design formula milk that best matches infant needs. Increasing knowledge in this area may help to regulate and adjust neonatal infant growth and improve neonatal and long-term health.

Author Contributions: Conceptualization: A.G.-D., S.M.A., M.A.M.-C., Á.L.L.d.P. and D.R.-C.; Methodology and Resources: A.G.-D., S.M.A., Á.L.L.d.P., D.R.-C.; Writing—Original Draft Preparation, A.G.-D., A.A. and D.R.-C.;

Supervision, S.M.A.; Writing—Review and Editing: A.G.-D., S.M.A., M.A.M.-C., Á.L.L.d.P., M.S.P. and D.R.-C.; Funding Acquisition: S.M.A. and M.A.M.-C.

Funding: This work was supported by Ministerio de Economia y Competitividad (grant number FEM2015-63631-R) to SMA and the Ministerio de Ciencia, Innovación y Universidades (Spain) (grant number RTI2018-097504-B-100) to SMA and MAM-C. Both grants were co-financed with FEDER funds.

Acknowledgments: We would like to thank Raquel Fuertes Ortega, a professional translator, for English editing.

Conflicts of Interest: The authors declare no conflict of interest.

References

1. Guaraldi, F.; Salvatori, G. Effect of Breast and Formula Feeding on Gut Microbiota Shaping in Newborns. *Front. Cell. Infect. Microbiol.* **2012**, *2*, 94. [CrossRef]
2. World Health Organization: Preterm Birth. Available online: https://bit.ly/2RWokG3 (accessed on 23 November 2018).
3. Blencowe, H.; Cousens, S.; Chou, D.; Oestergaard, M.; Say, L.; Moller, A.; Kinney, M. Born Too Soon: The Global Epidemiology of 15 Million Preterm Births. *Reprod. Health* **2013**, *10*, 1–14. [CrossRef] [PubMed]
4. Harrison, M.S.; Goldenberg, R.L. Global Burden of Prematurity. *Semin. Fetal Neonatal Med.* **2016**, *21*, 74–79. [CrossRef] [PubMed]
5. Iams, J.D.; Romero, R.; Culhane, J.F.; Goldenberg, R.L. Primary, Secondary, and Tertiary Interventions to Reduce the Morbidity and Mortality of Preterm Birth. *Lancet* **2008**, *371*, 164–175. [CrossRef]
6. Ion, R.; López Bernal, A. Smoking and Preterm Birth. *Reprod. Sci.* **2015**, *22*, 918–926. [CrossRef] [PubMed]
7. Al-gubory, K. Environmental Pollutants and Lifestyle Factors Induce Oxidative Stress and Poor Prenatal Development. *Reprod. Biomed. Online* **2014**, *29*, 17–31. [CrossRef] [PubMed]
8. Simón, L.; Pastor-Barriuso, R.; Boldo, E.; Fernández-Cuenca, R.; Ortiz, C.; Linares, C.; Medrano, M.J.; Galán, I. Smoke-Free Legislation in Spain and Prematurity. *Pediatrics* **2017**, *139*, e20162068. [CrossRef]
9. Blencowe, H.; Cousens, S.; Oestergaard, M.Z.; Chou, D.; Moller, A.-B.; Narwal, R.; Adler, A.; Garcia, C.V.; Rohde, S.; Say, L.; et al. National, Regional, and Worldwide Estimates of Preterm Birth Rates in the Year 2010 with Time Trends since 1990 for Selected Countries: A Systematic Analysis and Implications. *Lancet* **2012**, *379*, 2162–2172. [CrossRef]
10. Schanler, R.J.; Shulman, R.J.; Lau, C. Feeding Strategies for Premature Infants: Beneficial Outcomes of Feeding Fortified Human Milk Versus Preterm Formula. *Pediatrics* **1999**, *103*, 1150–1157. [CrossRef]
11. Meinzen-Derr, J.; Poindexter, B.; Wrage, L.; Morrow, A.L.; Stoll, B.; Donovan, E.F. Role of Human Milk in Extremely Low Birth Weight Infants' Risk of Necrotizing Enterocolitis or Death. *J. Perinatol.* **2009**, *29*, 57–62. [CrossRef]
12. Maayan-Metzger, A.; Avivi, S.; Schushan-Eisen, I.; Kuint, J. Human Milk Versus Formula Feeding Among Preterm Infants: Short-Term Outcomes. *Am. J. Perinatol.* **2012**, *29*, 121–126. [CrossRef] [PubMed]
13. Reiterer, F.; Scheuchenegger, A.; Resch, B.; Maurer-Fellbaum, A.; Avian, A.; Urlesberger, B. Outcomes of Very Preterm Infants with and without BPD Followed to Preschool Age. *Pediatr. Int.* **2019**, *61*, 381–387. [CrossRef] [PubMed]
14. Isaacs, E.B.; Fischl, B.R.; Quinn, B.T.; Chong, W.K.; Gadian, D.G.; Lucas, A. Impact of Breast Milk on Intelligence Quotient, Brain Size, and White Matter Development. *Pediatr. Res.* **2010**, *67*, 357–362. [CrossRef] [PubMed]
15. Parkinson, J.R.C.; Hyde, M.J.; Gale, C.; Santhakumaran, S.; Modi, N. Preterm Birth and the Metabolic Syndrome in Adult Life: A Systematic Review and Meta-Analysis. *Pediatrics* **2013**, *131*, e1240–e1263. [CrossRef]
16. Lapillonne, A.; Griffin, I.J. Feeding Preterm Infants Today for Later Metabolic and Cardiovascular Outcomes. *J. Pediatr.* **2013**, *162*, S7–S16. [CrossRef] [PubMed]
17. Howson, C.P.; Kinney, M.V.; Mcdougall, L.; Lawn, J.E. Born Too Soon: Preterm Birth Matters. *Reprod. Health* **2013**, *10*, 1–9. [CrossRef] [PubMed]
18. Castillo-Castañeda, P.C.; Gaxiola-Robles, R.; Méndez-Rodríguez, L.C.; Zenteno-Savín, T. Defensas Antioxidantes En Leche Materna En Relación Al Número De Gestas Y A La Edad De Las Madres. *Nutr. Hosp.* **2014**, *30*, 540–547. [PubMed]

19. Castillo-Castañeda, P.C.; García-González, A.; Bencomo-Alvarez, A.E.; Barros-Nuñez, P.; Gaxiola-Robles, R.; Celina Méndez-Rodríguez, L.; Zenteno-Savín, T. Micronutrient Content and Antioxidant Enzyme Activities in Human Breast Milk. *J. Trace Elem. Med. Biol.* **2019**, *51*, 36–41. [CrossRef] [PubMed]
20. Ballard, O.; Morrow, A.L. Human Milk Composition: Nutrients and Bioactive Factors. *Pediatr. Clin. N. Am.* **2013**, *60*, 49–74. [CrossRef] [PubMed]
21. World Health Organization: Exclusive Breastfeeding. Available online: https://bit.ly/2FaaEkE (accessed on 23 November 2018).
22. World Health Organization: Infant and Young Child Feeding. Available online: https://bit.ly/2VyOnRL (accessed on 5 March 2019).
23. American Academy of Pediatrics. Breastfeeding and the Use of Human Milk. *Pediatrics* **2012**, *129*, e827–e841. [CrossRef] [PubMed]
24. Singhal, A.; Sadaf Farooqi, I.; O'Rahilly, S.; Cole, T.J.; Fewtrell, M.; Lucas, A. Early Nutrition and Leptin Concentrations in Later Life. *Am. J. Clin. Nutr.* **2002**, *75*, 993–999. [CrossRef] [PubMed]
25. Lessen, R.; Kavanagh, K. Practice Paper of the Academy of Nutrition and Dietetics Abstract: Promoting and Supporting Breastfeeding. *J. Acad. Nutr. Diet.* **2015**, *115*, 450. [CrossRef]
26. Yuksel, S.; Yigit, A.A.; Cinar, M.; Atmaca, N.; Onaran, Y. Oxidant and Antioxidant Status of Human Breast Milk During Lactation Period. *Dairy Sci. Technol.* **2015**, *95*, 295–302. [CrossRef]
27. Singhal, A.; Cole, T.J.; Lucas, A. Early Nutrition in Preterm Infants and Later Blood Pressure: Two Cohorts after Randomised Trials. *Lancet* **2001**, *357*, 413–419. [CrossRef]
28. Cubero, J.; Sánchez, C.L.; Bravo, R.; Sánchez, J.; Rodriguez, A.B.; Rivero, M.; Barriga, C. Analysis of the Antioxidant Activity in Human Milk, Day Vs. Night. *Cell Membr. Free Radic. Res.* **2009**, *1*, 100–101.
29. De Vicente, B. 1 De Cada 5 Bebés no Recibe Leche Materna En Los Países Ricos. Available online: https://bit.ly/2RiYzeR (accessed on 21 January 2019).
30. Heiman, H.; Schanler, R.J. Nutrición Enteral En Prematuros: El Rol De La Leche Humana. *Rev. Enferm.* **2007**, *12*, 26–34.
31. Hanson, C.; Lyden, E.; Furtado, J.; Van Ormer, M.; Anderson-Berry, A. A Comparison of Nutritional Antioxidant Content in Breast Milk, Donor Milk, and Infant Formulas. *Nutrients* **2016**, *8*, 681. [CrossRef]
32. Mehta, R.; Petrova, A. Is Variation in Total Antioxidant Capacity of Human Milk Associated with Levels of Bio-Active Proteins? *J. Perinatol.* **2014**, *34*, 220–222. [CrossRef]
33. Savino, F.; Benetti, S.; Liguori, S.A.; Sorrenti, M.; Montezemolo, L.C. Di Advances on Human Milk Hormones and Protection Against Obesity. *Cell. Mol. Biol.* **2013**, *59*, 89–98.
34. Garwolińska, D.; Namieśnik, J.; Kot-Wasik, A.; Hewelt-Belka, W. Chemistry of Human Breast Milk—A Comprehensive Review of the Composition and Role of Milk Metabolites in Child Development. *J. Agric. Food Chem.* **2018**, *66*, 11881–11896. [CrossRef]
35. Mutinati, M.; Pantaleo, M.; Roncetti, M.; Piccinno, M.; Rizzo, A.; Sciorsci, R.L. Oxidative Stress in Neonatology: A Review. *Reprod. Domest. Anim.* **2014**, *49*, 7–16. [CrossRef] [PubMed]
36. Wilinska, M.; Borszewska-Kornacka, M.K.; Niemiec, T.; Jakiel, G. Oxidative Stress and Total Antioxidant Status in Term Newborns and Their Mothers. *Ann. Agric. Environ. Med.* **2015**, *22*, 736–740. [CrossRef] [PubMed]
37. Aceti, A.; Beghetti, I.; Martini, S.; Faldella, G.; Corvaglia, L. Oxidative Stress and Necrotizing Enterocolitis: Pathogenetic Mechanisms, Opportunities for Intervention, and Role of Human Milk. *Oxid. Med. Cell. Longev.* **2018**, *2018*, 1–7. [CrossRef] [PubMed]
38. Thibeault, D.W. The Precarious Antioxidant Defenses of the Preterm Infant. *Am. J. Perinatol.* **2000**, *17*, 167–182. [CrossRef] [PubMed]
39. Ahmad, P.; Jaleel, C.A.; Salem, M.A.; Nabi, G.; Sharma, S. Roles of Enzymatic and Nonenzymatic Antioxidants in Plants During Abiotic Stress. *Crit. Rev. Biotechnol.* **2010**, *30*, 161–175. [CrossRef] [PubMed]
40. Friel, J.K.; Martin, S.M.; Langdon, M.; Herzberg, G.R.; Buettner, G.R. Milk from Mothers of Both Premature and Full-Term Infants Provides Better Antioxidant Protection than Does Infant Formula. *Pediatr. Res.* **2002**, *51*, 612–618. [CrossRef]
41. Gutiérrez-Repiso, C.; Velasco, I.; Garcia-Escobar, E.; Garcia-Serrano, S.; Rodríguez-Pacheco, F.; Linares, F.; Ruiz De Adana, M.S.; Rubio-Martin, E.; Garrido-Sanchez, L.; Cobos-Bravo, J.F.; et al. Does Dietary Iodine Regulate Oxidative Stress and Adiponectin Levels in Human Breast Milk? *Antioxid. Redox Signal.* **2014**, *20*, 847–853. [CrossRef]

42. Ledo, A.; Arduini, A.; Asensi, M.A.; Sastre, J.; Escrig, R.; Brugada, M.; Aguar, M.; Saenz, P.; Vento, M. Human Milk Enhances Antioxidant Defenses Against Hydroxyl Radical Aggression in Preterm Infants. *Am. J. Clin. Nutr.* **2009**, *89*, 210–215. [CrossRef]
43. Friel, J.K.; Diehl-Jones, B.; Cockell, K.A.; Chiu, A.; Rabanni, R.; Davies, S.S.; Jackson Roberts, L. Evidence of Oxidative Stress in Relation to Feeding Type During Early Life in Premature Infants. *Pediatr. Res.* **2011**, *69*, 160–164. [CrossRef]
44. Quiles, J.L.; Ochoa, J.J.; Ramirez-Tortosa, M.C.; Linde, J.; Bompadre, S.; Battino, M.; Narbona, E.; Maldonado, J.; Mataix, J. Coenzyme Q Concentration and Total Antioxidant Capacity of Human Milk at Different Stages of Lactation in Mothers of Preterm and Full-Term Infants. *Free Radic. Res.* **2006**, *40*, 199–206. [CrossRef]
45. Zarban, A.; Taheri, F.; Chahkandi, T.; Sharifzadeh, G.; Khorashadizadeh, M. Antioxidant and Radical Scavenging Activity of Human Colostrum, Transitional and Mature Milk. *J. Clin. Biochem. Nutr.* **2009**, *45*, 150–154. [CrossRef] [PubMed]
46. Ankrah, N.A.; Appiah-Opong, R.; Dzokoto, C. Human Breastmilk Storage and the Glutathione Content. *J. Trop. Pediatr.* **2000**, *46*, 111–113. [CrossRef] [PubMed]
47. Hernández-Aguilar, M.T.; De La Torre, M.J.L.; Borja-Herrero, C.; Lasarte-Velillas, J.J.; Martorell-Juan, L. Antioxidant Properties of Human Milk. *J. Pediatr. Biochem.* **2013**, *3*, 161–167.
48. Hamprecht, K.; Goelz, R. Postnatal Cytomegalovirus Infection through Human Milk in Preterm Infants. *Clin. Perinatol* **2016**, *44*, 121–130. [CrossRef] [PubMed]
49. Allen, A.A.; Baquero-Artiago, F. Review and Guidelines on the Prevention, Diagnosis and Treatment of Post-Natal Cytomegalovirus Infection. *J. Pediatr (Barc)* **2011**, *74*, 52.e1–52.e13.
50. Păduraru, L.; Dimitriu, D.C.; Avasiloaiei, A.L.; Moscalu, M.; Zonda, G.I.; Stamatin, M. Total Antioxidant Status in Fresh and Stored Human Milk from Mothers of Term and Preterm Neonates. *Pediatr. Neonatol.* **2018**, *59*, 600–605. [CrossRef] [PubMed]
51. Xavier, A.M.; Rai, K.; Hegde, A.M. Total Antioxidant Concentrations of Breastmilk-An Eye-Opener to the Negligent. *J. Heal. Popul. Nutr.* **2011**, *29*, 605–611. [CrossRef]
52. Young, I.; Woodside, J. Antioxidants in Health and Disease. *J. Clin. Pathol.* **2001**, *54*, 176–186. [CrossRef]
53. Goyal, M.M.; Basak, A. Human Catalase: Looking for Complete Identity. *Protein Cell* **2010**, *1*, 888–897. [CrossRef]
54. Brigelius-Flohé, R. Tissue-Specific Functions of Individual Glutathione Peroxidases. *Free Radic. Biol. Med.* **1999**, *27*, 951–965. [CrossRef]
55. Groussard, C.; Rannou-Bekono, F.; Machefer, G.; Chevanne, M.; Vincent, S.; Sergent, O.; Cillard, J.; Gratas-Delamarche, A. Changes in Blood Lipid Peroxidation Markers and Antioxidants after a Single Sprint Anaerobic Exercise. *Eur. J. Appl. Physiol.* **2003**, *89*, 14–20. [CrossRef] [PubMed]
56. Silvestre, D.; Miranda, M.; Muriach, M.; Almansa, I.; Jareno, E.; Romero, F.J. Antioxidant Capacity of Human Milk: Effect of Thermal Conditions for the Pasteurization. *Acta Paediatr.* **2008**, *97*, 1070–1074. [CrossRef] [PubMed]
57. Ballatori, N.; Krance, S.M.; Notenboom, S.; Shi, S.; Tieu, K.; Hammond, C.L. Glutathione Dysregulation and the Etiology and Progression of Human Diseases. *Biol. Chem.* **2009**, *390*, 191–214. [CrossRef] [PubMed]
58. Valko, M.; Leibfritz, D.; Moncol, J.; Cronin, M.T.D.; Mazur, M.; Telser, J. Free Radicals and Antioxidants in Normal Physiological Functions and Human Disease. *Int. J. Biochem. Cell Biol.* **2007**, *39*, 44–84. [CrossRef] [PubMed]
59. Liebmann, P.M.; Wölfler, A.; Felsner, P.; Holfer, D.; Schauenstein, K. Melatonin and the Immune System. *Int. Arch. Allergy Immunol.* **1997**, *112*, 203–211. [CrossRef] [PubMed]
60. Reiter, R.J. Antioxidant Actions of Melatonin. *Adv. Pharmacol.* **1996**, *38*, 103–117.
61. Kennaway, D.J.; Wright, H. Melatonin and Circadian Rhythms. *Curr. Top. Med. Chem.* **2002**, *2*, 199–209. [CrossRef] [PubMed]
62. Blask, D.E.; Sauer, L.A.; Dauchy, R.T. Melatonin as a Chronobiotic/Anticancer Agent: Cellular, Biochemical, and Molecular Mechanisms of Action and Their Implications for Circadian-Based Cancer Therapy. *Curr. Top. Med. Chem.* **2002**, *2*, 113–132. [CrossRef]
63. Cuzzocrea, S.; Reiter, R.J. Pharmacological Actions of Melatonin in Acute and Chronic Inflammation. *Curr. Top. Med. Chem.* **2002**, *2*, 153–165. [CrossRef]
64. Guerrero, J.M.; Reiter, R.J. Melatonin-Immune System Relationships. *Curr. Top. Med. Chem.* **2002**, *2*, 167–179. [CrossRef]

65. Reiter, R.J.; Tan, D.X.; Cabrera, J.; D'Arpa, D.; Sainz, R.M.; Mayo, J.C.; Ramos, S. The Oxidant/Antioxidant Network: Role of Melatonin. *Biol. Signals Recept.* **1999**, *8*, 56–63. [CrossRef] [PubMed]
66. Reiter, R.J.; Tan, D.-X. What Constitutes a Physiological Concentration of Melatonin? *J. Pineal Res.* **2003**, *34*, 79–80. [CrossRef] [PubMed]
67. Tan, D.; Hardeland, R.; Manchester, L.C.; Poeggeler, B.; Lopez-Burillo, S.; Mayo, J.C.; Sainz, R.M.; Reiter, R.J. Mechanistic and Comparative Studies of Melatonin and Classic Antioxidants in Terms of Their Interactions with the ABTS Cation Radical. *J. Pineal Res.* **2003**, *34*, 249–259. [CrossRef] [PubMed]
68. Bonnefont-Rousselot, D.; Collin, F. Melatonin: Action as Antioxidant and Potential Applications in Human Disease and Aging. *Toxicology* **2010**, *278*, 55–67. [CrossRef] [PubMed]
69. Colella, M.; Biran, V.; Baud, O. Melatonin and the Newborn Brain. *Early Hum. Dev.* **2016**, *102*, 1–3. [CrossRef] [PubMed]
70. Melatonin. *Drugs and Lactation Database (LactMed)*; National Library of Medicine (US): Bethesda, MD, USA, 2006.
71. Shelby, R.D.; Cromeens, B.; Rager, T.M.; Besner, G.E. Influence of Growth Factors on the Development of Necrotizing Enterocolitis. *Clin. Perinatol.* **2019**, *46*, 51–64. [CrossRef]
72. Lawrence, R.M.; Pane, C.A. Human Breast Milk: Current Concepts of Immunology and Infectious Diseases. *Curr. Probl. Pediatr. Adolesc. Health Care* **2007**, *37*, 7–36. [CrossRef]
73. Loui, A.; Eilers, E.; Strauss, E.; Pohl-Schickinger, A.; Obladen, M.; Koehne, P. Vascular Endothelial Growth Factor (VEGF) and Soluble VEGF Receptor 1 (sFlt-1) Levels in Early and Mature Human Milk from Mothers of Preterm versus Term Infants. *J. Hum. Lact.* **2012**, *28*, 522–528. [CrossRef]
74. Hirai, C.; Ichiba, H.; Saito, M.; Shintaku, H.; Yamano, T.; Kusuda, S. Trophic Effect of Multiple Growth Factors in Amniotic Fluid or Human Milk on Cultured Human Fetal Small Intestinal Cells. *J. Pediatr. Gastroenterol. Nutr.* **2002**, *34*, 524–528. [CrossRef]
75. Munblit, D.; Abrol, P.; Sheth, S.; Chow, L.Y.; Khaleva, E.; Asmanov, A.; Lauriola, S.; Padovani, E.M.; Comberiati, P.; Boner, A.L.; et al. Levels of Growth Factors and Iga in the Colostrum of Women from Burundi and Italy. *Nutrients* **2018**, *10*, 1216. [CrossRef]
76. Funakoshi, H.; Nakamura, T. Hepatocyte Growth Factor: From Diagnosis to Clinical Applications. *Clin. Chim. Acta* **2003**, *327*, 1–23. [CrossRef]
77. Kobata, R.; Tsukahara, H.; Ohshima, Y.; Ohta, N.; Tokuriki, S.; Tamura, S.; Mayumi, M. High Levels of Growth Factors in Human Breast Milk. *Early Hum. Dev.* **2008**, *84*, 67–69. [CrossRef] [PubMed]
78. Min, J.-K.; Lee, Y.-M.; Kim, J.H.; Kim, Y.-M.; Kim, S.W.; Lee, S.-Y.; Gho, Y.S.; Oh, G.T.; Kwon, Y.-G. Hepatocyte Growth Factor Suppresses Vascular Endothelial Growth Factor-Induced Expression of Endothelial ICAM-1 and VCAM-1 by Inhibiting the Nuclear Factor-κB Pathway. *Circ. Res.* **2005**, *96*, 300–307. [CrossRef] [PubMed]
79. Wagner, C.L.; Taylor, S.N.; Johnson, D. Host factors in amniotic fluid and breast milk that contribute to gut maturation. *Clin. Rev. Allergy Immunol.* **2008**, *34*, 191–204. [CrossRef] [PubMed]
80. Chang, C.-J.; Chao, J.C.-J. Effect of human milk and epidermal growth factor on growth of human intestinal Caco-2 cells. *J. Pediatr. Gastroenterol. Nutr.* **2002**, *34*, 394–401. [CrossRef]
81. Radulescu, A.; Zhang, H.-Y.; Chen, C.-L.; Chen, Y.; Zhou, Y.; Yu, X.; Otabor, I.; Olson, J.K.; Besner, G.E. Heparin-Binding EGF-Like Growth Factor promotes intestinal anastomotic healing. *J. Surg. Res.* **2012**, *171*, 540–550. [CrossRef] [PubMed]
82. Gregory, K.E.; Walker, W.A. Immunologic factors in human milk and disease prevention in the preterm infant. *Curr. Pediatr. Rep. Online* **2013**, *1*, 222–228. [CrossRef]
83. Dvorak, B.; Fituch, C.C.; Williams, C.S.; Hurst, N.M.; Schanler, R.J. Increased Epidermal Growth Factor Levels in Human Milk of Mothers with Extremely Premature Infants. *Pediatr. Res.* **2003**, *54*, 15–19. [CrossRef]
84. Castellote, C.; Casillas, R.; Ramírez-Santana, C.; Pérez-Cano, F.J.; Castell, M.; Moretones, M.G.; López-Sabater, M.C.; Franch, A. Premature Delivery Influences the Immunological Composition of Colostrum and Transitional and Mature Human Milk. *J. Nutr.* **2011**, *110*, 1181–1187. [CrossRef]
85. Boesmans, W.; Gomes, P.; Janssens, J.; Tack, J.; Vanden Berghe, P. Brain-derived neurotrophic factor amplifies neurotransmitter responses and promotes synaptic communication in the enteric nervous system. *Gut* **2008**, *57*, 314–322. [CrossRef]
86. Ismail, A.M.; Babers, G.M.; El Rehany, M.A. Brain-Derived Neurotrophic Factor in Sera of Breastfed Epileptic Infants and in Breastmilk of Their Mothers. *Breastfeed. Med.* **2015**, *10*, 277–282. [CrossRef] [PubMed]

87. Li, R.; Xia, W.; Zhang, Z.; Wu, K. S100b protein, brain-derived neurotrophic factor, and glial cell line-derived neurotrophic factor in human milk. *PLoS ONE* **2011**, *6*, e21663. [CrossRef] [PubMed]
88. Blum, J.W.; Baumrucker, C.R. Colostral and milk insulin-like growth factors and related substances: Mammary gland and neonatal (intestinal and systemic) targets. *Domest. Anim. Endocrinol.* **2002**, *23*, 101–110. [CrossRef]
89. Milsom, S.R.; Blum, W.F.; Gunn, A.J. Temporal changes in insulin-like growth factors I and II and in insulin-like growth factor binding proteins 1, 2, and 3 in human milk. *Horm. Res.* **2008**, *69*, 307–311. [CrossRef] [PubMed]
90. Ozgurtas, T.; Aydin, I.; Turan, O.; Koc, E.; Hirfanoglu, I.M.; Acikel, C.H.; Akyol, M.; Erbil, M.K. Vascular endothelial growth factor, basic fibroblast growth factor, insulin-like growth factor-I and platelet-derived growth factor levels in human milk of mothers with term and preterm neonates. *Cytokine* **2010**, *50*, 192–194. [CrossRef] [PubMed]
91. Elmlinger, M.W.; Hochhaus, F.; Loui, A.; Frommer, K.W.; Obladen, M.; Ranke, M.B. Insulin-like growth factors and binding proteins in early milk from mothers of preterm and term infants. *Horm. Res.* **2007**, *68*, 124–131. [CrossRef] [PubMed]
92. Lenhartova, N.; Matasova, K.; Lasabova, Z.; Javorka, K.; Calkovska, A. Impact of early aggressive nutrition on retinal development in premature infants. *Physiol. Res.* **2017**, *66*, S215–S226. [PubMed]
93. Shimizu, H.; Ohsaki, A.; Oh-I, S.; Okada, S.; Mori, M. A new anorexigenic protein, nesfatin-1. *Peptides* **2009**, *30*, 995–998. [CrossRef] [PubMed]
94. Tapping, R.I.; Tobias, P.S. Soluble CD14-Mediated Cellular Responses to Lipopolysaccharide. *Chem. Immunol.* **2000**, *74*, 108–121. [PubMed]
95. Ulevitch, R.J.; Tobias, P.S. Receptor-Dependent Mechanisms of Cell Stimulation by Bacterial Endotoxin. *Annu. Rev. Immunol.* **1995**, *13*, 437–457. [CrossRef] [PubMed]
96. Munblit, D.; Peroni, D.G.; Boix-Amorós, A.; Hsu, P.S.; Land, B.V.; Gay, M.C.L.; Kolotilina, A.; Skevaki, C.; Boyle, R.J.; Collado, M.C.; et al. Human Milk and Allergic Diseases: An Unsolved Puzzle. *Nutrients* **2017**, *9*, 894. [CrossRef] [PubMed]
97. Khodayar-Pardo, P.; Mira-Pascual, L.; Collado, M.C.; Martínez-Costa, C. Impact of lactation stage, gestational age and mode of delivery on breast milk microbiota. *J. Perinatol.* **2014**, *34*, 599–605. [CrossRef] [PubMed]
98. Kalliomäki, M.; Ouwehand, A.; Arvilommi, H.; Kero, P.; Isolauri, E. Transforming growth factor-β in breast milk: A potential regulator of atopic disease at an early age. *J. Allergy Clin. Immunol.* **1999**, *104*, 1251–1257. [CrossRef]
99. Serrao, F.; Papacci, P.; Costa, S.; Giannantonio, C.; Cota, F.; Vento, G.; Romagnoli, C. Effect of early expressed human milk on insulin-like growth factor 1 and short-term outcomes in preterm infants. *PLoS ONE* **2016**, *11*, e0168139. [CrossRef] [PubMed]
100. Saso, A.; Blyuss, O.; Munblit, D.; Faal, A.; Moore, S.E.; Doare, K. Le Breast Milk Cytokines and Early Growth in Gambian Infants. *Front. Pediatr.* **2019**, *6*, 414. [CrossRef] [PubMed]
101. Savino, F.; Liguori, S.A.; Lupica, M.M. Adipokines in breast milk and preterm infants. *Early Hum. Dev.* **2010**, *86*, 77–80. [CrossRef] [PubMed]
102. Catli, G.; Anik, A.; Tuhan, H.Ü.; Kume, T.; Bober, E.; Abaci, A. The relation of leptin and soluble leptin receptor levels with metabolic and clinical parameters in obese and healthy children. *Peptides* **2014**, *56*, 72–76. [CrossRef] [PubMed]
103. Sagawa, N.; Yura, S.; Itoh, H.; Kakui, K.; Takemura, M.; Nuamah, M.A.; Ogawa, Y.; Masuzaki, H.; Nakao, K.; Fujii, S. Possible role of placental leptin in pregnancy: A review. *Endocrine* **2002**, *19*, 65–71. [CrossRef]
104. Savino, F.; Liguori, S.; Fissore, M.; Oggero, R. Breast Milk Hormones and Their Protective Effect on Obesity. *Int. J. Pediatr. Endocrinol.* **2009**, *2009*, 1–8. [CrossRef]
105. Smith-Kirwin, S.; O'Connor, D.; De Johnston, J.; Lancey, E.; Hassink, S.; Funanage, V. Leptin expression in human mammary epithelial cells and breast milk. *J. Clin. Endocrinol. Metab.* **1998**, *83*, 1810–1813. [CrossRef]
106. Casabiell, X.; Piñeiro, V.; Tomé, M.A.; Peinó, R.; Diéguez, C.; Casanueva, F.F. Presence of leptin in colostrum and/or breast milk from lactating mothers: A potential role in the regulation of neonatal food intake. *J. Clin. Endocrinol. Metab.* **1997**, *82*, 4270–4273. [CrossRef] [PubMed]
107. Yu, X.; Rong, S.S.; Sun, X.; Ding, G.; Wan, W.; Zou, L.; Wu, S.; Li, M.; Wang, D. Associations of breast milk adiponectin, leptin, insulin and ghrelin with maternal characteristics and early infant growth: A longitudinal study. *Br. J. Nutr.* **2018**, *120*, 1380–1387. [CrossRef] [PubMed]

108. Eilers, E.; Ziska, T.; Harder, T.; Plagemann, A.; Obladen, M.; Loui, A. Leptin determination in colostrum and early human milk from mothers of preterm and term infants. *Early Hum. Dev.* **2011**, *87*, 415–419. [CrossRef] [PubMed]
109. Houseknecht, K.L.; McGuire, M.K.; Portocarrero, C.P.; McGuire, M.A.; Beerman, K. Leptin is present in human milk and is related to maternal plasma leptin concentration and adiposity. *Biochem. Biophys. Res. Commun.* **1997**, *240*, 742–747. [CrossRef] [PubMed]
110. Miralles, O.; Sánchez, J.; Palou, A.; Picó, C. A physiological role of breast milk leptin in body weight control in developing infants. *Obesity* **2006**, *14*, 1371–1377. [CrossRef] [PubMed]
111. Uçar, B.; Kırel, B.; Bör, Ö.; Kılıç, F.S.; Doğruel, N.; Durmuş Aydoğdu, S.; Tekin, N. Breast Milk Leptin Concentrations in Initial and Terminal Milk Samples: Relationships to Maternal and Infant Plasma Leptin Concentrations, Adiposity, Serum Glucose, Insulin, Lipid and Lipoprotein Levels. *J. Pediatr. Endocrinol. Metab.* **2000**, *13*, 149–156. [CrossRef] [PubMed]
112. A Fields, D.; Demerath, E.W. Relationship of insulin, glucose, leptin, IL-6 and TNF-α in human Breast-Milk with Infant Growth and Body Composition. *Pediatr. Obes.* **2012**, *7*, 304–312. [CrossRef]
113. Resto, M.; O'Connor, D.; Leef, K.; Funanage, V.; Spear, M.; Locke, R. Leptin Levels in Preterm Human Breast Milk and Infant Formula. *Pediatrics* **2001**, *108*, 1–4. [CrossRef]
114. Dundar, N.O.; Anal, O.; Dundar, B.; Ozkan, H.; Cahskan, S.; Büyükgebiz, A. Longitudinal Investigation of the Relationship between Breast Milk Leptin Levels and Growth in Breast-fed Infants. *J. Pediatr. Endocrinol. Metab.* **2005**, *18*, 181–187. [CrossRef]
115. Larsson, M.W.; Lind, M.V.; Larnkjær, A.; Due, A.P.; Blom, I.C.; Wells, J.; Lai, C.T.; Mølgaard, C.; Geddes, D.T.; Michaelsen, K.F. Excessive weight gain followed by catch-down in exclusively breastfed infants: An exploratory study. *Nutrients* **2018**, *10*, 1290. [CrossRef]
116. Kubota, N.; Yano, W.; Kubota, T.; Yamauchi, T.; Itoh, S.; Kumagai, H.; Kozono, H.; Takamoto, I.; Okamoto, S.; Shiuchi, T.; et al. Adiponectin Stimulates AMP-Activated Protein Kinase in the Hypothalamus and Increases Food Intake. *Cell Metab.* **2007**, *6*, 55–68. [CrossRef] [PubMed]
117. Maeda, N.; Takahashi, M.; Funahashi, T.; Kihara, S.; Nishizawa, H.; Kishida, K.; Nagaretani, H.; Matsuda, M.; Komuro, R.; Ouchi, N.; et al. PPARγ Ligands Increase Expression and Plasma Concentrations of Adiponectin, an Adipose-Derived Protein. *Diabetes* **2001**, *50*, 2094–2099. [CrossRef] [PubMed]
118. Luoto, R.; Laitinen, K.; Nermes, M.; Isolauri, E. Impact of maternal probiotic-supplemented dietary counseling during pregnancy on colostrum adiponectin concentration: A prospective, randomized, placebo-controlled study. *Early Hum. Dev.* **2012**, *88*, 339–344. [CrossRef] [PubMed]
119. Martin, L.J.; Woo, J.G.; Geraghty, S.R.; Altaye, M.; Davidson, B.S.; Banach, W.; Dolan, L.M.; Ruiz-Palacios, G.M.; Morrow, A.L. Adiponectin is present in human milk and is associated with maternal factors. *Am. J. Clin. Nutr.* **2006**, *83*, 1106–1111. [CrossRef] [PubMed]
120. Wang, Y.; Zhang, Z.; Yao, W.; Morrow, A.; Peng, Y. Variation of maternal milk adiponectin and its correlation with infant growth. *Chin. J. Pediatr.* **2011**, *49*, 338–343.
121. Dündar, N.O.; Dündar, B.; Cesur, G.; Yilmaz, N.; Sütu, R.; Özgüner, F. Ghrelin and adiponectin levels in colostrum, cord blood and maternal serum. *Pediatr. Int.* **2010**, *52*, 622–625. [CrossRef]
122. Ozarda, Y.; Gunes, Y.; Tuncer, G.O. The concentration of adiponectin in breast milk is related to maternal hormonal and inflammatory status during 6 months of lactation. *Clin. Chem. Lab. Med.* **2012**, *50*, 911–917. [CrossRef]
123. Weyermann, M.; Brenner, H.; Rothenbacher, D. Adipokines in Human Milk and Risk of Overweight in Early Childhood. *Epidemiology* **2007**, *18*, 722–729. [CrossRef]
124. Newburg, D.S.; Woo, J.G.; Morrow, A.L. Characteristics and Potential Functions of Human Milk Adiponectin. *J. Pediatr.* **2010**, *156*, S1–S46. [CrossRef]
125. Savino, F.; Sorrenti, M.; Benetti, S.; Lupica, M.M.; Liguori, S.A.; Oggero, R. Resistin and leptin in breast milk and infants in early life. *Early Hum. Dev.* **2012**, *88*, 779–782. [CrossRef]
126. Ilcol, Y.O.; Hizli, Z.B.; Eroz, E. Resistin is present in human breast milk and it correlates with maternal hormonal status and serum level of C-reactive protein. *Clin. Chem. Lab. Med.* **2008**, *46*, 118–124. [CrossRef] [PubMed]
127. Briana, D.D.; Boutsikou, M.; Baka, S.; Gourgiotis, D.; Marmarinos, A.; Hassiakos, D.; Malamitsi-Puchner, A. Perinatal Changes of Plasma Resistin Concentrations in Pregnancies with Normal and Restricted Fetal Growth. *Neonatology* **2008**, *93*, 153–157. [CrossRef]

128. Ng, P.C.; Lee, C.H.; Lam, C.W.K.; Chan, I.H.S.; Wong, E.; Fok, T.F. Ghrelin in preterm and term newborns: Relation to anthropometry, leptin and insulin. *Pediatr. Res.* **2005**, *58*, 725–730. [CrossRef]
129. Van Der Lely, A.J.; Tschöp, M.; Heiman, M.L.; Ghigo, E. Biological, physiological, pathophysiological, and pharmacological aspects of ghrelin. *Endocr. Rev.* **2004**, *25*, 426–457. [CrossRef] [PubMed]
130. Aydin, S.; Ozkan, Y.; Erman, F.; Gurates, B.; Kilic, N.; Colak, R.; Gundogan, T.; Catak, Z.; Bozkurt, M.; Akin, O.; et al. Presence of obestatin in breast milk: Relationship among obestatin, ghrelin, and leptin in lactating women. *Nutrition* **2008**, *24*, 689–693. [CrossRef] [PubMed]
131. Aydin, S.; Aydin, S.; Ozkan, Y.; Kumru, S. Ghrelin is present in human colostrum, transitional and mature milk. *Peptides* **2006**, *27*, 878–882. [CrossRef] [PubMed]
132. Kierson, J.; Dimatteo, D.; Locke, R.; MacKley, A.; Spear, M. Ghrelin and cholecystokinin in term and preterm human breast milk. *Acta Paediatr. Int. J. Paediatr.* **2006**, *95*, 991–995. [CrossRef]
133. Aydin, S.; Geckil, H.; Karatas, F.; Donder, E.; Kumru, S.; Kavak, E.C.; Colak, R.; Ozkan, Y.; Sahin, I. Milk and blood ghrelin level in diabetics. *Nutrition* **2007**, *23*, 807–811. [CrossRef]
134. Cesur, G.; Ozguner, F.; Yilmaz, N.; Dundar, B. The relationship between ghrelin and adiponectin levels in breast milk and infant serum and growth of infants during early postnatal life. *J. Physiol. Sci.* **2012**, *62*, 185–190. [CrossRef]
135. Savino, F.; Petrucci, E.; Lupica, M.M.; Nanni, G.E.; Oggero, R. Assay of ghrelin concentration in infant formulas and breast milk. *World J. Gastroenterol.* **2011**, *17*, 1971–1975. [CrossRef]
136. Seim, I.; Walpole, C.; Amorim, L.; Josh, P.; Herington, A.; Chopin, L. The expanding roles of the ghrelin-gene derived peptide obestatin in health and disease. *Mol. Cell. Endocrinol.* **2011**, *340*, 111–117. [CrossRef] [PubMed]
137. Osakia, A.; Shimizu, H.; Ishizuka, N.; Suzuki, Y.; Mori, M.; Inoue, S. Enhanced expression of nesfatin/nucleobindin-2 in white adipose tissue of ventromedial hypothalamus-lesioned rats. *Neurosci. Lett.* **2012**, *521*, 46–51. [CrossRef] [PubMed]
138. Aydin, S. The presence of the peptides apelin, ghrelin and nesfatin-1 in the human breast milk, and the lowering of their levels in patients with gestational diabetes mellitus. *Peptides* **2010**, *31*, 2236–2240. [CrossRef] [PubMed]
139. Mesmin, C.; Fenaille, F.; Becher, F.; Tabet, J.C.; Ezan, E. Identification and characterization of apelin peptides in bovine colostrum and milk by liquid chromatography-Mass spectrometry. *J. Proteome Res.* **2011**, *10*, 5222–5231. [CrossRef] [PubMed]
140. Castan-Laurell, I.; Dray, C.; Attané, C.; Duparc, T.; Knauf, C.; Valet, P. Apelin, diabetes, and obesity. *Endocrine* **2011**, *40*, 1–9. [CrossRef] [PubMed]
141. Akcilar, R.; Turgut, S.; Caner, V.; Akcilar, A.; Ayada, C.; Elmas, L.; Özcan, T.O. The effects of apelin treatment on a rat model of type 2 diabetes. *Adv. Med. Sci.* **2015**, *60*, 94–100. [CrossRef] [PubMed]
142. Heinonen, M.V.; Purhonen, A.K.; Miettinen, P.; Pääkkönen, M.; Pirinen, E.; Alhava, E.; Åkerman, K.; Herzig, K.H. Apelin, orexin-A and leptin plasma levels in morbid obesity and effect of gastric banding. *Regul. Pept.* **2005**, *130*, 7–13. [CrossRef] [PubMed]
143. Lv, S.-Y.; Yang, Y.J.; Qin, Y.J.; Mo, J.R.; Wang, N.B.; Wang, Y.J.; Chen, Q. Central apelin-13 inhibits food intake via the CRF receptor in mice. *Peptides* **2012**, *33*, 132–138. [CrossRef]
144. Dinarello, C.A. Proinflammatory cytokines. *Chest* **2000**, *118*, 503–508. [CrossRef]
145. Agarwal, S.; Karmaus, W.; Davis, S.; Gangur, V. Immune markers in breast milk and fetal and maternal body fluids: A systematic review of perinatal concentrations. *J. Hum. Lact.* **2011**, *27*, 171–186. [CrossRef]
146. Watanabe, M.A.E.; de Oliveira, G.G.; Oda, J.M.M.; Ono, M.A.; Guembarovski, R.L. Cytokines in Human Breast Milk: Immunological Significance for Newborns. *Curr. Nutr. Food Sci.* **2012**, *8*, 2–7. [CrossRef]
147. Peroni, D.G.; Pescollderungg, L.; Piacentini, G.L.; Rigotti, E.; Maselli, M.; Watschinger, K.; Piazza, M.; Pigozzi, R.; Boner, A.L. Immune regulatory cytokines in the milk of lactating women from farming and urban environments. *Pediatr. Allergy Immunol.* **2010**, *21*, 977–982. [CrossRef] [PubMed]
148. Rajani, P.S.; Seppo, A.E.; Järvinen, K.M. Immunologically Active Components in Human Milk and Development of Atopic Disease, With Emphasis on Food Allergy, in the Pediatric Population. *Front. Pediatr.* **2018**, *6*, 6. [CrossRef] [PubMed]
149. Polat, A.; Tunc, T.; Erdem, G.; Yerebasmaz, N.; Tas, A.; Beken, S.; Basbozkurt, G.; Saldir, M.; Zenciroglu, A.; Yaman, H.; et al. Interleukin-8 and its receptors in human milk from mothers of full-term and premature infants. *Breastfeed. Med.* **2016**, *11*, 247–251. [CrossRef] [PubMed]

150. Aspinall, R.; Prentice, A.M.; Ngom, P.T. Interleukin 7 from maternal milk crosses the intestinal barrier and modulates T-cell development in offspring. *PLoS ONE* **2011**, *6*, e20812. [CrossRef] [PubMed]
151. Maheshwari, A.; Lu, W.; Lacson, A.; Barleycorn, A.A.; Nolan, S.; Christensen, R.D.; Calhoun, D.A. Effects of Interleukin-8 on the Developing Human Intestine. *Cytokine* **2002**, *20*, 256–267. [CrossRef] [PubMed]
152. Mehta, R.; Petrova, A. Very Preterm Gestation and Breastmilk Cytokine Content During the First Month of Lactation. *Breastfeed. Med.* **2011**, *6*, 21–24. [CrossRef] [PubMed]
153. Abdelhamid, A.E.; Chuang, S.L.; Hayes, P.; Fell, J.M.E. Evolution of in vitro cow's milk protein-specific inflammatory and regulatory cytokine responses in preterm infants with necrotising enterocolitis. *Pediatr. Res.* **2011**, *69*, 165–169. [CrossRef] [PubMed]
154. Branton, M.H.; Kopp, J.B. TGF-beta and fibrosis. *Microbes Infect.* **1999**, *1*, 1349–1365. [CrossRef]
155. Donnet-Hughes, A.; Duc, N.; Serrant, P.; Vidal, K.; Schiffrin, E. Bioactive molecules in milk and their role in health and disease: The role of transforming growth factor-β. *Immunol. Cell Biol.* **2000**, *78*, 74–79. [CrossRef] [PubMed]
156. Morita, Y.; Campos-Alberto, E.; Yamaide, F.; Nakano, T.; Ohnisi, H.; Kawamoto, M.; Kawamoto, N.; Matsui, E.; Kondo, N.; Kohno, Y.; et al. TGF-β Concentration in Breast Milk is Associated with the Development of Eczema in Infants. *Front. Pediatr.* **2018**, *6*, 1–6. [CrossRef] [PubMed]
157. Iyengar, S.R.; Walker, W.A. Immune factors in breast milk and the development of atopic disease. *J. Pediatr. Gastroenterol. Nutr.* **2012**, *55*, 641–647. [CrossRef] [PubMed]
158. Rogier, E.W.; Frantz, A.L.; Bruno, M.E.C.; Wedlund, L.; Cohen, D.A.; Stromberg, A.J.; Kaetzel, C.S. Secretory antibodies in breast milk promote long-term intestinal homeostasis by regulating the gut microbiota and host gene expression. *Proc. Natl. Acad. Sci. USA* **2014**, *111*, 3074–3079. [CrossRef] [PubMed]
159. Reed, J.R.; Schwertfeger, K.L. Immune cell location and function during post-natal mammary gland development. *J. Mammary Gland Biol. Neoplasia* **2010**, *15*, 329–339. [CrossRef] [PubMed]
160. Ustundag, B.; Yilmaz, E.; Dogan, Y.; Akarsu, S.; Canatan, H.; Halifeoglu, I.; Cikim, G.; Denizmen Aygun, A. Levels of cytokines (IL-1β, IL-2, IL-6, IL-8, TNF-α) and trace elements (Zn, Cu) in breast milk from mothers of preterm and term infants. *Med. Inflamm.* **2005**, *2005*, 331–336. [CrossRef] [PubMed]
161. Hawkes, J.S.; Bryan, D.L.; James, M.J.; Gibson, R.A. Cytokines (IL-1β, IL-6, TNF-α, TGF-β1, and TGF-β2) and prostaglandin E2 in human milk during the first three months postpartum. *Pediatr. Res.* **1999**, *46*, 194–199. [CrossRef] [PubMed]
162. Hrdý, J.; Novotná, O.; Kocourková, I.; Prokešová, L. Cytokine expression in the colostral cells of healthy and allergic mothers. *Folia Microbiol. (Praha)* **2012**, *57*, 215–219. [CrossRef]
163. Erbağci, A.B.; Çekmen, M.B.; Balat, Ö.; Balat, A.; Aksoy, F.; Tarakçioğlu, M. Persistency of high proinflammatory cytokine levels from colostrum to mature milk in preeclampsia. *Clin. Biochem.* **2005**, *38*, 712–716. [CrossRef]
164. Soto-Ramírez, N.; Boyd, K.; Zhang, H.; Gangur, V.; Goetzl, L.; Karmaus, W. Maternal serum but not breast milk IL-5, IL-6, and IL-13 immune markers are associated with scratching among infants. *Allergy Asthma Clin. Immunol.* **2016**, *12*, 399. [CrossRef]
165. Patki, S.; Kadam, S.; Chandra, V.; Bhonde, R. Human breast milk is a rich source of multipotent mesenchymal stem cells. *Hum. Cell* **2010**, *23*, 35–40. [CrossRef]
166. Molès, J.P.; Tuaillon, E.; Kankasa, C.; Bedin, A.S.; Nagot, N.; Marchant, A.; McDermid, J.M.; Van de Perre, P. Breastmilk cell trafficking induces microchimerism-mediated immune system maturation in the infant. *Pediatr. Allergy Immunol.* **2018**, *29*, 133–143. [CrossRef] [PubMed]
167. Cacho, N.T.; Lawrence, R.M. Innate immunity and breast milk. *Front. Immunol.* **2017**, *8*, 584. [CrossRef]
168. Järvinen, K.-M.; Soumalainen, H. Leucocytes in human milk and lymphocyte subsets in cow's milk-allergic infants. *Pediatr. Allergy Immunol.* **2002**, *13*, 243–254. [CrossRef] [PubMed]
169. Vass, R.A.; Kemeny, A.; Dergez, T.; Ertl, T.; Reglodi, D.; Jungling, A.; Tamas, A. Distribution of bioactive factors in human milk samples. *Int. Breastfeed. J.* **2019**, *14*, 1–10. [CrossRef] [PubMed]
170. Bendiks, M.; Kopp, M.V. The relationship between advances in understanding the microbiome and the maturing hygiene hypothesis. *Curr. Allergy Asthma Rep.* **2013**, *13*, 487–494. [CrossRef] [PubMed]
171. Perez, P.F.; Dore, J.; Leclerc, M.; Levenez, F.; Benyacoub, J.; Serrant, P.; Segura-Roggero, I.; Schiffrin, E.J.; Donnet-Hughes, A. Bacterial Imprinting of the Neonatal Immune System: Lessons from Maternal Cells? *Pediatrics* **2007**, *119*, e724–e732. [CrossRef]

172. Mcguire, M.K.; Mcguire, M.A. Human Milk: Mother Nature's Prototypical Probiotic Food? *Adv. Nutr.* **2015**, *6*, 112–123. [CrossRef] [PubMed]
173. Boix-Amorós, A.; Collado, M.C.; Mira, A. Relationship between milk microbiota, bacterial load, macronutrients, and human cells during lactation. *Front. Microbiol.* **2016**, *7*, 3389. [CrossRef]
174. Grönlund, M.; Gueimonde, M.; Laitinen, K.; Kociubinski, G.; Gronroos, T.; Salminen, S.; Isolauri, E. Maternal breast-milk and intestinal bifidobacteria guide the compositional development of the bifidobacterium microbiota in infants at risk of allergic disease. *Clin. Exp. Allergy* **2007**, *37*, 1764–1772. [CrossRef]
175. Martín, V.; Maldonado-Barragán, A.; Moles, L.; Rodriguez-Baños, M.; Del Campo, R.; Fernández, L.; Rodríguez, J.M.; Jiménez, E. Sharing of Bacterial Strains Between Breast Milk and Infant Feces. *J. Hum. Lact.* **2012**, *28*, 36–44. [CrossRef]
176. Gueimonde, M.; Sakata, S.; Kalliomäki, M.; Isolauri, E.; Benno, Y.; Salminen, S. Effect of maternal consumption of Lactobacillus GG on transfer and establishment of fecal bifidobacterial microbiota in neonates. *J. Pediatr. Gastroenterol. Nutr.* **2006**, *42*, 166–170. [CrossRef] [PubMed]
177. Barthow, C.; Wickens, K.; Stanley, T.; Mitchell, E.A.; Maude, R.; Abels, P.; Purdie, G.; Murphy, R.; Stone, P.; Kang, J.; et al. The Probiotics in Pregnancy Study (PiP Study): Rationale and design of a double-blind randomised controlled trial to improve maternal health during pregnancy and prevent infant eczema and allergy. *BMC Pregnancy Childbirth* **2016**, *16*, 1–14. [CrossRef] [PubMed]
178. Underwood, M.A.; Gaerlan, S.; De Leoz, M.L.A.; Dimapasoc, L.; Kalanetra, K.M.; Lemay, D.G.; German, J.B.; Mills, D.A.; Lebrilla, C.B. Human milk oligosaccharides in premature infants: Absorption, excretion, and influence on the intestinal microbiota. *Pediatr. Res.* **2015**, *78*, 670–677. [CrossRef] [PubMed]

© 2019 by the authors. Licensee MDPI, Basel, Switzerland. This article is an open access article distributed under the terms and conditions of the Creative Commons Attribution (CC BY) license (http://creativecommons.org/licenses/by/4.0/).

MDPI
St. Alban-Anlage 66
4052 Basel
Switzerland
Tel. +41 61 683 77 34
Fax +41 61 302 89 18
www.mdpi.com

Nutrients Editorial Office
E-mail: nutrients@mdpi.com
www.mdpi.com/journal/nutrients

www.ingramcontent.com/pod-product-compliance
Lightning Source LLC
LaVergne TN
LVHW070237100526
838202LV00015B/2139